Weather Hazards Handbook

Weather Hazards Handbook

Edited by Jonah Hedges

SYRAWOOD
PUBLISHING HOUSE
New York

Published by Syrawood Publishing House,
750 Third Avenue, 9th Floor,
New York, NY 10017, USA
www.syrawoodpublishinghouse.com

Weather Hazards Handbook
Edited by Jonah Hedges

International Standard Book Number: ISBN 978-1-68286-559-0 (Hardback)

Cataloging-in-Publication Data

Weather hazards handbook / edited by Jonah Hedges.
p. cm.
Includes bibliographical references and index.
ISBN 978-1-68286-559-0
1. Weather. 2. Climatic extremes. 3. Climatology. I. Hedges, Jonah.
QC981 .W43 2018
551.6--dc23

Printed in China.

TABLE OF CONTENTS

PREFACE

Weather hazards are defined as extreme weather phenomena that can cause damage to people as well as surroundings. Some examples of extreme weather events include tornadoes, floods, storms, droughts, etc. Most weather hazards are unseasonal and hence, add to its severity. Recent climatic phenomena can be attributed to long-term changes in the Earth's temperature and global warming. This book includes some of the vital pieces of work being conducted across the world, on various topics related to weather hazards. The readers would gain knowledge that would broaden their perspective about this topic.

This book has been the outcome of endless efforts put in by authors and researchers on various issues and topics within the field. The book is a comprehensive collection of significant researches that are addressed in a variety of chapters. It will surely enhance the knowledge of the field among readers across the globe.

It gives us an immense pleasure to thank our researchers and authors for their efforts to submit their piece of writing before the deadlines. Finally in the end, I would like to thank my family and colleagues who have been a great source of inspiration and support.

Editor

An Indirect Cue of Predation Risk Counteracts Female Preference for Conspecifics in a Naturally Hybridizing Fish *Xiphophorus birchmanni*

Pamela M. Willis[1]*, Gil G. Rosenthal[2,3], Michael J. Ryan[1]

1 Section of Integrative Biology, University of Texas at Austin, Austin, Texas, United States of America, **2** Department of Biology, Texas A & M University, College Station, Texas, United States of America, **3** Centro de Investigación Científica de las Huastecas "Aguazarca", Calnali, Hidalgo, Mexico

Abstract

Mate choice is context dependent, but the importance of current context to interspecific mating and hybridization is largely unexplored. An important influence on mate choice is predation risk. We investigated how variation in an indirect cue of predation risk, distance to shelter, influences mate choice in the swordtail *Xiphophorus birchmanni*, a species which sometimes hybridizes with *X. malinche* in the wild. We conducted mate choice experiments to determine whether females attend to the distance to shelter and whether this cue of predation risk can counteract female preference for conspecifics. Females were sensitive to shelter distance independent of male presence. When conspecific and heterospecific *X. malinche* males were in equally risky habitats (i.e., equally distant from shelter), females associated primarily with conspecifics, suggesting an innate preference for conspecifics. However, when heterospecific males were in less risky habitat (i.e., closer to shelter) than conspecific males, females no longer exhibited a preference, suggesting that females calibrate their mate choices in response to predation risk. Our findings illustrate the potential for hybridization to arise, not necessarily through reproductive "mistakes", but as one of many potential outcomes of a context-dependent mate choice strategy.

Editor: Vincent Laudet, Ecole Normale Supérieure de Lyon, France

Funding: This work was funded by grants to PMW from the Society for Integrative and Comparative Biology, the Southwestern Association of Naturalists, the American Livebearer Association, and the PADI Foundation grants. PMW was supported by a Zoology Scholarship Endowment for Excellence Fellowship from the University of Texas. The funders had no role in study design, data collection and analysis, decision to publish, or preparation of the manuscript.

Competing Interests: The authors have declared that no competing interests exist.

* E-mail: pamwillis@utexas.edu

Introduction

Mate choice is influenced by multiple factors, including mate preference and the current context in which mate choice decisions are made [1,2]. Internal and environmental conditions influence the costs and benefits individuals accrue from their choice of mate. As a result, mate choice often varies according to environmental factors such as the level of predation risk [3,4] or the density of potential mates [5,6], and attributes of the choosers themselves such as age, condition, or reproductive state [7–9]. For example, female túngara frogs become less choosy about potential mates as the time remaining for successful reproduction declines [10]. Factors that alter mate choice can in turn affect the strength of sexual selection acting on, or the targets of sexual selection in, the opposite sex. Context-dependent variation in mate choice is therefore important because it may influence evolution by sexual selection; for example, the evolution of novel male traits or the ease of speciation [1,11–12].

While the context-dependence of mate choice within species is well-recognized, less consideration has been given to the possibility that mate choice between species is similarly influenced by current internal and environmental conditions and can result in hybridization. There is evidence, however, that changing conditions can, by altering or constraining mate choice, make hybridization more likely [13–17] For example, in hybridizing populations of grebes (*Aechmophorus occidentalis* and *A. clarkii*), males increasingly pursue heterospecific females as conspecific partners become scarce over the breeding season [14]. Understanding the extent to which context-dependent mate choice can promote or inhibit hybridization is important because of its potential to affect the origin, loss, and fate of evolutionary lineages through interspecific gene flow (see [18–20]).

A prominent influence on mate choice is predation risk [1]. For many species, mate sampling increases the risk of being detected by predators (e.g., [21]) and individuals often become less choosy when predation risk is high (e.g., [4,22]). For example, female sand gobies (*Pomatoschistus minutus*) prefer large colorful males, but become indiscriminate around predators [3]. Predation risk might similarly decrease choosiness in potentially hybridizing individuals and thereby increase the chance of hybridization, although this hypothesis has not been tested experimentally.

Many animals rely heavily on indirect cues, such as the distance to cover or level of illumination, to assess the risk of predation [23,24], and perceived predation risk increases with distance to cover in many species, including fish (e.g., [25,26]). Here we investigate whether variation in the distance to shelter influences female mate choice in the swordtail fish *Xiphophorus birchmanni*. Hybridization and introgression occur between this species and its congener, *X. malinche*, in several tributaries of the Río Pánuco basin in Hidalgo, Mexico [27,28]. First generation hybrids occur at very low frequencies, with a greater preponderance of backcross individuals and later generation hybrids [28]. The two species inhabit shallow rocky streams subject to seasonal flooding and drought [29], and likely experience considerable variation in predation pressure, to which swordtails attend [30,31]. Males of each species differ morphologically in several ways [27,29]. Previous studies have shown that female *X. birchmanni* typically prefer the cues of conspecific males over those of *X. malinche* [32,33], but that their choice of mate can vary with environmental conditions [32,34]. *Xiphophorus birchmanni* is therefore an excellent model for investigating the effects of predation risk on female choice. We experimentally tested female sensitivity to predation risk, and whether it influences their choice of conspecifics over heterospecifics.

Methods

Ethics Statement

All research was conducted in compliance with the Guide for the Care and Use of Laboratory Animals. This study was approved by the Institutional Animal Care and Use Committee at The University of Texas, protocol number 07012201. All efforts were made to maximize animal welfare.

Fish Collection and Experimental Design

We collected *X. birchmanni* from Garces (20°56′24″N, 98°16′54″W), and *X. malinche* from Chicayotla (20°55′26″N, 98°34′35″W) [28], in 2008 and 2009. Subjects and stimuli, all sexually-mature, were either wild-caught or first-generation descendents. Females were isolated from males for at least two weeks before testing.

Water was conditioned with Prime (Seachem Laboratories Inc, Georgia, USA) and carbon-filtered for ≥24 hrs before use. Two filtered 500 W halogen lamps provided downwelling irradiance (UV and visible) comparable to that of natural *Xiphophorus* habitat, following [35]. The sides of the test aquarium (76×30×30 cm) were lined with Teflon and overlain with filter gels, providing diffused horizontal irradiance [35].

Fifteen females were individually offered the choice between conspecific and heterospecific (*X. malinche*) males, presented at opposite ends of the test tank behind clear, UV-transmittant, porous barriers (Fig. 1). The barriers prevent physical interactions between the sexes, while allowing female access to male visual and olfactory cues, both important in *X. birchmanni* mate choice [32,33]. Males were presented in randomly-assigned groups of three per side to allow females access to within-species phenotypic variation and to reduce male stress (as can occur when males are presented individually, PMW pers. obs.); this design also approaches the setting in the wild, where swordtails occur in large social groups and males court in the presence of other males [36]. Each pair of male groups served as stimuli for three to five females. One *X. malinche* male died during the experiment, and was replaced by a similarly-sized male.

In many species, perceived predation risk increases with distance to cover (e.g., [25,26]). Swordtails are reluctant to venture far from cover in both the wild [29] and in the lab, and seek shelter immediately when startled (PMW, pers. obs.). We used distance to shelter as an indirect cue of predation risk. Females spontaneously took cover underneath a central shelter (a 7.5 cm diameter sponge filter) upon introduction to the tank. Emergence from beneath the shelter, followed by visits to both sides of the tank, marked the onset of a 3 min acclimation period. We then conducted two consecutive, 5 min trials: one with the shelter equidistant from either male compartment, and one with it closer to the heterospecific side (Fig. 1). Trial order was alternated

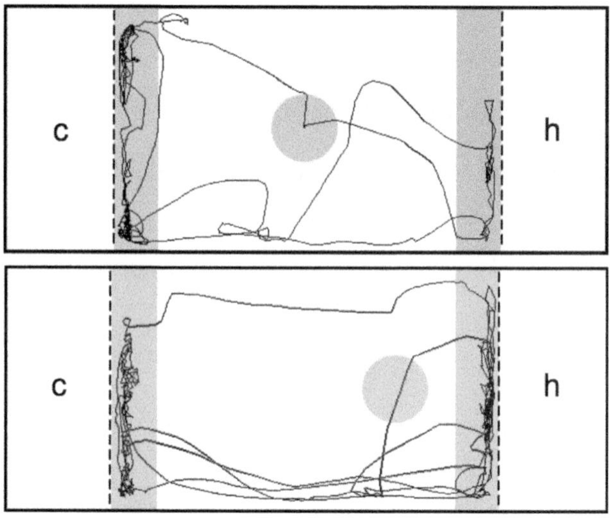

Figure 1. Experimental tank design. Shaded regions represent shelter (circles) and zones (rectangles) in which association time was recorded. Upper panel = shelter equidistant from either male compartment. Lower panel = shelter closer to the heterospecific compartment. c = conspecific male compartment, h = heterospecific male compartment. Drawn to scale. Representative paths of a subject female over two consecutive trials are shown.

between females. To minimize disturbance to the fish between trials, the shelter was moved remotely by overhead pulley very slowly over a ~30 sec period, without removing the fish from the tank. If the first trial was the one with the shelter equidistant from either side, we moved the shelter halfway to the offset position, and then returned it to the center, so as to introduce a comparable amount of shelter movement over the two trials, regardless of trial order. Trials in which the female either hid or was inactive for over half the trial were declared void. As a measure of mate choice we recorded association time with each male stimulus, which predicts mate choice and reproductive success in *Xiphophorus* (e.g., [37–39]). We collected data using the automated EthoVision XT video tracking system (version 5.0, Noldus Information Technology, Wageningen, Netherlands). Tanks were emptied, rinsed, and dried between females.

The shelter was visible to males as well as females, allowing for the possibility that males might (also) respond to changes in shelter position. If so, any treatment effect could reflect female responses to changes in male behavior, rather than the distance to shelter. We controlled for potential male-shelter interactions in two ways. First, we recorded two uncorrelated (Pearson $r = 0.04$) measures of male behavior: the number of aggressive events among males within the group (bite attempts and lateral displays, [40]) and overall group activity (total time during which at least one male is actively moving). This allowed us to account for any variation in female behavior arising due to changes in these (or other correlated) male behaviors (see 'Statistical analyses'). Second, we repeated the experiment without males. This allowed us to observe the influence of shelter position alone on female behavior.

Statistical Analyses

We used linear mixed models to examine the influence of shelter distance, male species (or side, for trials with no male stimuli), and their interaction on female association time. In the presence of a significant interaction, we performed nested contrasts to evaluate the difference in association time between male species (or side) within each treatment. We used a maximum likelihood protocol implemented by the *lme4* package of R. Female ID and male pair group were treated as random effects to account for the non-independence of within-subject and within-male-group measures. Treatment order was included as a covariate. For trials including male stimuli, male aggression and overall activity were also included as covariates. We used the second-order Akaike Information Criterion (AICc) for model selection [41], using the *MuMIn* package in R. As no competing models ranked highly (i.e. delta AICc<2), model averaging was unnecessary [41]. Model residuals were examined to ascertain assumptions of normality and homogeneity of variances were met.

Conventional significance testing of fixed effects in mixed models is a contentious issue, primarily because it is not clear how to calculate the appropriate degrees of freedom [42]. As null hypothesis testing is a familiar paradigm for many biologists, we used Markov chain Monte Carlo sampling (10000 samples) of the posterior distribution of the parameters to generate 95% posterior density credibility intervals and *p*-values [42] using the *pvals.fnc* function in the *languageR* library of R.

Results

Females preferentially associated with conspecific males when the shelter was equidistant from either species (nested contrast, $P_{MCMC} = 0.018$; Figure 2A). When the shelter was closer to the heterospecific side, females no longer exhibited a preference (nested contrast, $P_{MCMC} = 0.123$; linear mixed model, significant male-species-by-shelter-position interaction; Table 1, Figure 2A). A significant effect of male species was also detected, with females associating more with conspecifics overall (Table 1). The two measures of male behavior, aggression and activity, decreased model fit considerably ($\Delta AIC = 7.249$), and were not retained in the final model. No other significant effects were detected (Table 1).

In the absence of males, and with the shelter equidistant from either side, females spent their time equally on either side (nested contrast, $P_{MCMC} = 0.225$; Figure 2B). With the shelter offset, however, females spent more time on the side nearest the shelter (nested contrast, $P_{MCMC} = 0.005$; Figure 2B), resulting in a significant side-by-shelter-position interaction (Table 1). No other significant effects were detected (Table 1).

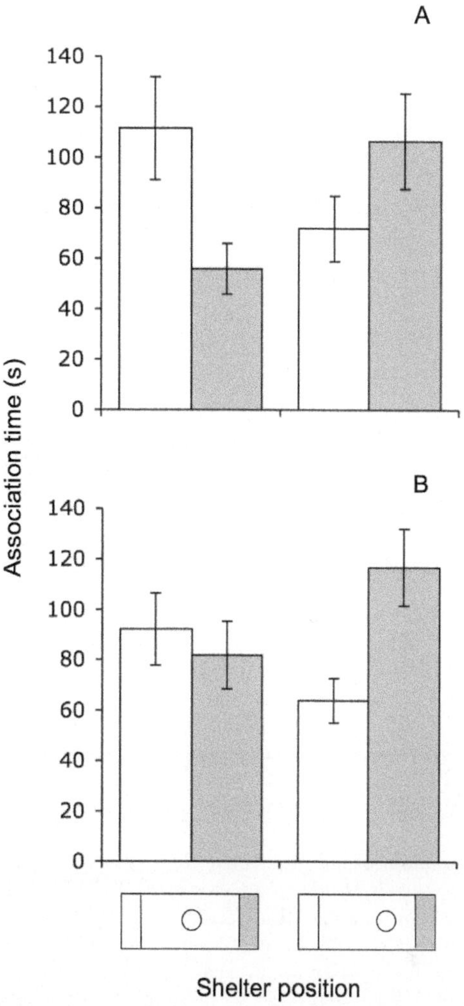

Figure 2. Female time in association zones as a function of male species and shelter position. Female time spent in either zone in the A) presence or B) absence of male stimuli, with the shelter either centered or offset within the tank. In A), white bars = conspecific male side, gray bars = heterospecific male side. In B), bar color designates opposite sides of the tank.

Discussion

In many species, searching for and choosing among potential mates can conflict with predator avoidance (e.g., [21]), causing individuals to adjust their mate choices to the level of risk [3–4,22]. We have shown that, when conspecific and heterospecific males are equally distant from shelter, female *X. birchmanni* prefer conspecifics. Such conspecific preferences are widespread in many taxa, and are important in limiting gene flow between many closely-related species in sympatry [43–45]. However, we have also shown that female *X. birchmanni* are sensitive to the perceived risk of predation, and adjust their mate choices accordingly. These findings provide the first experimental evidence that predation risk can override preferences for conspecifics among (actually or potentially) hybridizing species. Our results are significant because animal hybridization can be an important source of evolutionary change [18–20]. Predation risk may therefore be important, not only in shaping the form and strength of sexual selection within populations (e.g., [46]), but for introducing genetic novelty between them.

Mate sampling under threat of predation likely occurs In wild *Xiphophorus* [30,47]. Cichlid fishes co-occur with all northerm swordtail species and are likely important predators [30,48], as are birds (GGR, unpub. dat.). Other potential predators of *X. birchmanni* include the Mexican tetra (*Astyanax mexicanus*), eleotrid fishes and snakes [47,49]. Seasonal flooding and drought likely expose individuals to variation in predation risk. For example, drought conditions frequently isolate *Xiphophorus* fishes in small pools (e.g., [50–51]), potentially exposing individuals to increased risk of predation, particularly by birds. Conditions of elevated predation risk may constrain female mate choice and increase the likelihood of hybridization.

Although many other species of *Xiphophorus* live in sympatry with others and can hybridize in the lab, hybridization has ony been reported for a few species pairs, and in all cases appears rare (reviewed in [52]). In *X. birchmanni*, both organic pollution [32] and encounter rates with conspecific males [34] reduce female discrimination against *X. malinche* males, in addition to predation risk as reported here. Predation risk may therefore facilitate hybridization in the wild; however, it may be that a variety of conditions are necessary to overcome the strong prezygotic isolation observed in other sympatric species pairs.

Table 1. Best linear mixed models of effects on female association time, in the presence or absence of male stimuli.

Experiment	Parameter	Coefficient estimate	HPD_{lower}	HPD_{upper}	P_{MCMC}
Males present	Male species	−55.67	−102.37	−13.73	0.015
	Shelter position	−39.67	−85.26	2.79	0.073
	Treatment order	14.97	−17.86	47.75	0.372
	Male species * Shelter position	90.33	28.44	153.45	0.006
Males absent	Side	−10.21	−48.76	28.83	0.599
	Shelter position	−28.43	−65.40	9.82	0.138
	Treatment order	6.41	−23.14	35.39	0.677
	Side * Shelter position	63.12	10.26	116.46	0.020

Effect sizes (coefficients) with Bayesian 95% credibility intervals (HPD_{lower}, HPD_{upper}) and P-values.

Female preference for the sword is common in many *Xiphophorus* species, and this preference is considered ancestral [53]. However, this male ornament is also attractive to predators [47], and predation on sworded males can eliminate female preference for the sword [31]. Swords are both absent in *X. birchmanni* males and unattractive to female conspecifics [29,33], and the trait has failed to spread across hybrid zones [27]. In the present study, predation risk reduced female discrimination against (sworded) heterospecifics, suggesting that the relationships among female preference, sword presence and predation risk in this species may be complex. Futures studies of the role of the sword in the patterns reported here can help illuminate its complex role in the evolution of *Xiphophorus*.

Given that predation risk influences mate choice in a wide range of species [1], it seems likely that it can alter mate choice in other hybridizing taxa. For example, Enos Lake stickleback species underwent extensive hybridization and subsequent species collapse following the introduction of an omnivorous crayfish [54–55], the presence of which inhibits reproductive behavior in one of the parental species [56]; however, the exact mechanisms of this collapse are unknown. Identifying the consequences of hybridization is important for understanding the selective forces affecting reproductive isolation and speciation, and a well-developed literature addresses this topic (e.g., [43,57–59]); however, less often addressed are the behavioral causes of hybridization, which are important for the same reason. Our findings suggest that, in some cases, hybridization may arise as just one of the many potential outcomes of a context-dependent mate choice strategy. Studying the behavioral causes from this perspective may broaden our understanding of the processes reducing or increasing diversity.

Acknowledgments

Thanks to G. Calabrese, J. Cheng, F. Masood, S. Monsivais, J. Moran, and S. Stapleton for assistance with data collection. Felix Breden and several anonymous reviewers provided valuable feedback on earlier versions of the manuscript. We thank the Mexican federal government for permission to collect fish.

Author Contributions

Conceived and designed the experiments: PMW. Performed the experiments: PMW. Analyzed the data: PMW. Contributed reagents/materials/analysis tools: GGR MJR. Wrote the paper: PMW. Data interpretation: PMW GGR MJR. Article revision for intellectual content: GGR MJR.

References

1. Jennions MD, Petrie M (1997) Variation in mate choice and mating preferences: a review of causes and consequences. Biol Rev 72: 283–327.
2. Cotton S, Small J, Pomiankowski A (2006) Sexual selection and condition-dependent mate preferences. Current Biology 16: R755–R765.
3. Forsgren E (1992) Predation risk affects mate choice in a gobiid fish. Am Nat 140: 1041–1049.
4. Hedrick AV, Dill LM (1993) Mate choice by female crickets is influenced by predation risk. Anim Behav 46: 193–196.
5. Shelly TE, Bailey WJ (1992) Experimental manipulation of mate choice by male katydids: the effect of female encounter rate. Behav Ecol Sociobiol 30: 277–282.
6. Berglund A (1995) Many mates make male pipefish choosy. Behaviour 132: 213–218.
7. Gray DA (1999) Intrinsic factors affecting female choice in house crickets: time cost, female age, nutritional condition, body size, and size-relative reproductive investment. J Insect Behav 12: 691–700.
8. Kodric-Brown A, Nicoletto, PF (2001) Age and experience affect female choice in the guppy (*Poecilia reticulata*). Am Nat 157: 316–323.
9. Hebets EA, Wesson J, Shamble PS (2008) Diet influences mate choice selectivity in adult female wolf spiders. Anim Behav 76: 355–365.
10. Lynch KS, Rand AS, Ryan MJ, Wilczynski W (2005) Plasticity in female mate choice associated with changing reproductive states. Anim Behav 69: 689–699.
11. Gomulkiewicz R (1991) Limited female choice and the evolution of secondary sexual characters. Am Nat 138: 1396–1407.
12. Thibert-Plante X, Hendry AP (2011) Factors influencing progress towards sympatric speciation. J Evol Biol 24: 2186–2196.
13. Wilson DS, Hedrick A (1982) Speciation and the economics of mate choice. Evol Theory 6: 15–24.
14. Nuechterlein GL, Buitron D (1998) Interspecific mate choice by late-courting male western grebes. Behav Ecol 9: 313–321.
15. Wiley C, Fogelberg N, Sæther SA, Veen T, Svedin N, et al. (2007) Direct benefits and costs for hybridizing *Ficedula* flycatchers. J Evol Biol 20: 854–864.
16. Pfennig KS (2007) Facultative mate choice drives adaptive hybridization. Science 318: 965–967.
17. Gerlai R (2007) Mate choice and hybridization in Lake Malawi cichlids, *Sciaenochromis fryeri* and *Cynotilapia afra*. Ethology 113: 673–685.
18. Seehausen O (2004) Hybridization and adaptive radiation. Trends Ecol Evol 19: 198–207.
19. Mallet J (2007) Hybrid speciation. Nature 446: 279–283.
20. Seehausen O, Takimoto G, Roy D, Jokela J (2008) Speciation reversal and biodiversity dynamics with hybridization in changing environments. Mol Ecol 17: 30–44.
21. Pocklington R, Dill LM (1995) Predation on females or males: who pays for bright male traits? Anim Behav 49: 1122–1124.
22. Brooksmythe I, Detto T, Backwell PRY (2008) Female fiddler crabs settle for less: the travel costs of mate choice. Anim Behav 76: 1775–1781.
23. Hughes JJ, Ward D (1993) Predation risk and distance to cover affect foraging behaviour in Namib Desert gerbils. Anim Behav 46: 1243–1245.
24. Orrock JL, Danielson BJ, Brinkerhoff RJ (2004) Rodent foraging is affected by indirect, but not by direct, cues of predation risk. Behav Ecol 15: 433–437.
25. Dill LM, Fraser AHG (1984) Risk of predation and the feeding behavior of juvenile coho salmon (*Oncorhynchus kisutch*). Behav Ecol Sociobiol 16: 65–71.
26. Lima SL, Dill LM (1990) Behavioral decisions made under the risk of predation: a review and prospectus. Can J Zool 68: 619–640.

27. Rosenthal GG, De La Rosa Reyna XF, Kazianis S, Stephens MJ, Morizot DC, et al. (2003) Dissolution of sexual signal complexes in a hybrid zone between the swordtails *Xiphophorus birchmanni* and *Xiphophorus malinche* (Poeciliidae). Copeia. pp 299–307.
28. Culumber ZW, Fisher HS, Tobler M, Mateos M, Barber PH, et al. (2011) Replicated hybrid zones of *Xiphophorus* swordtails along an elevational gradient. Mol Ecol 20: 342–356.
29. Rauchenberger M, Kallman KD, Morizot DC (1990) Monophyly and geography of the Río Pánuco basin swordtails (genus *Xiphophorus*) with descriptions of four new species. Am Mus Novit 2975: 1–41.
30. Coleman SW, Rosenthal GG (2006) Swordtail fry attend to chemical visual cues in detecting predators and conspecifics. Plos ONE 1: e118.
31. Johnson JB, Basolo AL (2003) Predator exposure alters female mate choice in the green swordtail. Behav Ecol 14: 619–625.
32. Fisher HS, Wong BBM, Rosenthal GG (2006) Alteration of the chemical environment disrupts communication in a freshwater fish. Proc R Soc Lond B 273: 1187–1193.
33. Wong BBM, Rosenthal GG (2006) Female disdain for swords in a swordtail fish. Am Nat 167: 136–140.
34. Willis PM, Ryan MJ, Rosenthal GG (2011) Encounter rates with conspecific males influence female mate choice in a naturally hybridizing fish. Behav Ecol Advance Access published July 11, 2011.
35. Cummings ME, Rosenthal GG, Ryan MJ (2003) A private ultraviolet channel in visual communication. Proceedings of the Royal Society of London B 270: 897–904.
36. Fisher HS, Rosenthal GG (2007) Male swordtails court with an audience in mind. Biol Lett 3: 5–7.
37. Ryan MJ, Hews DK, Wagner WEJ (1990) Sexual selection on alleles that determine body size in the swordtail *Xiphophorus nigrensis*. Behav Ecol Sociobiol 26: 231–237.
38. Cummings M, Mollaghan D (2006) Repeatability and consistency of female preference behaviours in a northern swordtail, Xiphophorus nigrensis. Anim Behav 72: 217–224.
39. Walling CA, Royle NJ, Lindström J, Metcalfe NB (2010) Do female association preferences predict the likelihood of reproduction? Behav Ecol Sociobiol 64: 541–548.
40. Morris MR, Gass L, Ryan MJ (1995) Assessment and individual recognition of opponents in the pygmy swordtails *Xiphophorus nigrensis* and *X. multilineatus*. Behav Ecol Sociobiol 37: 303–310.
41. Burnham KP, Anderson DR (2002) Model selection and multimodel inference: a practical information-theoretic approach. New York: Springer-Verlag. 488 p.
42. Baayen RH, Davidson DJ, Bates DM (2008) Mixed-effects modeling with crossed random effects for subjects and items. J Mem Lang 59: 390–412.
43. Mayr E (1963) Animal species and evolution. Cambridge: Belknap Press.
44. Seehausen O, van Alphen JJM, Witte F (1997) Cichlid fish diversity threatened by eutrophication that curbs sexual selection. Science 277: 1808–1811.
45. Wirtz P (1999) Mother species-father species: unidirectional hybridization in animals with female choice. Anim Behav 58: 1–12.
46. Breden F, Stoner G (1987) Male predation risk determines female preference in the Trinidad guppy. Nature 329: 831–833.
47. Rosenthal GG, Martinez TYF, de Leon FJG, Ryan MJ (2001) Shared preferences by predators and females for male ornaments in swordtails. Am Nat 158: 146–154.
48. Basolo AL, Wagner WE (2004) Covariation between predation risk, body size and fin elaboration in the green swordtail, *Xiphophorus helleri*. Biol J Linn Soc 83: 87–100.
49. Fisher HS, Rosenthal, G G (2010) Relative abundance of *Xiphophorus* fishes and its effect on sexual communication. Ethology 116: 32–38.
50. Gordon M (1953) The ecological niche of the pygmy swordtail, *Xiphophorus pygmaeus*, in the Río Axtla, Mexico. Copeia 1953: 148–150. pp 148–150.
51. Tatarenkov A, Healey CIM, Avise JC (2010) Microgeographic population structure of green swordtail fish: genetic differentiation despite abundant migration. Mol Ecol 19: 257–268.
52. Kallman KD, SD Kazianis (2006) The genus *Xiphophorus* in Mexico and Central America. Zebrafish 3: 271–285.
53. Basolo AL (1995) Phylogenetic evidence for the role of a pre-existing bias in sexual selection. Proc R Soc B 259: 2223–2228.
54. Gow JL, Peichel CL, Taylor EB (2006) Contrasting hybridization rates between sympatric three-spined sticklebacks highlight the fragility of reproductive barriers between evolutionarily young species. Mol Ecol 15: 739–752.
55. Taylor EB, Boughman JW, Groenenbroom M, Sniatynski M, Schluter D, et al. (2006) Speciation in reverse: morphological and genetic evidence of the collapse of a three-spined stickleback (*Gasterosteus aculeatus*) species pair. Mol Ecol 15: 343–355.
56. Velema GJ (2010) Investigating the role of invase American signal crayfish (*Pacifastacus leniusculus*) in the collapse of the benthic-limnetic threespine stickleback species pair (*Gasterosteus aculeatus*) in Enos Lake, British Columbia. M.Sc. thesis, Vancouver: University of British Columbia. 81 p.
57. Hewitt GM (1988) Hybrid zones – natural laboratories for evolutionary studies. Trends Ecol Evol 3: 158–167.
58. Burke JM, Arnold ML (2001) Genetics and the fitness of hybrids. Ann Rev Genetics 35: 31–52.
59. Naisbit RE, Jiggins CD, Mallet J (2001) Disruptive sexual selection against hybrids contributes to speciation between *Heliconius cydno* and *Heliconius melpomene*. Proc R Soc B 268: 1849–1854.

2

Grassland Resistance and Resilience after Drought Depends on Management Intensity and Species Richness

Anja Vogel[1]*, Michael Scherer-Lorenzen[2], Alexandra Weigelt[3]

1 Institute of Ecology, Friedrich Schiller University Jena, Jena, Germany, **2** Michael Scherer-Lorenzen, Faculty of Biology – Geobotany, University of Freiburg, Freiburg, Germany, **3** Alexandra Weigelt, Institute of Biology, University of Leipzig, Leipzig, Germany

Abstract

The degree to which biodiversity may promote the stability of grasslands in the light of climatic variability, such as prolonged summer drought, has attracted considerable interest. Studies so far yielded inconsistent results and in addition, the effect of different grassland management practices on their response to drought remains an open question. We experimentally combined the manipulation of prolonged summer drought (sheltered vs. unsheltered sites), plant species loss (6 levels of 60 down to 1 species) and management intensity (4 levels varying in mowing frequency and amount of fertilizer application). Stability was measured as resistance and resilience of aboveground biomass production in grasslands against decreased summer precipitation, where resistance is the difference between drought treatments directly after drought induction and resilience is the difference between drought treatments in spring of the following year. We hypothesized that (i) management intensification amplifies biomass decrease under drought, (ii) resistance decreases with increasing species richness and with management intensification and (iii) resilience increases with increasing species richness and with management intensification.

We found that resistance and resilience of grasslands to summer drought are highly dependent on management intensity and partly on species richness. Frequent mowing reduced the resistance of grasslands against drought and increasing species richness decreased resistance in one of our two study years. Resilience was positively related to species richness only under the highest management treatment. We conclude that low mowing frequency is more important for high resistance against drought than species richness. Nevertheless, species richness increased aboveground productivity in all management treatments both under drought and ambient conditions and should therefore be maintained under future climates.

Editor: Jon Moen, Umea University, Sweden

Funding: The University of Jena funded the drought experiment within the ProChance program (www.uni-jena.de, grant to AW) and the University of Zurich gave additional support (www.uzh.de). The management experiment was supported by the Friedrich Schiller University of Jena and the ETH Zurich (funds from N. Buchmann). The Jena Experiment was funded by the Deutsche Forschungsgemeinschaft (FOR 456,www.dfg.de) with additional support from the Friedrich Schiller University of Jena, the Max Planck Society, the University of Zurich and the Swiss National Science Foundation (3100AQ-107531 to B. Schmid). The funders had no role in study design, data collection and analysis, decision to publish, or preparation of the manuscript.

* E-mail: anja.vogel@uni-jena.de

Introduction

There is agreement that the world's ecosystems will likely have to cope with future climatic changes, such as increased mean temperatures, a higher frequency of extreme weather events as well as changes in wind and precipitation patterns [1]. Among the different scenarios are projected decreases in summer precipitation and increases in autumn, winter and spring precipitation in subtropical and temperate regions (see also [2] that shows the same for the region of our study site). Along with the present and ongoing climate change, biodiversity is challenged by land-use changes to meet the growing demand for ecosystem services [3]. The consequences of those changes for ecosystem functioning, ecosystem services and human wellbeing have been the focus of research in the last few decades. It has been found that plant species diversity can have positive effects on multiple ecosystem processes [4] and if many times, places, functions and environmental changes were considered [5]. Aboveground plant biomass production, the most-studied process in biodiversity research, has been consistently found to rise in response to plant diversity in grasslands [6,7,8,9,10], an important finding for agricultural management. This positive relationship of species richness and productivity even holds under nutrient-rich conditions [11,12] and perturbations such as intense livestock grazing [13].

Beyond productivity itself, the temporal stability of biomass production has also been found to relate positively with species richness [14,15,16,17], meaning that temporal variability in productivity is lower in species-rich compared to species-poor communities. Besides the consistent results of diversity effects on temporal stability, the relationship between diversity and other aspects of stability, like resistance and resilience after perturbations (such as climatic changes) are mixed. Resistance, the degree of change after perturbations [18], is usually calculated as the difference of some performance measure between perturbed and unperturbed conditions and reflects the extent to which the mean of an ecosystem property changes after a single perturbation event. Resilience (in the sense of engineering resilience [19]), the time after perturbation until pre-perturbation levels are regained [18], is usually expressed as the rate of return of a variable at a given time after perturbation [20,21,22]. Both, resistance and resilience after perturbations were expected to increase with species richness [14]. In experimental studies it has been found to be true for resistance of biomass production against parasitism [23] and herbivore attack [24], but also neutral relationships between herbivore resistance and species richness have been reported [13,25]. The experimental results on resistance of grassland biomass production against drought show mostly negative or neutral relationships with species richness [21,22,26], depending on the level of productivity. It has been documented that more productive grasslands have a lower ability to withstand perturbation (lower resistance) than less productive communities, indicated by the higher loss of absolute aboveground biomass after a drought perturbation [22]. Thus, depending on the slope of the biodiversity-productivity relationship under unperturbed conditions, resistance either decreases with increasing species richness (negative slope) [21,22], increases with species richness (positive slope, which was only documented for a grassland experiment, where species richness was not independently manipulated) [26] or does not change (no significant relationship), when species richness has no effect on productivity like in the studies by Kahmen *et al.* [27] and Wang *et al.* [28]. It can thus be hypothesized that high species richness has a positive effect on the productivity of ecosystems but also results in a lower ability to resist future climatic changes. Besides the important ecosystem property to resist perturbations, a fast return to preperturbation levels (high resilience) would help to maintain ecosystem functioning. In very early studies it had been suggested that resistance and resilience are inversely correlated [29]. MacGillivray *et al.* [30] supported this prediction by measuring drought resistance and resilience in semi-natural grassland communities. It is also consistent with the work by van Ruijven *et al.* [22] who reported about experimental grasslands with decreasing resistance with species richness and increasing resilience. Although they confirmed that resistance rather depends on initial productivity of grasslands than on species richness *per se* (proportional resistance did not decrease with species richness), they found species richness to be a stronger predictor for recovery than productivity (recovery and proportional recovery increased with species richness).

Despite these findings, it remains unclear whether the idea that species richness promotes the productivity of grasslands at the consequence of lower resistance but higher resilience against perturbation, is applicable to a broad range of grasslands. For example, Gilgen and Buchmann [31] found that drought effects on biomass production vary between different grassland types along an altitudinal gradient. However, the explanatory factors of that study, namely environmental conditions and management strategies, were not independent and their individual effects could thus not be separated. In managed grasslands, management intensity has profound consequences on species richness and productivity. Fertilized grasslands, for example, have low species richness but a high productivity due to the high resource availability. Because of a low resistance with high productivity, one would expect that fertilization decreases the resistance of grasslands to perturbation. Indeed, Grime *et al.* [32] suggested that fertile grasslands would be less resistant compared to extensively managed grasslands because they contain more species with a high relative growth rate, which are highly susceptible to drought stress. However, higher growth rates should lead to faster regrowth after perturbation and we would expect a high resilience in fertilized grasslands.

We experimentally manipulated plant species richness, management intensity (amount of fertilizer and mowing frequencies, Figure 1) and prolonged summer drought separately to investigate the effects of biodiversity on resistance and resilience against drought perturbation in grasslands differing in their management intensity. We hypothesize that (i) biomass decreases under drought perturbation, especially under high management intensity due to the higher productivity under fertilization, (ii) resistance decreases with increasing species richness and with management intensification but proportional resistance (measure of resistance that is corrected for initial biomass productivity) does not change with species richness and (iii) resilience as well as proportional resilience increases with increasing species richness as well as with management intensity.

Results

The harvest at the end of the induced drought period showed significant increases in aboveground biomass with sown species richness in both years (Table 1, Figure 2). This was true for all management and drought treatments. Aboveground biomass was significantly lower under drought only in the frequently mown grasslands (M4F100, M4F200, Figure 2), i.e., the drought response of grasslands was affected by management intensity (Table 1).

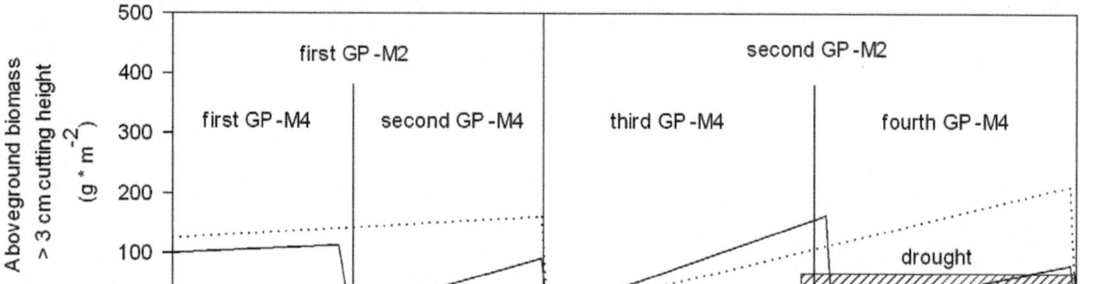

Figure 1. Time course of the growing season, including management and drought interventions of our study system. Stars indicate fertilization dates, vertical lines show mowing dates and dashed area represents our drought period. Frequently mown grassland types (M4, solid line) had four cuts per year and therefore four growth periods (first to fourth GP) previous to every cut. Normal mown grassland types (M2, dotted line) had two cuts per year and therefore two growth periods. Aboveground biomass is plotted as mean standing biomass above cutting height for each mowing grassland type.

Table 1. Summary of mixed effects models for aboveground biomass in August 2008 and 2009 to test for effects of diversity (numbers of initially sown species and functional group richness), management and drought treatments.

	Biomass August 2008					Biomass August 2009			
	df	AIC	L ratio	p		AIC	L ratio	p	
Nullmodel	11	1882.024				2058.232			
Block	14	1885.305	2.719	0.4370		2061.191	3.041	0.3854	
Species Richness (log-scale) = SR	15	1843.338	43.967	<0.0001	***	2026.597	36.595	<0.0001	***
Number of Functional Groups = FG	16	1845.331	0.007	0.9336		2026.316	2.280	0.1310	
Management	19	1815.300	36.031	<0.0001	***	1941.106	91.210	<0.0001	***
Drought	20	1789.755	27.545	<0.0001	***	1927.831	15.275	0.0001	***
Management × Drought	23	1786.773	8.982	0.0295	*	1930.126	3.705	0.2951	
Management × SR	26	1789.410	3.363	0.3390		1935.555	0.572	0.9029	
Management × FG	29	1793.380	2.030	0.5662		1937.716	3.838	0.2795	
Drought × SR	30	1793.947	1.433	0.2313		1939.645	0.072	0.7890	
Drought × FG	31	1794.484	1.463	0.2264		1941.613	0.032	0.8579	
SR × FG	32	1795.201	1.283	0.2574		1942.251	1.362	0.2431	

Models were fitted by stepwise inclusion of variables and likelihood ratio tests (L ratio) were applied to assess statistical significance of variables (p-values). Significance is given with * = $p<0.05$. ** = $p<0.01$. *** = $p<0.001$; df = degrees of freedom.

Resistance slightly decreased with increasing species richness in 2008 (Table 2, Figure 3 left) and with increasing mowing frequency in both years (Table 2, Figure 3). Proportional resistance was not affected by species richness but still decreased with increasing mowing frequency in 2008 (Table 2, Figure 3). We found no effect of the interaction of species diversity and management treatment (mowing frequency or amount of fertilizer) on resistance or proportional resistance. Resilience was positively related to species richness only in the most intensively managed grassland (significant interaction of SR and F in Table 3, Figure 4). Proportional resilience was not affected by species richness (Table 3).

Discussion

In our study we independently manipulated biodiversity loss, management intensity and drought. We were therefore able to distinguish the effects of all single treatments from their interactions on aboveground biomass. We found that the response of experimental grasslands to drought depends on management intensity. Aboveground biomass decreased after induced summer drought only in grasslands with frequent mowing (four times per year), not in grasslands with only two cuts per year. Differences in growth status due to mowing may explain our findings. The low canopy height after mowing generally increases soil surface evaporation through increased wind speed at ground level and low plant cover. Hu et al. [33] reported increased evaporation with decreased canopy density, measured as leaf area index (LAI), especially at LAI values lower than 2 $m^2 * m^{-2}$. After mowing, LAI in our experiment was close to 0 $m^2 * m^{-2}$ (data not shown). In addition to reduced precipitation this would mean a further decrease of soil moisture in the frequently mown (M4-) treatment compared to the normal mown (M2-) treatment. Consequently, a decrease of aboveground biomass only occurs if drought hits the communities at an early growth status, when soil is not sufficiently covered by plants. Thus, grasslands with high mowing frequency and hence frequently low LAI have to be considered as more sensitive to drought. In contrast, extensively mown grasslands would only suffer from drought if it occurs in the regrowth phase after mowing. This was not the case in both of the study years for our grasslands mown only twice a year (M2). They were well-advanced in height growth at the beginning of the drought treatment. The same might have been the case in the study of Jentsch et al. [34], where no effect on aboveground productivity could be detected when drought was induced during peak growing season in June [34]. In contrast Pfisterer and Schmid [21] found negative drought effects in their grassland with the same management as in our experiment (two cuts per year and drought induction before late cut). The longer drought period (and therefore shorter time for normal regrowth) compared to our experiment could be one explanation for contrasting results but also site specific (climatic conditions) and treatment differences have to be taken into account. This study site has mean annual precipitation amounts twice as high as in our site (Table 4) and rain shelters were adjusted close to vegetation, thus a stronger heat effect could be assumed.

Along with our second hypothesis we wanted to test, whether the observed relationships of resistance and species richness [22] change with management intensity. We only found slightly decreasing resistance with species richness and this effect of species richness did not change with management intensity. Instead we found a strong effect of management intensity itself on resistance because high mowing frequency decreased the resistance of grasslands against drought. High mowing frequency resulted in both lower absolute biomass accumulation (resistance) as well as lower relative biomass accumulation under drought (proportional resistance) compared to normal mown stands. Our results support the idea that species richness affects resistance due to increasing productivity with species number, and not due to number of species *per se* [22], because proportional resistance was not related to species richness. De Boeck *et al.* [35] found that more productive and species-rich communities have a higher evapotranspiration and water use efficiency compared to monocultures. They concluded that decreased aboveground biomass is one potential mechanism for saving water, because it reduces the transpirational surface of the canopy.

After the second drought in the following year, resistance was constant across the plant diversity gradient, meaning that there

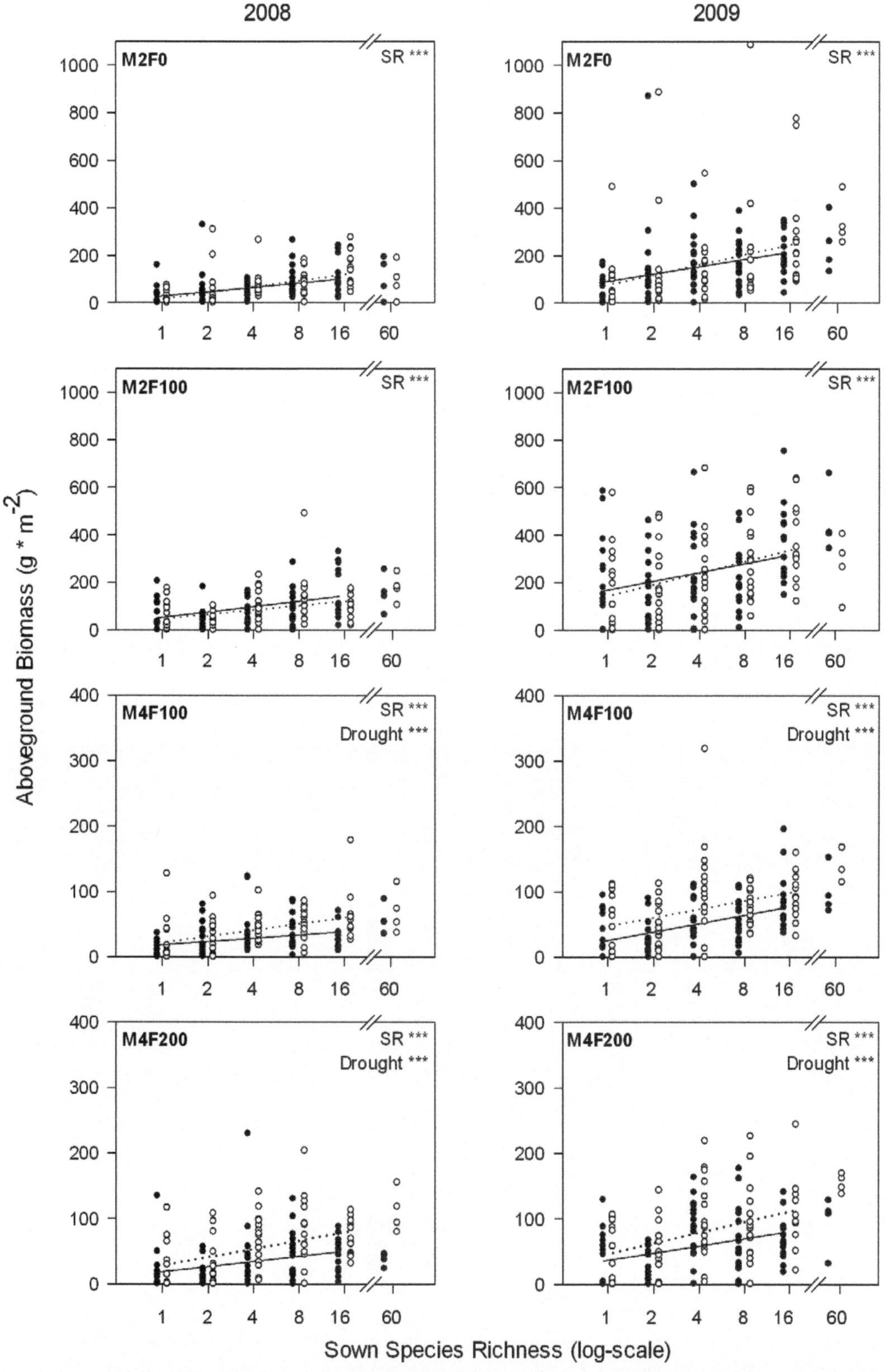

Figure 2. Aboveground biomass across sown species richness gradient for each management treatment. Aboveground biomass at the end of the induced drought period in August 2008 (left column) and 2009 (right column) measured regrowth since the last cut (M2-types: June; M4-types: July). The ambient treatment is given in open circles (dotted regression line) and drought treatment in closed circles (solid regression line). Significant effects obtained from mixed models for every single management treatment per year: SR = effect of sown species richness (linear), drought = difference in drought and ambient treatments, * = $p<0.05$, ** = $p<0.01$, *** = $p<0.001$.

Table 2. Summary of mixed effects models for resistance and proportional resistance of aboveground biomass after drought in August 2008 and 2009 to test for effects of diversity (numbers of initially sown species and functional group richness), management (separated into mowing and fertilizer amounts) and drought treatments.

		Resistance 2008				Resistance 2009					proportional Resistance 2008				proportional Resistance 2009		
	df	AIC	L ratio	*p*		AIC	L ratio	*p*		df	AIC	L ratio	*p*		AIC	L ratio	*p*
Nullmodel	3	1026.667				1158.463				6	914.183				1025.004		
Block	6	1028.312	4.355	0.2256		1157.890	6.573	0.0868		9	917.457	2.726	0.4358		1028.257	2.746	0.4324
Species richness (log-scale) = SR	7	1025.371	4.941	0.0262	*	1157.731	2.598	0.1417		10	918.179	1.278	0.2582		1030.178	0.080	0.7778
Number of functional groups = FG	8	1025.530	1.840	0.1749		1159.167	0.563	0.4529		11	919.132	1.047	0.3062		1032.016	0.161	0.6880
Mowing = M	9	1007.267	20.263	<0.0001	***	1145.509	15.658	0.0001	***	12	909.735	11.397	0.0007	***	1030.359	3.657	0.0558
M×SR	10	1007.980	1.287	0.2565		1147.369	0.141	0.7077		13	911.734	0.001	0.9757		1032.358	0.001	0.9788
M×FG	11	1008.535	1.445	0.2293		1149.119	0.250	0.6169		14	910.415	3.320	0.0685		1034.324	0.034	0.8534
Fertilizer amount = F	12	1010.209	0.325	0.5685		1150.820	0.298	0.5851		15	911.540	0.874	0.3498		1036.322	0.002	0.9626
F×SR	13	1011.601	0.608	0.4354		1151.880	0.940	0.3322		16	913.419	0.122	0.7272		1038.275	0.047	0.8290
F×FG	14	1010.700	2.901	0.0885		1152.796	1.084	0.2978		17	912.595	2.824	0.0929		1039.751	0.524	0.4689

Models were fitted by stepwise inclusion of variables and likelihood ratio tests (L ratio) were applied to assess statistical significance of variables (p-values). Significance is given with * = $p<0.05$, ** = $p<0.01$, *** = $p<0.001$; df = degrees of freedom.

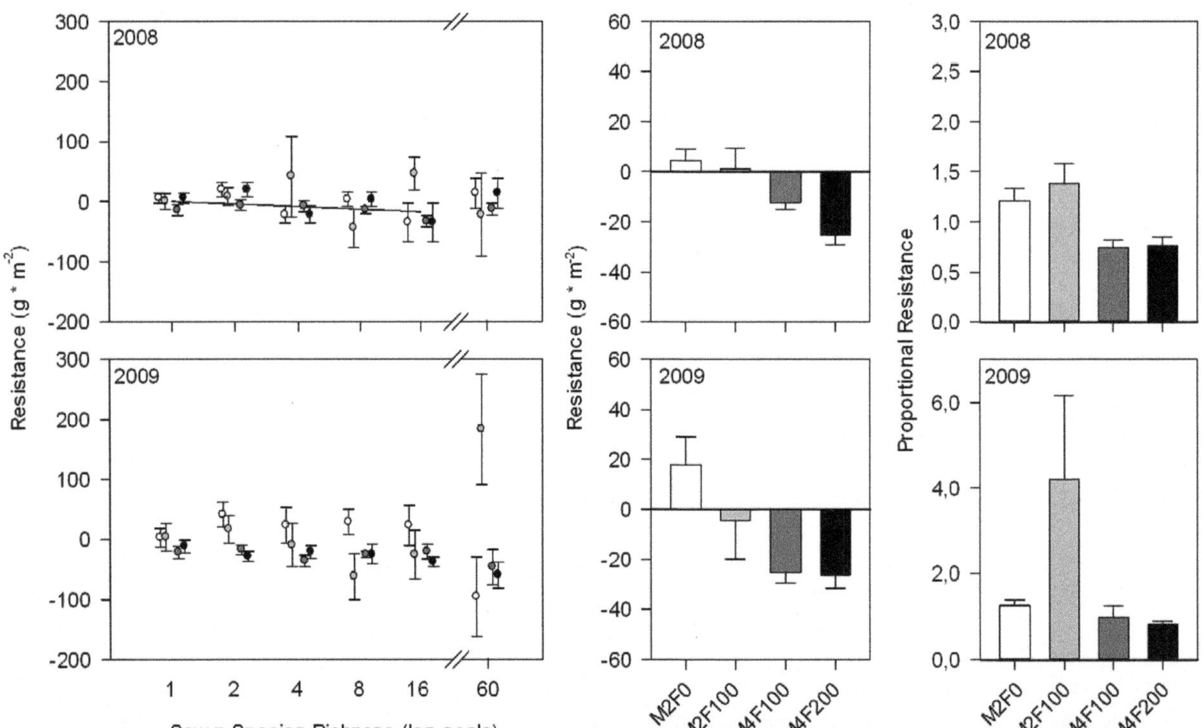

Figure 3. Resistance of biomass production for 2008 and 2009. Resistance was calculated as the difference of drought and corresponding ambient treatments at the end of the drought period in August and was plotted against species richness (left) and management treatments (middle). Regression lines are given for significant effects of species richness. Proportional resistance (ratio of drought to ambient treatment) was plotted against management treatment (right). Management treatments are shown in white (M2F0), gray (M2F100), dark gray (M4F100) and black (M4F200).

Table 3. Summary of mixed effects models for resilience and proportional resilience of aboveground biomass of the first cut in spring 2009 to test for effects of diversity (numbers of initially sown species and functional group richness), management (separated into mowing and fertilizer amounts) and drought treatments.

		Resilience Spring 2009				**proportional Resilience 2009**		
	df	**AIC**	**L ratio**	***p***		**AIC**	**L ratio**	***p***
Nullmodel	6	3898,242				1024,902		
Block	9	3903,177	1,065	0.7855		1026,706	4,196	0.2410
Species richness (log-scale) = SR	10	3903,771	1,406	0.2357		1026,624	2,081	0.1491
Number of functional groups = FG	11	3905,674	0,097	0.7557		1028,398	0,226	0.6345
Mowing = M	12	3905,225	2,449	0.1176		1029,770	0,628	0.4280
M×SR	13	3907,219	0,006	0.9395		1030,613	1,157	0.2821
M×FG	14	3909,208	0,011	0.9161		1032,446	0,167	0.6826
Fertilizer amount = F	15	3907,540	3,668	0.0555		1032,955	1,491	0.2221
F×SR	16	3904,600	4,939	0.0263	*	1033,658	1,297	0.2548
F×FG	17	3905,130	1,470	0.2253		1033,873	1,784	0.1816

Models were fitted by stepwise inclusion of variables and likelihood ratio tests (L ratio) were applied to assess statistical significance of variables (p-values). Significance is given with * = $p<0.05$, ** = $p<0.01$, *** = $p<0.001$; df = degrees of freedom.

was no stronger decrease in absolute amounts of aboveground biomass in the species rich compared to the species poor communities due to drought. Different precipitation patterns might explain this year-to-year changes, since rainfall just before the induced drought period was different in both years (Table 4).

The main rain events of 2008 occurred in the very wet April, whereas May and June were unusually dry. Soil moisture (volumetric water content measured in 8 cm depth of an unsheltered reference area) decreased to an average of 17.0% in summer. In contrast in 2009, rain events were regularly distributed

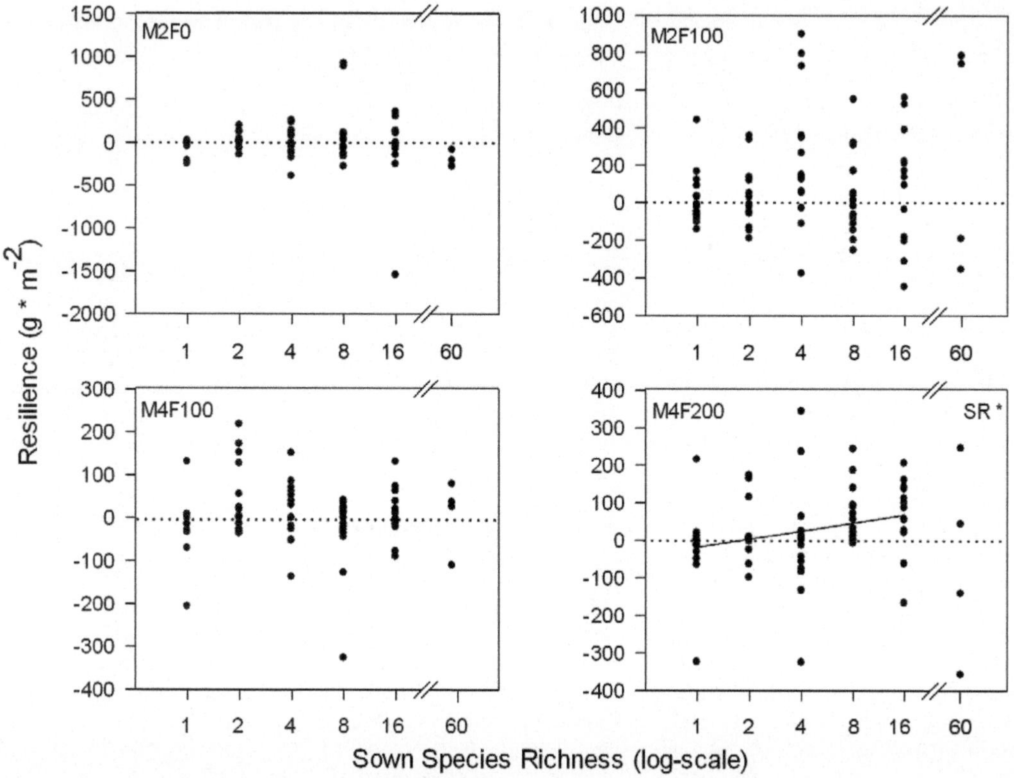

Figure 4. Resilience of biomass over sown species richness for each management treatment. Resilience of biomass was calculated as the difference of drought and corresponding ambient treatments for the first harvest in spring 2009 (M2-types: June, M4-types: April). Regression lines are given for significant effects obtained from linear models for every single management treatment per year: SR = effect of sown species richness (linear), drought = difference in drought and ambient treatments, * = p<0.05, ** = p<0.01, *** = p<0.001.

Table 4. Climatic parameters measured on field site during the two study years 2008 and 2009 with reference period 1961–1990 measured by the German Weather Service DWD in Jena, city center.

Month	Air temperature (°C)			Precipitation (mm)			Soil moisture (Vol%)	
	1961–90	2008	2009	1961–90	2008	2009	2008	2009
J	0.40	5.00	−3.09	37.00	24.50	9.00	37.29	22.11
F	1.40	3.76	1.15	34.00	20.40	33.70	37.15	33.31
M	4.80	5.11	5.04	43.00	55.80	42.50	37.97	37.01
A	8.60	7.90	11.58	57.00	91.80	73.70	37.35	31.39
M	13.40	14.05	13.89	62.00	22.00	62.60	25.62	31.05
J	16.70	17.13	15.01	75.00	54.40	52.90	21.74	28.58
J	18.20	18.51	18.34	52.00	40.60	85.10	17.75	31.29
A	17.40	17.90	18.59	63.00	58.60	14.60	16.61	22.34
S	14.20	12.05	14.56	42.00	50.00	53.60	21.71	23.67
O	9.80	9.13	8.42	39.00	55.30	47.30	26.30	28.70
N	5.00	5.60	8.06	41.00	19.90	68.30	28.16	34.00
D	1.70	1.28	0.64	42.00	38.60	80.00	30.59	36.22
Year	9.30	11.75	11.48	587.00	531.90	623.30	28.19	29.97

Values represent monthly means (temperature, soil moisture) or sums (precipitation).

and average soil moisture was at 25.3% in summer. A higher soil moisture at the beginning of the drought treatment in 2009 may have stimulated a better growth of the plants compared to the year before and might have more rapidly lowered the loss of soil water through evaporation. Species-rich communities are thought to be more water efficient [35,36], i.e. they produce more biomass per unit of water. Such communities could therefore benefit from higher soil moisture compared to species-poor communities. In consequence, they could be proportionally less affected from drought stress than species poor communities. Whether annual precipitation patterns would explain the different resistance-species richness patterns between years, can only be underpinned by long-term data.

Resilience increased with species richness as previously reported [22] but only under highest management intensity. In contrast, we found that species richness only affected absolute resilience, whereas the proportion of biomass increase in previously dried subplots compared to their ambient conditions (proportional resilience) did not change with species richness. Furthermore, resilience and species richness were only positively related in the very intensively managed grasslands (frequently mown and high fertilized). The positive relationship of resilience and species richness in the M4F200 management treatment was weak but significant and might be strongly due to the positive responses of the 8-species-mixtures. It is known that species richness increases shoot-root-ratios indicating a better resource use with species richness [37]. We can only speculate why this was more effective in the previously dried subplots compared to the corresponding ambient treatment under highest management intensity. It may be explained in part by higher fertilizer amounts affecting below-ground processes. It has been reported that drought did not necessarily increase root growth [38,39,40]. Together with a decreased aboveground biomass, drought may have decreased shoot-root-ratios especially in more diverse mixtures. The corresponding ambient treatments might have a much higher shoot-root-ratio due to the higher biomass and the well-known fact that fertilization decreases root growth relative to shoot growth. After drought, this lower shoot-root-ratio of the dried subplots may be a prerequisite for better aboveground biomass allocation in comparison to the ambient in the intensively fertilized treatment. Furthermore, the plasticity and intensity of growth responses of roots and shoots under drought is highly species-specific [39,41], and thus plant species composition is an important determinant of community root and shoot growth [31]. This could explain that especially the 8-species-mixtures had a high resilience, when intensively managed.

Our results indicate that management intensity affects the resistance of grasslands after drought, with growth phenology being the underlying cause: grasslands at the regrowth stage are more sensitive to decreased precipitation and loose more biomass, than grasslands at later stages with a more fully developed canopy. As a consequence, low mowing frequency enhances drought resistance because of a lower probability to face reduced precipitation during the regrowth stage. Nevertheless, species richness and aboveground biomass were positively related even under drought conditions, which shows that biomass yield is higher the more diverse a community is, no matter under which management intensity and climatic conditions the community grew. Thus high plant species diversity should be maintained under future climates.

Materials and Methods

Study site and experimental treatments

We used the gradient of plant species richness established in the Jena biodiversity experiment and superimposed a gradient of management intensity and a drought treatment. The field site is located in the floodplain of the river Saale in Jena (Thuringia, Germany, 50°55′N, 11°35′E, 130 m above sea level) with a mean annual air temperature of 9.3°C and precipitation amount of 587 mm measured during 1961–1990. The study site was used as a highly fertilized arable field before [42]. The soils are loamy Eutric Fluvisols. In 2002, 80 grassland plots of different plant species mixtures were established from a pool of 60 mesophilic grassland species from Molinio-Arrhenateretea meadows typical for the regional alluvial plains.

The gradient in plant species richness (1, 2, 4, 8, 16 and 60 species) in the Jena experiment is combined with a gradient in the number of functional groups (1, 2, 3 or 4 functional groups namely grasses, small herbs, tall herbs and legumes) with about four replicates per species richness×functional groups combination. Mixtures were arranged in a randomized block design to account for edaphic variations with increasing distance to the river Saale. Experimental plots were maintained by weeding blockwise in two annual weeding campaigns. For further details see [42].

The gradient in management intensity was established in 2006 with four subplots on every plot of the 80 plant mixtures. Management varied in mowing regime (M2: two cuts, M4: four cuts per year) and the amount of NPK-fertilizer application (F0: no fertilizer; F100: 100.0 kg N ha^{-1} a^{-1}, 43.6 kg P ha^{-1} a^{-1}, 83.0 kg K ha^{-1} a^{-1}; F200: 200 kg N ha^{-1} a^{-1}, 87.2 kg P ha^{-1} a^{-1}, 166.0 kg K ha^{-1} a^{-1}) and was combined as follows: M2F0, M2F100, M4F100, M4F200 [11]. All three fertilizer treatments were arranged randomly on an area of each 1.6 m×4 m within the main plots of 20×20 m, while M2F0 treatment was always located in the central core area of the plots, representing the standard management of the whole field site. Fertilization was done twice a year (31 March and 23 June 2008, 31 March and 16 June 2009) and mowing was done in spring (end of April, only M4-subplots), in early summer (beginning of June, all subplots), end of July (M4-subplots) and in late summer at the beginning of September (all subplots).

Drought was induced in 2008 and 2009 using transparent rain shelters during six weeks in summer previous to the last annual cut (25 July to 2 September 2008 and 16 July to 1 September 2009, Figure 1). Rain shelters were made of LDPE greenhouse film (www.dm-folien.com) in 2008 and of PVC sheets (www.paruschke-kunststoffe.de, product code: PVCSPK7018K10) in 2009 because of its higher durability. Rain shelters were inclined in a height of 1.3 to 1.5 m to enable ventilation and runoff of rain water in one direction 1 m away from our core area. Control subplots remained unsheltered and received ambient precipitation. We established one sheltered (hereafter named "drought" treatment) and one unsheltered subplot (named "ambient") of 1.6 m×2 m size for each management treatment in each of the 80 plots covering the whole diversity gradient. Measurements of the soil water content revealed a soil moisture decrease of about 17% in the drought treatment compared to the ambient treatment. The open-side construction of the rain shelters could not prevent a temperature increase of about 1.5–2.2°C on soil surface in the drought treatments, but no warming was detected at 20 cm height. The PVC sheets reduced photosynthetically active radiation (PAR) by 28% maximum. In 2009 we established an additional roof control in all plots, e.g. a sheltered subplot where we added collected rain water, to measure the pure roof effect (heat, altered light conditions) on our response variable. We found that the results of the roof control were similar to those of the ambient treatment (data not shown).

Since the Jena Experiment was established, climatic conditions were measured by a weather station directly on field site so that we were able to document weather data during our experimental phase 2008–2009 (Table 4). Rain shelters excluded 59.5 mm precipitation in 2008 (reduction of 40% compared to unroofed subplots during the summer July–September) and 53.7 mm in 2009 (reduction of 35%).

Data collection

All measurements were restricted to a central area of 1 m×1 m on every subplot to minimize edge effects (precipitation, varying soil nutrients, different height of neighboring vegetation). We clipped aboveground biomass of a 20 cm×50 cm area at 3 cm height above soil surface two days prior to every mowing event of a subplot. Biomass was sorted into sown species, unsown weeds and dead plant material, dried until constant weight (70°C, 48 h) and weighed. Here we present the results of the biomass of sown species.

Statistical analysis

We calculated resistance and resilience from our biomass data according to van Ruijven and Berendse [22]. In contrast to that study, which compared perturbed and unperturbed plots in two consecutive years, we were able to compare data from each drought treatment with its corresponding ambient treatment at the same time. Resistance of biomass production was calculated as the difference of biomass under perturbed and unperturbed conditions (drought - ambient) at the end of the drought period in August. Proportional resistance calculated as the ratio of drought to ambient treatment biomass was determined to account for productivity effects on resistance. Resilience determines the change in biomass production after perturbation and was calculated as difference of post-drought biomass and the corresponding ambient treatment from the first harvest after drought (M4-subplots: April 2009, M2-subplots: June 2009). Proportional resilience was calculated as the ratio of post-drought biomass and the corresponding ambient treatment from the first harvest after drought. The proportional values indicate, whether the ratio of biomass decrease or increase due to drought change.

We analyzed the data with mixed effects models using the *nlme*-package of R 2.8.1. to account for the nested design of our experiment (drought/ambient nested within management nested within plots of different diversity levels). Because we were not interested in effects of each species combination, we used plots as random factors in the model, as well as management and our drought treatment. We fitted a series of models by stepwise inclusion of fixed effects. First, we included block in our fixed term to account for all edaphic variation in the field and the blockwise management and data sampling. Then we included sown species richness and functional group richness as diversity factors and our experimental treatments (management, drought) and their interaction with diversity treatments stepwise in the fixed term of the models with the maximum likelihood method. We applied likelihood ratio tests for model comparison and estimating the significance of the fixed effects. We are aware that sown diversity and realized diversity can vary between the management treatments, because management intensification is expected to reduce species diversity [43]. We therefore fitted additional models with realized species richness and realized functional group richness instead of sown species and functional group richness. These models presented the same conclusions as with design variables. For better comparison with other experimental results, we present results of sown diversity effects in the paper and realized diversity effects in the (Table S1, S2, S3, Figure S1, S2, S3 and Methods S1 for information on data acquisition of realized species richness). To meet the assumptions of mixed effects models (normally distributed within group errors and random effects), biomass and resistance were log transformed. When data were heteroscedastic (in case of resilience), variance functions were included [44]. The 60 species mixtures merely serve as reference plots and are excluded from analysis as they are not fully compatible with the experimental design of the experiment.

Supporting Information

Methods S1 Realized species richness was recorded during drought period in August 2008 and 2009 in every subplot. We recorded presence and absence of every single sown species in 10 squares of 1 dm^2 size along one transect and repeated this three times within our study area of 1 m^2. The realized species number was the sum of all species that were present in at least one out of the 30 squares.

Figure S1 Aboveground biomass across realized species richness gradient for each management treatment. Aboveground biomass at the end of the induced drought period in August 2008 (left) and 2009 (right) for each management treatment since the last cut (M2-types: June; M4-types: July). The ambient treatment is given in open circles (dotted regression line) and drought treatment in closed circles (solid regression line). Significant effects obtained from mixed models for every single management treatment per year: SR = effect of realized species richness (linear), drought = difference in drought and ambient treatments, * = $p<0.05$, ** = $p<0.01$, *** = $p<0.001$.

Figure S2 Resistance in biomass production over realized species richness. Resistance was calculated as the difference of drought and corresponding ambient treatments. Realized species richness represents the mean of realized species numbers of drought and ambient treatment. Management treatments are shown in white (M2F0), gray (M2F100), dark gray (M4F100), black (M4F200).

Figure S3 Resilience of biomass over realized species richness for each management treatment. Resilience was calculated as the difference of drought and corresponding ambient treatments for the first harvest in spring 2009 (M2-types: June, M4-types: April). Realized species richness represents the mean of realized species numbers of drought and ambient treatment. Regression lines are given for significant effects obtained from linear models for every single management treatment per year: SR = effect of realized species richness (linear), drought = difference in drought and ambient treatments, * = $p<0.05$, ** = $p<0.01$, *** = $p<0.001$.

Table S1 Summary of mixed effects models for aboveground biomass in August 2008 and 2009 to test for effects of management, drought and diversity (realized numbers of species and functional groups) treatments.

Table S2 Summary of mixed effects models for resistance and proportional resistance of aboveground biomass after drought in August 2008 and 2009 to test for effects of diversity (realized numbers of species and functional groups) and management treatments (separated into mowing and fertilizer amounts).

Table S3 Summary of mixed effects models for resilience computed as the difference between previously drought and ambient treatment in aboveground biomass as well as for proportional resilience of the first cut in spring 2009 to test for effects of management (separated into mowing and fertilizer amounts) and diversity (realized numbers of species and functional groups) treatments.

Acknowledgments

We thank numerous student helpers, the gardeners Steffen Eismann, Steffen Ferber, Silke Hengelhaupt, Ute Köber, Katja Kunze, Heike Scheffler as well as Gerlinde Kratzsch, Victoria Stabrey, Stefan Lorenz, Kymbat Dikambaeva and interns Kyle Siefers and Kathrin Erfurt for their help in maintaining the field site and help during data collection. We especially thank Victor Malakhov for his help during installation and maintaining the rain shelters as well as help during biomass harvests. We also want to thank Bernhard Schmid for his support throughout the experiment as well as Dan Flynn and two anonymous reviewers for helpful comments on the manuscript.

Author Contributions

Conceived and designed the experiments: AW MSL. Performed the experiments: AV AW. Analyzed the data: AV. Contributed reagents/materials/analysis tools: AW MSL. Wrote the paper: AV AW MSL. Interpretation of results: AV AW MSL.

References

1. IPCC (2007) Climate Change 2007: Synthesis Report Intergovernmental Panel on Climate Change.
2. Jacob D, Göttel H, Kotlarski S, Lorenz P, Sieck K (2008) Klimaauswirkungen und Anpassung in Deutschland-Phase 1: Erstellung regionaler Klimaszenarien für Deutschland. In:, , Umweltbundesamt, editor (2008) Climate Change 11-08. Dessau-Roßlau: Umweltbundesamt.
3. Millennium Ecosystem Assessment M, editor (2005) Ecosystems and Human Well-being: Synthesis. Washington, DC: Island Press.
4. Hector A, Bagchi R (2007) Biodiversity and ecosystem multifunctionality. Nature 448: 188–191.
5. Isbell F, Calcagno V, Hector A, Connolly J, Harpole WS, et al. (2011) High plant diversity is needed to maintain ecosystem services. Nature 477: 199–U196.
6. Hector A, Schmid B, Beierkuhnlein C, Caldeira MC, Diemer M, et al. (1999) Plant diversity and productivity experiments in European grasslands. Science 286: 1123–1127.
7. Tilman D, Reich PB, Knops J, Wedin D, Mielke T, et al. (2001) Diversity and productivity in a long-term grassland experiment. Science 294: 843–845.
8. van Ruijven J, Berendse F (2003) Positive effects of plant species diversity on productivity in the absence of legumes. Ecology Letters 6: 170–175.
9. Marquard E, Weigelt A, Temperton VM, Roscher C, Schumacher J, et al. (2009) Plant species richness and functional composition drive overyielding in a six-year grassland experiment. Ecology 90: 3290–3302.
10. Reich PB, Knops J, Tilman D, Craine J, Ellsworth D, et al. (2001) Plant diversity enhances ecosystem responses to elevated CO2 and nitrogen deposition (vol 410, pg 809, 2001). Nature 411: 824–+.
11. Weigelt A, Weisser WW, Buchmann N, Scherer-Lorenzen M (2009) Biodiversity for multifunctional grasslands: equal productivity in high-diversity low-input and low-diversity high-input systems. Biogeosciences 6: 1695–1706.
12. Kirwan L, Luescher A, Sebastia MT, Finn JA, Collins RP, et al. (2007) Evenness drives consistent diversity effects in intensive grassland systems across 28 European sites. Journal of Ecology 95: 530–539.
13. Isbell FI, Wilsey BJ (2011) Increasing native, but not exotic, biodiversity increases aboveground productivity in ungrazed and intensely grazed grasslands. Oecologia 165: 771–781.
14. Yachi S, Loreau M (1999) Biodiversity and ecosystem productivity in a fluctuating environment: The insurance hypothesis. Proceedings of the National Academy of Sciences of the United States of America 96: 1463–1468.
15. Tilman D, Reich PB, Knops JMH (2006) Biodiversity and ecosystem stability in a decade-long grassland experiment. Nature 441: 629–632.
16. van Ruijven J, Berendse F (2007) Contrasting effects of diversity on the temporal stability of plant populations. Oikos 116: 1323–1330.
17. Eisenhauer N, Milcu A, Allan E, Nitschke N, Scherber C, et al. (2011) Impact of above- and below-ground invertebrates on temporal and spatial stability of grassland of different diversity. Journal of Ecology 99: 572–582.
18. Pimm SL (1984) The complexity and stability of ecosystems. Nature 307: 321–326.
19. Gunderson LH (2000) Ecological resilience - in theory and application. Annual Review of Ecology and Systematics 31: 425–439.
20. Tilman D (1996) Biodiversity: Population versus ecosystem stability. Ecology 77: 350–363.
21. Pfisterer AB, Schmid B (2002) Diversity-dependent production can decrease the stability of ecosystem functioning. Nature 416: 84–86.
22. van Ruijven J, Berendse F (2010) Diversity enhances community recovery, but not resistance, after drought. Journal of Ecology 98: 81–86.
23. Joshi J, Matthies D, Schmid B (2000) Root hemiparasites and plant diversity in experimental grassland communities. Journal of Ecology 88: 634–644.

24. Pfisterer AB, Diemer M, Schmid B (2003) Dietary shift and lowered biomass gain of a generalist herbivore in species-poor experimental plant communities. Oecologia 135: 234–241.
25. Scherber C, Heimann J, Kohler G, Mitschunas N, Weisser WW (2010) Functional identity versus species richness: herbivory resistance in plant communities. Oecologia 163: 707–717.
26. Tilman D, Downing JA (1994) Biodiversity and Stability in Grasslands. Nature 367: 363–365.
27. Kahmen A, Perner J, Buchmann N (2005) Diversity-dependent productivity in semi-natural grasslands following climate perturbations. Functional Ecology 19: 594–601.
28. Wang YF, Yu SX, Wang J (2007) Biomass-dependent susceptibility to drought in experimental grassland communities. Ecology Letters 10: 401–410.
29. Leps J, Osbornovakosinova J, Rejmanek M (1982) Community stability, complexity and species life-history strategies. Vegetatio 50: 53–63.
30. Macgillivray CW, Grime JP, Band SR, Booth RE, Campbell B, et al. (1995) Testing predictions of the resistance and resilience of vegetation subjected to extreme events. Functional Ecology 9: 640–649.
31. Gilgen AK, Buchmann N (2009) Response of temperate grasslands at different altitudes to simulated summer drought differed but scaled with annual precipitation. Biogeosciences 6: 2525–2539.
32. Grime JP, Brown VK, Thompson K, Masters GJ, Hillier SH, et al. (2000) The response of two contrasting limestone grasslands to simulated climate change. Science 289: 762–765.
33. Hu ZM, Yu GR, Zhou YL, Sun XM, Li YN, et al. (2009) Partitioning of evapotranspiration and its controls in four grassland ecosystems: Application of a two-source model. Agricultural and Forest Meteorology 149: 1410–1420.
34. Jentsch A, Kreyling J, Elmer M, Gellesch E, Glaser B, et al. (2011) Climate extremes initiate ecosystem-regulating functions while maintaining productivity. Journal of Ecology 99: 689–702.
35. De Boeck HJ, Lemmens C, Bossuyt H, Malchair S, Carnol M, et al. (2006) How do climate warming and plant species richness affect water use in experimental grasslands? Plant and Soil 288: 249–261.
36. Van Peer L, Nijs I, Reheul D, De Cauwer B (2004) Species richness and susceptibility to heat and drought extremes in synthesized grassland ecosystems: compositional vs physiological effects. Functional Ecology 18: 769–778.
37. Bessler H, Temperton VM, Roscher C, Buchmann N, Schmid B, et al. (2009) Aboveground overyielding in grassland mixtures is associated with reduced biomass partitioning to belowground organs. Ecology 90: 1520–1530.
38. Kreyling J, Beierkuhnlein C, Elmer M, Pritsch K, Radovski M, et al. (2008) Soil biotic processes remain remarkably stable after 100-year extreme weather events in experimental grassland and heath. Plant and Soil 308: 175–188.
39. Weisshuhn K, Auge H, Prati D (2011) Geographic variation in the response to drought in nine grassland species. Basic and Applied Ecology 12: 21–28.
40. Molyneux DE, Davies WJ (1983) Rooting pattern and water relations of 3 pasture grasses growing in drying soil. Oecologia 58: 220–224.
41. Foulds W (1978) Response to soil-moisture supply in 3 leguminous species.1. growth, reproduction and mortality. New Phytologist 80: 535–545.
42. Roscher C, Schumacher J, Baade J, Wilcke W, Gleixner G, et al. (2004) The role of biodiversity for element cycling and trophic interactions: an experimental approach in a grassland community. Basic and Applied Ecology 5: 107–121.
43. Cop J, Vidrih M, Hacin J (2009) Influence of cutting regime and fertilizer application on the botanical composition, yield and nutritive value of herbage of wet grasslands in Central Europe. Grass and Forage Science 64: 454–465.
44. Pinheiro JC, Bates DM, eds. Mixed-Effects Models in S and S-PLUS. New York: Springer Verlag. 528 p.

3

The Influence of Variable Rainfall Frequency on Germination and Early Growth of Shade-Tolerant Dipterocarp Seedlings in Borneo

Michael J. O'Brien[1]*, Christopher D. Philipson[1], John Tay[2], Andy Hector[1]

1 Institute of Evolutionary Biology and Environmental Studies, University of Zurich, Zurich, Switzerland, 2 School of International Tropical Forestry, University Malaysia Sabah, Kota Kinabalu, Sabah, Malaysia

Abstract

Climate change induced alterations to rainfall patterns have the potential to affect the regeneration dynamics of plant species, especially in historically everwet tropical rainforest. Differential species response to infrequent rainfall may influence seed germination and seedling establishment in turn affecting species distributions. We tested the role of watering frequency intervals (from daily to six-day watering) on the germination and the early growth of Dipterocarpaceae seedlings in Borneo. We used seeds that ranged in size from 500 to 20,000 mg in order to test the role of seed mass in mediating the effects of infrequent watering. With frequent rainfall, germination and seedling development traits bore no relationship to seed mass, but all metrics of seedling growth increased with increasing seed mass. Cumulative germination declined by 39.4% on average for all species when plants were watered at six-day intervals, and days to germination increased by 76.5% on average for all species from daily to six-day intervals. Final height and biomass declined on average in the six-day interval by 16% and 30%, respectively, but the percentage decrease in final size was greater for large-seeded species. Rooting depth per leaf area also significantly declined with seed mass indicating large-seeded species allocate relatively more biomass for leaf production. This difference in allocation provided an establishment advantage to large-seeded species when water was non-limiting but inhibited their growth under infrequent rainfall. The observed reduction in the growth of large-seeded species under infrequent rainfall would likely restrict their establishment in drier microsites associated with coarse sandy soils and ridge tops. In total, these species differences in germination and initial seedling growth indicates a possible niche axis that may help explain both current species distributions and future responses to climate change.

Editor: Bruno Hérault, Cirad, France

Funding: This work was funded by the Swiss National Science Foundation (Grant #31003A_125461). The funders had no role in study design, data collection and analysis, decision to publish, or preparation of the manuscript.

Competing Interests: The authors have declared that no competing interests exist.

* E-mail: mikey.j.obrien@gmail.com

Introduction

Germination and early seedling growth are highly susceptible to changes in climatic conditions such as temperature and water availability [1–4]. In aseasonal tropical forests, species have adapted to everwet conditions which may make these systems especially sensitive to changes in the rainfall regime [5,6]. Previous research on the effects of drought in tropical forests has mainly focused on total water deficits during periods of no rain associated with a dry season [7–10]. However, in moist tropical forests, the timing and variability in rainfall may have detrimental effects on regeneration regardless of total monthly rainfall [11,12].

In Borneo, rainfall varies greatly on daily, weekly and monthly timescales despite it being classified as a moist aseasonal climate [13]. Additionally, the forests of Borneo have evolved with El Niño Southern Oscillation (ENSO) events which are associated with drier conditions and more variable rainfall, defined as short-term rainless periods followed by extreme rainfall [13,14]. Seedling establishment in these forests may be especially susceptible to infrequent rainfall because trees from the family Dipterocarpaceae (dipterocarps) which dominate the primary forest canopy often have recalcitrant seeds (i.e. short viability with no soil seed banks due to desiccation sensitivity), and their seedling recruitment relies on the episodic mast fruitings, which commonly coincide with ENSO events [14,15]. Furthermore, the effect of rainfall variability on plant growth is increasingly relevant as global precipitation cycles are expected to intensify with climate change [10,16,17]. In the tropics, an increase in both frequency and intensity of ENSO events could have substantial impacts on tropical forest dynamics, especially at the sensitive seed and seedling stage [12,13,16,18–21].

Seed size is an important trait which determines species success to climatic stress [1,22]. Larger seeds often have deeper root extension beyond the drying soil profile and greater sugar reserves which provides a competitive advantage under stressful drought conditions [1,2,22–25]. For example, Daws et al. [4] demonstrated that larger-seeded species were able to germinate at lower water potentials, implying tolerance to drought-imposed desiccation. The success of seedling establishment under infrequent rainfall may be directly related to seed mass [22].

We germinated seeds under multiple different frequencies of water availability to examine the effects of rainfall variability on early seedling establishment. We used species spanning a range of more than two orders of magnitude in seed size to assess the importance of seed mass in mediating the potential negative impacts of infrequent rainfall. We examined the role of seed mass in seed germination and seedling growth by measuring both pre- and post-germination response.

Methods

Ethics Statement

Approval for research at this site was given by the Malaysian Economic Planning Unit (EPU Permit: #2738 UPE: 40/200/19/2640).

Study Site

The experiment was conducted between 4 August and 8 November 2010 at the Sabah Biodiversity Experiment (SBE; N05°05′20″ E117°38′32″; 102 MASL). This site is located ≈22 km north of Danum Valley Field Centre (DVFC) in the state of Sabah, Malaysia [26]. Mean annual rainfall (s.e.) from 1986–2010 from DVFC was 2848.5 (94.0) mm. This experiment was conducted *ex situ* under two layers of 70% shade-cloth and thin transparent polyethylene sheeting to exclude rainfall. The shade-houses at the SBE are raised with grated flooring, excluding large mammals but not small rodents and insects. The daily mean temperature (s.e.) under the polyethylene sheeting during the course of the experiment was 25.3°C (0.08) with a minimum of 21.5°C and a maximum of 34°C. The mean (s.e.) percent direct light was 4.7% (0.1) and a red:far-red ratio of 1.11 (0.01) (measured by simultaneous shade-house and open sky photosynthetically active radiation sensors (SKP 210 quantum sensor; Skye instruments LTD, Llandrindod Wells, Powys, UK)) which created a light environment similar to a small gap in the forest understory. A nearby forest gap (approximately 8 m^2) in the Malua Forest Reserve had 4.98% (0.002) direct light and a red:far-red ratio of 1.04 (0.002).

Seed Collection

Fruits of eight dipterocarps were collected from the Malua Forest Reserve surrounding the SBE during the masting event which began in late July 2010. Species were selected to ensure a variety of genera, seed size and morphological characteristics (Table 1). Visibly healthy seeds (i.e. free of fungus, decay and herbivore damage) were checked for maturity by opening five to ten seeds per species and ensuring that the radicle was fully formed. Additionally, seeds were checked for health in water (i.e. seeds that sank in a basin of water were deemed healthy). Fruit wings were removed, and individual seeds without wings were weighed for seed mass. An additional 50 seeds of every species were used to generate estimates of seed dry biomass. These seeds were weighed, dried at 64°C to a constant weight and reweighed. Regressions were developed for each species relating dry and wet mass.

Watering Frequency and Planting

Pots were filled with homogenized forest soil obtained from the Innoprise-FACE Foundation Rainforest Rehabilitation Project (http://www.face-thefuture.com) (pot sizes: 7.0×23.0 cm for large-seeded *S. macrophylla* to allow greater soil volume; 4.5×22.0 cm for all other species). The soil used was classified as clay comprised of 50% clay, 30% silt and 20% sand similar to upslope sites of the Malua Forest Reserve. Total monthly rainfall (30 day period) was set at 240 mm and dispensed at four frequencies to assess the role of rainfall pattern on germination and seedling development. The water quantity was calculated based on the area of the pots and the millimeters of water per day for each watering frequency. This quantity equated to approximately 13, 25, 51 and 63 ml pot^{-1} per watering for daily, two, four and six-day frequencies. This watering regime altered frequency while sustaining an equal monthly rainfall, and provided seedlings with similar total water as natural rainfall during that period (Figure 1). Watering frequencies were chosen based on analysis of monthly rainfall records at the DVFC. The daily frequency was based on the maximum number of days with rain in a 30 day period and six-day frequency was based on the minimum number of days with rain in a 30 day period from historical rainfall records (5 days in April 1998; [13]). The two and four-day frequencies provided approximately the mean and 25% quantile of rainfall days. Furthermore, soil matric potentials during the course of the experiment remained within the range of natural values recorded in the forest at DVFC (Figure 2; [5]).

One hundred seeds of each species were placed in pots and randomly assigned to one of the four treatments (25 seeds per species per treatment). Seeds were placed on their side on the soil surface to replicate natural seed position in the forest and allow for monitoring of germination. All pots were watered to saturation prior to planting (Figure 2).

Soil Matric Potential

Pots were measured for volumetric soil water content with an ML2x Theta Probe and HH2 moisture meter (Delta-T Devices, Burwell, Cambridge, UK) from day zero to eighty-four. Measurements were taken before and one hour after watering to determine minimum water content and the extent to which the soil was rehydrated. To calculate a drying curve for the clay soil used in the experiment, the relationship between soil matric potential and volumetric soil moisture was determined using the filter paper method [27]. A range of volumetric soil moisture from 1.5% to 46% was used to develop two linear equations between volumetric soil moisture and soil matric potential. Two equations were required because soil matric potential declined at a faster rate below 28% volumetric soil moisture (Figure S1).

Seed and Seedling Monitoring

Seeds were monitored daily for germination (radicle emergence) and mortality. Only seeds with visible signs of mortality were recorded (i.e. fungus, mammal browse, insect browse, or desiccation). Following germination, seedlings were monitored daily for mortality and leaf formation. All surviving seedlings were harvested on their 60th day. Roots were extracted and measured for length. Leaf photographs were taken for calculation of leaf area. Seedlings were dried to a constant temperature at 64°C, and leaves, stem and roots weighed.

Analysis

Soil matric potential was analyzed as a function of watering frequency, pre- and post-watering measurements, species, day and their interaction with a general least squares (gls; allowing for non-constant variance of different species using the varIdent function) in the nlme package for R version 2.13.2 [28,29]. Including a factor for species identity never improved the model fit, so it was dropped from the soil matric potential analysis.

Phenological and morphological variables (cumulative germination; number of days to germination; number of days to leaf formation; diameter (mm); height (mm); root length (mm); total dry biomass (g); leaf area (cm^2)) in the daily watering were compared using linear mixed effects models (with species identity as a random effect) to assess differences in baseline performance across the seed mass spectrum. Seed mass was $\log_{10}$ transformed

Table 1. Summary table of seed traits of dipterocarp species ordered from largest to smallest seed mass.

Species	Wet mass (mg)	Dry mass (mg)	Seed description[a]	Days to germination[a]
(Acronym)	Mean (95% CI)	Mean (95% CI)		
Shorea macrophylla	55229	21248	thick woody seed	14–37
Ashton (SM)	(54606–55853)	(20996–21500)	coat, recalcitrant	
Parashorea tomentella	6533	4228	thick woody seed	14–184
Meijer (PT)	(5910–7157)	(3976–4480)	coat, recalcitrant	
Dryobalanops lanceolata	6483	2714	green soft seed	7–13
Burck (DL)	(6042–6924)	(2536–2892)	coat, recalcitrant	
Parashorea malaanonan	3840	1760	thick woody seed	14–184
Merr. (PM)	(3217–4464)	(1508–2012)	coat, recalcitrant	
Hopea nervosa	2909	1272	thin hard seed	8–48
King (HN)	(2285–3533)	(1020–1523)	coat, recalcitrant	
Shorea beccariana	2756	944	thin hard seed	Unknown
Burck (SB)	(2133–3380)	(742–1245)	coat, recalcitrant	
Shorea parvifolia	853	514	thin hard seed	7–52
Dyer (SP)	(230–1477)	(262–766)	coat, recalcitrant	
Shorea argentifolia	809	485	thin hard seed	8–15
Symington (SA)	(230–1477)	(233–737)	coat, recalcitrant	

[a]Descriptions and germination range taken from [30] and recalcitrance taken from KEW database.

and a weighted variance for each species using the varIdent function was included for all analysis in order to meet assumptions of linearity. We analyzed the data both with and without *S. macrophylla* because a large number of the seeds of this species experienced mortality due to fungal infection. We chose to retain it in all analysis, as removing it did not qualitatively alter the results (see Figure S4, S5, S6, S7, S8, S9 without *S. macrophylla*).

The effect of water frequency on seedling variables was assessed with pooled means of all species in each watering frequency. We accounted for species variance around the pooled mean with a random effect of species varying within treatment. The lmer function in the lme4 package was used for this analysis due to the added complexity of the random effects.

In order to examine the tolerance to infrequent watering, the mean relative difference between daily and six-day watering of each species was calculated for cumulative germination, days to germination, final total biomass and height (eq. 1)

$$response = \frac{y_{i,six} - y_{i,daily}}{y_{i,daily}} \quad (1)$$

Where $y_{i,six}$ is the mean performance of variable i in the six-day watering frequency, and $y_{i,daily}$ is the mean performance of variable i in the daily frequency. Relative response equates the effect of infrequent watering for a given variable in proportion to its baseline performance [12]. Large differences from zero can be interpreted as a variable which was sensitive to infrequent watering where small differences would be a variable tolerant to infrequent watering. Generalized least square models compared these proportions with species seed mass to assess the relationship between seed mass and relative response to infrequent watering for germinating seeds and early seedling growth. We examined the change in biomass allocation metrics (i.e. shoot:root (g g^{-1}; SRR), root mass:total mass (g g^{-1}; RMR) and root depth:leaf area (mm cm^{-2}; RLA) ratios) with seed mass to explain trends in the relative response to infrequent watering using the same linear methods for assessing baseline variable differences.

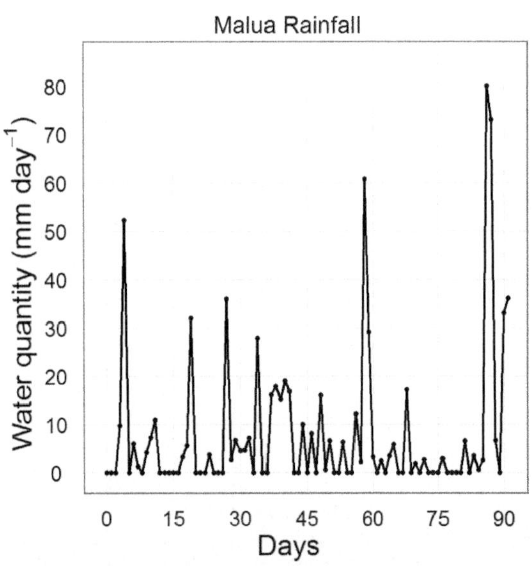

Figure 1. Local rainfall at Malua forest during the experiment. Daily rainfall over the period of the experiment. Total rainfall was approximately 724 mm over the course of 90 days with 44 near rainless days (<1 mm day^{-1}).

Results

Soil Matric Potential

Pre-watering soil matric potential was significantly lower at the end of each four and six-day watering frequency but recovered to that of the daily and two-day treatments after watering (Figure 2;

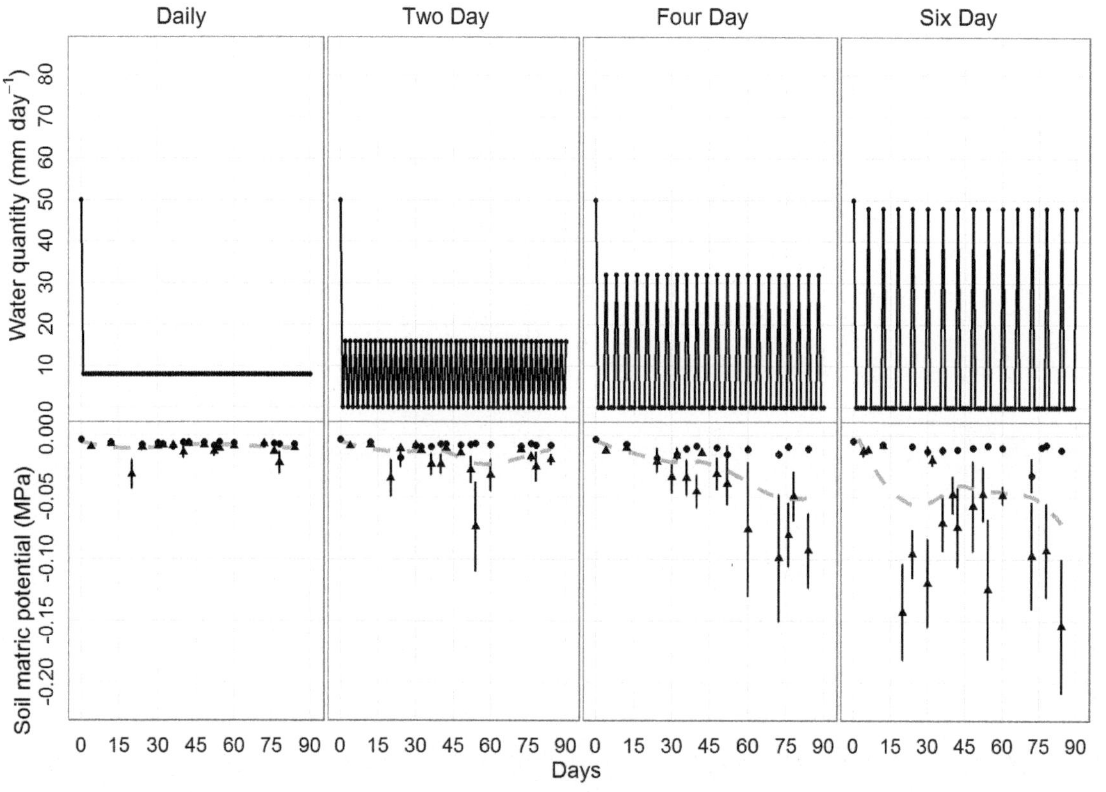

Figure 2. The effect of watering pattern on soil matric potential. The frequency of watering per treatment as a function of time. Each treatment received 720 mm of water. The lower panels show the change in soil matric potential with each treatment (pre- and post-watering measures are jittered for readability). The grey line represents a loess curve of change in the smoothed average through time.

Table S1 [significant pre-watering*treatment interaction]). The difference between pre- and post-watering soil matric potential increased throughout the course of the experiment and at different rates for each treatment (Figure 2; Table S1 [significant three-way interaction]). Initial mean pre-watering soil matric potential for all watering frequencies was −0.004 (±0.001 s.e.) MPa (Figure 2). At the final measure on the 84th day, mean pre-watering soil matric potential for daily, two, four and six-day watering was −0.01 (±0.007 s.e.), −0.02 (±0.005 s.e.), −0.09 (±0.03 s.e.) and −0.15 (±0.05 s.e.) MPa, respectively (Figure 2).

Baseline Performance

We found species-specific differences in phenological and morphological variables (Figure S2) with growth clearly related to seed size. However, germination variables did not correlate with seed mass (Figure S2). Height, diameter, root depth, biomass and leaf area increased with seed mass by 73.5 mm (95% CI: 15.3–131.7), 1.9 mm (95% CI: 1.3–2.5), 25.0 mm (95% CI: 3.5–46.5), 0.74 g (95% CI: 0.47–1.00) and 104.4 cm^2 (95% CI: 62.5–146.4) per $\log_{10}$ mg, respectively (Figure 3; Figure S2).

Water Frequency, Species Characteristics and Seed Mass

Altering water frequency from daily to six-day watering negatively affected most variables. Mean (daily – six-day) cumulative germination (23–14 seeds), height (158.4–118.4 mm), diameter (3.1–1.9 mm), root depth (93.7–83.1 mm), biomass (1.1–0.4 g) and leaf area (115.0–55.6 cm^2) declined with infrequent watering while days to germination (9–15 days) increased from daily to six-day watering, respectively (Figure 4; Figure S3). Timing of leaf formation and root depth per leaf area were not affected by infrequent watering (Figure S3).

There was no relationship between stress response and seed mass for cumulative germination (Figure 5). However, seed mass was related to response to infrequent water for days to germination, height, and total biomass (Figure 5). Days to germination increased by as much as 217% for *S. argentifolia* (the smallest seeded species) and as little as a 10.4% decrease for *S. macrophylla* (i.e. no effect for the largest seeded species) from daily to six-day frequency. On average a 76.5% (±23.1 s.e.) increase in days to germination was observed. The increase in days to germination declined with seed mass (slope = −0.89, 95% CI: −1.7– −0.04). Height decreased on average by 25.7% (±11.4 s.e.) for all species. Height decreased by as much as 36.9% for *P. tomentella* (for the largest seeded species surviving to 60 days) and 8.5% for *S. argentifolia* (the smallest seeded species) from daily to six-day frequency. Differences in height between treatments increased with seed mass (slope = −0.58, 95% CI: −0.79– −0.36). Biomass decreased by as much as 71.4% for *P. tomentella* and 7.9% for *S. parvifolia*, and on average decreased by 39.1% (±11.5 s.e.) for all species. Large-seeded species had greater differences in biomass between treatments (slope = −0.55, 95% CI: −0.78– −0.34). Not one seedling from the largest seeded species, *S. macrophylla*, survived to the end of the experiment in the six-day watering cycle. RLA declined with seed mass by −1.0 (95% CI: −1.53– −0.48) per $\log_{10}$ mg, but none of the alternative allocation measures showed a significant trend with seed mass (Figure 6; Figure S2).

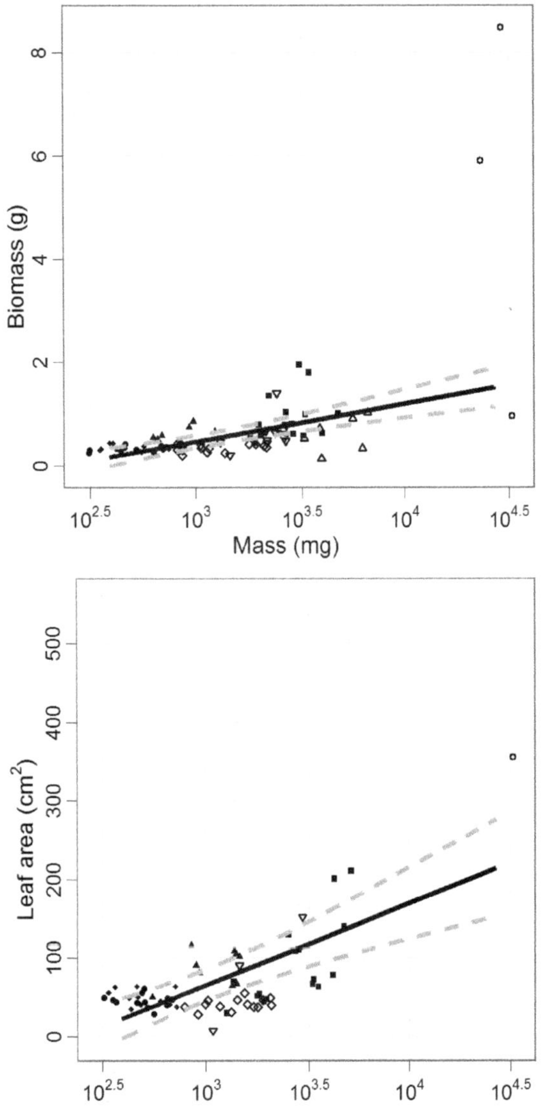

Figure 3. Relationship between growth and seed mass in daily watering. The relationship of baseline biomass and leaf area as a function of seed mass in the daily watering treatment (mean ±95% CI). Unequal variance was accounted for using a weighted variance for each species. Leaf area was based on the last harvest after 60 days. Points represent individual seedlings. (SM: open circle, PT: upward triangle, DL: closed square, PM: downward triangle, HN: open diamond, SB: closed triangle, SP: closed diamond, SA: closed circle).

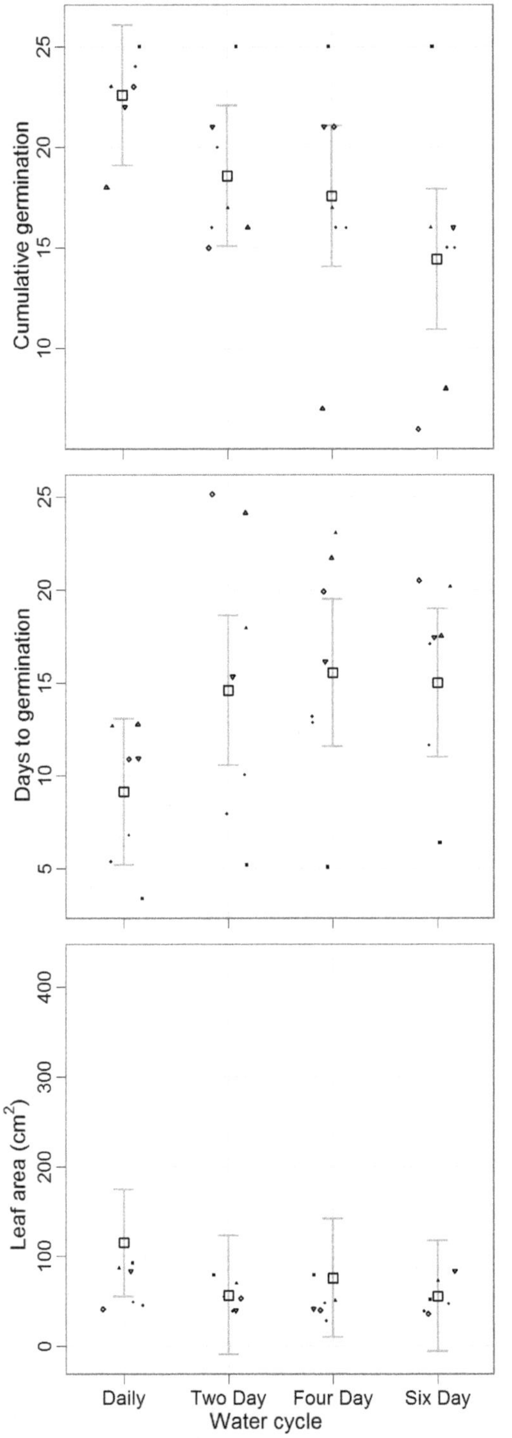

Figure 4. Effect of watering frequency on germination and growth. Cumulative germination, days to germination and final seedling leaf area calculated in each treatment. Variation in species and treatments was accounted for with a random effect for species in treatment. Leaf area was calculated from the last harvest after 60 days. The large open squares represent the pooled mean (±95% CI) of the eight Bornean climax species. All three traits were negatively affected by infrequent watering. Grey points represent species means for each variable in each treatment. The points are jittered for readability. (SM: open circle, PT: upward triangle, DL: closed square, PM: downward triangle, HN: open diamond, SB: closed triangle, SP: closed diamond, SA: closed circle).

Discussion

This study examined the role of infrequent rainfall on seedling establishment of Bornean dipterocarp species focusing on seed mass as a mediating trait. We found that infrequent watering altered the temporal distribution of seeds and size distribution of plants across a seed mass spectrum, with infrequent rainfall favoring large-seeded species before germination but favoring small-seeded species after germination. Under the high rainfall frequency characterizing the aseasonal tropics, all seed sizes took a similar number of days to germinate, and large-seeded species grew to a larger size than small-seeded species. However, infrequent rainfall disproportionately hindered large-seeded species growth compared with that of small-seeded species, which

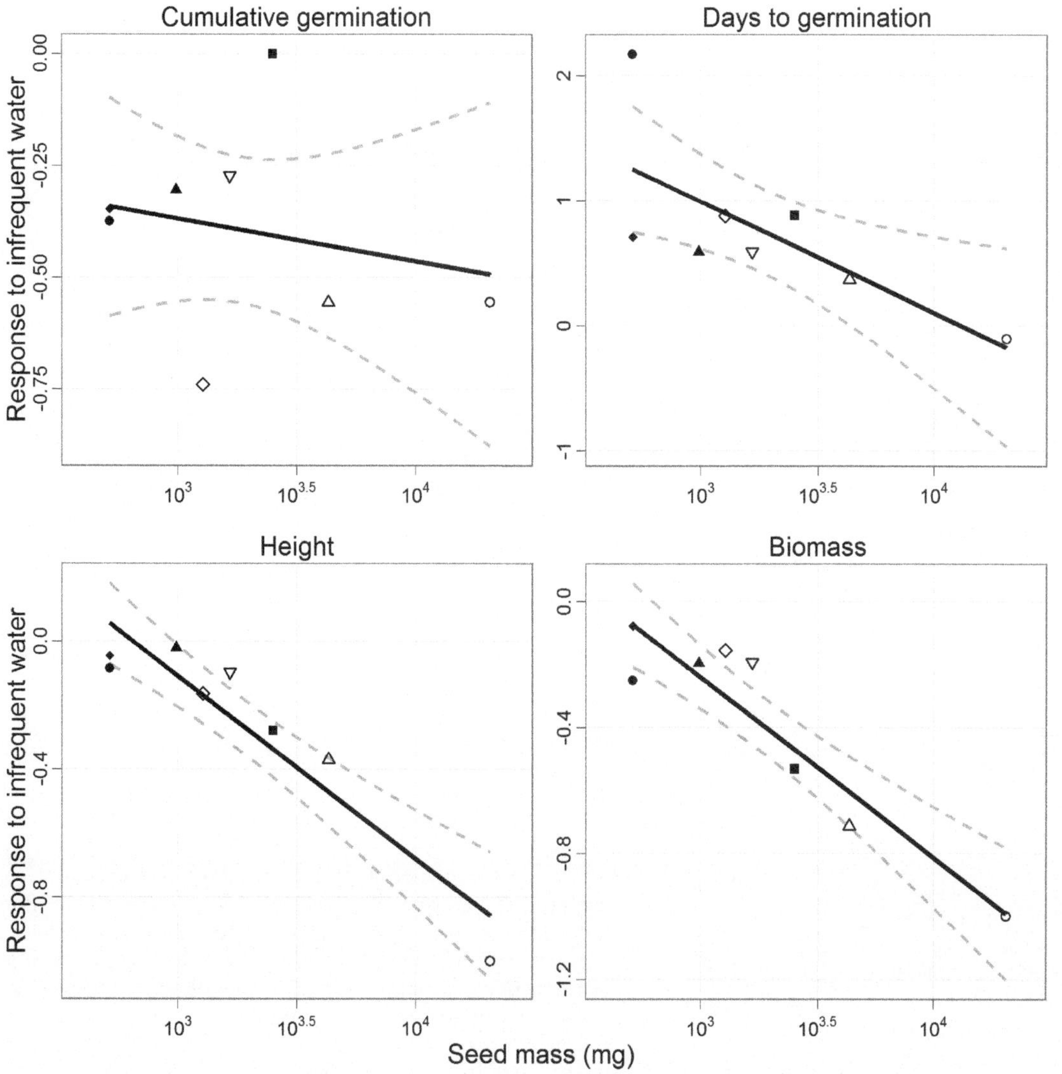

Figure 5. Relative response to infrequent water as a function of seed mass. Response to infrequent watering (i.e. the relative difference between daily and six-day watering) as a function of seed mass for cumulative germination, days to germination and seedling height and biomass (mean ±95% CI). No relationship existed between cumulative germination and seed mass. Germination of large seeds was more resistant to infrequent watering, but after germination large-seeded species had significantly greater declines in growth due to infrequent watering. Points represent mean values for each species. (SM: open circle, PT: upward triangle, DL: closed square, PM: downward triangle, HN: open diamond, SB: closed triangle, SP: closed diamond, SA: closed circle).

equalized final seedling size between large and small seeded species.

Germination

Days to germination showed no simple trends with seed mass suggesting other unmeasured seed traits were likely influential in germination. Seed coat thickness and toughness were two unmeasured traits that likely affected germination. The five species with the longest days to germination regardless of seed mass had the thickest and most lignified seed coats ([30]; Table 1; Figure 3). Alternatively, the species with the fewest days to germination (*D. lanceolata*) was the third largest seed but had a soft, green coat. Thick lignified seed coats would require a softening or breaking for the radicle to emerge. This would result from abiotic forces (e.g. the wetting and drying of the seed coat from water or diel temperature fluctuation) or biotic traits (e.g. increased emergence force from larger radicles associated with larger seeds) [25,31,32]. Species may also vary in response to environmental factors such as light quantity and quality and temperature that may account for unexplained variation in days to germination [3]. Therefore, the timing of germination is driven by complex interactions between seed traits and microsite conditions, which did not simply correlate with seed mass for dipterocarps [3,31].

Infrequent rainfall inhibited days to germination for small-seeded species while large-seeded species remained unaffected (Figure 5). This temporal germination advantage favors large seeds as they reached greater height and rooting depths which could potentially inhibit establishment and growth of small-seeded species [22]. This advantage for large seeds would be more pronounced under two scenarios. First, under canopy gaps, lower humidity and higher daytime temperatures would increase seed and soil surface drying associated with infrequent rainfall, possibly slowing small seed germination to a greater extent [2,3,8,12,32]. Second, current climate projections suggest that rainfall variability

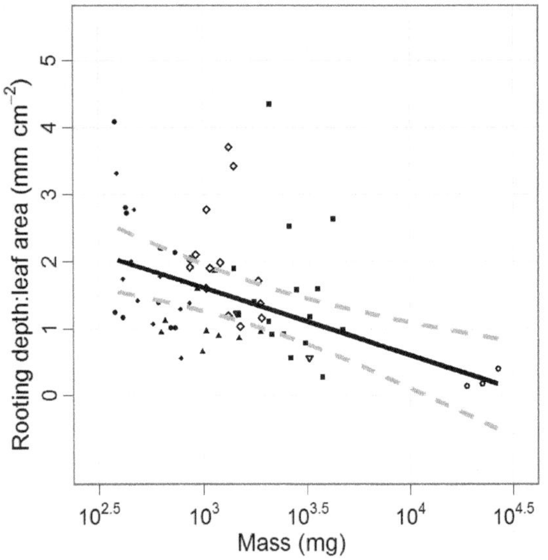

Figure 6. Rooting depth per leaf area as a function of seed mass. Root depth per leaf area (mean ±95% CI) as a function of seed mass. Unequal variance was accounted for using a weighted variance for each species. Large-seeded species produce much larger leaves increasing transpiration and water demand without proportionally larger rooting depth, leaving them susceptible to breaks in hydraulic conductivity and water stress. Points represent individual seedlings. (SM: open circle, PT: upward triangle, DL: closed square, PM: downward triangle, HN: open diamond, SB: closed triangle, SP: closed diamond, SA: closed circle).

will increase. Walsh and Newbery [13] have shown that consecutive months with short-term drought have been more frequent since the last decades of the 20th century. For example, from September 1997 to May 1998 there were four rainless periods of at least 10 days [13]. More recently, from February 2010 to April 2010 there were three rainless periods of at least 7 days (O'Brien, unpublished data). Under more frequent or longer rainless periods, large-seeded species would have a germination advantage [13,20]. However, dipterocarp seeds likely have a threshold to infrequent rainfall that, if reached during extreme ENSO events, may cause total regeneration failure, especially since many dipterocarps are desiccation sensitive [14,30].

Infrequent rainfall reduced cumulative germination, but the reduction was independent of seed mass (Figure 5). Natural rainfall patterns during the course of this experiment showed days of high rain and more than ten periods with consecutive no rain days, six of which were >2 days long (Figure 2). Overall seed germination in the six-day watering was reduced by 37% relative to the daily watering, which would have marginal impacts on seedling establishment under such ubiquitous seed production. However, greater declines should be expected under projected climate scenarios where extended dry periods (>10 days) are becoming more frequent [12,13].

Seedling Growth

Large-seeded species had greater growth under frequent rainfall (Figure 3). The greater stored non-structural carbohydrates often associated with larger seeds allows increased biomass growth and greater absolute height growth before leaf development [1,24,33]. Although absolute growth was greater for large-seeded species after 60 days, small-seeded species commonly have higher relative growth rate and may overtake large-seeded species given longer time periods [34,35].

All growth metrics showed a decline with infrequent watering. The magnitude of the effect was dependent on the metric with total biomass and leaf area having the greatest decline and root depth having the smallest. The overall decline in growth was likely due to either water limitation which inhibited cell expansion and division or due to the rapid shift from an aerobic to anaerobic environment [36]. Daily and two-day watering maintained soil matric potential at a constant level throughout the experiment while four and six-day watering caused levels more than ten times lower than daily watering. Soil drying also increased throughout the course of the experiment (Figure 2) because seedlings began establishing and taking up water. As root systems depleted soil water to meet growth and transpiration demands, drought stress increased, and growth was limited. However, soil matric potentials never dropped much below −0.2 MPa which is a relatively mild drought stress, and we hypothesize that drying would be more significant in the forest setting. Regardless of the extent of drought stress experienced by the seedlings, soil drying followed by rapid water saturation may alone be enough of a stress to cause internal water deficits [37,38].

Growth of large-seeded species was more inhibited by infrequent watering than growth of small-seeded species (Figure 5). After 60 days of growth, a difference of only 24 mm between *D. lanceolata* (a large seeded species) and *S. argentifolia* (the smallest seeded species) was observed under six-day watering versus 76 mm under daily watering. If this trend were to continue with longer rainless periods, it would alter the competitive rank of the species. The effect was even more pronounced for biomass in which the large-seeded *P. tomentella* (0.61 g –0.18 g) declined to below the mass of the smallest seeded *S. argentifolia* (0.34 g –0.25 g). Although soil matric potential did not show a significant difference between species, seedlings with greater water uptake could deplete soil water from their immediate soil rhizosphere causing a break in water conductivity and a more intense drought effect [39]. Additionally, the drying and wetting of the soil in the infrequent watering treatment, which would alter nutrient availability and soil texture, could have negative effects on seedling growth regardless of deficit [37,38]. We hypothesize that small-seeded species are less susceptible to these fluctuations in soil properties because of their finer roots relative to large-seeded species, which allows them to more easily alter root growth in the varying soil environment. However, research focusing directly on seedling root plasticity with a fluctuating water table is necessary to elucidate the mechanism driving this process.

The inverse relationship between rooting depth per leaf area and seed mass (Figure 6) may explain the increased effect on large-seeded species. In support of this result, Engelbrecht et al. [12] also found significantly larger rooting depth per leaf area of smaller seeded species. Greater leaf area per unit of root depth increases the risk of desiccation as high transpiration demands are not met leading to a drawdown of the soil water [39]. For a seedling to sustain the required hydraulic conductivity, it must access an increased soil volume [39]. Large-seeded species produced greater leaf area which equated to increased transpiration and subsequent water loss. At the early establishment phase, large seedlings would be at a disadvantage under infrequent rainfall especially in rapidly drying sites such as coarse textured soils and large canopy gaps [2,8,32,39]. We recognize that seed mass is a simplification of multiple traits into a single dimension, and physiological differences in water use efficiency or carbohydrate storage may provide further explanations for the sensitivity of species [40].

Our results have implications for species distributions in Bornean forests. Large-seeded species may preferentially establish in wetter areas which sustain consistent soil moisture such as lowland riparian zones, and small-seeded species may persist in sites with well-drained soils which have a more fluctuating water table. Recent work done in Lambir Hills National Park, Sarawak, Malaysia found an increase in mean seed mass along a soil moisture and fertility gradient with the largest seed masses found on fine clayey soil textures [41], which supports our seed size-sensitivity relationship.

Conclusions

This study demonstrates differential responses of large- and small-seeded species to watering frequency and temporal distribution of rainless periods. The timing of rainless periods (before or after germination) altered the competitive advantage between seed sizes. Short-term rainless periods delayed germination of small-seeded species which gave a temporal advantage to large-seeded species, but at the seedling establishment phase, large seedlings resulting from larger seeds were susceptible to fluctuations in water which inhibited growth. Current climate change projections of reduced and more variable rainfall could further alter species germination and establishment patterns which would have long-term effects on community composition and species distributions.

Supporting Information

Figure S1 The curves used to estimate matric potential from volumetric moisture. The matric potential declines at a much faster rate below approximately 28% volumetric moisture (5.477–0.1591 ×volumetric; $R^2 = 0.977$) than above (2.307–0.044 × volumetric; $R^2 = 0.981$).

Figure S2 Baseline variables in of seeds and seedlings in daily watering. The relationship of baseline phenological and morphological characteristics with seed mass for germinating seeds and seedlings in the daily watering treatment. Seedling growth and allocation variables were based on the last harvest after 60 days. Solid lines represent model predictions with 95% CIs. Points represent individual observations (SM: open circle, PT: upward triangle, DL: closed square, PM: downward triangle, HN: open diamond, SB: closed triangle, SP: closed diamond, SA: closed circle). Log transforming growth supported untransformed results and was therefore not used for the analysis.

Figure S3 Effect of watering frequency on seed and seedling variables. The effect of water frequency on seed and seedling variables for seven Bornean shade-tolerant species pooled. Seedling variables were based on the last harvest after 60 days. Most variables were negatively affected by infrequent watering. Individual species points were jittered for readability. Open squares represent model predictions with 95% CIs. The smaller points represent mean for each species in each treatment. (SM: open circle, PT: upward triangle, DL: closed square, PM: downward triangle, HN: open diamond, SB: closed triangle, SP: closed diamond, SA: closed circle)

Figure S4 Baseline variables in of seeds and seedlings in daily watering without SM. The relationship of baseline phenological and morphological characteristics with seed mass for germinating seeds and seedlings in the daily watering treatment. Seedling growth and allocation variables were based on the last harvest after 60 days. Solid lines represent model predictions with 95% CIs. Points represent individual observations (PT: upward triangle, DL: closed square, PM: downward triangle, HN: open diamond, SB: closed triangle, SP: closed diamond, SA: closed circle). Log transforming growth supported untransformed results and was therefore not used for the analysis.

Figure S5 Effect of watering frequency on seed and seedling variables without SM. The effect of water frequency on seed and seedling variables for seven Bornean shade-tolerant species pooled. Seedling variables were based on the last harvest after 60 days. Most variables were negatively affected by infrequent watering. Individual species points were jittered for readability. Open squares represent model predictions with 95% CIs. The smaller points represent mean for each species in each treatment. (PT: upward triangle, DL: closed square, PM: downward triangle, HN: open diamond, SB: closed triangle, SP: closed diamond, SA: closed circle)

Figure S6 Relationship between growth and seed mass in daily watering without SM. The relationship of biomass and leaf area as a function of seed mass in the daily watering treatment (mean ±95% CI). Unequal variance was accounted for using a weighted variance for each species. Leaf area was based on the last harvest after 60 days. Points represent individual seedlings. (PT: upward triangle, DL: closed square, PM: downward triangle, HN: open diamond, SB: closed triangle, SP: closed diamond, SA: closed circle)

Figure S7 Effect of watering frequency on germination and growth without SM. Cumulative germination, days to germination and final seedling leaf area calculated in each treatment. Variation in species and treatments was accounted for with a random effect for species in treatment. Leaf area was calculated from the last harvest after 60 days. The large open squares represent the pooled mean (±95% CI) of the eight Bornean climax species. All three traits were negatively affected by infrequent watering. Grey points represent species means for each variable in each treatment. The points are jittered for readability. (PT: upward triangle, DL: closed square, PM: downward triangle, HN: open diamond, SB: closed triangle, SP: closed diamond, SA: closed circle)

Figure S8 Relative response to infrequent water as a function of seed mass without SM. Response to infrequent watering (i.e. the relative difference between daily and six-day watering) as a function of seed mass for cumulative germination, days to germination and seedling height and biomass (mean ±95% CI). No relationship existed between cumulative germination and seed mass. Germination of large seeds was more resistant to infrequent watering, but after germination large-seeded species had significantly greater declines in growth due to infrequent watering. Points represent mean values for each species. (PT: upward triangle, DL: closed square, PM: downward triangle, HN: open diamond, SB: closed triangle, SP: closed diamond, SA: closed circle)

Figure S9 Rooting depth per leaf area as a function of seed mass without SM. Root depth per leaf area (mean ±95% CI) as a function of seed mass.). Unequal variance was accounted for using a weighted variance for each species. Large-seeded species produce much larger leaves increasing transpiration and water demand without proportionally larger rooting depth, leaving them susceptible to breaks in hydraulic conductivity and water stress. Points represent individual seedlings. (PT: upward triangle, DL: closed square, PM: downward triangle, HN: open diamond, SB: closed triangle, SP: closed diamond, SA: closed circle)

Table S1 Summary of significance for parameters explaining soil matric potential. Species was removed from the model as it was never significant nor did it improve the fit of the model.

Acknowledgments

We thank Dr. Sebastian Leuzinger and Alicia Argüello Garcia for thorough manuscript revisions. Additionally, we would like to thank David Burslem and two additional anonymous reviewers for their revisions. Great appreciation goes to our hard working staff at the Sabah Biodiversity Experiment, especially site manager Philip Ulok and SEARRP director Dr. Glen Reynolds. This research is Sabah Biodiversity Experiment manuscript number 9.

Author Contributions

Conceived and designed the experiments: MJO CDP. Performed the experiments: MJO. Analyzed the data: MJO CDP AH. Contributed reagents/materials/analysis tools: AH. Wrote the paper: MJO AH. Local collaborator for the approval of a work permit in Malaysia: JT.

References

1. Leishman M, Westoby M (1994) The role of seed size in seedling establishment in dry soil conditions-experimental evidence from semi-arid species. Journal of Ecology 82: 249–258.
2. Dalling J, Hubbell S (2002) Seed size, growth rate and gap microsite conditions as determinants of recruitment success for pioneer species. Journal of Ecology 90: 557–568.
3. Daws M, Burslem DFRP, Crabtree L, Kirkman P, Mullins CE, et al. (2002) Differences in seed germination responses may promote coexistence of four sympatric *Piper* species. Functional Ecology 16: 258–267.
4. Daws MI, Crabtree LM, Dalling JW, Mullins CE, Burslem DFRP (2008) Germination responses to water potential in neotropical pioneers suggest large-seeded species take more risks. Annals of Botany 102: 945–951.
5. Gibbons J, Newbery D (2002) Drought avoidance and the effect of local topography on trees in the understorey of Bornean lowland rain forest. Plant Ecology 164: 1–18.
6. Potts MD (2003) Drought in a Bornean everwet rain forest. Journal of Ecology 91: 467–474.
7. Condit R, Hubbell S, Foster R (1995) Mortality rates of 205 neotropical tree and shrub species and the impact of a severe drought. Ecological Monographs 65: 419–439.
8. Poorter L, Hayashida-Oliver Y (2000) Effects of seasonal drought on gap and understorey seedlings in a Bolivian moist forest. Journal of Tropical Ecology 16: 481–498.
9. Engelbrecht BMJ, Kursar TA (2003) Comparative drought-resistance of seedlings of 28 species of co-occurring tropical woody plants. Oecologia 136: 383–393.
10. Beier C, Beierkuhnlein C, Wohlgemuth T, Penuelas J, Emmett B, et al. (2012) Precipitation manipulation experiments-challenges and recommendations for the future. Ecology Letters 15: 899–911.
11. Blain D, Kellman M (1991) The effect of water supply on tree seed germination and seedling survival in a tropical seasonal forest in Veracruz, Mexico. Journal of Tropical Ecology 7: 69–83.
12. Engelbrecht BMJ, Dalling JW, Pearson TRH, Wolf RL, Gálvez DA, et al. (2006) Short dry spells in the wet season increase mortality of tropical pioneer seedlings. Oecologia 148: 258–269.
13. Walsh RP, Newbery DM (1999) The ecoclimatology of Danum, Sabah, in the context of the world's rainforest regions, with particular reference to dry periods and their impact. Philosophical Transactions of the Royal Society of London Series B, Biological sciences 354: 1869–1883.
14. Curran LM, Caniago I, Paoli GD, Astianti D, Kusneti M, et al. (1999) Impact of El Niño and logging on canopy tree recruitment in Borneo. Science 286: 2184–2188.
15. Ashton P, Givnish T, Appanah S (1988) Staggered flowering in the Dipterocarpaceae: new insights into floral induction and the evolution of mast fruiting in the aseasonal tropics. American Naturalist 132: 44–66.
16. Huntington TG (2006) Evidence for intensification of the global water cycle: review and synthesis. Journal of Hydrology 319: 83–95.
17. Heisler-White JL, Blair JM, Kelly EF, Harmoney K, Knapp AK (2009) Contingent productivity responses to more extreme rainfall regimes across a grassland biome. Global Change Biology 15: 2894–2904.
18. Foster S (1986) On the adaptive value of large seeds for tropical moist forest trees: a review and synthesis. Botanical Review 52: 260–299.
19. Fenner M (1987) Seedlings. New Phytologist 106: 35–48.
20. Timmerman A, Oberhuber J, Blacher A, Esch M, Latif M (1999) Increased El Niño frequency in a climate model forced by future greenhouse warming. Nature 398: 1996–1999.
21. Lewis SL, Brando PM, Phillips OL, van der Heijden GMF, Nepstad D (2011) The 2010 Amazon drought. Science 331: 554.
22. Muller-Landau HC (2010) The tolerance-fecundity trade-off and the maintenance of diversity in seed size. Proceedings of the National Academy of Sciences of the United States of America 107: 4242–4247.
23. Baker H (1972) Seed weight in relation to environmental conditions in California. Ecology 53: 997–1010.
24. Westoby M, Leishman M (1996) Comparative ecology of seed size and dispersal. Philosophical Transactions of the Royal Society of London Series B, Biological sciences 351: 1309–1318.
25. Daws MI, Ballard C, Mullins CE, Garwood NC, Murray B, et al. (2007) Allometric relationships between seed mass and seedling characteristics reveal trade-offs for neotropical gap-dependent species. Oecologia 154: 445–454.
26. Hector A, Philipson C, Saner P, Chamagne J, Dzulkifli D, et al. (2011) The Sabah Biodiversity Experiment: a long-term test of the role of tree diversity in restoring tropical forest structure and functioning. Philosophical Transactions of the Royal Society of London Series B, Biological sciences 366: 3303–3315.
27. Deka R, Wairiu M, Mtakwa PW, Mullins C, Veenendaal EM, et al. (1995) Use and accuracy of the filter-paper technique for measurement of soil matric potential. European Journal of Soil Science 46: 233–238.
28. Pineheiro J, Bates D (2000) Mixed-effects Models in S and S-PLUS. New York: Springer-Verlag.
29. R Development Core Team (2013) R: a language and environment for statistical computing. R Foundation for Statistical Computing, Vienna, Austria, ISBN 3–900051–07–0.
30. Ng FSP (1992) Manual of forest fruits, seeds and seedlings, Malindo Printers, Malaysia.
31. Vázquez-Yanes C, Orozco-Segovia A (1993) Patterns of seed longevity and germination in the tropical rainforest. Annual Review of Ecology and Systematics 24: 69–87.
32. Pearson T, Burslem DFRP, Mullins C, Dalling J (2002) Germination ecology of neotropical pioneers: interacting effects of environmental conditions and seed size. Ecology 83: 2798–2807.
33. Nakagawa M, Nakashizuka T (2004) Relationships between physical and chemical characteristics of dipterocarp seeds. Seed Science Research 14: 363–369.
34. Baraloto C, Hardy OJ, Paine CET, Dexter KG, Cruaud C, et al. (2012) Using functional traits and phylogenetic trees to examine the assembly of tropical tree communities. Journal of Ecology 100: 690–701.
35. Philipson C, Saner P, Marthews T (2012) Light-based regeneration niches: evidence from 21 dipterocarp species using size-specific RGRs. Biotropica 44: 627–636.
36. Hsiao T (1973) Plant responses to water stress. Annual Review of Plant Physiology 24: 519–570.
37. Parent C, Capelli N, Berger A, Crèvecoeur M, Dat J (2008) An overview of plant responses to soil waterlogging. Plant Stress 2: 20–27.
38. Poorter H, Niklas KJ, Reich PB, Oleksyn J, Poot P, et al. (2012) Biomass allocation to leaves, stems and roots: meta-analyses of interspecific variation and environmental control. New Phytologist 193: 30–50.
39. Sperry JS, Hacke UG, Oren R, Comstock JP (2002) Water deficits and hydraulic limits to leaf water supply. Plant, Cell & Environment 25: 251–263.
40. O'Brien MJ (2013) The Role of Functional-traits in the Response of Tropical Forests to Global Change. Zurich, Switzerland: University of Zurich.
41. Katabuchi M, Kurokawa H, Davies SJ, Tan S, Nakashizuka T (2012) Soil resource availability shapes community trait structure in a species-rich dipterocarp forest. Journal of Ecology 100: 643–651.

4

Spatial Pattern of Standing Timber Value across the Brazilian Amazon

Sadia E. Ahmed*, Robert M. Ewers

Department of Life Sciences, Imperial College London, Silwood Park Campus, Ascot, Berkshire, United Kingdom

Abstract

The Amazon is a globally important system, providing a host of ecosystem services from climate regulation to food sources. It is also home to a quarter of all global diversity. Large swathes of forest are removed each year, and many models have attempted to predict the spatial patterns of this forest loss. The spatial patterns of deforestation are determined largely by the patterns of roads that open access to frontier areas and expansion of the road network in the Amazon is largely determined by profit seeking logging activities. Here we present predictions for the spatial distribution of standing value of timber across the Amazon. We show that the patterns of timber value reflect large-scale ecological gradients, determining the spatial distribution of functional traits of trees which are, in turn, correlated with timber values. We expect that understanding the spatial patterns of timber value across the Amazon will aid predictions of logging movements and thus predictions of potential future road developments. These predictions in turn will be of great use in estimating the spatial patterns of deforestation in this globally important biome.

Editor: Jerome Chave, Centre National de la Recherche Scientifique, France

Funding: The work was funded by The Grantham Institute for Climate Change (http://www3.imperial.ac.uk/climatechange) and Microsoft Research (http://research.microsoft.com/en-us/labs/cambridge/) through Ph.D. funding. The funders had no role in study design, data collection and analysis, decision to publish, or preparation of the manuscript.

Competing Interests: This work was funded through Microsoft Research and the Grantham Institute via Ph.D. funding.

* E-mail: sadia.ahmed05@imperial.ac.uk

Introduction

The Amazon is the largest remaining area of tropical forest [1], containing half of the world's tropical forest biome [2], covering an area of nearly 5 million km^2 [3] and accounting for approximately 10% of the Earth's terrestrial net primary productivity and biomass [4–5]. It is a highly biodiverse system [6] housing a quarter of all global biodiversity [2]. However it is also a system under threat, with an average of 19,500 km^2 of forest cleared each year between 1996 and 2005 [7]. The spatial patterns of deforestation are determined largely by the patterns of roads that open access to frontier areas, leaving them susceptible to colonisation and further development [8–13]. Kirby *et al.* [14] showed that distance from roads is in fact the strongest predictor of deforestation in the Amazon, and Southworth *et al.* [15] reported that deforestation patterns often closely mirror the pattern of the road network with deforestation rates falling with distance from main roads.

The expansion of the secondary road network in the Amazon is primarily driven by the logging sector [16] and logging is a huge industry in the Amazon, with an estimated US$ 2.5 billion of timber extracted each year [16]. Estimates suggest that the amount of forest that is clear cut each year is matched with an equal area being selectively logged each year; approximately 10,000–15,000 km^2/year [17–18]. Even selectively logged forests can lose more than 40% canopy cover through damage to surrounding trees during extraction and increased fire risks [17]. Additionally, roads used by loggers to access timber also serve to open up frontier areas to colonists who further degrade and deforest [11–13].

A key goal of the logging sector, indeed any economically driven sector, is to maximise profits. This forms the basis of many land cover/land use change models, which assume a desire to maximise profit and use profit maximisation to determine potential land uses (e.g. [19]). There are two key aspects to profit; revenue and costs. In the logging industry the amount and value of timber extracted determines revenue. To accurately model deforestation driven by logging, it is, then, important to know the spatial distribution of timber values. The location of valuable timber is important because extraction is usually selective [20–21] and loggers ideally want to harvest areas that yield the highest profits, i.e. are situated on high density, high value timber and that are accessible with the least cost.

Knowing the economic value of forests is important for conservation as well. For example, Verissimo *et al.* [20] suggested locations for sustainable 'Flona' (national forests that allow sustainable logging) by combining data on protected areas, human occupation and forest value. They identified locations that would be economically viable to harvest but that would also provide biodiversity protection. The suggested a set of locations that covered 34% of Amazonian forest, of which 38% was also of high conservation priority. Further, the Amazon offers a host of ecosystem services, from climate regulation, water regulation and carbon storage, to forest products [1,22–27]. In order to calculate a true cost-benefit analysis of logging in these forests, the value of the ecosystem services provided by the standing forest needs to be quantified and compared with the timber values obtained by felling.

There are various estimates of the timber value in the Amazon; some specific e.g. \$15.4 billion [28], and some less specific e.g. 'several trillion dollars' [29]. However, timber value across the Amazon is difficult to predict because extensive surveys are labour and cost intensive, thus timber value estimates are often based on modelling. Often when looking at forest value, timber values are estimated in terms of net value (profit). For example, Stone [30] modelled the net value of timber using three price classes (valuable ~>\$300/m^3, medium ~200–300, and low value ~100–175 US\$/m^3) as a decaying function of greater distances from sawmills reducing net value, with the assumption that loggers will extract valuable timber from further away. This approach was also used by Verissimo *et al.* [20] and Merry *et al.* [28] who built on the work of Stone [30], making their spatial models more detailed in terms of industry behaviour.

Although spatial profitability has been modelled across the Amazon, it is interesting to know how much the forest is worth in terms of standing timber value. Therefore we have modelled timber value across the Amazon such that valuable tree stands are shown as valuable irrespective of extraction costs. We used ordinary kriging to generate a value map from RADAMBRAZIL survey data for 11 tree genera that are economically important. Kriging methods have been used before in the Amazon region to estimate the spatial distribution of tree diversity [31], tree species distribution [32] and timber density [33]. Our approach extends that of Arima *et al.* [33] who used kriging to estimate total timber density as a way of determining locations for logging road destinations. Here, we combine data on timber density and value to generate a map of potential timber revenue for the Amazon.

Analysis

The RADAMBRAZIL [34] survey is a selection of surveys carried out between 1968 and 1978 which aimed to map the natural resources of Brazil. Among the data collected on soils, geology and potential land uses, an extensive survey of vegetation was also carried out. The RADAMBRAZIL data set contains information on a total of 89 families and 513 genera of trees, recording the timber volume of individual trees within plots of known location. We aggregated species data by genus for 2465 RADAMBRAZIL forest plots across the Brazilian Amazon. Timber properties, and thus value, are inevitably heterogeneous within and between groupings; genera within families, species within genera and individuals within species will show variation. For example, within the *Tabebuia* genus, which is generally known for desirable hardwood timber such as *T. guayacan*, one may also find medium weight wood species such as *T. roseo-alba* or even light weight wood species such as *T. cassinoides* [35]. However, it was felt that genus was an appropriate level at which to carry out analyses relating to timber values because 71% [36] to 74% [37] of variation in wood density measures among species is explained by genus affiliation, whereas only 25% to 34% is explained by family affiliation.

A list of the top 15 Amazonian tree genera harvested in terms of volume exported (in 1000 m^3) and the average price per cubic metre, in US\$, was obtained from ITTO (International Tropical Timber Organisation [38]). The average value for each genus (US\$/m^3) was obtained by averaging the total export value (US\$/m^3) over the two years for which data were available (2006 and 2007). We were able to obtain data from both RADAMBRAZIL and ITTO for a total of 11 genera (*Dicorynia, Goupia, Hymenaea, Hymenolobium, Manilkara, Mora, Nectandra, Peltogyne, Swartzia, Swietenia, Tabebuia*), which were used in all subsequent analyses. We multiplied the average export value by the total volume of timber recorded at each RADAMBRAZIL location to obtain a timber value for that genus in that plot. RADAMBRAZIL forest plots were 1 ha in area, so our resulting timber values are in units of US\$.ha^{-1}. Genus value estimates were summed across all 11 genera in each plot to obtain a total timber value (US\$.ha^{-1}). Total value data was log-transformed to make it normally distributed (all analyses were done with log-transformed total value data, unless otherwise stated). Timber value data from ITTO was for processed sawn wood, whereas the RADAMBRAZIL timber volume data was for standing timber. Timber volume is lost at many stages of the logging process, meaning that 1 m^3 of timber represents a different 'volume' in the ITTO and RADAMBRAZIL datasets. Measurements of timber volume loss in sawmill operations based in the Eastern Amazon state of Para suggest a conversion efficiency of 35% from raw timber to sawn wood [39]. To account for this loss of timber volume and to standardise 'volume' measurements among the two datasets, we multiplied all values derived from RADAMBRAZIL data by a scaling factor of 0.35.

Ordinary kriging was carried out in ArcGIS 9.3 using the Geostatistical package suite of tools. We initially modelled each of the 11 genera separately (Figure 1), before combining values from all genera within plots (Figure 2) to interpolate total value across the Brazilian Amazon. Output values were anti-logged to give true US\$/ha values of timber, after which a forest mask was overlaid on the output map to remove extraneous timber values from areas known to be non-forest and/or previously deforested. We also calculated and present the spatial pattern of error in the kriged values. Because our model was conducted on log-transformed data, we present log-transformed timber values and the standard error associated with that value (Figure 2c,d). An assumption of kriging analysis is that values in locations that are closer in space are more similar than those that are further away. To test for this spatial autocorrelation, we calculated Moran's I in R 2.10.1 (R Development Core Team [40]) using the 'pgirmess' library [41] for total timber value across the Amazon. Moran's I statistics showed that locations up to ~200 km were strongly positively correlated (Moran's $I = 0.26$–0.19, $P<0.001$), after which the correlation coefficient drops to <0.04 and is generally non-significant. Furthermore, we validated the un-scaled kriging predictions by using linear regression to model predicted against observed timber value. There was a significant, positive relationship ($r^2 = 0.31$, $p<0.001$, $df = 2463$) and the regression line did not differ significantly from a 1:1 relationship (slope = 1.02, 95% confidence interval 0.95–1.09), indicating that the predictions of timber value by kriging were robust.

The 11 genera contributed very different amounts to the overall timber value, and exhibited genus-specific spatial patterns of timber value (Figure 1 & Table 1). The genus *Hymenaea* contributed the most to the total values (mean value US\$394 - per ha; 95% confidence interval 376–2136), whereas *Dicorynia* contributed the least (mean value US\$5 per ha 95% confidence interval 5–10). The mean total timber revenue predicted was US\$813 per hectare (95% confidence interval 477–4161), with a distribution showing a few locations with particularly high values of >US\$1500 (Figure 2, inset a). There was, however, one area known to be of relatively low value along the eastern edge of the study area that was predicted to be high value. This was likely a result of the kriging interpolation, as there were no survey locations in this area to moderate the predictions. Standard errors on the kriging interpolation (Figure 2, inset d) reflect this, showing a band of high error along the eastern edge of the study area.

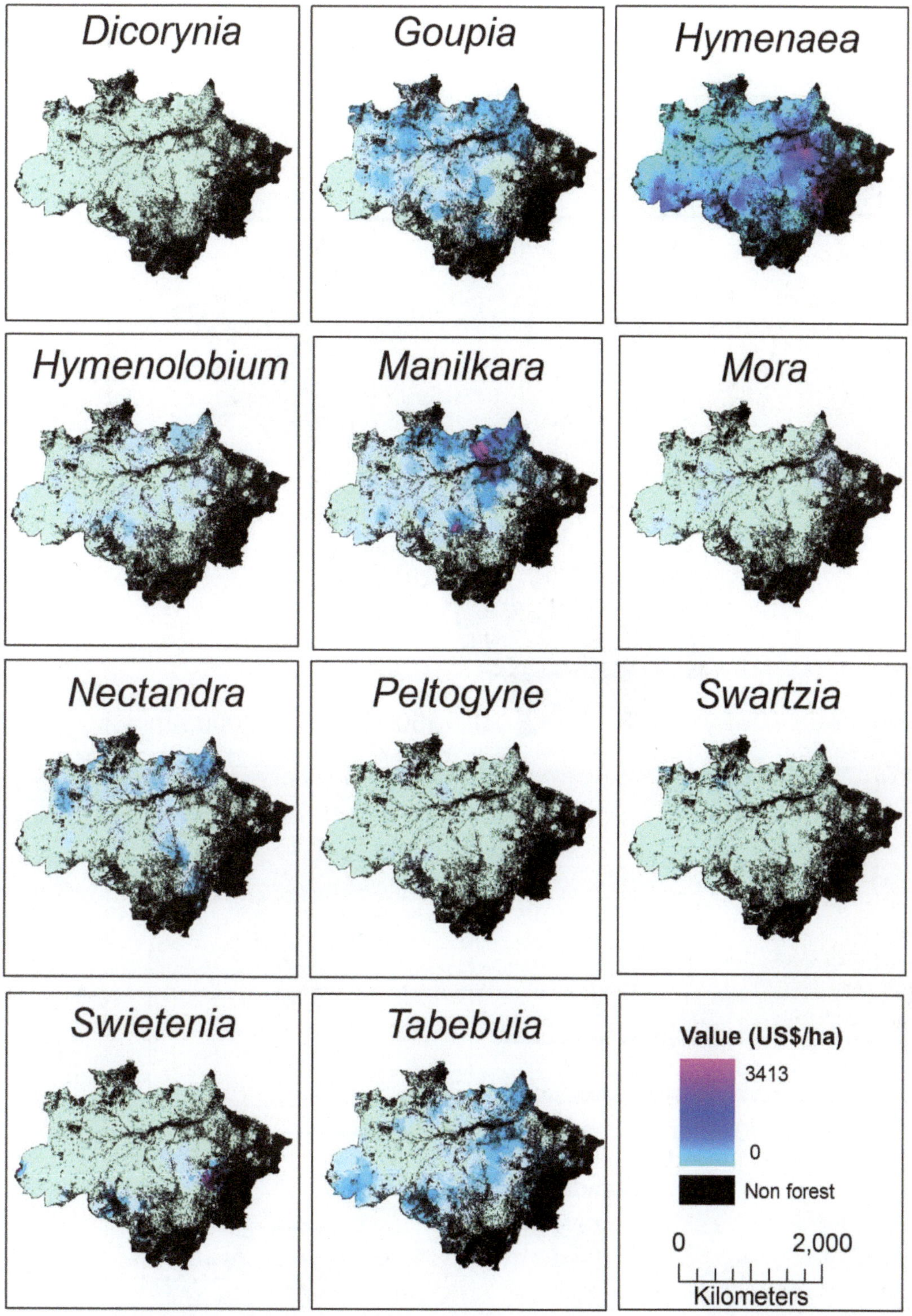

Figure 1. Predicted timber value across the Amazon by genus. Each of the 11 high-value genera, show genus-specific spatial patterns in the distribution of timber value and differences in total value (US$.ha^{-1}). *Hymenaea* appears to consistently have the highest value across the Amazon. Other genera, such as *Manilkara, Nectandra* and *Tabebuia* show 'hotspots' of high values in relatively restricted areas of the Amazon. Some genera, such as *Dicorynia* and *Peltogyne*, show lower values that do not vary much across space.

Discussion

The predicted timber revenue values are comparable to other estimates of timber value in the Amazon. For example, a report by Nepstad *et al.* [42] predicts a maximum *net* value of timber in the Amazon to be US$550, whereas we predict a *gross* maximum value of US$3150 and mean of US$813 per hectare. Once the costs of converting standing timber into sawn wood are taken into account,

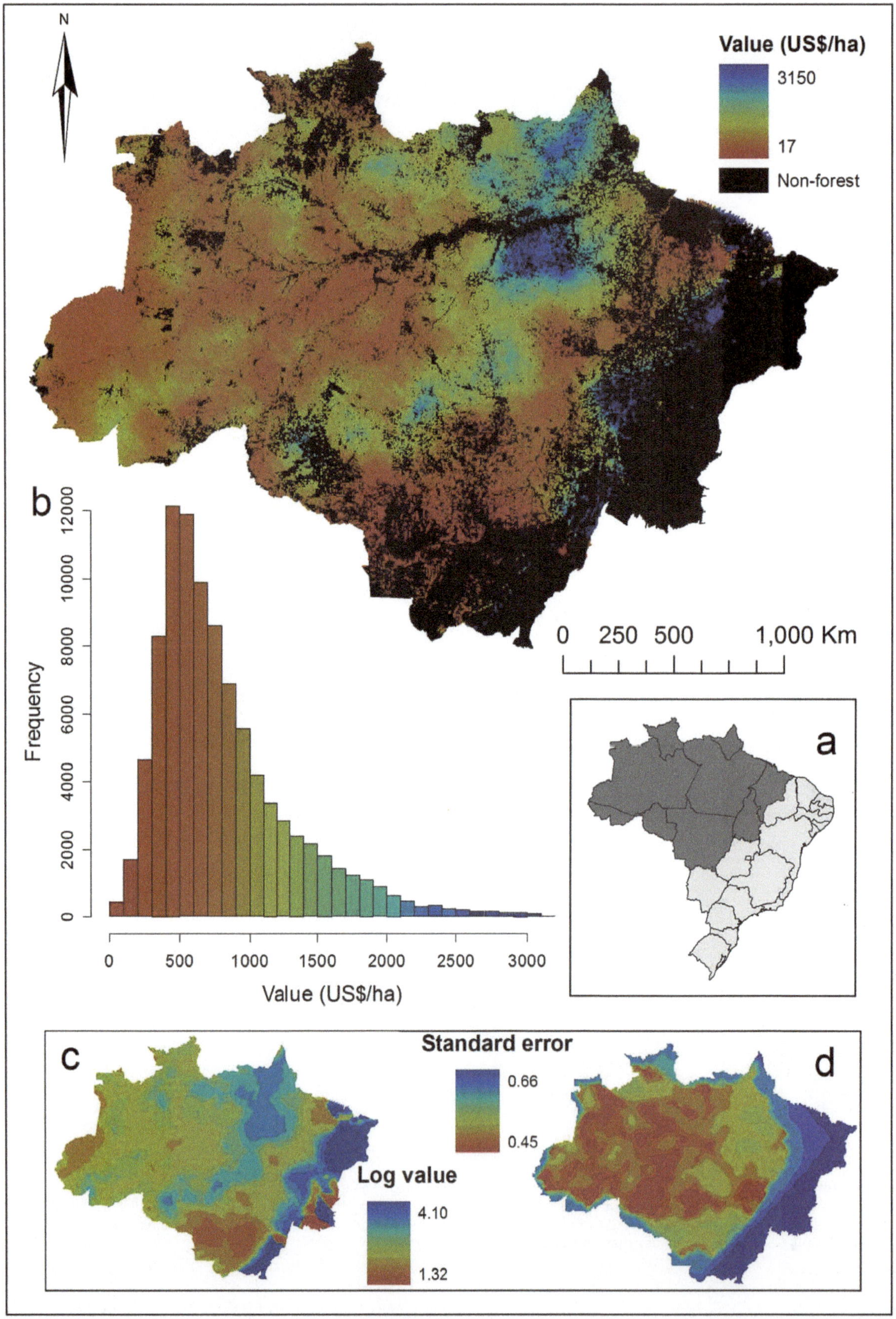

Figure 2. Map of timber value in the Amazon. Values range from low (US$17 per ha) to high (US$3150 per ha). Insets: (a) dark shading shows the spatial extent of the Brazilian Amazon within Brazil, including the state boundaries; (b) frequency distribution of timber values (US$.ha^{-1}) in the

Brazilian Amazon, calculated over 151,073,784 equal-area grid squares of area 0.25 km^2; (c) log- transformed timber value across the Amazon without non-forest areas removed (the raw outputs from the kriging analysis). As non-forest areas are not removed the maximum predicted value in panel (c) is 4.10, equating to US$4400 per ha, is higher than the maximum value of US$3150 per ha in panel (a). This discrepancy arises because the maximum values in panel (c) occur along the eastern margins of the Amazon where, in fact, there is little forest standing. (d) standard error of predicted log-transformed timber value.

our figures are comparable. However spatial patterns of high value timber differ between our predictions and those of Nepstad *et al.* [42], primarily because they considered net value (i.e. profit) while we consider gross value. So, while we find the highest value areas to be concentrated in the north east, they reported that high value areas were concentrated around transport systems that offer cost effective access to the forest.

The differences among genera in their predicted values can be attributed to several factors, of which the first is variation in the spatial distributions of the genera themselves. For example, *Dicorynia* is mainly recorded in the north-west of the Brazilian Amazon and was only present in 21 of 2465 locations. Other genera, for instance *Hymenaea*, *Nectandra* and *Swartzia* were well represented, being present in 945, 1100 and 909 of 2465 locations respectively. They are also relatively evenly distributed across the Brazilian Amazon rather than being clustered in a small region, which influences kriging predictions. Second the value per cubic metre also varies considerably between genera, for example the average value of *Swartzia* species is US$105 per ha whereas the average value of *Tabebuia* is US$317 per ha, further influencing the different predicted values of the 11 genera. Third, differences in the abundance (i.e. volume) of each genus influenced the predicted value, for instance *Dicorynia* had a total recorded volume of just 101.24 m^3 whereas *Hymenaea* had a total recorded volume of 4528.30 m^3. As a result of these three differences among genera, when their individual values are combined the genus-specific patterns are masked and the total values across the Amazon average out. Thus, simply adding the 11 separate genus level kriging predictions does not produce the same results as kriging the total value directly. This averaging effect explains why the maximum genus-specific value obtained of US$3143 per ha is slightly higher than the maximum total value which is US$3150 per ha.

There is a clear spatial pattern in timber value across the Amazon (Figure 1), with the most valuable timber in the north eastern region. Various studies have established that there is an east to west gradient in average wood density/wood specific gravity, with high wood densities occurring in the east [36–37,43–44]. This pattern in wood density is also associated with gradients of increasing seed mass in the east [43], higher above ground live biomass (AGLB) in the northeast and central Amazon [45], and a threefold variation in coarse wood production with higher production in the west [44,46].

The northeast region of the Amazon has old, nutrient poor, well drained soils and a moderately seasonal climate that is occasionally subject to drought [47]. By contrast, the less valuable areas sit in the west and to the south, where there are richer soils, a more seasonal climate and a more dynamic environment in terms of individual tree turnover [43,47]. These edaphic (soil) and climatic conditions help to explain the emergent pattern of timber value we have presented. ter Steege *et al.* [43] identified two primary gradients in tree composition in the Amazon; the first gradient parallels a major gradient in soil fertility, and the second composition gradient is related to climate, specifically dry season length. They found that in the east the most abundant genera are legumes, yet none of the most abundant genera in the western Amazon are legumes. Thus, unsurprisingly, the poor soils of the east appear then to favour species that are able to cope with low nutrient levels, these species tend to be long-lived trees with slow growth rates but high wood density. Conversely, the more fertile soils in the west are associated with higher growth rates [46], lower wood densities [36], high productivity and a high turnover of individuals [43,48], with an average stem turnover rate of 2.6%.yr^{-1} in the west/south-west compared to a rate of just 1.35%.yr^{-1} in the northeast [49–50]. Additionally, ENSO (El-

Table 1.

Genus	Mean value (US$.$ha^{-1}$)	Lower 95% CI (US$.$ha^{-1}$)	Upper 95% CI (US$.$ha^{-1}$)	Number of locations	Value (US$.$m^{-3}$)
Dicorynia	5	5	10	21	158
Goupia	221	221	587	824	257
Hymenaea	394	376	2136	945	569
Hymenolobium	18	18	26	261	96
Manilkara	208	208	1438	598	281
Mora	14	14	27	98	216
Nectandra	179	173	523	1100	292
Peltogyne	20	20	34	255	143
Swartzia	42	42	57	909	105
Swietenia	40	40	250	57	1096
Tabebuia	107	107	232	757	317
Total	813	477	4161	2465	

Mean predicted value of timber (to the nearest US$.$ha^{-1}$, plus 95% confidence intervals) across the Brazilian Amazon by genus, as calculated over 151,073,784 equal-area grid squares of 0.25 km^2 (data presented in Figures 1 and 2). In some cases the lower 95% CI is the same as the mean, reflecting the heavily left-skewed kriging predictions (i.e. Figure 1b). The number of RADAMBRAZIL locations each genus was recorded at and the average export value (US$.$m^{-3}$) are also shown.

Nino Southern Oscillation) causes episodic droughts in the eastern and central Amazon [51], however it has little affect on rainfall in the south-west. High wood density species often have a lower vulnerability to drought stress [37] and are thus less affected by episodic drought than their faster growing light wood counterparts. Mean stand level wood specific gravity was found to be 15.8% higher in eastern and central Amazonia compared to western Amazonia [36]. Given that high value trees often have a high wood density and take a long time to grow (mean wood density is inversely correlated with wood productivity [52]), the highly productive, high turnover areas of the west are not ideal for slow growth trees. However, the poor soils of the east that competitively favour legumes (family, *Fabaceae*), which includes seven of the 11 valuable genera considered in this study (*Swietenia, Mora, Swartzia, Peltogyne, Hymenolobium, Hymenaea, Dicorynia*), are more suited to slow growing, stress resistant species that are out competed on richer soils.

We were not able to include all economically valuable timber genera in this study, omitting genera such as *Carapa* that have high average prices. It was necessary to select genera which were present both in the RADAMBRAZIL data set and for which export value (US.$$m^{-3}$) data was available. It is reasonable to assume that inclusion of other economically important timber genera could alter the absolute values emerging from our analyses. However, given the ecological similarities among many economically important timber trees, we feel that the general spatial patterns would remain the same, with the highest value tree stands being located in the northeastern region of the Brazilian Amazon. Another point to note is the possibility that allied genera were confused in the field surveys, with genera such as *Swartzia* and *Bocoa* sometimes misidentified. However, we again feel that the general trends found here would be robust to minor errors in the field data, partly because no single genera contributes more than 19.66% of all individuals analysed.

We have shown that the patterns of standing timber value in the Amazon reflect known, large-scale ecological gradients extending across the Amazon, determining the spatial distribution of functional traits of trees which are, in turn, correlated with timber values. We expect that understanding the spatial patterns of timber value across the Amazon will aid predictions of logging movements and thus predictions of potential future road developments. These predictions in turn will be of great use in estimating the spatial patterns of deforestation in this globally important biome.

Acknowledgments

We would like to thank an anonymous reviewer for comments. We would also like to thank IMAZON (Instituto do Homem e Ambiente da Amazônia) and IBGE for providing data.

Author Contributions

Analyzed the data: SEA RME. Wrote the paper: SEA RME.

References

1. Foley JA, Asner GP, Costa MH, Coe MT, Defries R, et al. (2007) Amazonia revealed, forest degradation and the loss of ecosystem goods and services in the Amazon basin, Frontiers in Ecology and Environment, 5: 25–32.
2. Betts RA, Malhi Y, Roberts JT (2008) The future of the Amazon: new perspectives from climate, ecosystem and social services, Philosophical Transactions of the Royal Society B, 363: 1729–1735.
3. Moran EF (1993) Deforestation and land use in the Brazilian Amazon, Human Ecology, 21: 1–21.
4. Melillo JM, Houghton RA, Kicklighter DW, McGuire AD (1996) Tropical deforestation and the global carbon budget, Annual Review Energy and Environment, 2: 293–310.
5. Malhi Y, Grace J (2000) Tropical forests and atmospheric carbon dioxide, Trends in Ecology and Evolution, 15: 332–337.
6. Dirzo R, Raven PH (2003) Global state of biodiversity and loss, Annual Review of Environment and Resources, 28: 137–167.
7. Nepstad D, Soares-Filho BS, Merry F, Lima A, Moutinho P, et al. (2009) The end of deforestation in the Brazilian Amazon, Science, 326, 1350–1351.
8. Fearnside PM (1987) Deforestation and international economic development projects in Brazilian Amazonia, Conservation Biology, 1: 214–221.
9. Verissimo A, Barreto P, Tarifa R, Uhl C (1995) Extraction of a high-value natural resource in Amazonia: the case of mahogany, Forest Ecology and Management, 72: 39–60.
10. Mertens B, Poccard-Chapuis R, Piketty MG, Lacques AE, Venturieri A (2002) Crossing spatial analyses and livestock economics to understand deforestation processes in the Brazilian Amazon: the case of Sao Felix do Xingu in south Para, Agricultural Economics, 27: 269–294.
11. Caldas MM, Simmons C, Walker R, Perz S, Aldrich S, et al. (2010) Settlement formation and land cover use change: A case study in the Brazilian Amaon, Journal of Latin Amerian Geography, 9: 125–144.
12. Laurance WF, Goosem M, Laurance GW (2009) Impacts of roads and linear clearings on tropical forests, Trends in Ecology and Evolution, 24: 659–669.
13. Perz SG, Caldas MM, Arima E, Walker RJ (2007) Unofficial road building in the Amazon: socioeconomic and biophysical explanations, Development and Change, 38: 529–551.
14. Kirby KR, Laurance WF, Albernaz AK, Schroth G, Fearnside PM, et al. (2006) The future of deforestation in the Brazilian Amazon, Futures, 38: 432–453.
15. Southworth J, Marsik M, Qiu Y, Perz S, Cumming G, et al. (2011) Roads as drivers of change: trajectories across the Tri-national frontier in MAP, the Southwestern Amazon, Remote Sensing, 3: 1047–1066.
16. Arima EY, Walker RT, Perz SG, Caldas M (2005) Loggers and forest fragmentation: behavioural models of road building in the Amazon Basin, Annals of the Association of American Geographers, 95: 525–541.
17. Nepstad DC, Verissiomo A, Alencar A, Nobre C, Lima E, et al. (1999) Large scale impoverishment of Amazonian forests by logging and fire, Nature, 398, 505–508.
18. Laurance WF, Albernaz AKM, Schroth G, Fearnside PM, Bergen S, et al. (2002) Predictors of deforestation in the Brazilian Amazon, Journal of Biogeography, 29, 737–748.
19. Evans TP, Manire A, De Castro F, Brondizio E, McCracken S (2001) A dynamic model of household decision making and parcel level landcover change in the eastern Amazon, Ecological Modelling, 143: 95–113.
20. Verissimo A, Cochrane MA, Souza JR C, Salomao R (2002) Priority areas for establishing national forests in the Brazilian Amazon, Conservation Ecology, 6: 4–13.
21. Asner GP, Keller M, Silvas JNM (2004a) Spatial and temporal dynamics of forest canopy gaps following selective logging in the Eastern Amazon, Global Change Biology, 10: 765–783.
22. Fearnside PM (2005) Deforestation in Brazillian Amazonia, history, rates and consequences, Conservation Biology, 19: 680–688.
23. Bradshaw CJA, Sodhi NS, Peh KSH, Brook BW (2007) Global evidence that deforestation amplifies flood rish and severity in t he developing world, Global Change Biology, 13: 2379–2395.
24. Asner GP, Nepstad D, Cardinot G, Ray D (2004b) Drought stress and carbon uptake in an Amazon forest measured with spaceborne imaging spectroscopy, PNAS, 101: 6039–6044.
25. Bonan GB (2008) Foresta and climate change, forcings, feedbacks and the climate benefits of forests, Science, 320: 1444–1449.
26. Malhi Y, Roberts TJ, Betts RA, Killeen TJ, Li W, et al. (2008) Climate change, deforestation and te fate of the Amazon, Science, 319, 169–172.
27. Montagnini F, Jordan CF (2005) *Tropical Forest Ecology*, Springer, Berlin.
28. Merry F, Soares-Filho B, Nepstad D, Amacher G, Rodrigues H (2009) Balancing conservation and economic sustainability: the future of the Amazon timber industry, Environmental Management, 44, 395–407.
29. Uhl C, Barreto P, Verissimo A, Vidal E, Amaral P, et al. (1997) Natural resource management in the Brazilian Amazon, BioScience, 47, 160–168.
30. Stone SW (1998) Using a geographic information system for applied policy analysis: the case of logging in the Eastern Amazon, Ecological Economics, 27, 43–61.
31. ter Steege H, Pitman NCA, Phillips OL, Chave J, Sabatier D, et al. (2003) A spatial model of tree α-diversity and tree density for the Amazon, Biodiversity and Conservation, 12: 2255–2277.
32. Prates-Clark CDC, Saatchi SS, Agosti D (2008) Predicting geographical distribution models of high value timber trees in the Amazon basin using remotely sensed data, Ecological Modelling, 211: 309–323.
33. Arima EY, Walker RT, Sales M, Souza JR C, Perz SG (2008) The fragmentation of space in the Amazon basin: emergent road networks, Photogrammetric Engineering & Remote Sensing, 74: 699–709.
34. IBGE website. Available: ftp://geoftp.ibge.gov.br/mapas/banco_dados_georeferenciado_recursos_naturais/Accessed 2011 Jan, 03.

35. Gentry AH (1992) A synopsis of Bignoniaceae ethnobotany and economic botany, Annals of the Missouri Botanical Garden, 79, 53–64.
36. Baker TR, Phillips OL, Malhi Y, Almeida S, Arroyo L, et al. (2004) Variation in wood density determines spatial patterns in Amazonian forest biomass, Global Change Biology, 10, 545–562.
37. Chave J, Muller-Landau HC, Baker TR, Easdale TA, ter Steege H, et al. (2006) Regional and phylogenetic variation of wood density across 2456 neotropical tree species, Ecological applications, 16, 2356–2367.
38. ITTO international Tropical Timber Organisation website. Available: http://www.itto.int/Accessed: 2011 Mar, 15.
39. Gerwing JJ, Johns JS, Vidal E (1996) Reducing waste during logging and log processing: forest conversion in eastern Amazonia, Unasylva, 187: article 3. Available online at: http://www.fao.org/docrep/w2149e/w2149e00.htm Accessed: 2012 Mar 18.
40. R Development Core Team (2009) R, A language and environment for statistical computing. R Foundation for Statistical Computing, Vienna.
41. Giraudoux PR (2011) pgirmess: Data analysis in ecology. R package version 1.5.1. http://CRAN.R-project.org/package=pgirmess Accessed: 2012 Apr 2.
42. Nepstad D, Merry F, Rodrigues HO, Schwartzman S (2007) The costs and benefits of reducing carbon emissions from deforestation and forest degradation in the Brazilian Amazon, REDD, United Nations Framework Convention on Climate Change (UNFCCC), Conference of the Parties (COP), Thirteenth Session. Available online at: http://www.whrc.org/policy/pdf/cop13/WHRC_Amazon_REDD.pdf Accessed: 2012 Apr 2.
43. ter Steege H, Pitman NCA, Phillips OL, Chave J, Sabatier D, et al. (2006) Continental-scale patterns of canopy tree composition and function across Amazonia, Nature, 443, 444–447.
44. Baker TR, Phillips OL, Laurance WF, Pitman NCA, Almeida S, et al. (2009) Do species traits determine patterns of wood production in Amazonian forests? Biogeosciences, 6, 297–307.
45. Saatchi SS, Houghton RA, Dos Santos Alvala RC, Soares JV, Yu Y (2007) Distribution of aboveground live biomass in the Amazon Basin, Global Change Biology, 13, 816–837.
46. Malhi Y, Baker T, Wright J, Phillips OL, Almeida S, et al. (2004) The above ground coarse wood productivity of 104 neotropical forest plots, Global Change Biology, 10, 563–591.
47. Malhi Y, Phillips OL, Lloyd J, Baker T, Wright J, et al. (2002) An international network to monitor the structure, composition and dynamics of Amazonian forests (RAINFOR), Journal of Vegetation Science, 13, 439–450.
48. Phillips OL, Baker L, Arroyo L, Higuchi N, Killeen TJ, et al. (2004) Pattern and process in Amazon tree turnover 1976–2001, Philosophical Transactions of the Royal Society London B, 359, 381–407.
49. Stephenson NL, Van Mantgem PJ (2005) Forest turnover rates follow global and regional patterns of productivity, Ecology Letters, 8, 524–531.
50. Quesada CA, Lloyd J, Schwarz M, Baker T, Phillips OL, et al. (2009) Regional and large scale patterns in Amazon forest structure and function are mediated by variations in soil physical and chemical properties, Biogeosciences Discussions, 6, 3993–4057.
51. Malhi Y, Wright J (2004) Spatial patterns and recent trends in the climate of tropical regions, Philosophical Transactions of the Royal Society London B, 359, 311–329.
52. Malhi Y, Wood D, Baker T, Wright J, Phillips OL, et al. (2006) The regional variation of aboveground live biomass in old-growth Amazonian forests, 12, 1107–1138.

5

Recent Weather Extremes and Impacts on Agricultural Production and Vector-Borne Disease Outbreak Patterns

Assaf Anyamba[1,4]*, Jennifer L. Small[1,5], Seth C. Britch[2], Compton J. Tucker[1], Edwin W. Pak[1,5], Curt A. Reynolds[1], James Crutchfield[3], Kenneth J. Linthicum[2]

1 National Aeronautics and Space Administration, Goddard Space Flight Center, Biospheric Sciences Laboratory, Greenbelt, Maryland, United States of America, **2** United States Department of Agriculture, Agricultural Research Service, Center for Medical, Agricultural, & Veterinary Entomology, Gainesville, Florida, United States of America, **3** United States Department of Agriculture, Foreign Agricultural Service, International Production & Assessment Division, Washington, District of Columbia, United States of America, **4** Universities Space Research Association, Columbia, Maryland, United States of America, **5** Science Systems and Applications Incorporated, Lanham, Maryland, United States of America

Abstract

We document significant worldwide weather anomalies that affected agriculture and vector-borne disease outbreaks during the 2010–2012 period. We utilized 2000–2012 vegetation index and land surface temperature data from NASA's satellite-based Moderate Resolution Imaging Spectroradiometer (MODIS) to map the magnitude and extent of these anomalies for diverse regions including the continental United States, Russia, East Africa, Southern Africa, and Australia. We demonstrate that shifts in temperature and/or precipitation have significant impacts on vegetation patterns with attendant consequences for agriculture and public health. Weather extremes resulted in excessive rainfall and flooding as well as severe drought, which caused ~10 to 80% variation in major agricultural commodity production (including wheat, corn, cotton, sorghum) and created exceptional conditions for extensive mosquito-borne disease outbreaks of dengue, Rift Valley fever, Murray Valley encephalitis, and West Nile virus disease. Analysis of MODIS data provided a standardized method for quantifying the extreme weather anomalies observed during this period. Assessments of land surface conditions from satellite-based systems such as MODIS can be a valuable tool in national, regional, and global weather impact determinations.

Editor: Tetsuro Ikegami, The University of Texas Medical Branch, United States of America

Funding: This work was made possible by funding from USDA Foreign Agricultural Service towards the Global Agricultural Monitoring project, DoD Armed Forces Health Surveillance Center's Global Emerging Infections Surveillance and Response System (AFHSC/GEIS) under the Human Febrile and Vector -Borne Illnesses (FVBI) Program and USDA Agricultural Research Service. The funders had no role in study design, data collection and analysis, decision to publish, or preparation of the manuscript.

Competing Interests: The authors have declared that no competing interests exist.

* E-mail: assaf.anyamba@nasa.gov

Introduction

Severe drought and flooding events occurred in the 2010–2012 period with pronounced impacts on agriculture and public health at regional and national scales. Such extreme weather events frequently result from factors driving interannual climate variability, such as the *El Niño*/Southern Oscillation (ENSO) phenomenon and related tropical and extra-tropical teleconnections [1–3]. Importantly, these extremes are increasingly being amplified by long-term shifts in climate and global warming [4,5]. Extreme weather patterns associated with ENSO have varying impacts on regional ecosystems, agriculture, and health [5–10], and these impacts were very evident during the 2010–2012 period. Anomalously wet or dry conditions can lead to ecological settings favouring emergence or re-emergence of disease vectors and pathogens of global public health relevance such as West Nile virus, malaria, dengue virus, cholera, Murray Valley encephalitis virus, and Rift Valley fever virus among others [11–16]. Extremes in temperature and rainfall can affect crop yields, crop pests, pasture productivity, and create hazardous fire conditions [17–21]. Heavy floods damage infrastructure and crops, and wash away productive topsoil. While socioeconomic impacts of such events vary regionally, depending on existing resources for mitigation, in general they result in increased volatility in agricultural production and commodity prices, as well as higher costs for transport, infrastructure repair, and public health services.

Differences in assessment capability have made it difficult to quantify the global extent and impact of large-scale extreme weather events; however, the 2010–2012 period presents an opportunity to quantify impacts by analysis of satellite-based indicators of surface conditions during these years. We utilized 2000–2012 normalized difference vegetation index (NDVI) and land surface temperature (LST) measurements from the Moderate-Resolution Imaging Spectroradiometer (MODIS) instrument on-board NASA's Earth Observing System Terra and Aqua satellites [22] to illustrate the impacts of weather extremes on agriculture and vector-borne disease outbreak patterns for selected regional locations around the world shown in Figure 1. The anomalous conditions during 2010–2012 are the most extreme weather events in the Terra MODIS 12 year record, and the timing and unique intensity of these events is corroborated by analyses using longer term climate data sets [5,23–24]. We postulate that because both severe drought or flooding may create ecological conditions for the emergence of disease vectors and may substantially affect agricultural production, then under such extreme weather events as documented for 2010–2012 we should observe outbreaks of vector-borne diseases and sharp declines, in the cases of drought, and anomalous increases, in the cases of sufficient above-normal rainfall, in agricultural production. Our aim in this manuscript is to map the locations of documented 2010–2012 extremes at selected regional locations around the world using various satellite datasets, and describe and illustrate the impacts of these extremes by analyzing agricultural production data and vector-borne disease location data for the selected regions.

Materials and Methods

We analyzed two standard products from the MODIS instrument, the NDVI and LST, in combination with rainfall, crop production data, and vector-borne disease occurrence data extracted from an array of resources to investigate the impacts of weather extremes on agriculture and public health for selected regions of study shown in Figure S1 in File S1. The regions of study were selected based on a number of metrics and observations including assessment of areas of defined geographic anomalies in vegetation index data using the Global Agricultural Monitoring System at http://glam1.gsfc.nasa.gov/ and the Crop Explorer of major agricultural regions at http://www.pecad.fas.usda.gov/cropexplorer/. In addition, we monitored and evaluated vector-borne disease outbreak information from the Program for Monitoring Emerging Diseases (ProMED-mail), World Organization for Animal Health (OIE) (http://www.oie.int/) and US Centres for Disease Control and Prevention data records, available at http://diseasemaps.usgs.gov/wnv_us_human.html and information gathered through collaborations with the Global Emerging Infections Surveillance and Response System (GEIS) division of the DoD Armed Forces Health Surveillance Center (http://www.afhsc.mil/geisPartners). The selected regions and period of study experienced temporally persistent, spatially coherent anomalies and are unique in exemplifying clusters of agricultural production shortfalls and/or increase in vector-borne disease outbreaks associated with background weather anomalies taking place in each region (SI). Where any changes in agricultural production or a vector-borne disease outbreak were associated with extremes resulting from changes in the phase of ENSO, we referred to documentation from the NOAA Climate Prediction Center for information. Details of the data sets and analyses are given in the SI Materials and Methods (File S1), including Table S1 and Figures S1 to S4 in File S1. The NDVI quantifies the photosynthetic capacity of vegetation and how this varies in time [25,26], and LST provides a measure of the temperature of the land surface [27]. Both data sets can be used to monitor and assess agricultural growing conditions [28] and also infer ecological conditions that lead to the emergence and increase in abundance of disease vectors [16]. To gauge the impact of extreme weather events on regional ecosystems, we calculated monthly NDVI and LST anomalies by expressing differences between the monthly values and long-term mean 2000–2011 NDVI or LST values [29]. We also derived seasonal anomaly metrics by calculating cumulative three month anomalies, for example June, July, August (JJA) [SI]. The LST and NDVI monthly and seasonal anomalies were examined over the growing seasons for the selected six global agricultural regions worldwide (Figure 1 and Figure S1 in File S1) and were contrasted with agricultural production data extracted from the USDA Production, Supply, and Distribution Online (PSD) electronic database (http://www.fas.usda.gov/psdonline/) for the focal regions. Seasonal NDVI is particularly relevant to agriculture, as cumulative NDVI is linearly related to gross primary production, and hence highly correlated to agricultural production [30]. Accordingly temperature (LST) during the growing season regulates crop growth [7,18,19] and in general vegetation, which in this case can be observed through NDVI time series. Anomaly calculations using MODIS data provided unbiased and spatially continuous representations of the impacts of extreme weather conditions on regional vegetation for the study

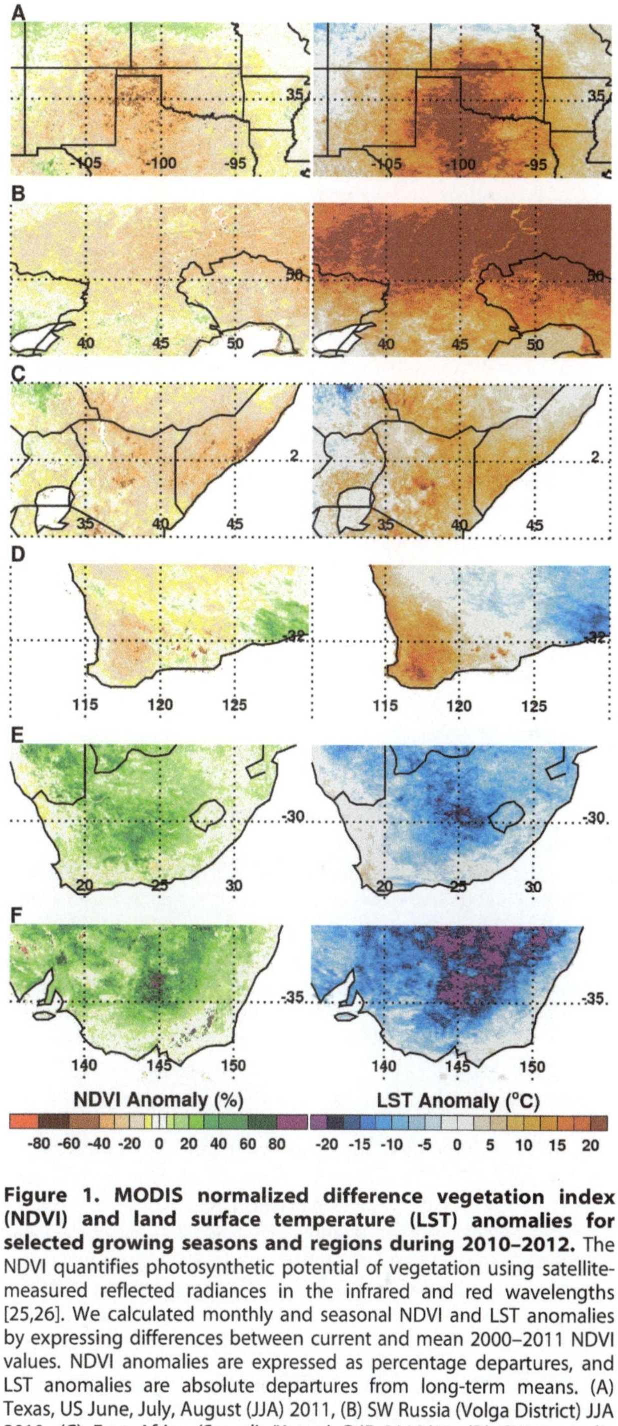

Figure 1. MODIS normalized difference vegetation index (NDVI) and land surface temperature (LST) anomalies for selected growing seasons and regions during 2010–2012. The NDVI quantifies photosynthetic potential of vegetation using satellite-measured reflected radiances in the infrared and red wavelengths [25,26]. We calculated monthly and seasonal NDVI and LST anomalies by expressing differences between current and mean 2000–2011 NDVI values. NDVI anomalies are expressed as percentage departures, and LST anomalies are absolute departures from long-term means. (A) Texas, US June, July, August (JJA) 2011, (B) SW Russia (Volga District) JJA 2010, (C) East Africa (Somalia/Kenya) DJF 2010/11, (D) SW Australia (Western Australia) SON 2010, (E) South Africa (Free State/North West) DJF 2010/11, (F) SE Australia (New South Wales) SON 2010.

period and were augmented with rainfall data extracted and derived for the focal regions from the Global Precipitation Climatology Project [31] (Table 1). Various publicly-available databases were surveyed to develop maps of outbreak locations of dengue, Rift Valley fever, Murray Valley encephalitis, and West Nile virus disease (SI), and these maps were juxtaposed with the focal anomalous regions identified in this study. We also compared the seasonal distribution of LST and NDVI values for the focal regions in the study period to gauge the direction of shifts of the variables from long-term means resulting from extreme conditions. Historical ENSO patterns and timing were determined from NOAA data portals (http://www.cpc.ncep.noaa.gov/products/MD-index.shtml) to establish that extreme weather conditions associated with the cold *La Niña* phase of ENSO and associated teleconnections resulted either in severely depressed vegetation conditions/drought and above-average LST, or excessive rainfall, denser than average vegetation, and cooler than normal LST, with a heterogeneity of impacts on agricultural production and vector-borne disease transmission in the various focal regions around the world.

Ethics Statement

All disease case data analyzed were anonymized. We used GPS latitude-longitude coordinates to map approximate case locations, and we did not handle or deal with any human or animal specimens.

Results

Regions Affected by Drought: Below-Normal Rainfall and Above-Normal Temperatures

In 2010 and 2011, severe drought affected four regions: the southern United States, western Russia, Western Australia, and East Africa (Figure 1A–D); in 2012, the entire continental US was affected (Figure S2 in File S1). In all cases, drought reduced crop yields and led to increased wild fires. The southern half of the contiguous US experienced extreme and persistent drought from mid-2010 through September 2011, with the greatest impact in Texas (Figure 1A). The Texas drought epicenter had historic rainfall lows of up to ~66% below normal (Table 1), coupled with one of the hottest summers on record in 2011 [5,24,32], with monthly LSTs up to 8°C above normal (Figure 2A), and seasonal LST as high as 20°C above normal (Figure 1A). As a result of high temperatures, vegetation photosynthetic capacity was reduced by 40–60% below average (Figures 1A and 2A). Drought and scorching temperatures led to declines in crop conditions and abandonment of fields. In particular, this was because the Texas drought of 2011 was of historical proportions and classified as the most extreme rainfall deficit year in over 100 years [24,32]. The severity of the drought overwhelmed the agricultural system with high evapo-transpiration rates, rationing of ground water resources for irrigation and other uses. Although producers switched irrigation from corn to more drought-tolerant cotton in some areas, cotton production still declined by 50% to 771 kilotons below the 2000–2011 average (Figure 3A). Extreme water deficits had reduced the capacity of the cotton crop to carry fruit as a result of lower rates of leaf photosynthesis and severely reduced the development of floral buds thus reducing yields [33]. Direct losses from the drought approached $10 billion in the state of Texas alone in 2011 [34,35]. Hot, dry conditions also resulted in a very active fire season across the southern US costing more than $1 billion [36]. Unprecedented 100-year climate conditions of extreme high temperatures (Figure 4A,E) and lack of rainfall were linked to the highest period of West Nile virus activity on record in Texas and the rest of the continental US: the 2012 epidemic of West Nile virus disease across the continental US (Figure 5) was the largest such outbreak since the introduction of West Nile virus into the country in 1999, and the spike in human West Nile virus disease cases in 2012 can in part be associated with extreme drought [12] and anomalously high positive shift in summer mean temperatures from ~30°C to 33°C (Figure 4A). Elevated temperatures increase the efficiency of transmission of West Nile virus by both *Culex pipiens* and *Cx. tarsalis* mosquitoes, and have positive effects on mosquito population development and survival, biting rates, and viral replication within these mosquito species [37–39].

Across Eurasia, the summer drought of June-August 2010 was centered in western Russia (Figure 1B) with the drought area extending to Belarus, Poland, Germany, Ukraine, and Kazakhstan [23,40]. Cumulative seasonal LSTs reached as high as 20°C above normal with declines in NDVI of up to 40% below normal (Figures 1B and 2B). Severe and persistent drought sharply reduced agricultural production of major grains in Russia's Central, Volga, and Ural Districts. Estimated wheat production for 2010 was just over 41,500 kilotons, approximately 14% below the 2000–2011 average (Figure 3B), and barley production for 2010 was estimated at 8,350 kilotons, which was 51% below the 2000–2011 average and the lowest in that time span. The combination of severe drought and high temperatures led to the abandonment of approximately 13.3 million hectares of cropland, including half the grain area in the Volga District. The extreme heat resulted in more than 15,000 deaths in Russia during the 2010 summer [23,40]. As in the southern US, the drought caused extreme fire conditions. From July-September 2010 more than 1.25 million hectares burned, causing an estimated $630 million in damages [41].

Table 1. Total seasonal rainfall, long-term means, and anomalies for 2010–2011 extracted from the Global Precipitation Climatology Project [31] for regions presented in Figure 1.

Region	Season	Total (mm)	Mean (mm)	Anomaly (%)
US (Texas)	June–August 2011	59.40	174.11	−65.88
SW Russia (Volga District)	June–August 2010	35.48	133.18	−73.36
East Africa (Somalia/Kenya)	December 2010–February 2011	7.68	51.86	−85.19
SW Australia (Western Australia)	September–November 2010	25.84	90.08	−71.31
South Africa (Free State/North West)	December 2010–February 2011	363.92	253.69	43.45
SE Australia (New South Wales)	September–November 2010	255.27	102.93	148.00

Anomalies are calculated based on the 1979–2011 climatology period.

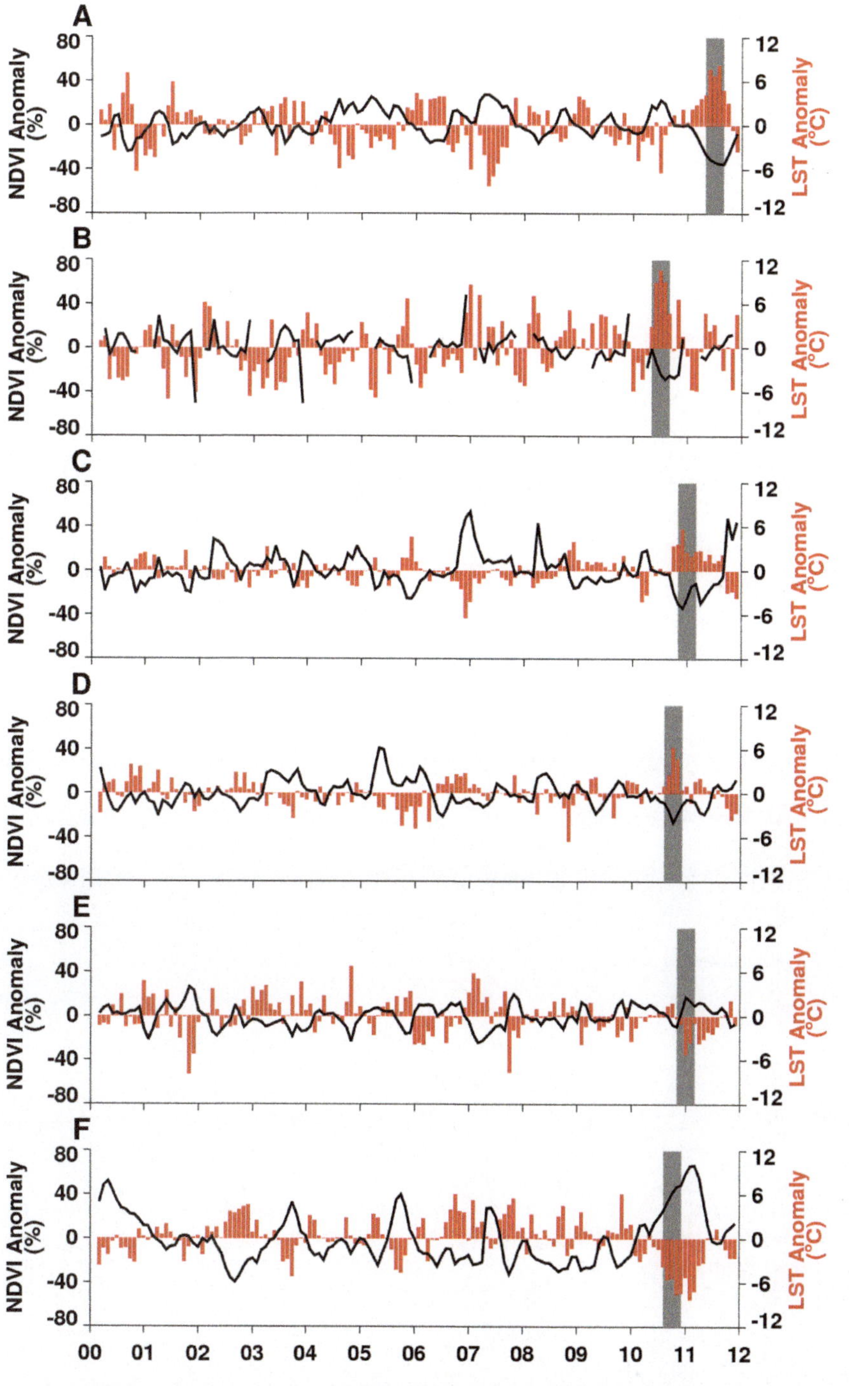

Figure 2. Monthly time series of 2000–2012 MODIS normalized difference vegetation index (NDVI; black line) and land surface temperature (LST; vertical red bars) anomalies for agricultural regions shown in Figure 1 and Figure S1 in File S1. Grey bars indicate selected periods of focused impact studies for each region.

East Africa experienced below-normal rainfall from early 2010 through mid-2011. Rainfall for December 2010 – February 2011 was ~85% below normal (Table 1) and seasonal LSTs were between 5–20°C above normal. The stress of these conditions on vegetation and pastures is illustrated by NDVI departures of 60% below normal (Figure 1C). The peak of the East Africa drought was during the 2010–2011 *La Niña* period (Figure 2C) and led to sharp declines in agricultural production, in particular sorghum, a staple food in this semi-arid region. For Somalia, the total sorghum production for 2011 was 25 kilotons, more than 80% below normal and the lowest for the last decade (Figure 3C). The drought diminished the productivity of pastures and caused widespread famine and high mortality in livestock throughout the region. Although it is difficult to gauge the total human impact of the East African famine, the United Nations estimated more than 13 million people required humanitarian assistance there by December 2011 [42]. Persistent above-normal temperatures were associated with the first known large-scale outbreak of dengue in

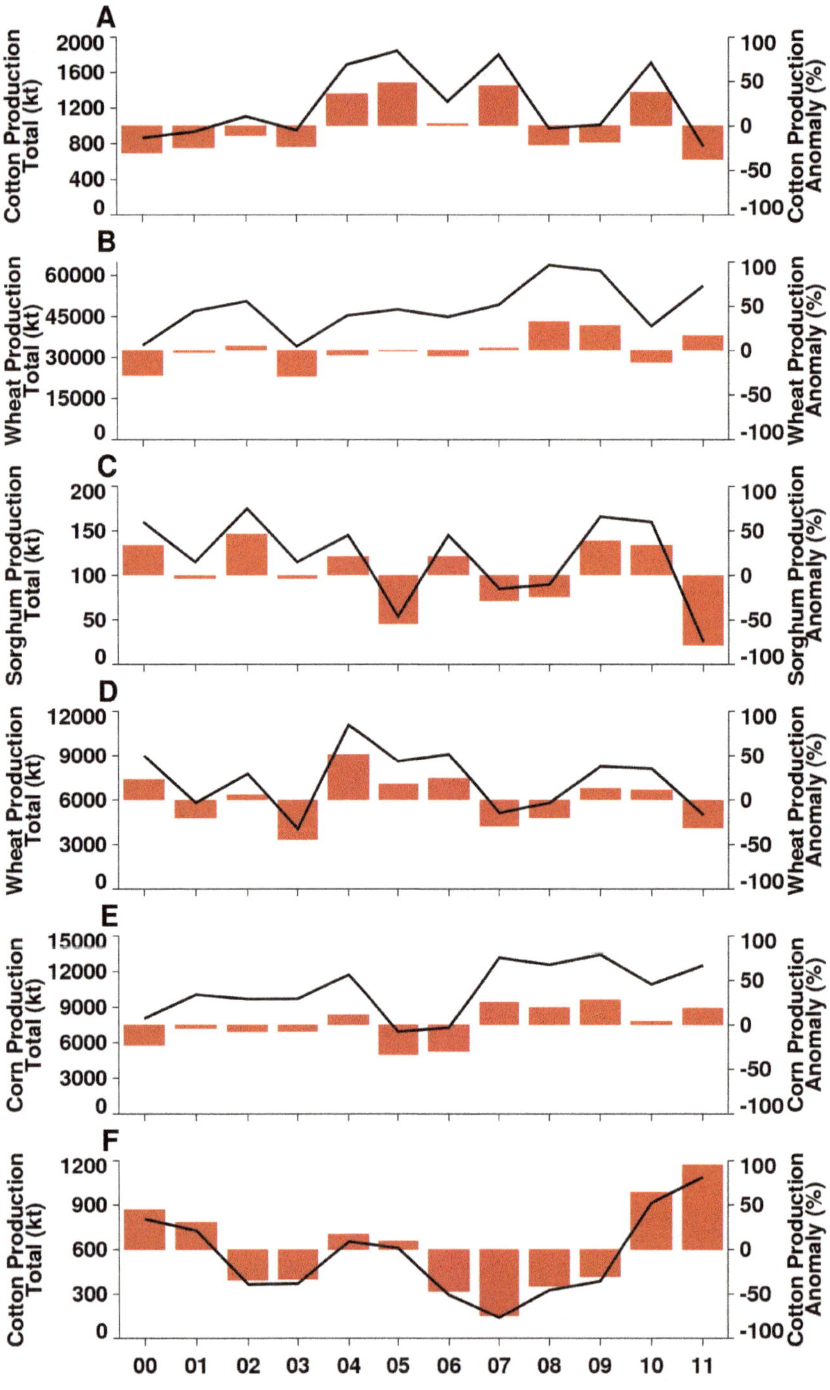

Figure 3. Crop production time series for selected regions in 2000–2011 expressed as annual totals (represented by the black line) and as annual anomalies, i.e., departures from the decadal mean (represented by the red bar chart).

East Africa. Dengue is a mosquito-borne disease associated with infrequent replenishment of stored water supply around households during hot, dry climatic conditions in densely populated areas [9,15,43], which has been shown to increase populations of *Aedes aegypti*, the primary mosquito vector of dengue [44]. Severe drought coupled with higher temperatures increased the abundance of container-breeding dengue virus vector mosquitoes in urban settings leading to the dengue outbreak that persists up to the present [45]. Outbreaks were centered in Mogadishu, Somalia, and Mandera, Kenya (Figure 5), compounding the effects of famine on already vulnerable populations.

Australia experienced both extremes: drought over the western quarter and extreme wet conditions over the eastern half of the continent. Drought and floods in Australia are linked with ENSO cycles, as seen in 1916, 1917, 1950, 1954–1956, 1973–1975, and 2010–2011 [46]. In western Australia, drought prevailed during the 2010 growing season: seasonal LSTs were up to 20°C above normal, vegetation conditions were 30–60% below normal

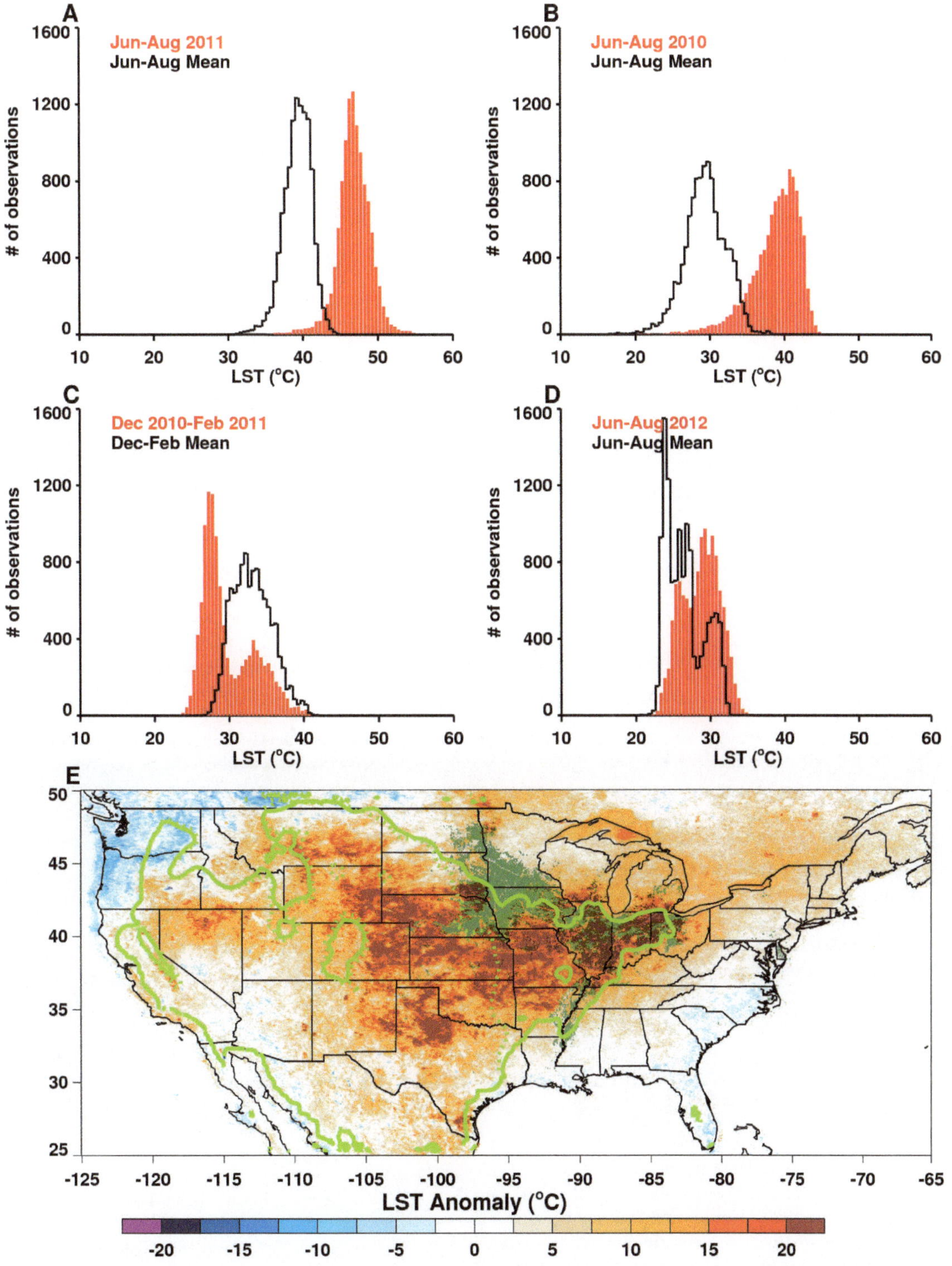

Figure 4. (A–D) Histograms of MODIS growing season land surface temperature (LST) distributions (red bars) compared against the long-term means (black lines) for selected regions of study. (E) Spatial pattern of LST anomalies for the US during June, July, August (JJA) 2012. The area shaded in olive green shows the dominant corn and soy growing region; the solid neon green line delineates the JJA 2012 seasonal 30°C isocline which encompasses the majority of the US agricultural region; and the dotted neon green line shows the JJA 2000–2011 long-term mean 30°C isocline. Persistent temperatures above 30°C destroy most crops and reduce yields [6,17,57].

(Figures 1D and 2D), and total rainfall for September – November 2010 was only ~26 mm, or ~71% below normal (Table 1). Regional wheat production was more than 30% below normal at just under 5,000 kilotons (Figure 3D). The persistent drought, low soil moisture, and above-normal temperatures centered on Western Australia's central agricultural region resulted in the worst overall agricultural production in 40 years [47].

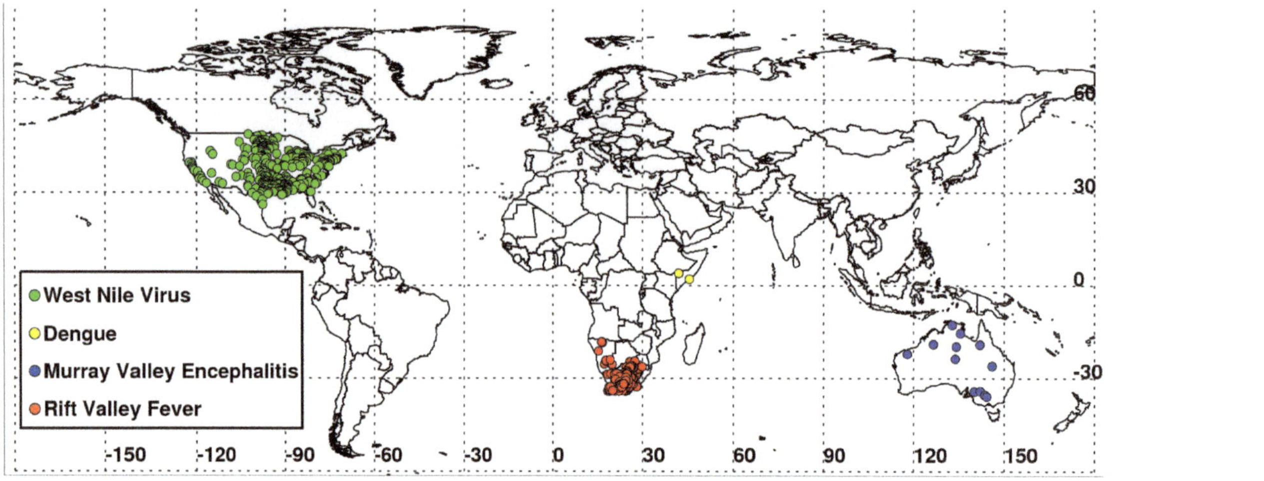

Figure 5. Global distribution of epidemics/epizootics of vector-borne disease outbreaks during 2010–2012 associated with weather extremes, showing the outbreak locations of West Nile virus disease (US, 2012), dengue (East Africa, 2011), Rift Valley fever (Southern Africa, 2011), and Murray Valley encephalitis (Australia, 2011).

Regions Affected by Excessive Rainfall and Below-Normal Temperatures

Three regions experienced excessive rain, floods, and below-normal temperatures during the September 2010 to May 2011 *La Niña* event [48]: the northern US and southern Canada; southeastern Australia; and southern Africa. In North America, heavy spring rains saturated fields, causing flooding and hampering field work in portions of the Corn Belt, Ohio Valley, Tennessee Valley, Missouri Valley, Pacific Northwest, and the northern Great Plains extending into southern Canada. The northern and central Great Plains and much of the Northeast and Atlantic Coast States accumulated rainfall totaling 200% or more above normal and experienced prolonged cool temperatures during this period (Figure S3 in File S1). Above-average precipitation also fell in the Pacific Northwest and rains triggered widespread flooding in the Mississippi and Missouri basins, putting thousands of acres of farmland under water and causing more than $5 billion in damage. Agricultural losses in southern Canada amounted to $1 billion, but impacts on US agricultural production were mixed: although states like Iowa reported bumper harvests, most agricultural states suffered reduced harvests from delayed planting of crops in spring 2011 because waterlogged fields were inaccessible [48,49].

The main corn producing region of South Africa, known as "The Maize Triangle" received heavy rainfall during the 2010–2011 *La Niña* [48], with December 2010-February 2011 total rainfall more than 360 mm, or ~43% higher than normal (Table 1). Abundant rainfall combined with cooler temperatures (Figures 1E and 2E) created favorable growing conditions throughout the country and resulted in the fifth successive year of 25% above-average maize production (Figure 3E). Cool, wet weather led to rapid green-up of vegetation (Figure 1E) and flooding of low-lying areas, or *dambos/pans*, creating ideal ecological conditions for hatching *Aedes* mosquito eggs infected with the Rift Valley fever virus. These conditions in association with the downward shifts in December-February 2010/2011 mean seasonal temperatures from ~40°C to 30°C in South Africa (Figure S4C in File S1) were conducive not only to increased mosquito populations, but increased virus infection in mosquitoes [50,51] and subsequent Rift Valley fever virus transmission. This resulted in the most extensive and widespread outbreak of Rift Valley fever in the region since the 1970s (Figure 5), negatively impacting the livestock industry and human health in southern Africa [51,52]. Overall, MODIS time series NDVI and LST anomalies during the 2010/11 *La Niña* were the most persistent and extreme anomalies for the southern Africa region for the 12 year record.

Southeastern Australia likewise experienced extreme wet conditions, with September-November 2010 rainfall total of ~255 mm, or 148% above normal (Table 1). High rainfall and cooler temperatures (Figure 1F) resulted in increased agricultural yields, putting Australia on track for one of the largest cotton crops in the last decade. The USDA estimated 2011 total cotton production in the region at 1,088 kilotons, approximately 95% higher than the 2000–2011 average (Figure 3F). Most cotton growing regions in New South Wales and Queensland received above-normal winter and spring rainfall, creating excellent planting conditions and sufficient irrigation resources. Overall, 2011 area planted and production estimates were at decadal highs and represent a significant recovery from severe droughts experienced during 2006–2009 (Figure 2F). However, the wet conditions also had negative impacts. The large accumulation of biomass in 2011 created heavy fuel loads that resulted in an intense fire season in early 2012, with large brush fires occurring in Victoria, south Australia, and New South Wales [53,54]. Rains also caused moderate to severe flooding in parts of eastern Australia, especially central-southern Queensland and northern New South Wales. At the peak of the *La Niña* event from November 2010 to January 2011, floods damaged crops in low-lying areas. The combined persistent and heavy rainfall, cooler temperatures, and vegetation growth created ideal conditions for an increase in populations of mosquito vectors of Murray Valley encephalitis virus, leading to outbreaks over northern and eastern Australia [55] (Figure 5). Murray Valley encephalitis mosquito vectors, primarily *Culex annulirostris*, favor cooler temperatures associated with heavy rainfall periods in the tropics and sub-tropics [56]. The downward shift from ~40°C to 30°C in December-January 2010/2011 mean seasonal temperatures compared to the long-term mean distribution for eastern Australia (Figure S4E in File S1) during this epidemic period shows that a cooler environment was conducive to increased mosquito populations and virus amplification, and increased virus transmission to humans.

Discussion

The above regional snapshots exemplify the contrasting patterns and impacts of extreme weather events and demonstrate that changing weather will vary geographically in both magnitude and extent. For example, Figure 4 shows seasonal LSTs against long-term mean seasonal LST distributions for selected regions during the study period. June-August LST distributions for both Texas, US (2011, Figure 4A) and Volga District, Russia (2010, Figure 4B) show significantly warmer and drier conditions compared to long term June-August means, with temperatures as high as ~45°C. Importantly, seasonal minimum temperatures for both regions were also ~10°C higher than normal, indicating an overall increase in seasonal LST, not simply increased LST variability. Temperatures above 30°C increase stress on vegetation and damage pastures, a variety of crops [6,7,17–20,57], and here led to sharp declines in agricultural production, such as a 50% decline in cotton production in Texas (Figure 3A) and a 25% decline in Volga District wheat (Figure 3B). In contrast, Figure 4C shows seasonal LSTs for December 2010-February 2011 over South Africa's Maize Triangle. Compared to Texas and Volga District, the weather in South Africa was cooler and wetter, with average temperatures in the range of 19–28°C that have been associated with increased corn yields [7,18,19,57]. Indeed, above-average corn production was observed over South Africa during the last five years (Figure 3E).

The 2011 drought effects in Texas were similar to the Iowa corn-growing area a year later (Figure 4D), where the 2012 seasonal minimum, mean, and maximum temperatures as high as 35°C, were all above normal, with higher monthly and long-term mean LSTs. The 2011–2012 US droughts were associated with negative public health impacts in addition to large scale agricultural impacts such as reduction in production of major commodities including corn and cotton. The spatial pattern in Figure 4E shows that in 2012 the drought epicenter had shifted to the central US unlike in 2011 when the drought was centered in Texas (Figure 1A). This shows the tendency for drought to persist, expand, and propagate over large areas once initiated. Extreme 2012 summer temperatures affected ~70% of the US corn growing area and led to extensive impoverishment of pasture and rangeland, plus record reductions in corn production. The widespread and persistent nature of drought and high temperatures also contributed to the large-scale epidemic of West Nile virus disease cases across the continental US [38] (Figure 5; SI).

Droughts of the magnitude we observed in the US, Russia, and southeastern Australia have produced recent volatility in agricultural commodity prices, increased costs for feed and water in the livestock industry, decreased supply of water resources, increased fire risk, and contributed to other detrimental societal impacts [17]. The 2012 growing season marked the third consecutive year of weather extremes that affected the northern hemisphere and resulted in depleted commodity stocks. The major 2010–2012 droughts were the most persistent and extreme for these regions in the 2000–2012 MODIS record. The 2011 drought in eastern Africa was also the most extreme event in the MODIS record for the region. Conversely, *La Niña* rains in Australia and South Africa had largely positive impacts on agriculture, but associated vector-borne disease outbreaks negatively impacted public health. Our findings show that extreme seasonal shifts in weather conditions, regardless of direction, favor different vector-borne disease systems in different ways and may lead to increased risk of vector-borne disease outbreaks. Transmission of different vector-borne pathogens such as West Nile virus or Rift Valley fever virus may be enhanced by opposite weather extremes because they are different viruses with completely different ecologies of transmission, different hosts, different mosquito vectors, and different optimal habitats. For instance, anomalously hot and dry conditions can lead to increased storage of water around households and consequent increases in populations of container-breeding mosquitoes like *Aedes aegypti* that may transmit dengue virus. On the other hand, anomalously prolonged wet conditions and flooding, aside from triggering the large-scale emergence of disease-vector mosquitoes, can enhance production and sustainment of vegetation which directly favors the survival of mosquitoes and thereby their vectorial capacity, that is, their capacity to become infected by viruses, such as Rift Valley fever virus, and then transmit them. Such environmentally enhanced outbreaks will vary globally depending on the virus and its transmission ecology and the geographic location and baseline condition of disease endemism and seasonality [58], and could favor the globalization of such pathogens. Our analysis of temperature and vegetation conditions provides a method for quantifying weather extremes consistently from region to region, and demonstrates the value of satellite data in monitoring and mapping the magnitude and extent of such events. As extreme weather events become more common under a more variable climate, nations will face costly adaptation. Systematically-collected satellite data such as those we describe here can be a valuable contribution to national, regional, and global weather impact determinations.

Acknowledgments

Holly Riebeek and Amy Houghton provided useful suggestions for improvement of the manuscript.

Author Contributions

Conceived and designed the experiments: AA KJL. Performed the experiments: AA JLS SCB. Analyzed the data: JLS AA SCB EWP JC. Contributed reagents/materials/analysis tools: AA JLS SCB CAR JC. Wrote the paper: AA SCB KJL JLS CJT JC CR EWP. Contributed knowledge on ecology and temperature effects on mosquito vectors: KJL SCB. Contributed knowledge on weather extreme impacts on vegetation and ecosystems: AA CJT. Contributed knowledge on extreme weather on agricultural production: CAR CR. Wrote data analysis software: JLS EWP.

References

1. Glantz MH (1991) Introduction. In: Glantz MH, Katz RW, Nicholls N, editors. Teleconnections Linking Worldwide Climate Anomalies. New York: Cambridge University Press. pp. 1–2.
2. Lyon B, Barnston AG (2005) ENSO and the spatial extent of interannual precipitation extremes in tropical land areas. J Clim 18: 5095–5109.
3. Hoerling MP, Kumar A (2000) Understanding and predicting extratropical teleconnections related to ENSO. In: Diaz HF, Markgraf V, editors. *El Niño* and the Southern Oscillation: Multiscale Variability and Global and Regional Impacts. New York: Cambridge University Press. pp. 57–88.
4. Hansen J, Sato M, Ruedy R (2012) Perception of climate change. Proc Natl Acad Sci U S A 109: 14726–14727, E2415–E2423.

5. Blunden J, Arndt DS (2013) State of the climate in 2012. Bull Am Meteorol Soc 94: S1–S258.
6. Battisti DS, Naylor RL (2009) Historical warnings of future food insecurity with unprecedented seasonal heat. Science 323: 240–244.
7. Legler DM, Bryant KJ, O'Brien JJ (1999) Impact of ENSO-related climate anomalies on crop yields in the US. Clim Change 42: 351–375.
8. Caviedes CN, Fik TJ (1992) The Peru-Chile eastern Pacific fisheries and climatic oscillation. In: Glantz MH, editor. Climate Variability, Climate Change, and Fisheries. Cambridge: Cambridge Univ Press. pp. 355–375.
9. Epstein PR (2005) Climate change and human health. N Engl J Med 353: 1433–1436.
10. Bouma MJ, Dye C (1997) Cycles of malaria associated with *El Niño* in Venezuela. J Am Med Assoc 278: 1772–1774.
11. Gubler DJ (2002) The global emergence/resurgence of arboviral diseases as public health problems. Arch Med Res 33: 330–342.
12. Epstein PR, Defilippo C (2001) West Nile virus and drought. Global Change Human Health 2: 105–107.
13. Pascual M, Rodó X, Ellner SP, Colwell R, Bouma MJ (2000) Cholera dynamics and *El Niño*-Southern Oscillation. Science 289: 1766–1769.
14. Nicholls NA (1986) A method for predicting Murray Valley encephalitis in southeast Australia using the Southern Oscillation. Aust J Exp Biol Med Sci 64: 587–594.
15. Chretien J-P, Anyamba A, Bedno SA, Breiman RF, Sang R, et al. (2007) Drought-associated chikungunya emergence along coastal East Africa. Am J Trop Med Hyg 76: 405–407.
16. Linthicum KJ, Anyamba A, Tucker CJ, Kelley PW, Myers MF, et al. (1999) Climate and satellite indicators to forecast Rift Valley fever epidemics in Kenya. Science 285: 397–400.
17. Rosenzweig CE, Iglesias A,Yang XB, Epstein PR, Chivian E (2001) Climate change and extreme weather events: Implications for food production, plant diseases, and pests. Global Change Human Health 2: 90–104.
18. Semenov MA, Porter JR (1995) Climatic variability and the modeling of crop yields. Agric Forest Meteorol 73: 265–283.
19. Schlenker W, Roberts MJ (2009) Nonlinear temperature effects indicate severe damages to US crop yields under climate change. Proc Natl Acad Sci U S A 106: 15594–15598.
20. Tubiello FN, Soussana JF, Howden SM (2007) Crop and pasture response to climate change. Proc Natl Acad Sci U S A 104: 19686–19690.
21. Mueller-Dombois D, Goldammer JG (1990) Fire in tropical ecosystems and global environmental change. In: Goldammer JG, editor. Fire in the Tropical Biota: Ecosystem Processes and Global Challenges. Heidelberg: Springer-Verlag. pp. 1–10.
22. Justice CO, Townshend JRG, Vermote EF, Masuoka E, Wolfe RE, et al. (2002) An overview of MODIS Land data processing and product status. Rem Sens Env 83: 3–15.
23. Trenberth KE, Fasullo JT (2012) Climate extremes and climate change: The Russian heat wave and other climate extremes of 2010. J Geophys Res 117: D17103.
24. Hoerling M, Kumar A, Dole R, Nielse-Gammon JW, Eischeid J, et al. (2013) Anatomy of an extreme event. J Climate 26: 2811–2832. doi: http://dx.doi.org/10.1175/JCLI-D-12-00270.1
25. Tucker CJ (1979) Red and photographic infrared linear combinations for monitoring vegetation. Remote Sens Environ 8: 127–150.
26. Myneni RB, Hall FG, Sellers PJ, Marshak AL (1995) The interpretation of spectral vegetation indexes. IEEE Trans Geosci Rem Sens 33: 481–486.
27. Wan Z (2008) New refinements and validation of the MODIS land-surface temperature/emissivity products. Rem Sens Env 112: 59–74.
28. Becker-Reshef I, Justice C, Sullivan M, Vermote E, Tucker CJ, et al. (2010) Monitoring global croplands with coarse resolution earth observations: The Global Agriculture Monitoring (GLAM) Project. Remote Sens 2: 1589–1609.
29. Anyamba A, Tucker CJ (2005) Analysis of Sahelian vegetation dynamics using NOAA-AVHRR NDVI data from 1981–2003. J Arid Environ 63: 596–614.
30. Prince SD, Tucker CJ (1986) Satellite remote sensing of rangelands in Botswana II: NOAA AVHRR and herbaceous vegetation. Int J Rem Sens 7: 1555–1570.
31. Adler RF, Huffman GJ, Chang A, Ferraro R, Xie P-P, et al. (2003) The version-2 Global Precipitation Climatology Project (GPCP) monthly precipitation analysis (1979-present). J Hydrometeorol 4: 1147–1167.
32. Winters KE (2013) A Historical Perspective on Precipitation, Drought Severity, and Streamflow in Texas during 1951–56 and 2011. Scientific Investigations Report 2013–5113. US Department of the Interior, US Geological Survey. Available: http://pubs.usgs.gov/sir/2013/5113/pdf/sir20135113.pdf. Accessed 2014 January 28.
33. Turner N, Heam AB, Begg JE, Constable GA (1986) Cotton (*Gossypium hirsutum* L.): physiological and morphological responses to water deficits and their relationship to yield. Field Crops Res 14: 153–170.
34. Anderson DP, Welch JM, Robinson J (2012) Agricultural impacts of Texas's driest year on record. Choices 27: 1–3.
35. Nielsen-Gammon JW (2012) The 2011 Texas drought. Texas Water J 3: 59–95.
36. NOAA National Climatic Data Center (2011) State of the Climate: Wildfires for Annual 2011. Available: http://www.ncdc.noaa.gov/sotc/fire/2011/13. Accessed 2012 November 27.
37. Kilpatrick AM, Meola MA, Moudy RM, Kramer LD (2008) Temperature, viral genetics, and the transmission of West Nile virus by *Culex pipiens* mosquitoes. PLoS Pathog 4: e1000092.
38. Johnson BJ, Sukhdeo MVK (2013) Drought-induced amplification of local and regional West Nile virus infection rates in New Jersey. J Med Entomol 50: 195–204.
39. Moudy RM, Meola MA, Morin LL, Ebel GD, Kramer LD (2007) A newly emergent genotype of West Nile virus is transmitted earlier and more efficiently by *Culex* mosquitoes. Am J Trop Med Hyg 77: 365–370.
40. Matsueda M (2011) Predictability of Euro-Russian blocking in summer of 2010. Geophys Res Lett 38: L06801.
41. Munich RE (2010) Geo Risks Research Heat wave, drought, wildfires in Russia (Summer 2010). MR Touch Natural Hazards – Event report. Available: https://www.munichre.com/app_pages/www/@res/pdf/NatCatService/catastrophe_portraits/Event_report_HW_DR_WF_Russia_touch_en.pdf. Accessed 2013 April 10.
42. Congressional Research Service (2012) Horn of Africa: The Humanitarian Crisis and International Response. Congressional Research Service Report for Congress, R42046.
43. Padmanabha H, Soto E, Mosquera M, Lord CC, Lounibos LP (2010) Ecological links between water storage behaviors and *Aedes aegypti* production: implications for dengue vector control in variable climates. Ecohealth 7: 78–90.
44. Subra R (1983) The regulation of preimaginal populations of *Aedes aegypti* L. (Diptera: Culicidae) on the Kenya coast. I. Preimaginal population dynamics and the role of human behavior. Ann Trop Med Parasitol 77: 195–201.
45. IRIN (2011) Kenya: Medics overwhelmed as dengue fever spreads. IRIN Humanitarian News and Analysis. Available: http://www.irinnews.org/Report/93848/KENYA-Medics-overwhelmed-as-dengue-fever-spreads. Accessed 2013 April 10.
46. Nicholls N, Lavery B, Frederiksen C, Drosdowsky W, Torok S (1996) Recent apparent changes in relationships between the *El Niño*-Southern Oscillation and Australian rainfall and temperature. Geophys Res Letts 23: 3357–3360.
47. USDA Foreign Agricultural Service (2011) Australia: Grain and Feed Update February 2011. Global Agricultural Information Network (GAIN) Report Number AS1101. Available: http://gain.fas.usda.gov/Recent%20GAIN%20Publications/Grain%20and%20Feed%20Update_Canberra_Australia_1-20-2011.pdf. Accessed 2013 April 19.
48. NOAA/NWS (2011) Climate Diagnostics Bulletin (2010, 2011). Available: http://www.cpc.ncep.noaa.gov/products/CDB/CDB_Archive_pdf/pdf_CDB_archive.shtml. Accessed 2013 March 30.
49. USDA Foreign Agricultural Service (2010) Canada: Grain and Feed Update November Quarterly 2010. USDA Foreign Agricultural Service Global Agricultural Information Network (GAIN) Report Number CA0043. Available: http://gain.fas.usda.gov/Recent%20GAIN%20Publications/Grain%20and%20Feed%20Update_Ottawa_Canada_11-01-2010.pdf. Accessed 2013 April 19.
50. Turell MJ (1993) Effects of environmental temperature on the vector competence of *Aedes taeniorhynchus* for Rift Valley fever and Venezuelan equine encephalitis viruses. Am J Trop Med Hyg 49: 672–670.
51. Grobbelaar AA, Weyer J, Leman PA, Kemp A, Paweska JT, et al. (2011) Molecular epidemiology of Rift Valley fever virus. Emerg Infect Dis 17: 2270–2276.
52. Métras RT, Porphyre DU, Pfeiffer A, Kemp PN, Thompson, et al. (2012) Exploratory space-time analyses of Rift Valley fever in South Africa in 2008–2011. PLoS Negl Trop Dis 6: e1808.
53. Australian Government Bureau of Meteorology Queensland Climate Services Centre. (2010) Monthly Weather Review: Queensland December 2010. Brisbane: Commonwealth of Australia. Available: http://bit.ly/jcdZLt. Accessed 2014 January 15.
54. Chanson H (2011) The 2010–2011 Floods in Queensland (Australia): Photographic Observations, Comments and Personal Experience. Hydraulic Model Report No. CH82/11, School of Civil Engineering, The University of Queensland, Brisbane, Australia. Available: http://espace.library.uq.edu.au/view/UQ:239732. Accessed 2014 January 15.
55. Knox J, Cowan RU, Doyle JS, Ligtermoet MK, Archer JS, et al. (2012) Murray Valley encephalitis: a review of clinical features, diagnosis and treatment. Med J Aust 196: 322–326.
56. Van Den Hurk AF, Craig SB, Tulsiani SM, Jansen CC (2010) Emerging tropical diseases in Australia. Part 4. Mosquito-borne diseases. Ann Trop Med Parasit 8: 623–640.
57. Porter JR, Semenov MA (2005) Crop responses to climatic variation. Phil Trans R Soc B 360: 2021–2035.
58. Anyamba A, Linthicum KJ, Small J, Collins K, Tucker CJ, et al. (2012) Climate teleconnections and recent patterns of human and animal disease outbreaks. PLoS Negl Trop Dis 6: e1465. doi:10.1371/journal.pntd.0001465.

An Integrated Model of Environmental Effects on Growth, Carbohydrate Balance, and Mortality of *Pinus ponderosa* Forests in the Southern Rocky Mountains

Christina L. Tague[1]*, Nathan G. McDowell[2], Craig D. Allen[3]

1 Bren School of Environmental Science and Management, University of California Santa Barbara, Santa Barbara, California, United States of America, **2** Earth and Environmental Sciences Division, Los Alamos National Laboratory, Los Alamos, New Mexico, United States of America, **3** United States Geological Survey, Fort Collins Science Center, Jemez Mountains Field Station, Los Alamos, New Mexico, United States of America

Abstract

Climate-induced tree mortality is an increasing concern for forest managers around the world. We used a coupled hydrologic and ecosystem carbon cycling model to assess temperature and precipitation impacts on productivity and survival of ponderosa pine (*Pinus ponderosa*). Model predictions were evaluated using observations of productivity and survival for three ponderosa pine stands located across an 800 m elevation gradient in the southern Rocky Mountains, USA, during a 10-year period that ended in a severe drought and extensive tree mortality at the lowest elevation site. We demonstrate the utility of a relatively simple representation of declines in non-structural carbohydrate (NSC) as an approach for estimating patterns of ponderosa pine vulnerability to drought and the likelihood of survival along an elevation gradient. We assess the sensitivity of simulated net primary production, NSC storage dynamics, and mortality to site climate and soil characteristics as well as uncertainty in the allocation of carbon to the NSC pool. For a fairly wide set of assumptions, the model estimates captured elevational gradients and temporal patterns in growth and biomass. Model results that best predict mortality risk also yield productivity, leaf area, and biomass estimates that are qualitatively consistent with observations across the sites. Using this constrained set of parameters, we found that productivity and likelihood of survival were equally dependent on elevation-driven variation in temperature and precipitation. Our results demonstrate the potential for a coupled hydrology-ecosystem carbon cycling model that includes a simple model of NSC dynamics to predict drought-related mortality. Given that increases in temperature and in the frequency and severity of drought are predicted for a broad range of ponderosa pine and other western North America conifer forest habitats, the model potentially has broad utility for assessing ecosystem vulnerabilities.

Editor: Ben Bond-Lamberty, DOE Pacific Northwest National Laboratory, United States of America

Funding: This study was funded by Ecosystems and the Climate and Land Use programs of the United States Geological Survey (http://www.usgs.gov/contracts/index.html). The funders had no role in study design, data collection and analysis, decision to publish, or preparation of the manuscript.

Competing Interests: The authors have declared that no competing interests exist.

* E-mail: ctague@bren.ucsb.edu

Introduction

Drought-related forest mortality is a widespread phenomenon and is expected to increase with continued climate warming [1], [2], [3]. Drought is also an important driver of forest productivity and longer-term patterns of growth and forest structure and composition [4]. Models provide tools for investigating productivity and mortality risk under climate change scenarios and are a key source of information for forest management planning and impact assessment [5], [6], [7]. Modeling approaches range from fully mechanistic physiological representations to envelope models that use historic climate-productivity relationships to estimate future responses. While fully mechanistic approaches are better able to disentangle the impact of multiple interacting processes, intermediate complexity carbon cycling models are valuable for estimating watershed to regional scale net primary production (NPP) and have been validated by comparisons with flux-tower and remote-sensing estimates of NPP [8], [9], [10]. Historically, however, forest carbon cycling models do not explicitly represent the physiological mechanisms involved in forest mortality [11].

Recent research has focused on mechanisms that underlie drought-related tree mortality and has identified both endogenous factors such as carbon starvation and hydraulic failure, and exogenous factors such as attack by insect and disease, that lead to death [12], [13], [11]. Identifying the specific mechanisms by which these factors individually or in combination lead to tree death remains an active area of research. Earlier work positioned hydraulic failure and carbon starvation as distinct mortality mechanisms, but more recent investigations suggest mortality may occur through interactions between these and other mechanisms [11], [14]. There is mounting evidence that death can occur prior to complete depletion of non-structural carbohydrate (NSC) reserves, suggesting that accurately modeling within plant transport of NSC may be necessary to predict mortality related to a depletion of energy reserves. Explicit representation of the mortality mechanisms is likely to require representing a combination of drought effects on NSC depletion, within plant transport of NSC, controls on the use of NSC reserves (or sink effects), and their interactions [15], [13], [14]. While models of within plant hydraulics and transport exist [16], the complexity of these models limits their applicability at watershed scales where forest behavior also depends on forest structure and composition patterns, climate gradients, and soil water holding and drainage characteristics.

Coupled models of ecosystem carbon cycling and hydrology offer intermediate complexity in representing plant physiology as well as plot to watershed scale controls on moisture and energy exchange. The challenge is to adapt these models to account for mortality risk without full representation of within plant mortality mechanisms until more mature (and efficiently parameterized) detailed mechanistic models are available [11]. We tested the addition of a simple two-parameter model of NSC storage dynamics to an existing coupled carbon cycling and hydrology model, the Regional Hydro Ecological Simulation System (RHESSys). While NPP declines with water stress, the relationship between daily or seasonal NPP and mortality also depends on energy reserves that can be used for maintenance, repair and defense. Several carbon cycling models include a dynamic NSC pool, but utilize all carbohydrates each day [17] or by the end of each year [18] (see [11] for a review).

In this paper, we explore the seasonal and inter-annual dynamics of the NSC pool, its relationship with climate and water availability, and potential linkages with NPP and mortality. We then use RHESSys with the inclusion of the NSC pool dynamics (RHESSys-NSC) to estimate the likelihood of mortality, and compare results with an observed mortality event at the lower ecotone of a ponderosa pine (*Pinus ponderosa*) distributional gradient [19], [20]. Simulating the tree and stand level NSC pool is only a proxy for more complex mortality mechanisms such as insect attack or hydraulic failure, but photosynthetic and NSC dynamics are strongly associated with each of these processes and with mortality [21], [22], [23], [24]. We use ten years of stem growth data from this site and two higher ponderosa pine sites across an 800 m elevational gradient to evaluate model estimates of productivity, and then use sensitivity analyses to explore linkages between NSC parameters, the likelihood of mortality, and their interactions with site characteristics across the elevational gradient.

Methods

Overview

Our approach was to first develop a sub-model of NSC pool dynamics and link it to RHESSys model simulations of photosynthesis and carbon allocation, test the model against observational data of growth and mortality, and then conduct a sensitivity analysis to determine optimal parameterization and the influence of climate and site characteristics on productivity and survival. To test if the modeled carbon cycle was reasonable, we compared predictions of inter-annual variation in NPP with observations of growth as basal area increment (BAI) for each of the three sites through time. We also compare model estimates of cross-site difference in biomass and leaf area index (LAI) with observations of stand structure for each site. While these comparisons are only semi-quantitative since we do not have direct measurements of NPP or LAI trajectories through time, they provide an indication of whether the model can accurately represent spatial differences in site productivity prior to the drought and the relative likelihood of surviving the drought.

Once we have developed our modeling approach, we then used RHESSys-NSC to answer two questions: 1) what is the relationship between estimates of minimum NSC during drought and the fraction of gross primary production (GPP) allocated to NSC; and 2) how do differences in precipitation and temperature between stands along an 800 m elevation gradient influence the NSC pool and ultimately the likelihood of mortality? To answer the first question, we examine the sensitivity of model estimates of minimum NSC during drought to the two parameters that control the fraction of GPP allocated to NSC. We use this sensitivity analysis to identify parameters that produce patterns of minimum NSC that correlate with observations of mortality.

We perform this sensitivity analysis for two different model implementations: 1) a stand level implementation at each site that accounts for site specific variation in pre-drought biomass, climate, and soil characteristics; and 2) a tree level implementation in which we assume the same average biomass per unit area for all three sites. In the tree level implementation all sites have the same biomass and only climate inputs and soil parameters differ. The tree level implementation examines the impact of drought on a "tree" or a given biomass per unit area at each of the three sites. The stand level implementation, on the other hand, includes spatial differences in current biomass that result from historic, decadal effects of spatial variation in environmental drivers (e.g. the cooler-wetter climate at the high elevation site supports greater biomass prior to the drought). Considering both a stand (e.g. average effects of the drought given stand characteristics that evolved prior to the drought) and a tree level (e.g. effects for the same biomass per unit area at each of the different sites) implementation allows us to disentangle the effect of pre-drought biomass from spatial differences in the severity of drought as a result of gradients in precipitation, temperature and differences in soils. Our second question then examines the physical drivers that give rise to spatial differences in forest mortality risk. To answer the second question, we examine how model predictions of NSC, productivity, and mortality vary between high and low elevation sites for stand level implementations given different temperature and precipitation drivers.

Model Implementation

RHESSys is a process-based model of coupled hydrologic and ecosystem processes, spatially modeling fluxes both within and between model unit grid cells of user-defined size. RHESSys has successfully modeled both hydrologic and carbon cycling behavior such as evapotranspiration, streamflow, net ecosystem CO_2 exchange, and NPP in mountainous regions in North America and Europe [25], [26], [27], [28], [29], [30].

RHESSys models the vertical flux of water between the atmosphere and canopy, and the litter and soil layers [31]. Modeled hydrologic processes include infiltration, vertical drainage between unsaturated and saturated stores, and lateral and vertical redistribution of shallow groundwater to deeper groundwater stores. Spatial patterns of radiation, and atmospheric drivers such as temperature, precipitation, humidity, and wind can be provided as inputs or estimated using empirical functions (i.e. [32]). For each model unit grid cell - energy, wind, and water are attenuated through the aboveground canopy as a function of LAI. RHESSys distinguishes between canopy interception of diffuse and direct radiation and sunlit and shaded leaves [33]. Snowmelt is estimated using a combination of an energy budget approach for radiation-driven melt and a temperature index-based approach for latent heat-driven melt processes. Transpiration from the canopy and evaporation of intercepted water and soil and litter moisture are computed using the Penman–Monteith equation [34], where stomatal conductance is computed using species-specific multiplicative models [35] of radiation, vapor pressure deficit, rooting zone soil moisture, CO_2, and air temperature. GPP is computed using the Farquhar equation [36]. Respiration is computed as a function of temperature and plant component nitrogen content [37], modified to account for adaptation to ambient temperature during growth [38]. Carbon and nutrient cycling sub-models are similar to those used in BIOME BGC [18] and Century N-Gas [39], respectively. RHESSys determines the fraction of net photosynthesis (gross photosynthesis minus maintenance and growth respiration of foliar, woody, and fine root biomass) allocated to leaves and stems based on stand age following Dickinson [40]. In this approach, the proportion of net photosynthesis allocated to leaves declines as LAI increases. Allocation to fine roots follows allocation to leaves and the remainder of net photosynthesis is allocated to stems. Leaf, fine-root, and branch turnover occur as a fixed proportion of the associated biomass components. Turnover rates and allocation parameters for ponderosa pine were taken from the RHESSys parameter database.

We extended the current carbon cycling sub-model within RHESSys to explicitly model a within-tree NSC pool (Figure 1). This NSC pool maintains the net balance across years and includes two new parameters to control the non-structural carbohydrate allocation and consumption. *NSC/NPP* is the fraction of annual NPP allocated to non-structural carbohydrate storage until total NSC reaches a threshold percentage of plant structural carbon (20% of total plant biomass). Our 20% of total plant biomass threshold is based on field measurements as well as plant structural considerations [41], [42], [43], [44], [45]. An additional scheme was required for NSC allocation to respiration versus growth, particularly during periods of low GPP. Our approach was to assume trees utilize NSC reserves preferentially to maintain respiration, and once respiration requirements are met, the remaining NSC is used to replace fine roots and leaf turnover given that excessive reduction in these compartments during drought accelerates future mortality. To account for this use of NSC to replace leaf and fine root turnover, we examined a range of values of *MinL/AGC,* or the threshold minimum ratio of leaf carbon stores to total above ground carbon. If the leaf to above ground carbon falls below the threshold value defined by the *MinL/AGC* parameter, the tree will access NSC reserves after maintenance respiration requirements are met, to restore leaf biomass. Limited data on the allocation of NSC to fine roots versus foliage forced us to assume that fine root biomass is assumed to follow leaf biomass - such that allocation to leaves from NSC reserves also triggers allocation to fine roots. We are exploring additional flexibility in leaf/fine root allocation ratios in response to stress via subsequent work. We also examine model sensitivity to values of the *MinL/AGC* parameter ranging from 0 to 25%. The 25% upper limit for *MinL/AGC* is based on field measurements of leaf:aboveground biomass ratio for mature ponderosa pine stands [46]. We refer to this model as RHESSys-NSC throughout the remainder of the paper.

We track the time series of model estimates of NSC reserves and argue that mortality is increasingly likely to occur in conditions where our estimated NSC reserves approach zero. We use NSC here as an indicator of mortality risk and argue that low NSC reserve estimates in our carbon cycling model will correlate with a number of more complex physiological mechanisms that have been linked to drought related mortality including carbohydrate starvation, declines in the plant's ability recover from cavitation and hydraulic failure, and reduction in allocation to defense mechanisms. Ultimately, we need to define a threshold NSC reserve, below which the likelihood of mortality increases. While the critical values of NSC below which mortality likelihood begins to increase remains uncertain, we assume that values below 0.5% of total above ground biomass are both physiologically realistic and will only occur under stress conditions that substantially reduce net accumulation of assimilates over time. For this paper we do not produce quantitative likelihood estimates for mortality, but rather use a fuzzy classification such that for stands with NSC below the 0.5% threshold the risk of mortality is high, for those near the 0.5% threshold the risk is moderate, and for those with NSC above the 0.5% threshold the likelihood of stand level mortality events is low. To test this use of NSC as an indicator of the relative likelihood of mortality, we compute NSC estimates for stands along an elevational gradient and compare the NSC estimates for sites that survived the 2000's drought with those in which 100% mortality occurred.

RHESSys Implementation for the New Mexico Study Watershed

We tested the new RHESSYs-NSC using ponderosa pine stands in the Jemez Mountains of northern New Mexico. This is a semi-arid continental climate region characterized by a bimodal precipitation pattern of summer and winter peaks, with strong elevational gradients of temperature and precipitation. We focus on three sites located across an 800 m elevation gradient within the Frijoles watershed of Bandelier National Monument, in the Jemez Mountains of northern New Mexico (Table 1 provides additional information on site characteristics. Sites have similar slope and aspect). The low elevation site is a relatively open stand of ponderosa pine mixed with one-seed juniper (*Juniperus monosperma*) and piñon pine (*Pinus edulis*) (Table 2 provides stand basal area (BA)). The mid-elevation site is closed canopy, even-aged ponderosa pine, and the high-elevation site is a relatively dense canopy site of mixed ponderosa pine, Douglas-Fir (*Pseudotsuga menziesii*) and aspen (*Populus tremuloides*). Observed ponderosa pine basal area was similar for the upper and mid site and substantially smaller for the low elevation site (Table 2). Although ponderosa pine BA is similar for high and low elevation sites, overall forest stand BA is greater for the high elevation site due to the presence of Douglas-fir and aspen trees. All stands are mature with ages between 95 and 300 years. In the simulation we modeled the high site stand as homogeneous ponderosa pine.

At each site, dendrometer bands have monitored weekly stem growth for ten ponderosa pine trees per site since 1991, from which basal area increment (BAI) was calculated [20]. In late 2002, after three years of drought, all monitored ponderosa pine trees at the low elevation site died. In contrast, while mid and high elevation sites showed BAI declines during the drought, these pines did not experience significant mortality. BAI, mortality and survival are the key observed variables we used to test the RHESSys-NSC model performance. While methods are available to translate BAI into stem-wood growth for comparison to modeled stemwood accumulation, we argue that uncertainties in allometric parameters are high, since no on-site allometrics are available, ultimately reducing the information content of such a comparison. Correlation between model estimates of NPP and BAI however, demonstrates whether or not the model generally captures site and inter-annual differences in productivity.

Daily maximum and minimum air temperature and precipitation inputs to RHESSys are derived from the nearby Los Alamos meteorology station (elevation 2262m). To estimate precipitation inputs for each of the sites across the elevation gradient, we apply an elevation-based precipitation adjustment (Precip-scalar in Table 1) to meteorology station records. We derive this precipitation adjustment from historic records available at two additional local stations located at 2896 m and 3231 m (Quemazon snotel and Bandelier fire tower stations, respectively). Records for these sites are shorter than for the primary station so they could not be used as input time series. Final precipitation scalars used are summarized in Table 1. Note that while a standard environmental

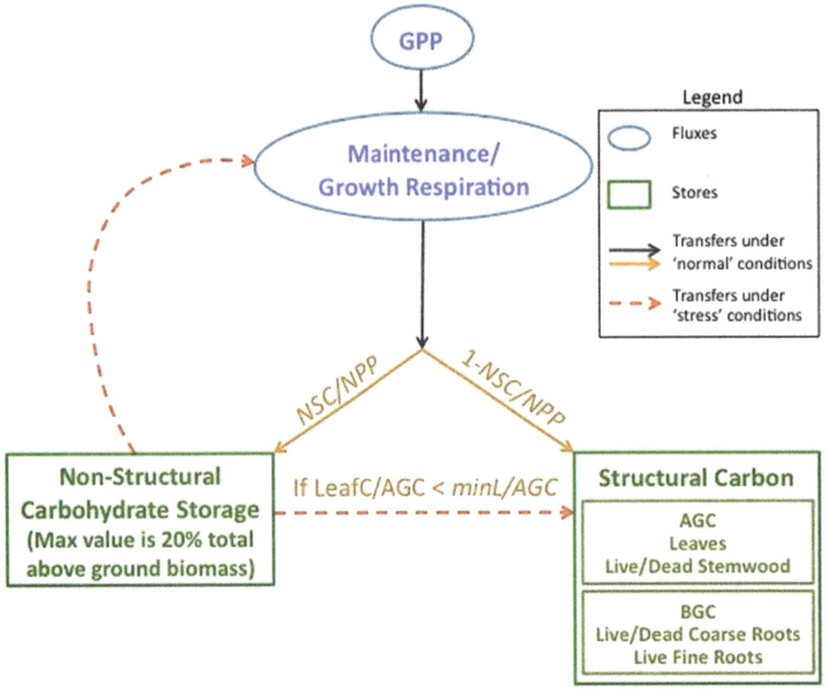

Figure 1. RHESSys submodel of allocation to NSC (non structural carbohydrate), with associated parameters. AGC is above ground plant structural carbon and BGC is below ground plant structural carbon. GPP is gross primary productivity. *NSC/NPP* is the fraction of annual NPP allocated to non-structural carbohydrate storage. *MinL/AGC* is defined as a threshold minimum for the ratio of leaf carbon (LeafC) to above ground carbon and determines when NSC will be depleted to maintain AGC.

lapse rate for temperature (6.5 °C/km) is commonly used in hydrologic modeling, recent studies show that in the mountainous West, lapse rates are often substantially less than environmental lapse rates and typically differ for maximum and minimum temperatures [47]. To account for this, temperature inputs are also adjusted for elevation using relationships among the three meteorology station records. Computed lapse rates for this site are −5.7°C per km and −5.3°C per km for maximum and minimum temperatures, respectively.

Soils at the upper site are Pachic Argiustolls (Lucito loam on 1–8% slopes and Mapache gravelly loam on 10–25% slopes), at the mid site the soils are mixed Vitrandic Haplustolls and Typic Paleustalfs (Adornado-Letrado complex on 5–20% slopes) and typic Haplustalfs (Jemez sandy loam on 2–8% slopes), and at the low site the soils are Vitrandic Haplustolls (Adornado very paragravelly ashy loamy coarse sand on 8–15% slopes). RHESSys soil parameters for each site (Table1) were derived from recent soil survey data [48]. Soils at the upper site have a slower drainage rate and greater water holding capacity relative to the lower site. The mid site soils have the highest water holding capacity but have drainage rates intermediate between upper and lower sites.

To initialize forest structural carbon (in leaves, roots and stems), we ran a 100-year spinup simulation. Meteorology data for the spinup period is synthesized by repeating the available 50-year (water year 1950 to 2000) Los Alamos meteorologic station record, using adjustments for elevation as discussed above. Modeling unit resolution for these simulations is a 30 m grid cell. Thus, at each site stand carbon cycling dynamics are modeled as per unit area carbon store and fluxes averaged over a 30 m grid cell with mean elevation, climate and soil characteristics found at each of the three sites. Use of the historic spinup allows the model to account for spatial differences in stand characteristics at each site, assuming similar ages since disturbance. In these stand-level simulations we include a 5% background whole tree (stem, leaf, stem and coarse root) turnover [49]. Subsequent simulations based on this spinup include the effect of a-priori differences in stand biomass at each site in response to climate variation and drought. This means, for example, that the upper elevation stand may have greater total biomass, and thus typically utilize more water when aggregated

Table 1. Site characteristics and soil parameters.

Site	Elev (m)	Precip scaler (dim*)	Soil maximum saturated hydraulic conductivity (m/day)	Soil decay of saturated hydraulic conductivity with depth (m)	Soil Ψ air entry (m)	Soil pore size index (dim)
High	2767	1.27	764	8.75	0.159	0.16
Mid	2308	1.03	1500	4	0.175	0.168
Low	2002	0.85	3667	0.38	0.125	0.139

*dim = dimensionless, Precip scalar is the multiplier on input daily precipitation data to adjust for cross site differences in rainfall inputs.

Table 2. Modeled and observed stand characteristics prior to the drought.

Site	LAI after 100 year spinup (averaged over 20 years) (m^2/m^2)	Pearson Correlation Coef (mean across all parameters) NPP vs. BAI	Observed Mean BAI (prior to 2000) (cm^2/yr)	Observed Stand Basal Area (BA) (m^2/ha)
High	2.4	0.45 (0.37–0.53)	35.3	29
Mid	2.0	0.75 (0.71–0.82)	19.6	32
Low	1.6	0.73 (0.68–0.83)	15.1	8.8

LAI is modeled leaf area index. BAI is measured annual basal area increment. NPP is model estimate of net primary productivity.

over a 30 m modeling unit. To assess performance of the carbon-cycling model across the elevational gradient, we compared predictions of above ground biomass (total leaf and stem) for the three sites to stand measurements of BA and LAI. We also compared modeled NPP and observed BAI for all sites to assess model ability to capture response to inter-annual variation in climate conditions.

Sensitivity analyses

Given the relative lack of empirical information regarding NSC dynamics during drought and prior to mortality, we performed sensitivity analyses of model growth predictions (NPP, change in LAI) and NSC to assess how much the different assumptions of NSC dynamics influenced predictions. Specifically, we varied *NSC/NPP* and *MinL/AGC* parameters to examine the sensitivity of model predictions of NSC and NPP. We began simulations in 1985 to account for any transient dynamics associated with shifting to a specific NSC parameter set. We then examine the model predictions for water years 1992 to 2007. In particular we compare the minimum NSC obtained for 1992–2007 to indicate likelihood of carbon starvation across the elevational gradient and across different values for storage parameters.

We perform sensitivity analysis for two different scenarios; one to estimate stand-level responses using spatially variable structural carbon (based on model spinup of leaves, stems and roots as described above) and a second to estimate tree-specific responses where initial biomass in leaf, stem and root compartments are set to be uniform across sites. We do this to disentangle differences in drought stress response that include the effect of apriori difference in biomass and biomass partitioning into leaves, stems and roots (stand-level response), and those due only to differences in site meteorology and soil conditions (tree-specific response). For tree-specific scenarios we use biomass (leaves, stems, and roots) components from the spinup for the mid elevation site and normalized the results by stand biomass. For tree-specific scenarios we assume no turnover of stemwood to reflect tree-based rather than stand-based estimates.

We used the results from the sensitivity analyses to select reasonable values of *NSC/NPP* and *MinL/AGC* parameters. We choose parameters that: 1) are consistent with cross-site and inter-annual patterns of observed BAI; 2) result in minimum values of NSC that are less than 0.5% of biomass for the low elevation site during drought; and 3) are physiologically and ecologically realistic. While the critical values of NSC below which mortality is likely to occur are uncertain, we assume that values below 0.5% indicate a high likelihood of stress related mortality. We then use this parameter set to run additional simulations, where we vary precipitation and temperature inputs and soil parameters to explore interactions among these controls on carbon cycling dynamics along the elevational gradient in more detail. For simplicity, we focus only on comparison between the high and low elevation sites. To compare the relative importance of temperature and precipitation dynamics, we contrast productivity estimates for simulations using spatially uniform structural carbon for the 1992 through 2007 period using a) the original meteorological dataset, b) simulations where all sites receive high site precipitation, c) simulations where all sites receive low site precipitation, d) simulations where all precipitation is assumed to fall as rain and temperature is unchanged, e) simulations where all sites use high site temperature, and f) simulations where all sites use low site temperature.

Results

Assessment of RHESSys biomass and productivity predictions across three sites

Model spinup leads to relatively stable biomass after approximately 50 years (Figure 2). The spinup estimates of LAI, biomass, and NPP are consistent with observed biomass and BAI across the elevation transect of the study site (Tables 2, and Figures 2, 3bc). The accuracy of the spinup simulations, along with the convergence of above and below ground biomass estimates to a relatively stable value prior to the drought, suggests that simulations are not likely to be sensitive to initial conditions or the age assumed at the beginning of the sensitivity analysis. The model estimates of biomass, NPP and LAI show increasing values with elevation that are consistent with observed patterns (Table 2, Figure 3c).

Model estimates show clear elevational differences in estimates of snow accumulation and melt, transpiration, and NPP (Figure 3). The high elevation site receives considerably more of its precipitation as snow relative to mid and low elevation sites (Table 1, Figure 3a). There is higher spring transpiration and NPP

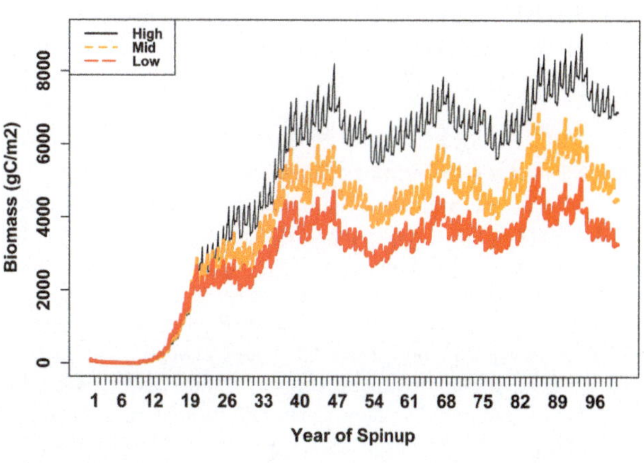

Figure 2. RHESSys estimates of total stand biomass for spinup period of 100 years.

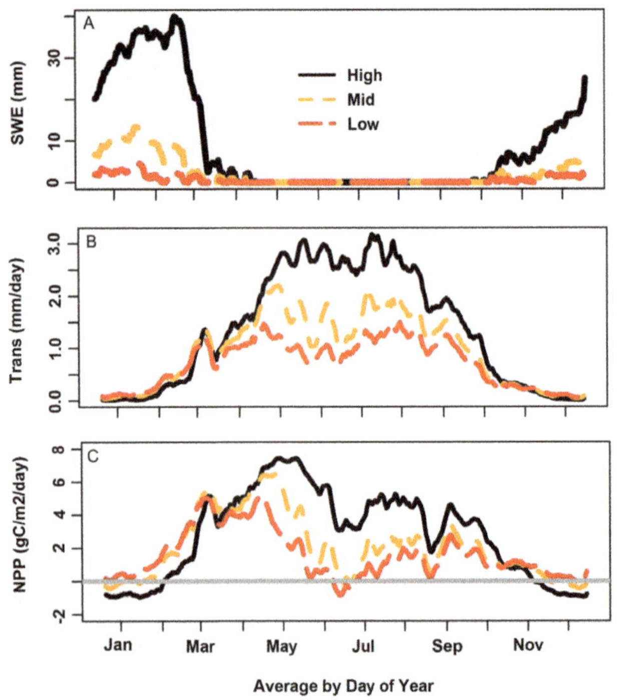

Figure 3. RHESSys estimates ecohydrologic fluxes for simulation period Graphs show a) snow water equivalent (per area of 30 m grid cell) for high, mid and low elevation site for water year 1992–2007 simulation period b) daily transpiration per grid cell, averaged by day of water year for pre-drought (1985–2000) period and c) daily net primary productivity (NPP) per grid cell, averaged by water year for pre-drought (1985–1999) period.

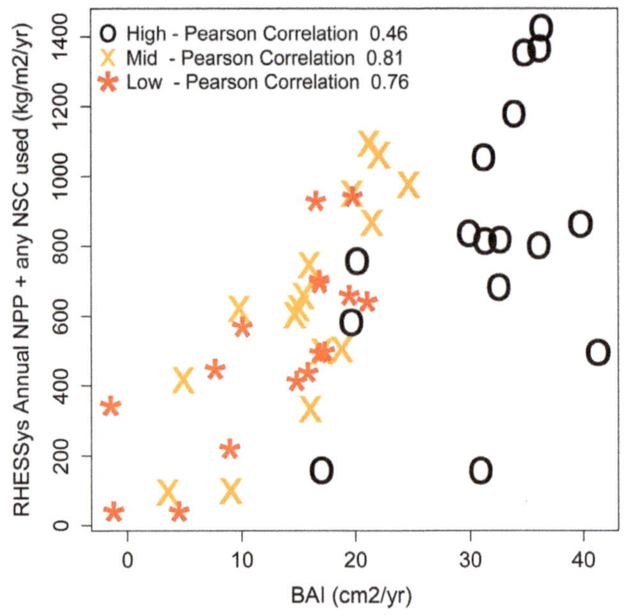

Figure 4. Comparison between observed basal area increment (BAI) and RHESSys estimates of annual net primary productivity (NPP). Points compare observed BAI and model NPP estimates for each year for the 1992–2007 period. Note in the NPP estimate we include the addition of any NSC that was removed from long term storage in a given year and used for growth. Here we report Pearson correlation coefficients using 0.3 and 0.05 for *NSC/NPP* and *MinL/AGC*, respectively. All correlation coefficients have p-value < 0.05. Table 2 provides mean and range of correlation coefficients for all NSC parameter values.

in the low elevation site, but lower summer transpiration and NPP - reflecting elevation differences in timing and magnitude of soil water and evaporative demand (Figure 3bc).

The relatively strong (>0.5) correlations between observed inter-annual BAI and modeled growth as NPP, plus any depletion of NSC pool for growth, suggest that major controls on inter-annual variation in productivity are captured (Table 2, Figure 4). These correlations remain above 0.4 regardless of assumptions of the parameters used for NSC (discussed below) and suggest that estimates of NPP are robust across this parameter uncertainty. In other words, uncertainty in NSC-related parameters does not preclude the use of the model to obtain realistic estimates of year-to-year variability in NPP for pre-drought years.

Sensitivity Analyses: Stand level simulations with cross-site differences in biomass

For stand-level simulations, minimum NSC over the simulation period (1992 to 2007) varies with NSC allocation parameters for all sites (Figure 5). Minimum NSC occurs during the early 2000's drought for all sites and parameter sets. There is a tradeoff between allocation to maintain carbohydrate storage reserves, which reduces drought vulnerability, and allocating to growth, which increases photosynthetic capacity. Thus, lower values for the *MinL/AGC* parameter allow greater loss of LAI during drought but maintain greater reserves and ultimately reduce mortality risk associated with low NSC values. A value of 0 for *MinL/AGC* reflects the scenario where trees use non-structural carbohydrate reserves to maintain respiration costs only, and do not produce foliage or fine roots from these reserves. Greater allocation to NSC storage (the *NSC/NPP* parameter) lead to higher values for the minimum NSC parameter, although for all three sites there is a threshold value for *NSC/NPP* above which increases in *NSC/NPP* do not lead to higher values of minimum NSC (Figure 5). This upper threshold results from interactions with the minimum leaf area (the *MinL/AGC* parameter) requirement and maximum allowable NSC (20% of total biomass), which also limits allocation to NSC when reserves are high.

Sensitivity of NSC to the *MinL/AGC* parameter varies with *NSC/NPP*. In general, lower values of *MinL/AGC* lead to higher NSC. The greatest sensitivity of minimum NSC estimates to *MinL/AGC* occurs at intermediate values of *NSC/NPP*. For lower *NSC/NPP* values, minimum NSC is strongly controlled by plant NPP, and differences due to the *MinL/AGC* parameter become negligible. At higher values of NSC/NPP, NSC becomes large enough to support a range of *MinL/AGC* requirements. For intermediate values of NSC/NPP, sensitivity in the *MinL/AGC* parameter reflects tradeoffs between maintaining leaf area and carbohydrate storage. Figure 6 demonstrates this trade-off for the upper elevation site. For values of *MinL/AGC* above 0.15, the minimum LAI over the full 1992–2007 period increases until a threshold is reached. Above this threshold, NSC allocation to leaves in good years ultimately leads to greater vulnerability to drought, and results in lower minimum LAI and minimum NSC. Higher, and probably physiologically unrealistic values of *MinL/AGC* (> 0.2), lead to mortality even in the high elevation site as the tree rapidly depletes NSC to maintain leaf area.

The minimum predicted NSC increases with elevation for all simulations except when *NSC/NPP* is set extremely low, in which

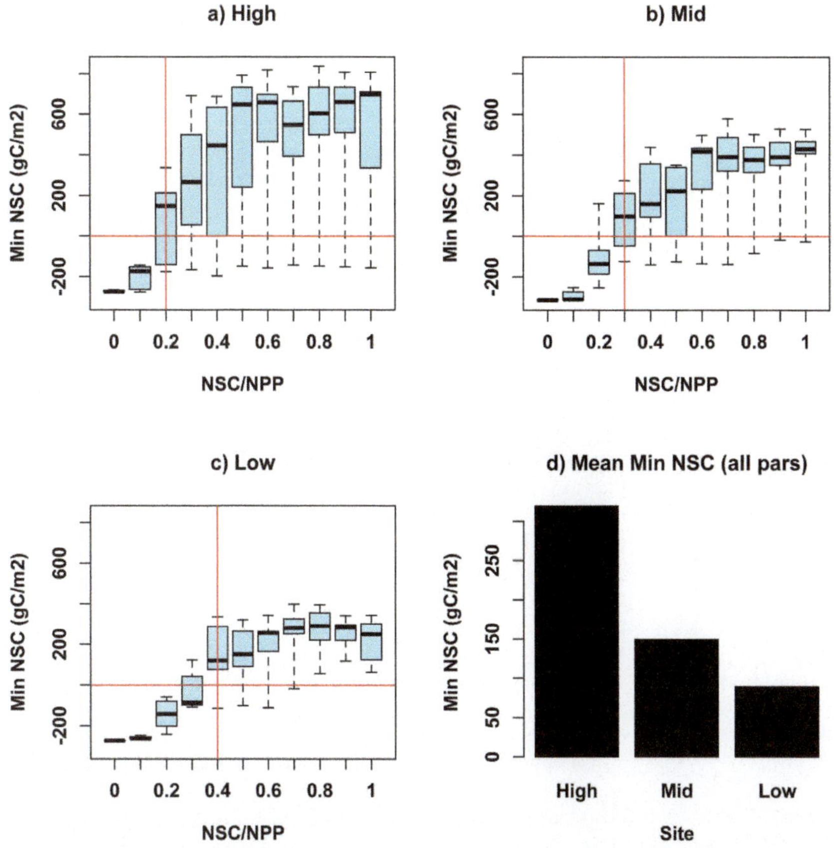

Figure 5. Estimates of minimum NSC over 1992–2007 simulation period for high, mid and low elevation stands. Graphs show variation in minimum NSC estimates across variation in parameters: *NSC/NPP* (x-axis) and *MinL/AGC* (box widths). Horizontal lines denote values of NSC below which mortality is likely. Vertical line denotes value of *NSC/NPP* parameter for which minimum NSC is greater than zero for most values (3rd quartile) of *MinL/AGC*. This value of *NSC/NPP* is 0.28, 0.32, 0.38 for high, mid and low sites respectively. Bottom right graph shows average value of minimum NSC across all parameters for high, mid and low elevation sites. Results are based on simulations using carbon storage initialized through 100 yr spinup.

case all sites achieve zero NSC. Thus, for most of our NSC allocation assumptions, the model estimates greater vulnerability to drought stress for the lower elevation sites (Figure 5). This spatial trend in minimum NSC is maintained after normalizing minimum NSC by total biomass. Here we note that if we assume a critical value for NSC of 0.5% of biomass as indicative of mortality, values for the *NSC/NPP* parameter between 0.24 and 0.36 correctly suggest mortality at the low elevation, but not mid to high elevation sites. This suggests that parameter values within this range provide the most realistic assumptions about NSC storage dynamics. The strong correlations of modeled NPP and observed BAI (Table 2) for most parameter sets suggests simulation of growth is robust across parameter uncertainty. Thus, while most NSC parameter sets correctly captured inter-annual and spatial variation in growth, only *NSC/NPP* values near 0.3 captured differences in the likelihood of drought related mortality.

Sensitivity Analyses: Tree level simulations with uniform biomass for all three sites

Results from the sensitivity analysis for tree-level simulations, where initial above and below ground structural carbon are spatially uniform, show similar patterns to those obtained for stand level simulations when structural carbon is varied across sites (Figure 7). As with spatially variable biomass simulations, estimates of minimum NSC stores are consistently less for the low elevation site across all parameter sets, with greater minimum NSC for mid rather than high elevation sites (Figure 7). Again, the threshold value of *NSC/NPP* above which NSC does not fall below our critical mortality threshold (0.5%), is lower for mid and high elevation sites, suggesting these stands can survive with relatively less allocation to NSC than the low site (Figure 7). The greatest cross-site differences in the attribution of mortality (storage < 0.5%) are found for *NSC/NPP* values of 0.2 to 0.4. For these tree-level simulations we used an initial biomass consistent with the mid-elevation site. In general, greater biomass (of individual trees or total stand) would be expected to increase respiration costs and water use and potentially increase the rate of NSC depletion during periods of drought and low GPP. Consistent with this, stand level simulations (Figure 5) for the high elevation site, where initial structural carbon values are greater, do show slightly lower minimum NSC relative to tree level simulations (Figure 7) for the same *NSC/NPP* parameter value. This difference, however, is relatively small and suggests that site condition differences (soil, meteorology) rather than pre-drought differences in stand structure, are the dominant controls on cross-site patterns of NSC.

In summary, our sensitivity analysis suggests values for the *NSC/NPP* parameter between 0.25 and 0.35 correctly predict the timing and magnitude of high mortality likelihood at the low elevation site while still allowing growth variation in the mid and high elevation sites to track observed values. For the *MinL/AGC* parameter, a

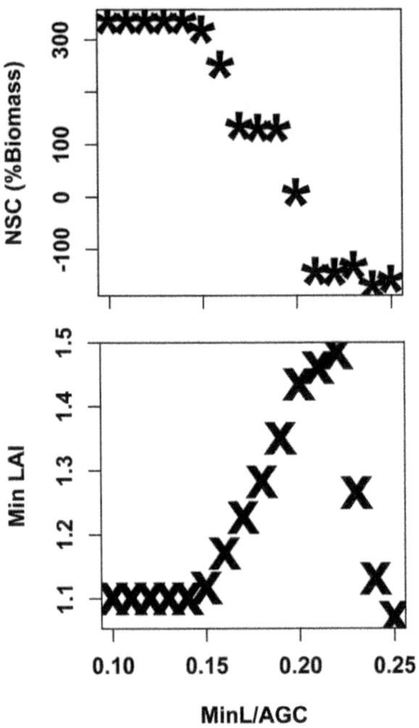

Figure 6. Estimates of a) minimum NSC and b) minimum LAI over 1992–2007 simulation period for high elevation stands. Graphs show variation in minimum NSC and LAI across variation in *MinL/AGC* with *NSC/NPP* set equal to 0.25.

wide range of values give reasonable NSC and growth predictions when combined with NSC/NPP parameter values in the 0.25 to 0.35 range. Field studies, however, suggest that observed minimum values for the ratio of leaf to above ground biomass are close to 0.05 (5%) [46], thus we set *MinL/AGC* to this value. While we do not have measurements of LAI and NSC throughout the drought, we argue that the trajectory of LAI and NSC using *NSC/NPP* of 0.25 and *MinL/AGC* of 0.05 (Figure 8ab) are consistent with observations that the mortality occurred only at the low elevation site and that productivity, prior to the drought, increased with elevation (as indicated by increasing mean measured BAI (Table 2)). Model estimates of NPP also show a good correlation between observed and modeled BAI (Table 2, Figure 4). Results confirm that our model and parameter sets are both reasonable from a physiological perspective and produce simulations that correctly capture observed differences in the likelihood of drought stress mortality during the 2000's drought.

Impact of differences in precipitation and temperature across sites on NSC and mortality patterns

To examine climate controls on cross-site differences in growth and survival we run simulations using parameter values for *NSC/NPP* of 0.25 and *MinL/AGC* of 5%. Differences in simulation results between the high and low elevation sites reflect complex interactions between air temperature and precipitation on productivity and NSC (Table 3). Despite the expected shifts in NPP by applying low elevation precipitation to the high site and vice versa, we observed that these changes in precipitation do not account fully for differences in NPP. Using air temperature from the low elevation site for the high elevation site only slightly decreases NPP relative to the baseline scenario. We note that temperature influences the model estimates of productivity directly through impacts on respiration and photosynthesis sub-models as well as indirectly though impacts on snow accumulation and melt and evaporative forcing. The combination of greater water holding capacity in the soils and greater precipitation at the high elevation site supports greater productivity at higher temperatures. For the low elevation site, using cooler temperatures leads to relatively small declines in productivity. This site is more clearly water limited, even when temperatures from the high elevation site are used. A shift from snow to rain at the high elevation site only slightly reduces NPP.

NSC and associated estimates of mortality risk show a different response pattern to temperature changes than the productivity patterns (Figure 9). NSC drops close to the mortality threshold for the high elevation site if either low elevation precipitation or low elevation temperature is used (dotted blue and purple lines in Figure 9a). The impact of snow versus rain is relatively small but can be seen early in the 2000 drought for the high elevation site where there is a steeper decline in NSC if all precipitation is assumed to fall as rain. For the low elevation site (Figure 9b), using high site precipitation delays, but does not prevent, mortality in the 2000 drought years; using the high elevation temperature, however, does. For the low elevation site, if both high precipitation and temperature are used, NSC remains above the mortality threshold; although it is still below high site values reflecting the impact of lower soil water holding capacity at the low elevation site.

This distinction between NSC and NPP behavior can be further explored by looking at model estimates of the relationship between NSC or NPP and annual precipitation (Figure 10). For NPP, upper and lower sites show a similar near linear relationship with annual precipitation until a threshold precipitation is reached (this threshold only occurs for the high elevation site). NSC integrates productivity over several years and results in a less direct relationship with precipitation.

Discussion

Our objective was to determine if a simple NSC model embedded within a coupled carbon-hydrologic model (RHESSys-NSC) can capture observed mortality during drought, and if successful, what does this tell us about how climate controls influence mortality along a relatively fine-scale (800 m) elevation gradient? []Our results demonstrate a successful implementation of a NSC model into RHESSys such that the model accurately predicts both inter-annual and spatial variation in productivity and likelihood of mortality during drought. Model estimates of GPP, respiration, and biomass components are also similar to observations for ponderosa pine in the Western US [50], [51]. Minimum NSC during drought showed a greater sensitivity to temperature differences than NPP. Notably, the model scenario in which the warmer low elevation temperatures drive the high elevation site showed only a slight decrease in NPP, but NSC would have fallen close to the mortality threshold during the drought. This has implications for predicting forest mortality risk in a changing climate where future droughts are likely to occur under warmer temperatures [52], [53], [54], [49].

Including a *NSC/NPP* ratio of 0.25 resulted in accurate simulations of both growth (based on correlations above 0.4 between observed BAI and modeled NPP) and low elevation site mortality across the 800 m elevational transect (Figure 8). We emphasize that resulting values for *NSC/NPP* also make physiological sense., There must be some allocation to NSC to maintain respiration so *NSC/NPP* must be greater than zero, however too much allocation to NSC (*NSC/NPP* or values much greater than 0.4) is less competitive given tradeoffs between using carbon to support storage versus growth. Thus, our sensitivity analysis shows

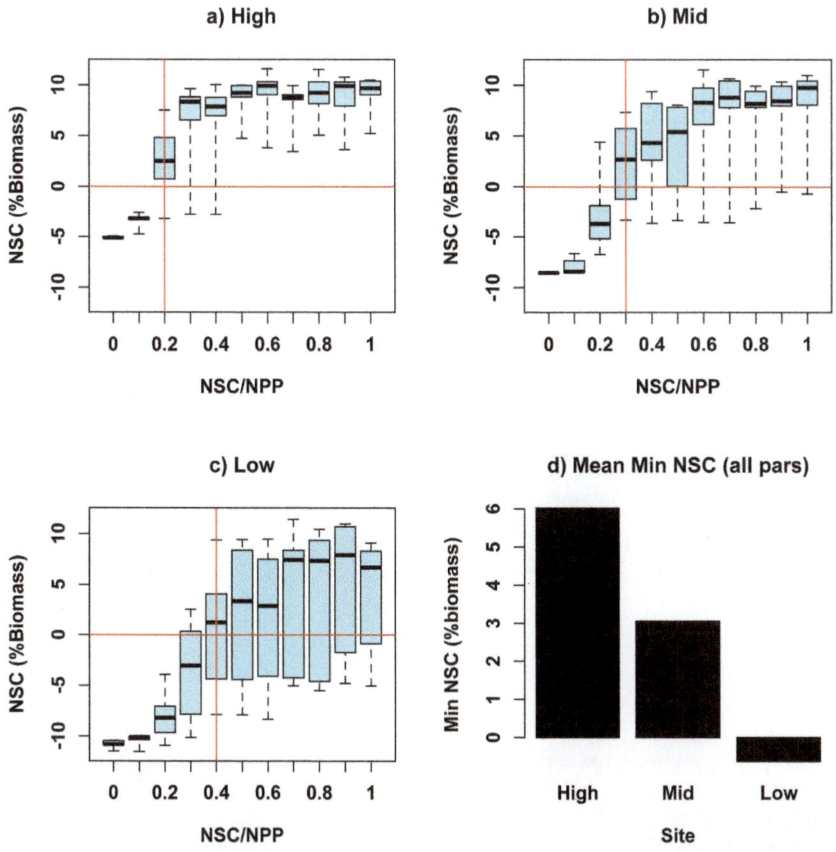

Figure 7. Estimates of minimum NSC as a percentage of biomass over 1992–2007 simulation period. Estimates of minimum NSC are shown for a tree of similar size for high, mid and low elevation sites. Horizontal lines denote values of NSC below which mortality is likely. Vertical lines denote the value of the *NSC/NPP* parameter for which minimum NSC is greater than zero for most values (3rd quartile) of *MinL/AGC*. Graphs show variation in minimum NSC across variation in parameters: *NSC/NPP* (x-axis) and *MinL/AGC* (box widths). Bottom right graph shows average value of minimum NSC across all parameters for high, mid and low elevation site. Initial structural carbon values are the same for all three sites.

that the best performing parameter sets are consistent with physically realistic NSC storage dynamics. Further, our results are consistent with a recent study in this watershed that concluded carbon assimilation was a major component of the growth and mortality response of ponderosa pine across these same three sites [20].

Our model results do not prove that carbon starvation (e.g. [20]) was the last step in mortality of these particular trees because hydraulic failure, attack by bark beetles (*Ips* species) or other factors may have triggered the final phases of mortality [55], [11], [56]. However, it is likely that declining NSC increases risks of non-recoverable cavitation [24] and promotes susceptibility to biotic mortality agents, as less carbon is available to allocate to defensive compounds [57], [58], [11]. Given this general support for energy-limitation as a key factor in mortality, we argue that this simple two-parameter NSC model provides a good approximation of mortality risk. Simplicity is important in the context of modeling for forest management and scenario development at watershed to landscape scales. While more detailed physiological-based models of mortality are emerging, they will be challenging to parameterize and apply over complex heterogeneous landscapes.

The modeled differences in NSC and NPP responses across the elevational transect with drought are not necessarily surprising, but illustrate the necessity of using a storage term (NSC in our case) to track conditions under which mortality is likely to occur. Previous models of tree scale mortality have similarly shown that short time scale estimates of deficits in net assimilation (photosynthesis minus respiration) occur frequently and do not correlate well with mortality [59]. We note that a large number of experimental datasets on conifers are now revealing that tree death is associated with severe depletions in NSC [23]. Recent experimental work, however, has also shown that the ability of forest to use available carbon resources or NSC can also be limited by temperature [60] and drought [61]. In these cases, NSC may not be fully depleted prior to death. In this study, we posit NSC as an index for more complex mechanisms. We argue that this general pattern of energy reserves and their depletion leading to mortality provides a first-order approximation to mortality risk, and is consistent with the hypothesis that NSC storage is either directly or indirectly related to mortality through its interdependence on hydraulics and biotic attack [11]. Even if the actual causes of mortality are physiological limitations not directly reflected by NSC depletion, our NSC index may serve as an indicator of the degree of stress. We argue that our relatively simple approach may be a useful one because the linkage between a carbon cycling model with a simple NSC storage term and a hydrology model can integrate the effect of environmental controls on both the longer term development of plant structure (including biomass, rooting depth, and photosynthetic capacity) and short term stress due to physical controls on moisture availability (e.g. vapor pressure deficit, soil water availability).

Future model developments that could benefit improved predictions, as well as enable investigation of physiological

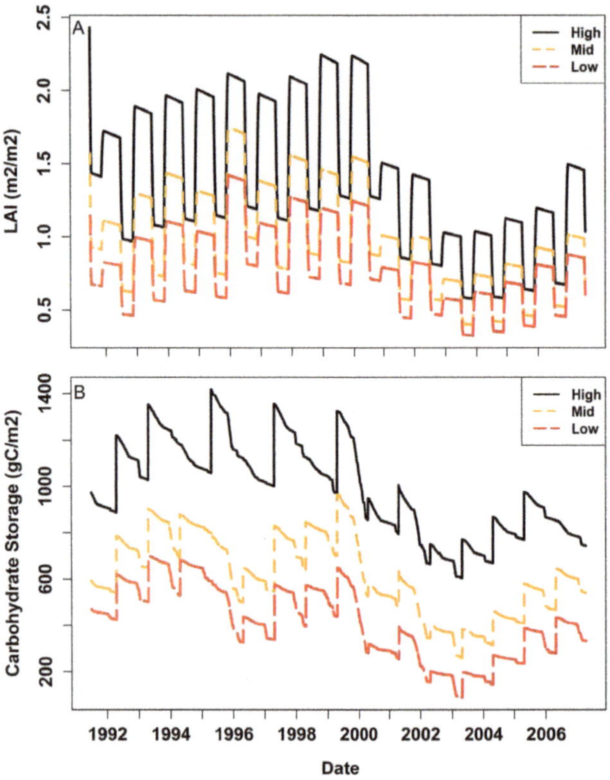

Figure 8. Trajectory of a) LAI and b) Non Structural Carbohydrate Storage. Trajectories of LAI and NSC are shown for high, mid and low elevation sites for 1992–2007 period. *NSC/NPP* and *MinL/AGC* are set to 0.3 and 0.05 respectively. We note that for the low elevation site, we predict that mortality would occur during the drought based on low NSC values. Based on this we could trigger a reduction of LAI to zero. We show predicted LAI without this mortality trigger to illustrate the response of the carbon cycling model and to point out that simply tracking needle loss would not be sufficient to capture the mortality event.

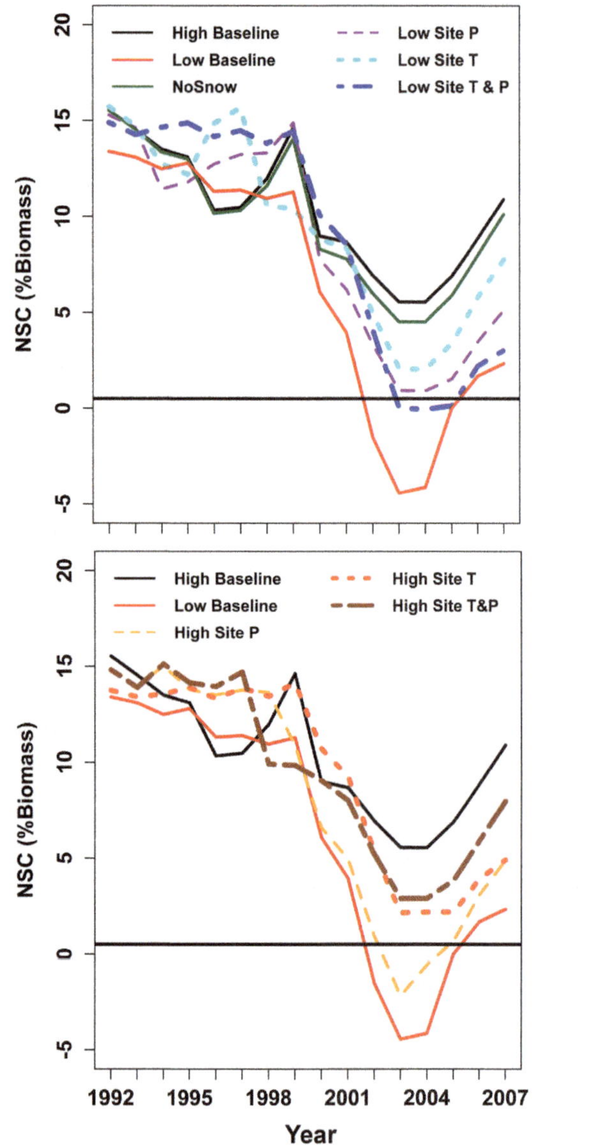

Figure 9. Trajectories of NSC through the 1992–2004 simulation period. Graphs of NSC trajectories are shown for both the pre-drought and drought years. *NSC/NPP* and *MinL/AGC* are set to 0.3 and 0.05 respectively and structural carbon estimates are initialized to be uniform across all 3 sites. Results are shown for the high site for 5 scenarios: baseline, all precipitation falls as rain (no snow), precipitation from low elevation site, temperature from low elevation site and both temperature and precipitation from the low elevation site. For the low elevation site 4 scenarios are shown: baseline, precipitation from high elevation site, temperature from high elevation site and both temperature and precipitation from the high elevation site.

mechanisms, include temporally variable *NSC/NPP* (i.e. increased ratio during drought), as well as other proposed mechanisms of mortality, such as hydraulic failure, phloem transport failure, and overwhelming biotic attack [11], leaf shedding [62] and responses related to the incorporation of NSC into plant tissues [13], [61]. We note that a key uncertainty in the model is in the strategies used for allocation of net photosynthesis to different plant components including leaves and roots. While the assumptions used in this study (age-based reduction in allocation to leaves and fine-roots) result in biomass and NPP estimates that are consistent with observations, different allocation strategies may alter results

Table 3. Change in estimated NPP with climate forcing conditions.

NPP (gC/m2/yr) Averaged for Water Years 1992–2004				
		Exchange of forcing data between High and Low site		
Site	**Baseline**	**Precipitation fromother site**	**Temperature fromother site**	**Precipitation andTemperature from other site**
High	2.00	1.30	2.10	1.27
Low	1.01	1.80	1.05	1.82

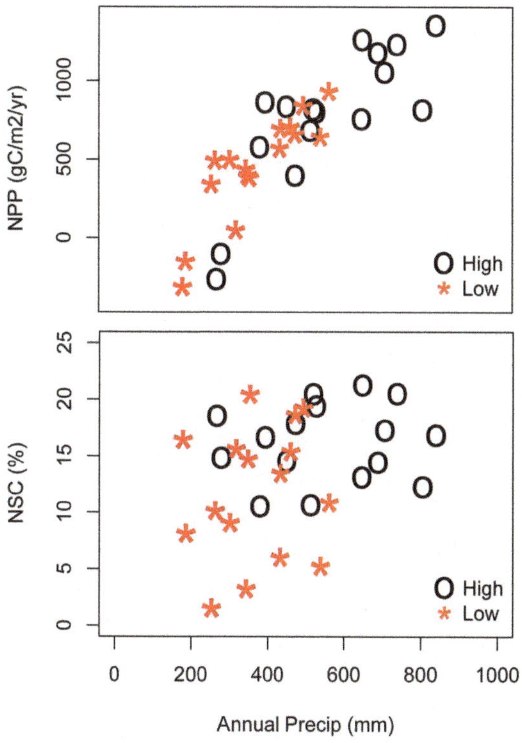

Figure 10. NPP and NSC against precipitation for high and low elevation sites over 1992-2007 simulation period.

and ultimately the sensitivity of model estimates to drought. Understanding plant allocation strategies remains an active area of research within the plant physiology literature [63]. Future work will examine the sensitivity of RHESSys-NSC drought responses to leaf and root allocation and senescence strategies.

Sensitivity analyses in this paper demonstrate the importance of within-watershed patterns of temperature and precipitation as controls on productivity and mortality. Thus, while representing plant physiological responses to water stress is important, combining these estimates with reasonable representation of local energy and moisture conditions also is a key characteristic of a useful model. Scenarios using our coupled model show that productivity/mortality risk may not always respond in parallel – such that increasing temperatures may increase both NPP and mortality risk – and that even within a watershed, temperature and moisture gradients combine in complex ways to shape responses.

An important component of the RHESSys modeling framework is the coupling between a hydrologic model that accounts for spatial difference in precipitation, snow accumulation and melt, and evapotranspiration/soil water controls on ecosystem carbon cycling and ultimately mortality. Modeling in this study focused on within watershed spatial patterns for three sites within a 20 km distance along an 800 m elevation gradient. We note that the ability to capture these within-watershed scale patterns is dependent on accurate resolution of temperature and precipitation patterns. Spatial differences in temperature and precipitation (as well as radiation and other atmospheric drivers) between these sites typically would not be resolved by global climate model output, and capturing high resolution (<50 km) spatial patterns in mountain environments remains a challenge for regional models. Techniques to improve downscaling of regional climate models remain an active area of research [64], [65], [66]. Our results demonstrate that accounting for these within-watershed spatial differences in microclimate clearly have important implications for estimates of watershed scale patterns of carbon cycling and water use.

Climate-induced tree mortality is emerging as an increasing concern for forest managers around the world [3], [49], [67], [68], and climate models project significant rises in global temperatures and increasing drought frequency and severity for many regions in this century, including much of the Southwest US [54], [69]. Models that link these climate projections with spatially explicit changes in forest productivity and tree mortality are needed to support adaptation planning and actions by forest managers [70], [71]. Results from this study demonstrate a successful application of a coupled ecohydrologic model in predicting forest behavior at scales relevant to park and local forest managers. Future work will use the model to estimate mortality risk for a broad range of climate and climate change scenarios for the Jemez Mountains region and elsewhere. Application of models like RHESSys that account for within-watershed spatial variation in interactions among climate, hydrology and carbon can be important tools in watershed-scale assessments directed at supporting climate change adaptation.

Acknowledgments

We gratefully acknowledge support from the Ecosystems and the Climate and Land Use programs of the U.S. Geological Survey (through the Western Mountain Initiative). Any use of trade, firm, or product names is for descriptive purposes only and does not imply endorsement by the U.S. Government.

Author Contributions

Conceived and designed the experiments: CT NM CA. Performed the experiments: CT. Analyzed the data: CT NM CA. Contributed reagents/materials/analysis tools: CT NM CA. Wrote the paper: CT NM CA. Designed software used in analysis: CT. Contributed data for model input: NM CA.

References

1. Seager R, Tzanova A, Nakamura J (2009) Drought in the Southeastern United States: causes, variability over the last millennium, and the potential for future hydroclimate change. J Climate 22: 5021–5045.
2. van Mantgem PJ, Stephenson NL, Byrne JC, Daniels LD, Franklin JF, et al. (2009) Widespread Increase of Tree Mortality Rates in the Western United States. Science 323: 521–524, doi:10.1126/science.1165000.
3. Allen CD, Macalady AK, Chenchouni H, Bachelot D, McDowell N, et al.(2010) A global overview of drought and heat-induced tree mortality reveals emerging climate change risks for forests. Forest Ecology and Management 259: 660–684.
4. Williams AP, Allen CD, Millar CI, Swetnam TW, Michaelsen J, et al. (2010) Forest responses to increasing aridity and warmth in the southwestern United States, Proc. Natl Acad. Sci. USA 107: 21289–21294, doi: 10.1073/pnas.0914211107.
5. Monserud RA (2003) Evaluating forest models in a sustainable forest management context. FBMIS 1: 35–47. Available: http://cms1.gre.ac.uk/conferences/iufro/fbmis/A/3_1_MonserudR_1.pdf.
6. Boisvenue C, Running SW (2006) Impacts of climate change on natural forest productivity – evidence since the middle of the 20th century. Glob Chang Biol 12: 862–882, doi: 10.1111/j.1365-2486.2006.01134.x.
7. Scheller RM, Mladenoff DJ (2007) An ecological classification of forest landscape simulation models: tools and strategies for understanding broad-scale forested ecosystems. Landsc Ecol 22: 491–505, doi: 10.1007/s10980-006-9048-4.
8. Nightingale JM, Phinn SR, Held AA (2004) Ecosystem process models at multiple scales for mapping tropical forest productivity. Progress in Physical Geography 28: 241–281.

9. Friend AD, Arneth A, Kiang NY, Lomas M, Ogee J, et al. (2007) FLUXNET and modeling the global carbon cycle. Global Change Biology 13: 610–633.
10. Randerson JT, Hoffman FM, Thornton PE, Mahowald NM, Lindsay K, et al. (2009) Systematic assessment of terrestrial biogeochemistry in coupled climate-carbon models. Glob Chang Biol 15: 2462–2484, doi:10.1111/j.1365-2486.2009.01912.x.
11. McDowell NG, Beerling DJ, Breshears DD, Fisher RA, Raffa KF, et al. (2011) The interdependence of mechanisms underlying climate-driven vegetation mortality. Trends Ecol Evol 26: 523–532.
12. Adams HD, Guardiola-Claramonte M, Barron-Gafford GA, Villegas JC, Breshears DD, et al.(2009) Temperature sensitivity of drought-induced tree mortality portends increased regional die-off under global-change-type drought. Proc Natl Acad Sci USA 106: 7063–7066, doi:10.1073/pnas.0901438106.
13. Sala A, Piper F, Hoch G (2010) Physiological mechanisms of drought-induced tree mortality are far from being resolved. New Phytol 186: 274–281.
14. Anderegg WRL, Berry JA, Smith DD, Sperry JS, Anderegg LDL, et al. (2012) The roles of hydraulic and carbon stress in a widespread climate-induced forest die-off. Proc. Natl Acad. Sci. USA 109: 233–237.
15. McDowell NG, Sevanto S (2010) The mechanisms of carbon starvation: how, when, or does it even occur at all? New Phytol 186: 264–266.
16. Hölttä T, Mencuccini M, Nikinmaa E (2009) Linking phloem function on structure: analysis with a coupled xylem-phloem transport model. Journal of Theoretical Biology 259: 325–337.
17. Williams M, Law BE, Anthoni PM, Unsworth MH (2001) Use of a simulation model and ecosystem flux data to examine carbon–water interactions in ponderosa pine. Tree Physiol 21: 287–298.
18. Thornton PE, Law BE, Gholz HL, Clark KL, Falge E, et al. (2002) Modeling and measuring the effects of disturbance history and climate on carbon and water budgets in evergreen needleleaf forests. Agricultural and Forest Meteorology 113: 185–222.
19. Allen CD (1989) Changes in the Landscape of the Jemez Mountains, New Mexico. Ph.D. dissertation, Department of Forestry and Natural Resources, University of California, Berkeley, CA., 346 p.
20. McDowell NG, Allen CD, Marshall L (2010) Growth, carbon-isotope discrimination, and drought-associated mortality across a Pinus ponderosa elevational transect. Glob Change Biol 16: 399–415, doi: 10.1111/j.1365-2486.2009.01994.x.
21. Stamp N (2003) Out of the quagmire of plant defense hypotheses. Q Rev Biol 78: 23–55
22. Guérard N, Maillard P, Brechet C, Lieutier F, Dreyer E (2007) Do trees use reserve or newly assimilated carbon for their defense reactions? A 13C labeling approach with young Scots pines inoculated with a bark-beetle-associated fungus (Ophiostoma brunneo ciliatum). Ann Sci 64: 601–608.
23. Galiano L, Martinez-Vilalta J, Lloret F (2011) Carbon reserves and canopy defoliation determine the recovery of Scots pine 4 yr after a drought episode. New Phytol 190: 750–759, doi: 10.1111/j.1469-8137.2010.03628.x
24. McDowell NG (2011) Mechanisms linking drought, hydraulics, carbon metabolism, and vegetation mortality. Plant Physiology 155: 1051–1059, doi: 10.1104/pp.110.170704.
25. Hartman MD, Baron JS, Lammers RB, Cline DW, Band LE, et al. (1999) Simulations of snow distribution and hydrology in a mountain basin. Water Resour Res 35: 1587–1603.
26. Mackay DS, Samanta S, Nemani RR, Band LE (2003) Multi-objective parameter estimation for simulating canopy transpiration in forested watersheds. Journal of Hydrology 277: 230–247.
27. Tague C, Grant GE (2009) Groundwater dynamics mediate low-flow response to global warming in snow-dominated alpine regions. Water Resour Res 45: W07421, doi: 10/1029/2008WR007179.
28. Zierl B, Bugmann H, Tague CL (2007) Water and carbon fluxes of European ecosystems: an evaluation of the ecohydrological model RHESSys. Hydrol Process 21: 3328–3339.
29. Hwang T, Band L, Hales TC (2009) Ecosystem processes at the watershed scale: Extending optimality theory from plot to catchment. Water Resour Res 45: W11425, doi: 10.1029/2009WR007775.
30. Tague C, Heyn K, Christensen L (2009) Topographic controls on spatial patterns of conifer transpiration and net primary productivity under climate warming in mountain ecosystems. Ecohydrology 2: 541–554, doi: 10.1002/eco.88.
31. Tague CL, Band LE (2004) RHESSys: Regional Hydro-Ecologic Simulation System—An object-oriented approach to spatially distributed modeling of carbon, water, and nutrient cycling. Earth Interactions 8: 1–42.
32. Hungerford RD, Nemani RR, Running SW, J Coughlan C (1989) MTCLIM – A mountain microclimate simulation-model. USDA Forest Service Intermountain Research Station Research Paper 414: 1–52.
33. Chen JM, Liu J, Cihlar J, Goulden ML (1999) Daily canopy photosynthesis model through temporal and spatial scaling for remote sensing applications. Ecol Model 124: 99–119, doi: 10.1016/S0304-3800(99)00156-8.
34. Monteith JL (1965) Evaporation and environment. In: Fogg GE, editor. Symposium of the Society for Experimental Biology, The State and Movement of Water in Living Organisms. New York: Academic Press, Inc. pp.205–234.
35. Jarvis PG, Mansfield TA (1981) Stomatal Physiology. In: Jarvis PD, Mansfield TA, editors. Society for Experimental Biology Seminar Series. New York: Cambridge University Press. 403p.
36. Farquhar GD, Von Caemmerer S (1982) Modeling of photosynthetic response to environmental conditions. Encyclopedia of plant physiology 12: 549–587.
37. Ryan MG (1991) A simple method for estimating gross carbon budgets for vegetation in forest ecosystems. Tree Physiol 9: 255–266.
38. Atkin OK, Tjoelker MG (2003) Thermal acclimation and the dynamic response of plant respiration to temperature. Trends Plant Sci 8: 343–351, doi: 10.1016/S1360-1385(03)00136-5.
39. Parton WJ, Mosier AR, Ojima DS, Valentine DW, Schimel DS, et al. (1996) Generalized model for N-2 and N2O production from nitrification and denitrification. Global Biogeochem Cycles 10: 401–412.
40. Dickinson RE, Shaikh M, Bryant R, Graumlich L (1998) Interactive canopies for a climate model. Journal of Climate 11: 2823–2836.
41. Barbaroux C, Breda N, Dufrene E (2003) Distribution of above-ground and below-ground carbohydrate reserves in adult trees of two contrasting broad-leaved species (*Auercus petraea and Fagus sylvatica*). New Phytol 157: 605–615, doi: 10.1046/j.1469-8137.2003.00681.x.
42. Wurth MKR, Pelaez-Riedl S, Wright SJ, Korner C (2005) Non-structural carbohydrate pools in a tropical forest. International Association for Ecology 143: 11–24.
43. Genet H, Breda N, Dufrene E (2009) Age-related variation in carbon allocation at tree and stand scales in beech (*Fagus sylvatica L.*) and sessile oak (*Quercus petraea (Matt.) Liebl.*) using a chronosequence approach. Tree Physiol 30: 177–192, doi: 10.1093/treephys/tpp105.
44. Gruber A, Pirkebner D, Florian C, Oberhuber W (2011) No evidence for depletion of carbohydrate pools in Scots pine (*Pinus sylvestris L.*) under drought stress. Plant Biol 14: 142–148, doi: 10.1111/j.1438-8677.2011.00467.x.
45. Fisher JB, Sitch S, Malhi Y, Fisher RA, Huntingford C, et al. (2010) Carbon cost of plant nitrogen acquisition: a mechanistic, globally applicable model of plant nitrogen uptake, retranslocation, and fixation. Global Biogeochemical Cycles 24: GB1014, doi: 10.1029/2009GB003621.
46. Callaway RM, Delucia EH, Schlesinger WH (1994) Biomass allocation of montane and desert ponderosa pine – an analog for response to climate-change. Ecology 75: 1474–1481.
47. Lundquist JD, Cayan DR (2007) Surface temperature patterns in complex terrain: Daily variations and long-term change in the central Sierra Nevada, California. J Geophys Res 112: D11124, doi: 10.1029/2006JD007561.
48. Schoeneberger PJ, Wysocki DA, Benham EC, Broderson WD, editors (2002) Field book for describing and sampling soils, Version 2.0. Natural Resources Conservation Service, National Soil Survey Center, Lincoln, NE.
49. Williams AP, Allen DC, Macalady AK, Griffin D, Woodhouse CA, et al. (2013) Temperature as a potent driver of regional forest-drought stress and tree mortality. Nature Climate Change 3: 292–297, doi: 10.1038/nclimate1693.
50. Law BE, Sun OJ, Campbell J, Van Tuyl S, Thornton P (2003) Changes in carbon storage and fluxes in a chronosequence of ponderosa pine. Glob Chang Biol 9: 510–524.
51. Anderson-Teixeira KJ, Delong JP, Fox AM, Brese DA, Litvak ME (2011) Differential responses of production and respiration to temperature and moisture drive the carbon balance across a climatic gradient in New Mexico. Glob Chang Biol 17: 410–424, doi:10.1111/j.1365-2486.2010.02269.x.
52. Breshears DD, Cobb NS, Rich PM, Price KP, Allen CD, et al. (2005) Regional vegetation die-off in response to global-change-type drought. Proc Natl Acad Sci USA 102, 15144–15148, doi: 10.1073/pnas.0505734102.
53. IPCC (2007) Synthesis Report, Contribution of Working Groups I, II and III to the Fourth Assessment Report of the Intergovernmental Panel on Climate Change. Cambridge, UK: Cambridge University Press. 104p.
54. Seager R, Vecchi GA (2010) Greenhouse warming and the twenty-first-century hydroclimate of southwestern North America. Proc Natl Acad Sci of the USA 107: 21277–21282.
55. Raffa KF, Aukema BH, Bentz BJ, Carroll AL, Hicke JA, et al. (2008) Cross-scale drivers of natural disturbances prone to anthropogenic amplification: dynamics of bark beetle eruptions. BioScience 58: 501–17.
56. Sturrock RN, Frankel SJ, Brown AV, Hennon PE, Kliejunas JT, et al. (2011) Climate change and forest diseases. Plant Pathol 60: 133–149.
57. Suarez ML, Ghermandi L, Kitzberger T (2004) Factors predisposing episodic drought-induced tree mortality in Nothofagus–site, climatic sensitivity and growth trends. J Ecol 92: 954–966.
58. Kane JM, Kolb TE (2010) Importance of resin ducts in reducing ponderosa pine mortality from bark beetle attack. Oecologica 164: 601–609, doi: 10.1007/s00442-010-1683-4.
59. Güneralp B, Gertner G (2007) Feedback loop dominance analysis of two tree mortality models: relationship between structure and behavior. Tree Physiology 27: 269–280.
60. Hoch G, Körner C (2005) Growth, demography and carbon relations of *Polylepis* trees at the world's highest tree line. Functional Ecology 19: 941–951.
61. Piper FI (2011) Drought induces opposite changes in the concentration of non-structural carbohydrates of two evergreen *Nothofagus* species of differential drought resistance. Ann For Sci 68: 415–424.
62. Breda N, Huc R, Granier A, Dreyer E (2006) Temperate forest trees and stands under severe drought: a review of ecophysiological responses, adaptation processes and long-term consequences. Ann For Sci 63: 625–644
63. Franklin O, Johansson J, Dewar RC, Dieckmann U, McMurtrie RE, et al. (2012) Modeling carbon allocation in trees: a search for principles. Tree Physiol 32: 648–666, doi: 10.1093/treephys/tpr138.

64. Fowler HJ, Blenkinsop S, Tebaldi C (2007) Linking climate change modeling to impacts studies: recent advances in downscaling techniques for hydrological modeling. International Journal of Climatology 27: 1547–1578.
65. Wood AW, Leung LR, Sridhar V, Lettenmaier DP (2004) Hydrologic implications of dynamical and statistical approaches to downscaling climate model outputs. Clim Change 62: 189–216.
66. Rummukainen M (2010) State-of-the-art with regional climate models, Wiles Interdisciplinary reviews, Clim Change 1: 82–96.
67. Breshears DD, Myers OB, Meyer CW, Barnes FJ, Zou CB, et al. (2009) Tree die-off in response to global change-type drought: mortality insights from a decade of plant water potential measurements. Front Ecol Environ 7: 185–189, doi: 10.1890/080016.
68. Carnicer J, Coll M, Ninyerola M, Pons X, Sánchez G, et al. (2011) Widespread crown condition decline, food web disruption, and amplified tree mortality with increased climate change-type drought. Proc Natl Acad Sci USA 108: 1474–1478, doi: 10.1073/pnas.1010070108.
69. IPCC (2007) Climate Change 2007: The physical science basis. In: Solomon S, Qin D, Manning, Chen Z, Marquis M, et al., editors. Contribution of working group I to the fourth assessment report of the intergovernmental panel on climate change. New York: Cambridge University Press. 996 p.
70. Joyce LA, Blate GM, Littell JS, McNulty J, Millar S, et al., editors. Preliminary review of adaptation options for climate-sensitive ecosystems and resources: a report by the U.S. Climate Change Science Program and the Subcommittee on Global Change Research. Washington, DC: US Environmental Protection Agency. pp. 3:1 – 3:127.
71. Peterson DL, Allen CD, Baron JS, Fagre DB, McKenzie D, et al. (2011) Response of Western mountain ecosystems to climatic variability and change: a collaborative research approach. In: Beever EA, Belant JL, editors. Ecological Consequences of Climate Change: Mechanisms, Conservation, and Management. Boca Raton, FL: CRC Press. pp. 163–190.

7

Water Regime Influences Bulk Soil and Rhizosphere of *Cereus jamacaru* Bacterial Communities in the Brazilian Caatinga Biome

Vanessa Nessner Kavamura[1,2]*, Rodrigo Gouvêa Taketani[1], Milena Duarte Lançoni[1,2], Fernando Dini Andreote[2], Rodrigo Mendes[1], Itamar Soares de Melo[1]

1 Laboratory of Environmental Microbiology, Embrapa Environment, Jaguariúna, São Paulo, Brazil, **2** Department of Soil Science, "Luiz de Queiroz" College of Agriculture, University of São Paulo, Piracicaba, São Paulo, Brazil

Abstract

We used the T-RFLP technique combined with Ion Torrent (PGM) sequencing of 16S rRNA and multivariate analysis to study the structure of bulk soil and rhizosphere bacterial communities of a cactus, *Cereus jamacaru,* from the Brazilian Caatinga biome, which is unique to Brazil. The availability of water shapes the rhizosphere communities, resulting in different patterns during the rainy and dry seasons. Taxonomic approaches and statistical analysis revealed that the phylum Actinobacteria strongly correlated with the dry season, while samples from the rainy season exhibited a strong correlation with the phylum Proteobacteria for rhizosphere samples and with the phyla Bacteroidetes, Firmicutes, Lentisphaerae, and Tenericutes for bulk soil samples. The STAMP software also indicated that the phylum Bacteroidetes, as well as two classes in the Proteobacteria phylum (γ and δ), were the most significant ones during the rainy season. The average abundance of the phylum Actinobacteria and the genus *Bacillus* was significantly greater during the dry season. Some significant genera found during the dry season might reflect their tolerance to the extreme conditions found in the Caatinga biome. They may also indicate the ecological function that microorganisms play in providing plants with some degree of tolerance to water stress or in assisting in their development through mechanisms of growth promotion. Alterations in microbial communities can be due to the different abilities of native microorganisms to resist and adapt to environmental changes.

Editor: Bas E. Dutilh, Radboud University Medical Centre, NCMLS, Netherlands

Funding: This work was supported by Embrapa project number: 02.09.01.008.00.03. The funders had no role in study design, data collection and analysis, decision to publish, or preparation of the manuscript.

Competing Interests: The authors have declared that no competing interests exist.

* E-mail: van_nessner@yahoo.com.br

Introduction

Drought is a complex and natural phenomenon that affects several parts of the world, with negative impacts on society, the economy and the environment. Bacteria living in the soil and rhizosphere of plants in dry environments may have undergone a selective pressure in order to survive. In addition to having tolerance to the extreme conditions they are exposed to, these bacteria may confer a certain level of tolerance to plants, as well as other functions such as plant growth-promotion [1] and soil maintenance due to their function and strategy in the ecosystem [2].

As suggested in other studies [3,4,5], we hypothesize that the availability of water affects the structure of microbial communities, resulting in different patterns during the dry and rainy seasons in the semi-arid climate of the Brazilian biome called Caatinga.

This semi-arid climate of northeast Brazil (3–17°S to 35–45°W) is determined by rainfall index, aridity index and drought risk [6] and covers approximately 8% of the country, an area of approximately 900,000 km^2 [7]. There are two well-defined rainy and dry seasons. Rainfall can be concentrated in December and January, March and April, or May and June (unpublished data). During the dry season, there is a predominance of hot and extremely dry weather, with temperatures up to 45°C.

This exclusive, understudied Brazilian biome, the Caatinga, harbors trees and shrubs that are highly adapted to this climate. These trees and shrubs include members of the Cactaceae family, which have developed adaptive features, such as succulent tissues for water storage and long spines to reduce water loss [8]. The species *Cereus jamacaru,* known as mandacaru, is well-distributed. This biome can be considered an extreme environment due to its high temperatures, long and irregular drought periods, low water availability and high ultraviolet radiation [9,10]. This type of environment likely contains extremophiles that are adapted to drought and high temperatures [11], but thus far there is a lack of studies on the microbial communities associated with *C. jamacaru.* Recent work by our group reported the culturable bacteria associated with *C. jamacaru* and two other cacti also present in this biome [1].

In the present study, we evaluated the structure of bulk soil and rhizosphere bacterial communities of *C. jamacaru* during dry and rainy seasons using five different sampling points and qPCR, T-RFLP, Ion Torrent (PGM) sequencing and multivariate analyses.

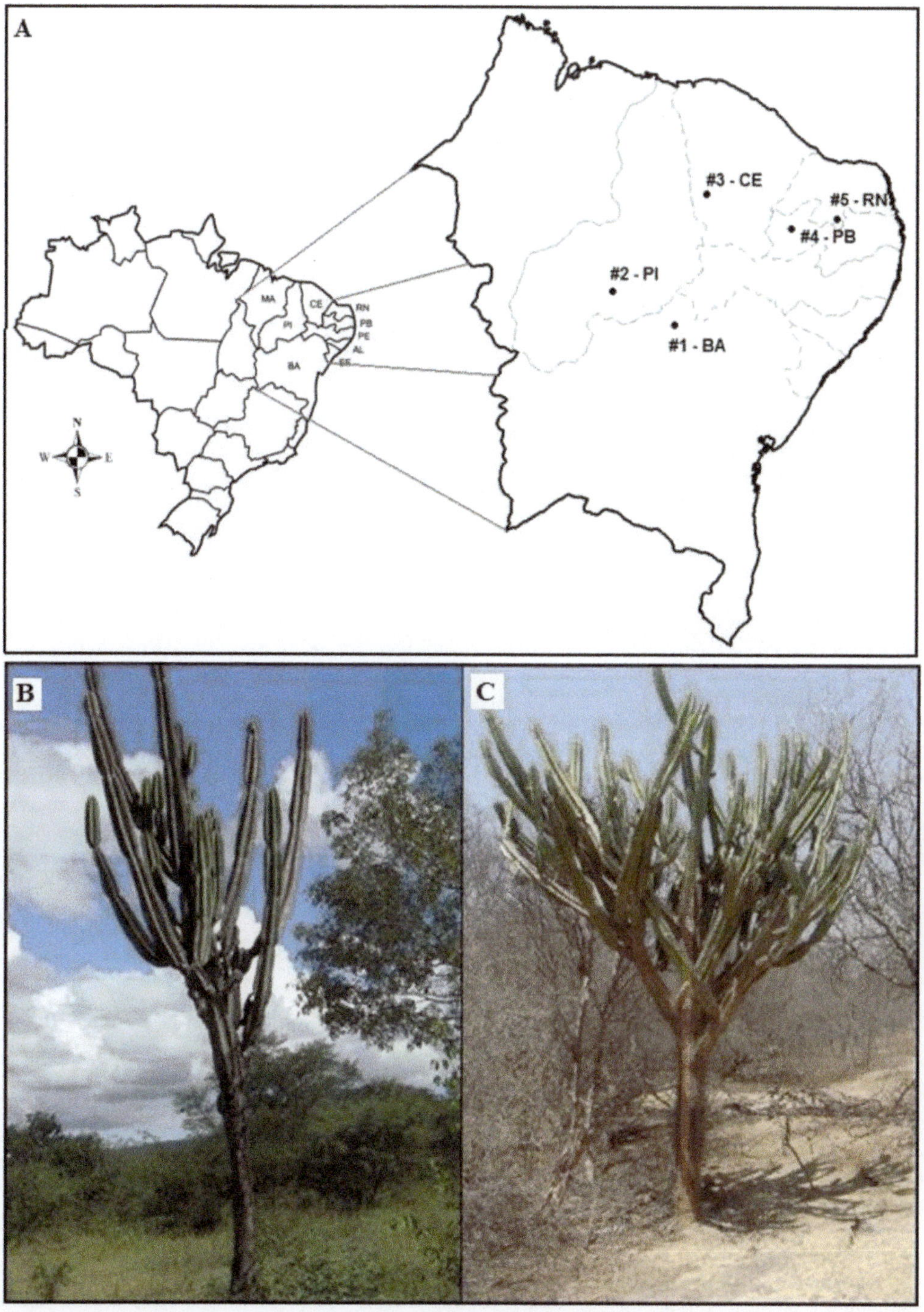

Figure 1. Sampling in the Caatinga biome in the semi-arid region of northeast Brazil. A - Map of Brazil in detail showing the distribution of the sampling sites comprising five states (#1 - Bahia (BA), S09°13′24.8′′; W41°05′11.4′′; #2 - Piauí (PI), S08°50′01.6′′, W42°33′13.3′′; #3 - Ceará (CE), S06°27′37.1′′, W40°44′50.5′′; #4 - Paraíba (PB), S06°42′44.2′′, W38°15′08.2′′; #5 - Rio Grande do Norte (RN), S06°39′15.6′′, W37°29′33.4′′). B and C - Pictures of *Cereus jamacaru*, a cactus found in the Caatinga biome of Brazil, during rainy (B) and dry seasons (C). (Pictures taken by the first author, 2009 and 2010).

Materials and Methods

Ethics Statement

This project was conducted with the authorization of the Institute of Environment and Renewable Natural Resources (IBAMA), process number 02001.004527/2011-90, and it did not involve endangered or protected species.

Caatinga Sampling Sites

Bulk soil (SL) and rhizosphere (RZ) samples of *Cereus jamacaru* were collected from five different sites in the Caatinga biome of

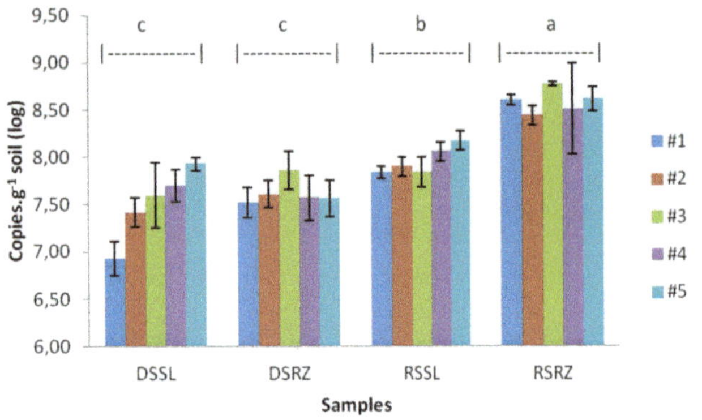

Figure 2. Quantification by qPCR of bacterial 16S rRNA gene copies in samples obtained during the rainy (RS) and dry (DS) seasons for both bulk soil (SL) and rhizosphere (RZ). Values indicate the average of three replications, and bars represent standard errors.

semi-arid northeast Brazil (figure 1A) during two distinct seasons (rainy (RS) (figure 1B) and dry (DS) (figure 1C)). The sampling of all sites was allowed by the regulatory body concerned with the protection of wildlife and protected areas (IBAMA).

Three replicates of each bulk soil and rhizosphere sample were mixed and sub-sampled for microbial community analysis and soil chemistry characterization. The soil analysis was performed in the Center for Research and Development of Soil and Environmental Resources of the Agronomic Institute of Campinas (IAC), with macronutrient and micronutrient determination as described previously [12]. The first sampling was performed during the rainy season (May, 2009), and the second one was performed during the dry season (October, 2010). The temperature of the soil and environment ranged from 28°C to 37°C during the rainy season. During the dry season, the temperature of the soil ranged from 42°C to 50°C, and the temperature of the environment varied from 33°C to 45°C. Samples were kept in plastic bags and stored at room temperature until processing.

DNA Extraction and Yield

Total community DNA from bulk soil and rhizosphere samples of *C. jamacaru* was extracted using a Power Soil™ DNA Isolation Kit (MOBIO Laboratories, Inc., Carlsbad, CA, USA) according to the manufacturer's protocol. Altogether, there were forty samples: two replicates each from the five sites taken during two different seasons (dry and rainy) and taken from bulk soil or rhizosphere material. The total yield of DNA varied from 10.4 $\mu g.\mu l^{-1}$ to 36.3 $\mu g.\mu l^{-1}$ across all samples, with an average of 25.94±6.58 $\mu g.\mu l^{-1}$ for bulk soil samples obtained during the rainy season, 22.26±5.62 $\mu g.\mu l^{-1}$ for bulk soil samples obtained during the dry season, 25.35±5.25 $\mu g.\mu l^{-1}$ for rhizosphere samples obtained during the rainy season and 23.08±7.03 $\mu g.\mu l^{-1}$ for rhizosphere samples obtained during the dry season.

Quantification of Total Bacterial Community by qPCR

The 16S rRNA gene was used to quantify bacterial community abundance. The quantification was performed in triplicate according to Taketani et al. 2009 [13] using an ABI Prism 7300 Cycler (Applied Biosystems). The specificity of the purified amplification products was confirmed by a melting curve analysis in which a unique peak was observed. The size of the amplicons was also checked using a 1.5% agarose gel stained with ethidium bromide. Standard curves were obtained by serially diluting a pool of bacterial 16S rRNA gene amplicons. The standard was diluted from 10^2 to 10^9 gene copies per microliter. The primer sets selected were P1 (5′-CCT ACG GGA GGC AGC AG-3′) and P2 (5′-ATT ACC GCG GCT GCT GG-3′) [14]. All quantitative PCR (qPCR) reactions were carried out in a 10-µl PCR mixture that contained 5 µl of Platinum SYBR Green qPCR Super Mix - UDG w/ROX (Invitrogen); 0.05 µl of each primer, 3.9 µl of Milli-Q water and 1 µl (approximately 30 ng) of environmental DNA. The PCR reactions were carried out using cycling times of 5 min at 94°C, followed by 30 cycles of 94°C for 30 sec, 57°C for 45 sec as the annealing temperature, and 72°C for 1 min. Only samples with specific melting curves and the expected amplicon sizes, as analyzed by agarose gel electrophoresis, were used for the quantification of the target group. The results were compared by Analysis of Variance and the Tukey's test at a significance level of 5%.

Bacterial Community Structure Analysis by T-RFLP

Amplification of 16S rRNA gene was performed with the primers 1492R (5′-TAC GGY TAC CTT GTT ACG ACT-3′) and bac27F (5′-AGA GTT TGA TCC TGG CTC AG – FAM-3′) in the following reaction: 2 µl of Dream Taq buffer, 1.2 µl of $MgCl_2$, 1.6 µl of dNTP (2.5 mM), 0.07 µl of each primer (5 pmol), 0.2 µl of Dream Taq, 2 µl of metagenomic DNA, and ultrapure sterilized water (Milli-Q) to a final volume of 20 µl. Amplification was evaluated by electrophoresis in agarose gels 1% (w/v). The amplification products were cleaved with the restriction enzyme *Hha*I (Fermentas, Life), according to the manufacturer's instructions. Reactions were performed using a thermal cycler (Applied Biosystems) at 37°C for 90 min, followed by an increase in temperature to 65°C for 30 sec. After the restriction reaction, the products were precipitated by the addition of 2 µl of EDTA (125 mM), 2 µl of sodium acetate (3 M) and 50 µl of ethanol (100%) according to the method suggested by the manual of BigDye® Terminator v3.1 Cycle Sequencing Kit. Samples were stored at −20°C until they were analyzed.

Terminal Restriction Fragments (T-RFs) analysis was performed using ABI PRISM 3500 Genetic Analyzer (Applied Biosystems). Data obtained from sequencing were analyzed with Gene Mapper v.4.1 (Applied Biosystems) software. For T-RFs, a limit baseline of 50 fluorescence units was used to discriminate "true" peaks from the background noise in this technique. We examined T-RFs that were >50 bp and <800 bp. Peak heights were transformed to relative data (percentage of detection) [15]. T-RFLP profiles were compared among different samples using T-RF relative abundance (>1%), in which each T-RF was considered to be one Operational Taxonomic Unit (OTU). Analysis of Similarity (ANOSIM) was performed with Bray-Curtis, offering a value ranging from −1 to 1. Values close to 0 indicate no differences among groups, while values close to 1 indicate a distinction among groups [16]. Non-metric Multidimensional Scaling (NMDS) graphs indicate the relative similarity of samples through distance ordering, in which similar samples are found to be very close [16]. Analyses were performed using Past 2.12 [17] software.

Bacterial Community Analysis using Ion Torrent (PGM) Sequencing

Each sample of DNA was amplified using the primers 967F (5′-CAA CGC GAA GAA CCT TAC C-3′) e 1046R (5′-CGA CAG CCA TGC ANC ACC T-3′) for the V6 region of 16S rRNA gene [18]; however, a different primer, 967F, with an additional tag of five nucleotides was synthesized for each sample (http://vamps.

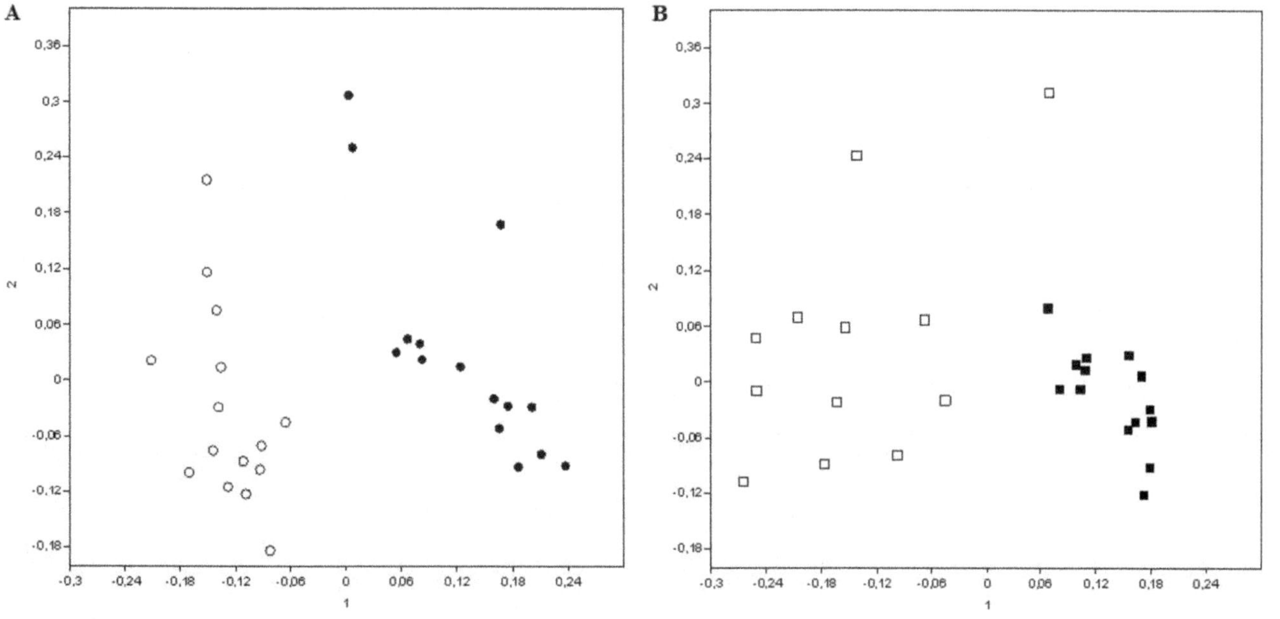

Figure 3. Non-metric multidimensional scaling (NMDS) of bacterial communities determined by T-RFLP, showing the seasonal variation among bulk soil (A) and rhizosphere (B) samples. Bulk soil samples are represented by circles and rhizosphere samples are represented by squares. Rainy season samples are represented by the white color and dry season samples are represented by the black color.

mbl.edu/). PCR reactions were carried out according to Sogin et al. [18]. After the amplification, each reaction was purified using Agencourt® AMPure® XP Reagent according to the manual provided by the manufacturer (Ion Amplicon Library Preparation (Purify the amplicon libraries), Life Technologies). At the beginning of the amplification process, we had forty samples; after the purification step, we ended with twenty samples because the replicates were joined. Each purified amplicon library was quantified using NanoDrop (Thermo Scientific), and the concentration of each sample was adjusted to prepare an equimolar pool (26 pM); 18 μL of this pool was used for emulsion PCR according to the Ion PGM™ 200 Xpress™ Template Kit manual from Life Technologies. After the recovery of ion spheres and enrichment, samples were prepared to be loaded on a 316 chip for posterior sequencing (Ion Sequencing Kit User Guide v2.0, Life Technologies) using the Ion Personal Genome Machine™ (PGM™) (Ion Torrent, Life Technologies).

Sequence Analyses

The initial handling of the sequences was performed using the online platform Galaxy (https://main.g2.bx.psu.edu/root), in which the raw data obtained from the sequencer Ion Personal Genome Machine™ (PGM™) (Ion Torrent, Life Technologies) were converted to FASTQ format using FASTQ Groomer [19] and were filtered by quality (95% of bases with Q20); in addition, reverse primers and adapters were removed on FASTX-toolkit. The classification of sequences was performed with mothur [20] following the recommendations of the SOP tutorial (http://www.mothur.org/wiki/Schloss_SOP) with adaptations from the Sogin tutorial (http://www.mothur.org/wiki/Sogin_data_analysis) to adjust to the short read from the V6 region. The removal of sequences (barcodes and forward primers) and library separation were performed using the trim.seqs command (settings, maxambig = 0, maxhomop = 6, bdiffs = 1, pdiffs = 2, minlength = 56, keepfirst = 60). Afterwards, the number of sequences was optimized using the unique.seqs command. Sequences were then aligned against the mothur version of the greengenes alignment by the align.seqs command using needleman as the alignment method (settings, ksize = 9 and gapopen = −1). The resulting alignment was passed through the screen.seqs command (settings, start = 4655, optimize = end, criteria = 95) to remove sequences that were outside of the desired range (i.e., the V6 region). The resulting alignment was then filtered using the filter.seqs command to remove any columns in the alignment that did not contain data. The number of sequences was once again optimized using the unique.seqs command followed by the pre.cluster command (settings, diffs = 2). The optimized alignment was used to check for chimeras using the chimera.uchime command. The resulting sequences were classified by Greengenes taxonomy using the classify.seqs command (cutoff = 60). The dist.seqs command was then used to generate a distance matrix (default settings). The sequences were then clustered into OTUs with the cluster command (average neighbor method). Richness estimators and diversity indexes were obtained by the summary.single command. The files *.groups, *.taxomony and *.names were exported to STAMP (Statistical Analysis of Metagenomic Profiles) [21] for further analysis. The comparison among multiple groups was performed using the software STAMP, which performs a statistical analysis of the sequences obtained in the samples, indicating the most abundant groups statistically and showing relevant differences in communities. Multivariate analyses were performed in Canoco 4.5 [22].

Sequence Accession Number

The 16S rRNA reads are available at: https://main.g2.bx.psu.edu/u/vannessner/h/16s-rrna-libraries-brazilian-caatinga-biome.

Results and Discussion

Soil Chemistry Characteristics

All soil samples were found to be acidic, with site #2 presenting the lowest pH (table S1). Most soil samples from the five sampling

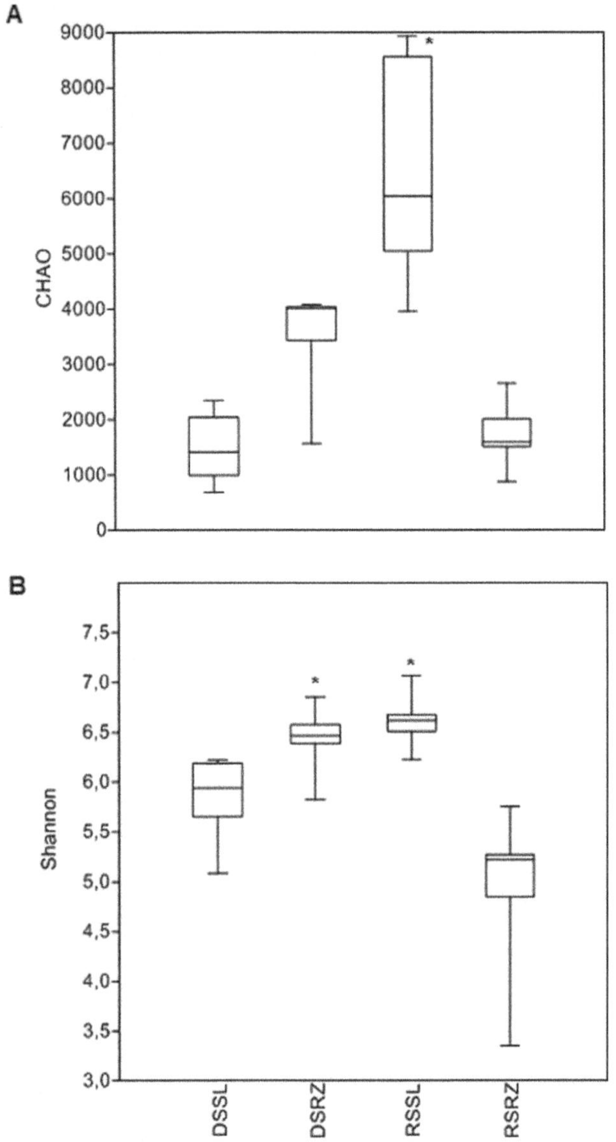

Figure 4. Estimation of total species richness obtained by the Chao index (A) and diversity using the Shannon-Wiener index (B) for samples from bulk soil during the dry season (DSSL), rhizosphere during the dry season (DSRZ), bulk soil during the rainy season (RSSL) and rhizosphere during the rainy season (RSRZ). Significant values according to Tukey's test ($p<0.01$) are indicated by an *.

sites presented significantly different values (according to Tukey's test at 5%) for most of the measured parameters. There is a high variability among the sampling sites, as observed in the clustering analysis using the unweighted pair-group average (UPGMA) with Euclidean similarity measure (figure S1), in which no pattern was found across different locations or seasons.

Abundance of Bacterial Communities by Quantitative PCR

The number of bacterial 16S rRNA gene copies identified in samples shows a general trend of increased abundance of bacteria in samples obtained during the rainy season, as compared to samples obtained during the dry season. Tukey's test confirms this result, with rhizosphere samples during the rainy season (RSRZ) being significantly more abundant than others (figure 2).

Samples from the rhizosphere during the rainy season have higher numbers of copies of the 16S rRNA gene compared to all other samples ($p<0.01$). In all sampling sites for the dry season, the rhizosphere (DSRZ) sample values ranged from 7.52 to 7.86 log of 16S rRNA gene copies per gram of soil. For rhizosphere samples obtained during the rainy season, the values are higher, ranging from 8.44 to 8.77 log of gene copies per gram of soil. Samples obtained from bulk soil ranged from 6.93 to 7.93 log for the dry season (DSSL) and from 7.84 to 8.17 log for the rainy season (RSSL). The transition from the rainy season to the dry season caused a variation in the rhizosphere samples, with a decrease of 9.8 to 12.6% in the numbers of copies per gram of soil. In the bulk soil samples, the decrease ranged between 2.9 and 11.6%, a more marked variation.

Bacterial Community Patterns

After observing a higher abundance of bacteria during the rainy season, we used T-RFLP to test whether the structure of the bacterial community varied. Non-metric multidimensional scaling (NMDS) was used with ANOSIM. The spatial variation between the five different sampling points was not significant, yielding a value of R close to zero ($R = -0.002$, $p = 0.456$). There is no clear division between the sampling sites (figure S2A).

The variation between bulk soil and rhizosphere samples produced a very low value of R ($R = 0.054$, $p<0.05$). There is a clearer separation between samples obtained during the rainy season (figure S2B), whereas bulk soil and rhizosphere samples obtained during the dry season slightly mix with each other (figure S2C), indicating a greater similarity between them. The selective pressure exerted by the dry season eliminated the rhizosphere effect, thus forming a tighter cluster.

Figure 3 shows a significant seasonal variation, with $R = 0.626$ ($p<0.001$). There is a clear separation between the rainy and dry seasons, both for bulk soil (figure 3A) and rhizosphere samples (figure 3B). Thus, despite the high variability found across the sampling sites, the distinction between seasons is evident.

These results suggest a change in the structure of the bacterial community according to seasonal variation (rainy vs. dry), as reported by several authors [5,23,24,25,26]. Several factors contribute to the design of the composition of bacterial communities [27]. In this analysis, it is evident that the community structure is shaped primarily by the presence or absence of water. The same result has been found for the rhizosphere of two leguminous trees (*Mimosa tenuiflora* and *Piptadenia stipulacea*) growing in the same Caatinga biome [28].

Bacterial Diversity Based on 16S rRNA Sequencing

We used a high-resolution technique to determine which groups were present in each season. A total of 590,043 sequences was obtained by Ion Torrent (PGM) Sequencing. Mothur classified 127,348 sequences as belonging to the domain Bacteria, of which 55.33% were grouped into thirty phyla and the remaining 44.67% remained unclassified. Twelve phyla showed relative frequencies greater than 1% for most samples.

Clustering analysis using Ward's method clearly distinguishes samples obtained during the different seasons, corroborating previous observations (figure S3). Samples obtained during the dry season form one group, while samples obtained during the rainy season form two distinct groups, separating the rhizosphere samples from the bulk soil samples.

The total species richness estimated by the Chao index was significantly higher ($p<0.01$) for bulk soil samples obtained during

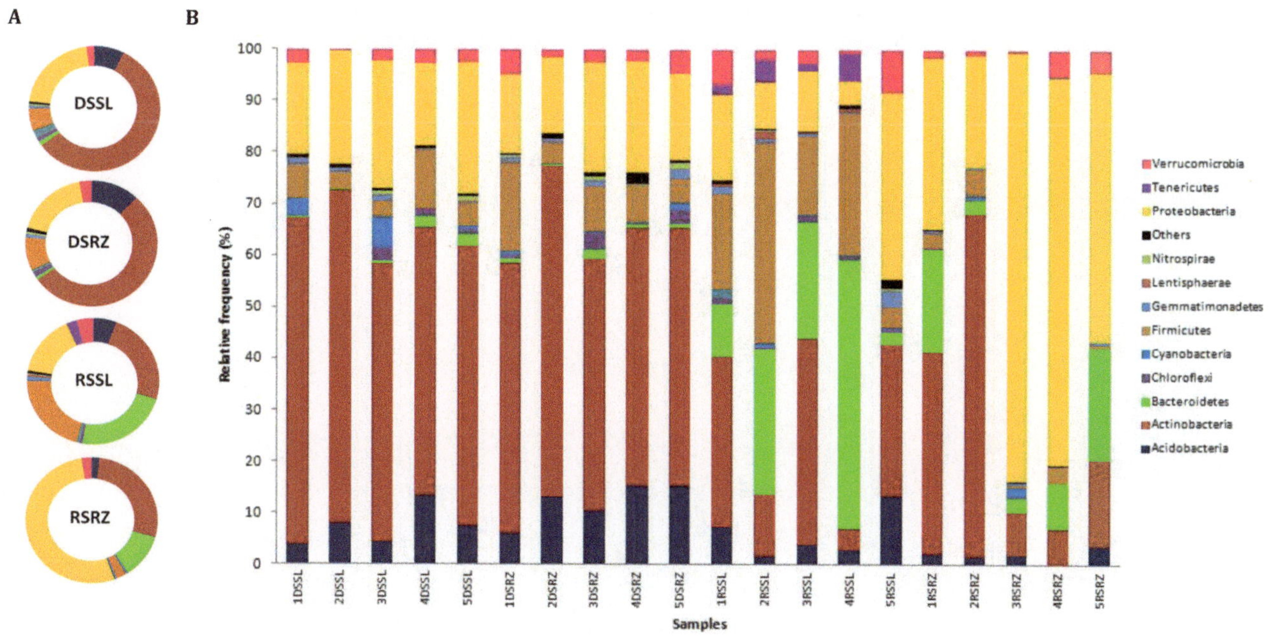

Figure 5. Classified sequences divided by phyla with relative frequencies higher than 1% (for the majority of the samples). A - Distribution of phyla, divided by samples. B - Samples obtained for the five different sampling sites from bulk soil during the dry season (DSSL), rhizosphere during the dry season (DSRZ), bulk soil during the rainy season (RSSL) and rhizosphere during the rainy season (RSRZ). "Others" include eighteen phyla with relative frequencies lower than 1%: AD3, Armatinonadetes, Candidatus Poribacteria, CCM11b, Chlamydiae, Elusimicrobia, Fibrobacteres, Fusobacteria, NC10, Planctomycetes, SC3, SC4, SPAM, Spirochaetes, Synergistetes, TM7, WPS-2 and WS3.

the rainy season (figure 4A), and diversity was also significantly higher ($p<0.01$) for rhizosphere samples obtained during the dry season and for bulk soil samples obtained during the rainy season (figure 4B).

In general, Actinobacteria was the dominant phylum, especially for samples obtained during the dry season (figure 5A). This phylum was also found in great abundance in rhizosphere samples obtained from a Mexican semi-arid cactus [29]. Bacteroidetes and Proteobacteria were detected with higher frequency during the rainy season, with Proteobacteria being more frequent in rhizosphere samples (figure 5B). The number of sequences affiliated with the phylum Acidobacteria was also higher in samples obtained during the dry season, comprising 7% and 12% of the reads from the bulk soil and the rhizosphere, respectively. Although being quite abundant in soil samples [30] and also in semi-arid environments [31], this phylum was not detected with high frequency in this study.

Principal Component Analysis (PCA) based on the relative frequencies of different phyla confirmed the results obtained from T-RFLP analysis, showing a clear separation between samples obtained from different seasons (figure 6). Samples obtained during the rainy season are more different from each other than samples obtained during the dry season. Based on the size of the arrows, samples from the dry season showed a stronger correlation to the phylum Actinobacteria and slightly correlated with the phylum Acidobacteria. The other phyla of Bacteroidetes, Lentisphaerae, Firmicutes, Proteobacteria and Tenericutes correlated with samples obtained during the rainy season, with rhizosphere samples correlating strongly with Proteobacteria, and the other four phyla correlating with bulk soil samples. To assess these differences statistically, comparisons were made among multiple groups using the software STAMP based on the criterion of the season. Significant differences ($p<0.05$) in the percentage of sequences were observed for some phyla. The numbers of representatives of Acidobacteria, Actinobacteria, and Nitrospirae were higher in the dry season; Bacteroidetes and Tenericutes ($p = 0.05$) had the largest representation during the rainy season. These data corroborate the data of two other works that reported that the abundance of representatives of the phylum Actinobacteria tends to be lower in damper soils [32,33]. Actinobacteria were found to increase in abundance during an experiment of throughfall exclusion [34]. Correspondingly, Proteobacteria and Bacteroidetes showed increased abundance during the rainy season, along with a drop in the proportion of Actinobacteria during the same period [35].

These differences might be due to the strategies adopted by these microorganisms, in which abundant groups during drought have a strategy of slow growth as oligotrophs, while favored groups during the rainy season might have quick responses to high resource availability, indicating a copiotrophic group [2].

Influence of Season in the Distribution of Microorganisms and their Possible Role in the Environment

At a finer phylogenetic resolution (that is, family and genus level), twenty-one families were significantly more abundant ($p<0.05$) during the dry season and twelve were significantly more abundant during the rainy season. Figure 7 shows the twenty-three most significant families. The Bacillaceae family, belonging to the phylum Firmicutes, class Bacilli, and order Bacillales, comprises Gram-positive endospore-forming microorganisms. The endospores are specialized resistance structures that allow the survival of microorganisms for extended periods in dry soils [36] and also give protection against several environmental stressors [37]. In a study conducted on the rhizosphere of the cactus *Mammillaria carnea* during rainy and dry seasons, the phylum Firmicutes had increased its abundance during the dry period due

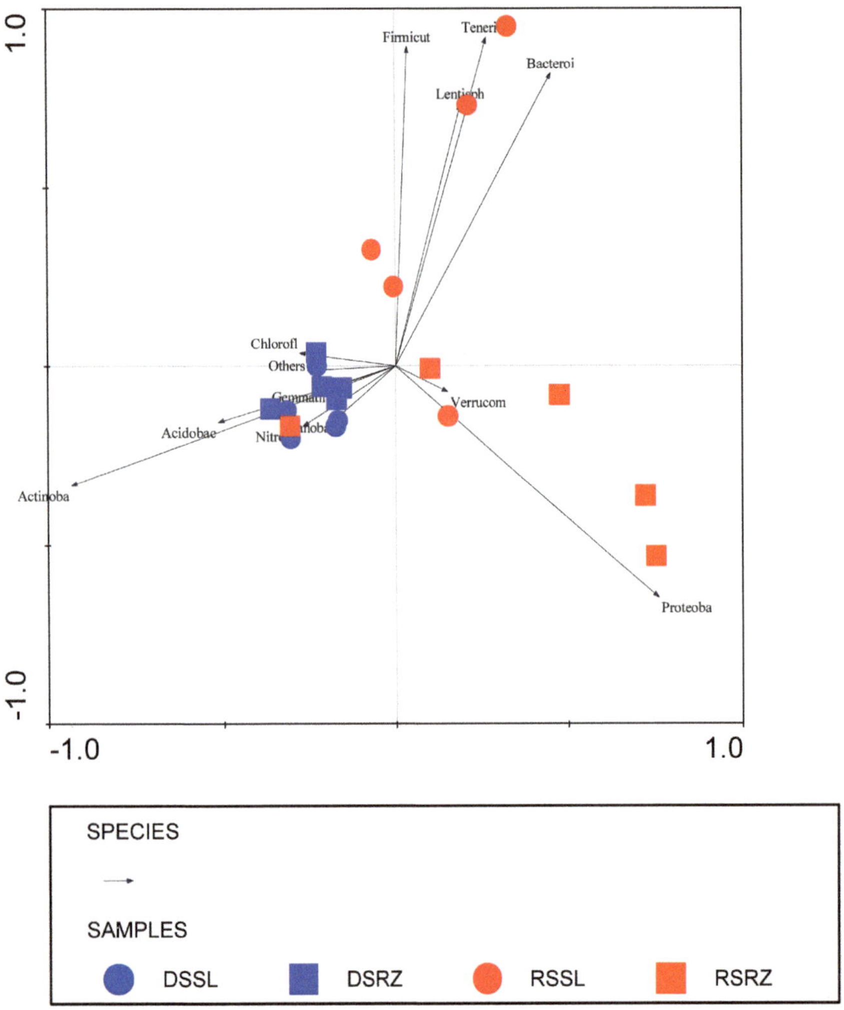

Figure 6. PCA of the 16S libraries based on taxonomic affiliation of reads determined by mothur. Samples are bulk soil during the rainy season (RSSL), bulk soil during the dry season (DSSL), rhizosphere during the rainy season (RSRZ) and rhizosphere during the dry season (DSRZ). The most common phyla are: Acidobacteria (Acidobac), Actinobacteria (Actinoba), Bacteroidetes (Bacteroi), Chloroflexi, Cyanobacteria (Cyanoba), Firmicutes (Firmicut), Gemmatimonadetes (Gemmati), Lentisphaerae (Lentisph), Nitrospirae (Nitro), Others, Proteobacteria (Proteoba), Tenericutes (Tener) and Verrucomicrobia (Verrucom).

primarily to the class Clostridia [5], which also has endospore-forming bacteria [38].

At the genus level, figure 8 shows the seventeen most significant genera, with thirteen more frequent during the dry season and four more frequent during the rainy season. The significant presence of one genus in one season over another does not exclude it from the other season, and vice-versa. For example, Bacilli that were identified at a significantly greater proportion during the dry season are also present during the rainy season but are enriched under the dry season. Because the sampling season was the major factor for differences in genera, the most representative genera might have developed mechanisms of adaptation to the unfavorable conditions imposed by water stress and other factors. Such adaptations might involve tolerance to high temperatures [39], desiccation tolerance genes [40], pigment production for protection against UV radiation [41], production of thermostable enzymes [42] and production of intracellular osmolytes [43]. We believe that the presence of a large number of sequences belonging to the genus *Bacillus* can be explained by the fact that endospore-forming bacteria might be favored in these types of environments due to resistance to heat and desiccation [44]. Six out of thirteen genera belong to the phylum Actinobacteria. They are also able to withstand harsh environments by producing spores resistant to desiccation and extreme heat [45]. Unlike the *Bacillus* genus, Gram-positive members with a high G+C content (Actinobacteria) are not capable of forming endospores. Some form spores similar to those produced by the genus *Streptomyces*, which are different from endospores and serve as a specialized reproductive structure that germinates under favorable conditions [46]. When spore production is not observed, they enter a dormant state similar to *Mycobacterium* characterized by low metabolic activity [47].

It is known that different plant species host specific microbial communities because plants are able to shape their rhizosphere microbiome [48,49]. The higher detection of some genera exclusively in the dry season can be due to the ecological role that some microorganisms play in the environment, such as providing a degree of tolerance against drought stress for plants they associate with [50]. Drought tolerance may come from the improvement of soil physical properties, such as soil aggregation through the production of exopolysaccharides and biofilm or

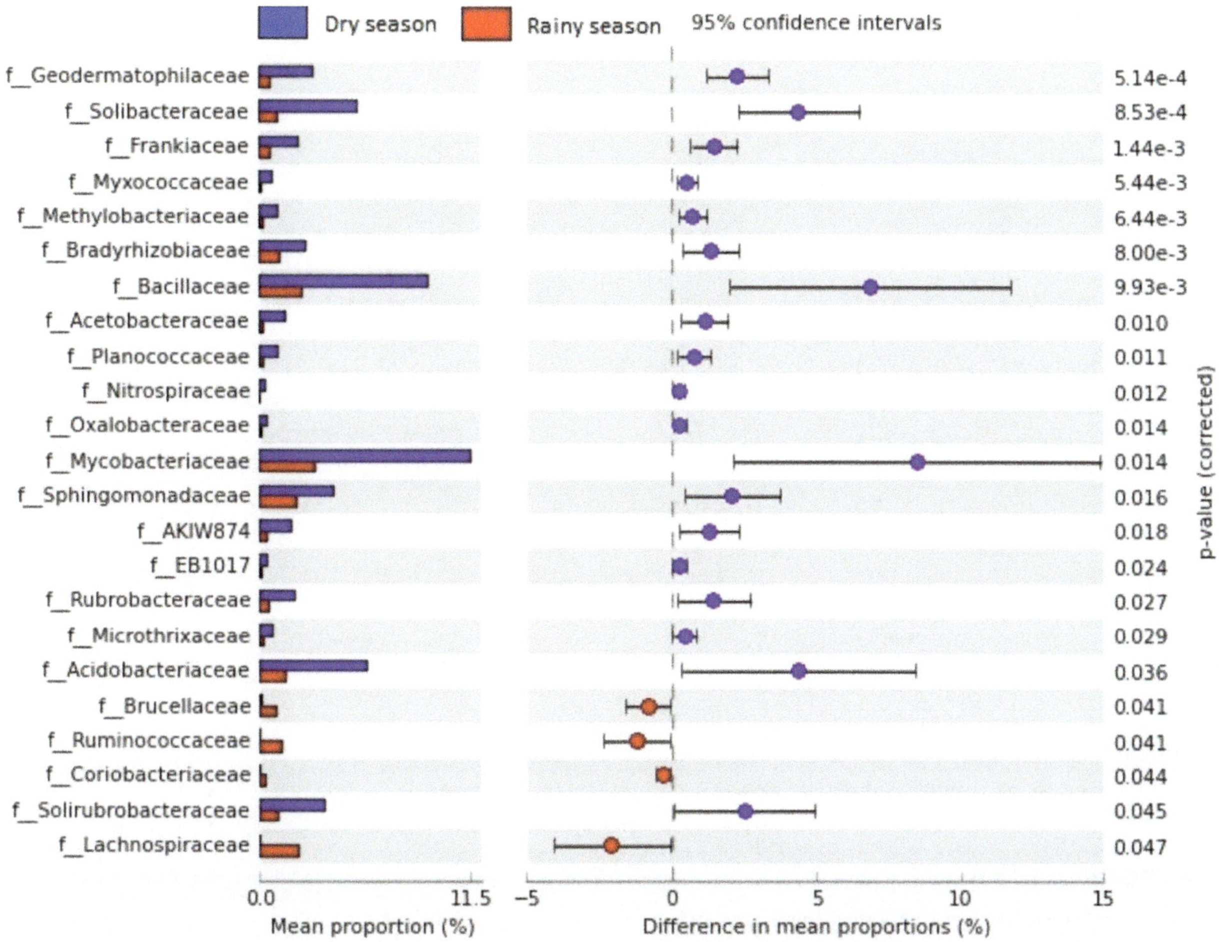

Figure 7. The twenty-three most significant families (p<0.05) with respect to the season.

protection by the production of osmolytes [51,52]. More details on how microorganisms help plants tolerate abiotic stress, i.e., drought, and even promote plant growth are provided by Kavamura et al. [1].

Conclusion and Future Perspectives

In general, qPCR and T-RFLP analysis combined with Ion Torrent (PGM) sequencing allowed us to confirm that the bulk soil and rhizosphere bacterial community of *C. jamacaru* showed a clear alteration between the rainy and dry seasons.

During the rainy season, a higher proportion of Gram-negative bacteria, especially those represented by Bacteroidetes and some classes of Proteobacteria, were observed. These include several microorganisms in soil that are important not only for nutrient cycling [53] but also for the whole ecosystem [54]. In the dry season, a higher proportion of Gram-positive bacteria, represented by the phylum Actinobacteria and the genus *Bacillus*, was observed.

We believe that the presence of dominant groups during the dry season, which are also present to a lesser extent during the rainy season, suggests that in soil, there are drought tolerant microorganisms. When conditions become unfavorable, sensitive groups decrease in proportion and the tolerant groups stand out due to the above-mentioned tolerance mechanisms. When the rainy season comes, drought-sensitive microorganisms are capable of growing rapidly, and they reestablish the community [55].

Changes in microbial communities can be observed due to the different abilities of native microorganisms to resist and adapt to environmental changes. However, these skills require a lot of energy and vary according to each microorganism [56]. Microorganisms fall into four categories depending on their ability to resist or adapt to environmental changes: i) microorganisms that have no mechanism for acclimatization; ii) microorganisms that only have acclimatization mechanisms; iii) microorganisms that possess inherent resistance; and iv) microorganisms with inherent resistance plus acclimatization mechanisms [57].

Gram-negative bacteria tend to fall into category ii, and Gram-positive bacteria tend to fall into categories iii or iv.

Thus, Gram-positive bacteria can be more resistant to rain/drought events, which is confirmed in this study. This result highlights a certain level of selection of microorganisms with effective tolerance mechanisms because bacterial communities that regularly suffer from stress episodes appear to be more tolerant than those that suffer from these episodes sporadically.

This work represents the first effort to better understand the bacteria associated with a widely distributed cactus in the Caatinga semi-arid biome of Brazil. Our group is now focusing on a complete description of the metagenome of this environment describing the metabolic pathways of the microorganisms inhabiting this biome to enable further understanding of how microbial communities remain active during the dry season, how they recover and the dynamics underlying these changes.

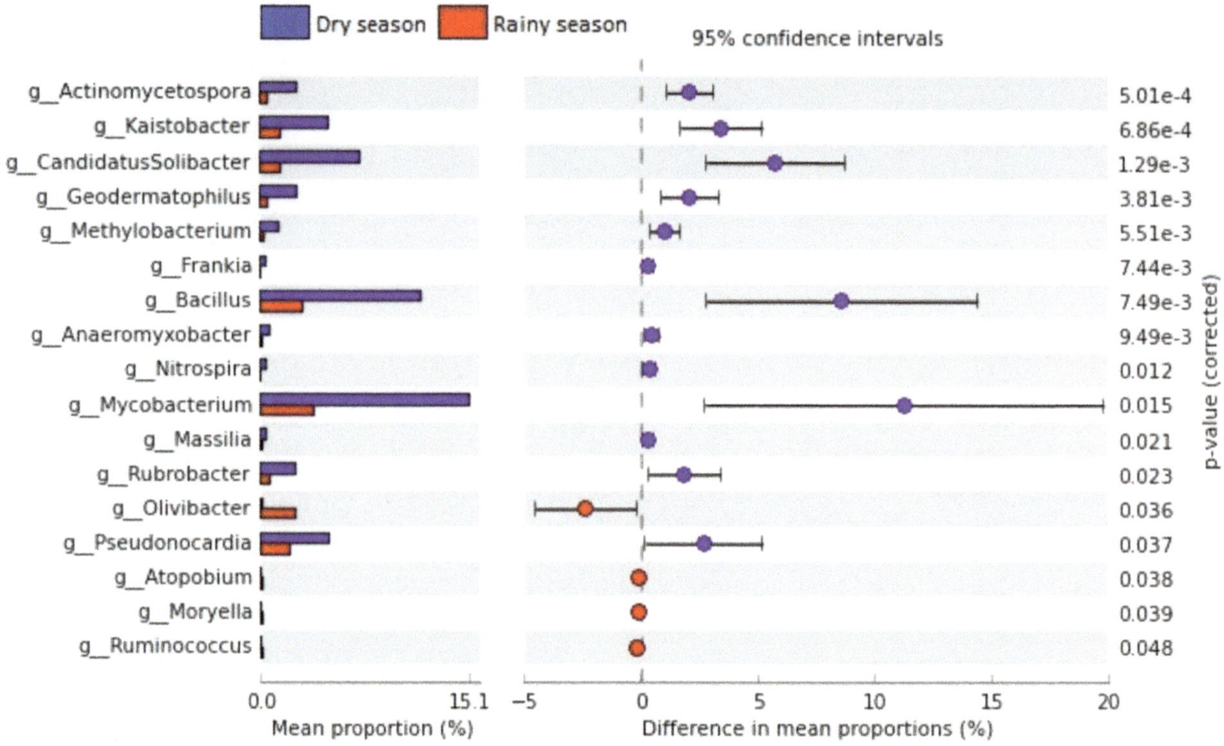

Figure 8. The seventeen most significant genera ($p<0.05$) with respect to the season.

Supporting Information

Figure S1 Clustering analysis of sampling sites, using UPGMA with Euclidian distance. Comparison during the rainy season (RS) and dry season (DS) for bulk soil samples (SL) for the five different sampling points: #1, #2, #3, #4 and #5.

Figure S2 Non-metric multidimensional scaling (NMDS) of bacterial communities determined by T-RFLP, showing the spatial variation (A) and source variation (B and C). Bulk soil samples are represented by circles and rhizosphere samples are represented by squares. Rainy season is represented by the white color and dry season is represented by the black color. The five different sampling points are represented by a cross (#1), a white triangle (#2), an x (#3), a black triangle (#4), and a rectangle (#5).

Figure S3 Clustering analysis using Ward's method obtained from 16S rRNA sequences showing a clear division according to the season. Dry season samples (DS) form one cluster, while rainy season samples (RS) form two separate clusters, one including bulk soil samples (SL) and the other including rhizosphere samples (RZ).

Table S1 Soil chemical features of each site for both seasons. Values are presented as the average ± standard errors (n = 3). In each line, values followed by the same letters do not differ statistically according to Tukey's test at 5%.

Acknowledgments

The authors thank the support of Cosme Corrêa dos Santos (UEFS) on identifying the plant species *Cereus jamacaru.* The authors also thank João Luiz da Silva and Suikinai Nobre Santos for their help with the Caatinga expeditions.

Author Contributions

Conceived and designed the experiments: VNK RGT FDA ISM. Performed the experiments: VNK MDL. Analyzed the data: VNK RGT MDL. Contributed reagents/materials/analysis tools: VNK RGT RM ISM. Wrote the paper: VNK.

References

1. Kavamura VN, Santos SN, Silva JL, Parma MM, Ávila LA, et al. (2013) Screening of Brazilian cacti rhizobacetria for plant growth promotion under drought. Microbiol Res 168: 183–191.
2. Fierer N, Bradford MA, Jackson RB (2007) Toward an ecological classification of soil bacteria. Ecology 88(6): 1354–1364.
3. Diallo MD, Martens M, Vloemans N, Cousin S, Vandekerckhove TTM, et al. (2004) Phylogenetic analysis of partial bacterial 16 S rDNA sequences of tropical grass pasture soil under *Acacia tortilis* subsp. *Raddiana* in Senegal. System Appl Microbiol 27: 238–252.
4. LaMontagne MG, Schimel JP, Holden PA (2003) Comparison of subsurface and surface soil bacterial communities in California grassland as assessed by Terminal Restriction Fragment Length Polymorphisms of PCR-Amplified 16S rRNA genes. Microb Ecol 46: 216–227.
5. Torres-Cortéz G, Millán V, Fernández-González AJ, Aguirre-Garrido JF, Ramírez-Saad HC, et al. (2012) Bacterial community in the rhizosphere of the cactus species *Mammillaria carnea* during dry and rainy seasons assessed by deep sequencing. Plant Soil 357(1–2): 275–288.
6. Ministério da Integração Nacional/SDR (2005) BRASIL 2005 Relatório final do grupo de trabalho interministerial para redelimitação do semi-árido nordestino e do polígono das secas. Brasília: Ministério da Integração Nacional/SDR. 118 p.

7. Giulietti AM, Harley R, Queiroz LP, Rapini A (2006) To set the scene. In: Queiroz LP, Rapini A, Giulietti AM, editors. Towards greater knowledge of the Brazilian semi-arid biodiversity. Brasília: Ministério de Ciência e Tecnologia. 11–15.
8. Gorelick R (2009) *Brasilicereus*, *Cipocereus*, and *Pilosocereus* in Eastern Brazil. Cactus Succul J 81(3): 126–137.
9. Desmarchelier C, Romão RL, Coussio J, Ciccia G (1999) Antioxidant and free radical scavenging activities in extracts from medicinal trees used in the 'Caatinga' region in northeast Brazil. J Ethnopharmacol 67: 69–77.
10. Santos SN, Kavamura VN, Silva JL, Melo IS, Andreote FD (2011) Plant growth promoter rhizobacteria in plants inhabiting harsh tropical environments and its role in agricultural improvements. In: Maheshwari DK, editor. Plant Growth and Health Promoting Bacteria. Berlin: Springer-Verlag. 251–272.
11. MacElroy RD (1974) Some comments on the evolution of extremophiles. Biosyst 6: 74–75.
12. Van Raij B, Andrade JC, Cantarella H, Quaggio JA (2001). Análise química para avaliação da fertilidade de solos tropicais. Campinas: Instituto Agronômico, 285p.
13. Taketani RG, dos Santos HF, van Elsas JD, Rosado AS (2009) Characterization of the effect of a simulated hydrocarbon spill on diazotrophs in mangrove sediment mesocosm. Anton van Leeuwenhoek 96: 343–354.
14. Muyzer G, Dewaal EC, Uitterlinden AG (1993) Profiling of complex microbial populations by denaturing gradient gel electrophoresis analysis of polymerase chain reaction-amplified genes coding for 16S rRNA. Appl Environ Microbiol 59: 695–700.
15. Culman SW, Gauch HG, Blackwood CB, Thies JE (2008) Analysis of T-RFLP data using analysis of variance and ordination methods: A comparative study. J Microbiol Meth 75: 55–63.
16. Bennett LT, Kasel S, Tibbits J (2008) Non-parametric multivariate comparisons of soil fungal composition: sensitivity to thresholds and indications of structural redundancy in T-RFLP data. Soil Biol Biochem 40: 1601–1611.
17. Hammer Ø, Harper DAT, Ryan PD (2001) Past: Paleontological Statistics Software package for education and data analysis. Palaeont Elec 4(1): 1–9.
18. Sogin ML, Morrison HG, Huber JA, Welch DM, Huse SM, et al. 2006 Microbial diversity in the deep sea and the underexplored "rare biosphere". PNAS 103(32): 12115–12120.
19. Blankenberg D, Gordon A, Von Kuster G, Coraor N, Taylor J, et al. (2010) Galaxy Team. Manipulation of FASTQ data with Galaxy. Bioinformatics. 26(14): 1783–1785.
20. Schloss PD, Westcott SL, Ryabin T, Hall JR, Hartmann M, et al. (2009) Introducing mother: open-source platform-independent, community-supported software for describing and comparing microbial communities. Appl Environ Microbiol 75(3): 7537–7541.
21. Parks DH, Beiko RG (2010) Identifying biologically relevant differences between metagenomic communities. Bioinformatics 26: 715–72.
22. Ter Braak CJF, Smilauer P (2002) CANOCO. Reference manual and CanoDraw for Windows user's guide: Software for Canonical Community Ordination (version 4.5). Ithaca: Microcomputer Power. 500 p.
23. Alonso-Sáez L, Balagué V, Sá EL, Sánchez O, González JM, et al (2007) Seasonality in bacterial diversity in North-west Mediterranean coastal waters: assessment through clone libraries, fingerprinting and FISH. FEMS Microbiol Ecol 60: 98–112.
24. Collignon C, Uraz S, Turpault MP, Frey-Klett P (2011) Seasons differently impact the structure of mineral weathering bacterial communities in beech and spruce stands. Soil Biol Biochem 43(10): 2012–2022.
25. Tatti E, Decorosi F, Viti C, Giovannetti L (2012) Despite long-term compost amendment seasonal changes are main drivers of soil fungal and bacterial population dynamics in a Tuscan vineyard. Geomicrobiol J 29(6): 506–519.
26. Castro HF, Classen AT, Austin EE, Norby RJ, Schadt CW (2010) Soil microbial community responses to multiple experimental climate change drivers. Appl Environ Microbiol 76(4): 999–1007.
27. Marschner P, Crowley D, Yang CH (2004) Development of specific rhizosphere bacterial communities in relation to plant species, nutrition and soil type. Plant Soil 261(1/2): 199–208.
28. Lançoni MD, Taketani RG, Kavamura VN, Melo IS (2013) Microbial community geographic patterns in the rhizosphere of two Brazilian semi-arid leguminous trees. World J Microbiol Biotechnol. In press.
29. Aguirre-Garrido JF, Montiel-Lugo D, Hernández-Rodríguez CC, Torres-Cortes G, Millán V, et al. (2012) Bacterial community structure in the rhizosphere of three cactus species from semi-arid highlands in central Mexico. A van Leeuw 101(4): 891–904.
30. Janssen PH (2006) Identifying the dominant soil bacterial taxa in libraries of 16S rRNA and 16S rRNA genes. Appl Environ Microbiol 72(3): 1719–1728.
31. Bachar A, Al-Ashhab A, Soares MIM, Sklarz MY, Angel R, et al. (2010) Soil microbial abundance and diversity along a low precipitation gradient. Microb Ecol 60: 453–461.
32. Alekhina LK, Dobrovol'Skaya TG, Pochatkova TN, Zvyagintsev DG (2001) Evaluation of bacterial diversity in soil microcosms at different moisture contents. Microbiology 70(6): 731–737.
33. Goodfellow M, Williams ST (1983) Ecology of actinomycetes. Annu Rev Microbiol 37: 189–216.
34. Bouskill NJ, Lim HC, Borglin S, Salve R, Wood TE, et al. (2013) Pre-exposure to drought increases the resistance of tropical forest soil bacterial communities to extended drought. ISME J 7: 384–394.
35. Cruz-Martínez K, Suttle KB, Brodie EL, Power ME, Andersen GL, et al. (2009) Despite strong seasonal responses, soil microbial consortia are more resilient to long-term changes in rainfall than overlying grassland. ISME J 3: 738–744.
36. Chen M, Alexander M (1973) Survival of soil bacteria during prolonged desiccation. Soil Biol Biochem 5: 213–221.
37. Griffiths BS, Philippot L (2012) Insights into the resistance and resilience of the soil microbial community. FEMS Microbiol Rev 37(2): 112–129.
38. Onyenwoke RU, Brill JA, Farahi K, Wiegel J (2004) Sporulation genes in members of the low G+C Gram-type-positive phylogenetic branch (*Firmicutes*). Arch Microbiol 182: 182–192.
39. Nazina TN, Tourova TP, Poltaraus AB, Novikova EV, Grigoryan AA, et al. (2001) Taxonomic study of aerobic thermophilic bacilli: descriptions of *Geobacillus subterraneus* gen. nov., sp. nov. and *Geobacillus uzenensis* sp. nov. from petroleum reservoirs and transfer of *Bacillus stearothermophilus*, *Bacillus thermocatenulatus*, *Bacillus thermoleovorans*, *Bacillus kaustophilus*, *Bacillus thermoglucosidasius* and *Bacillus thermodenitrificans* to *Geobacillus* as the new combinations *G. stearothermophilus*, *G. thermocatenulatus*, *G. thermoleovorans*, *G. kaustophilus*, *G. thermoglucosidasius* and *G. thermodenitrificans*. Int J Syst Evol Microbiol 51: 433–446.
40. Ward NL, Challacombe JF, Janssen PH, Henrissat B, Coutinho PM, et al. (2009) Three genomes from the phylum *Acidobacteria* provide insight into the lifestyles of these microorganisms in soils. Appl Environ Microbiol 75(7): 2046–2056.
41. Stankovic N, Radulovic V, Petkovic M, Vuckovic I, Jadranin M, et al. (2012) *Streptomyces* sp. JS520 produces exceptionally high quantities of undecylprodigiosin with antibacterial, antioxidative, and UV-protective properties. Appl Microbiol Biotechnol 96(5): 1217–1231.
42. Gtari M, Essoussi I, Maaoui R, Sghaier H, Boujmil R, et al. (2012) Contrasted resistance of stone-dwelling *Geodermatophilaceae* species to stresses known to give rise to reactive oxygen species. FEMS Microbiol Ecol 80: 566–577.
43. Lentzen G, Schwarz T (2006) Extremolytes: natural compounds from extremophiles for versatile applications. Appl Microbiol Biotechnol 72: 623–634.
44. Nicholson WL, Munakata N, Horneck G, Melosh HJ, Setlow P (2000) Resistance of *Bacillus* endospores to extreme terrestrial and extraterrestrial environments. Microbiol Mol Biol Rev 64(3): 548–572.
45. Wawrik B, Kutliev D, Abdivasievna UA, Kukor JJ, Zylstra GJ, et al. (2007) Biogeography of actinomycete communities and type II polyketide synthase genes in soils collected in New Jersey and Central Asia. Appl Environ Microbiol 73: 2982–2989.
46. Flärdh K, Buttner MJ (2009) *Streptomyces* morphogenetics: dissecting differentiation in a filamentous bacterium. Nature Rev Microbiol 7: 36–49.
47. Gengenbacher M, Kaufmann SHE (2012) *Mycobacterium tuberculosis:* success through dormancy. FEMS Microbiol Rev 36: 514–532.
48. Berendsen RL, Pieterse CMJ, Bakker PAHM (2012) The rhizosphere microbiome and plant health. Trends Plant Sci 17(8): 478–486.
49. Mendes R, Krujit M, Brujin I, Dekkers E, Van Der Voort M, et al. (2011) Deciphering the rhizosphere microbiome for disease-suppressive bacteria. Science 332: 1097–1100.
50. Grover M, Ali SZ, Sandhya V, Rasul A, Venkateswarlu B (2011) Role of microorganisms in adaptation of agriculture crops to abiotic stresses. World J Microbiol Biotechnol 27: 1231–1240.
51. Kaci Y, Heyraud A, Barakat M, Heulin T (2005) Isolation and identification of an EPS-producing *Rhizobium* strain from arid soil (Algeria): characterization of its EPS and the effect of inoculation on wheat rhizosphere soil structure. Res Microbiol 156: 522–531.
52. Vardharajula S, Ali AZ, Grover M, Reddy G, Bandi V (2011) Drought-tolerant plant growth promoting *Bacillus* spp.: effect on growth, osmolytes, and antioxidant status of maize under drought stress. J Plant Interact 6(1): 1–14.
53. Kersters K, De Vos P, Gillis M, Swings J, Vandamme P, et al. (2006) Introduction to the Proteobacteria. In: Dwarkin M, Falkow S, Rosenberg E, Schleifer KH, Stackebrandt E, editors. The Prokaryotes. New York: Springer. 3–37.
54. Spain AM, Krumholz LR, Elshhed S (2009) Abundance, composition, diversity and novelty of soil *Proteobacteria*. ISME J 3: 992–1000.
55. Van Gestel M, Merckx R, Vlassak K (1993) Microbial biomass response to soil drying and rewetting: the fate of fast- and slow-growing microorganisms in soils of different climates. Soil Biol Biochem 25: 109–123.
56. Schimel J, Balser TC, Wallenstein M (2007) Microbial stress-response physiology and its implications for ecosystem function. Ecology 88(6): 1386–1394.
57. Harris RF (1981) Effect of water potential on microbial growth and activity. In: Parr JF, Gardner WR, Elliott LF, editors. Water potential relations in soil microbiology. Madison, WI: American Society of Agronomy. 23–95.

8

Separating Drought Effects from Roof Artifacts on Ecosystem Processes in a Grassland Drought Experiment

Anja Vogel[1]*, Thomas Fester[2], Nico Eisenhauer[1], Michael Scherer-Lorenzen[3], Bernhard Schmid[4], Wolfgang W. Weisser[5], Alexandra Weigelt[1,6]

1 Institute of Ecology, Friedrich-Schiller-University Jena, Jena, Germany, **2** Department Umweltmikrobiologie, Helmholtz-Zentrum für Umweltforschung – UFZ, Leipzig, Germany, **3** Faculty of Biology, Geobotany, University of Freiburg, Freiburg, Germany, **4** Institute of Evolutionary Biology and Environmental Studies, University of Zurich, Zurich, Switzerland, **5** Terrestrial Ecology Research Group, Department of Ecology and Ecosystem Management, Center for Food and Life Sciences Weihenstephan, Technische Universität München, Freising, Germany, **6** Institute of Biology, University of Leipzig, Leipzig, Germany

Abstract

1: Given the predictions of increased drought probabilities under various climate change scenarios, there have been numerous experimental field studies simulating drought using transparent roofs in different ecosystems and regions. Such roofs may, however, have unknown side effects, called artifacts, on the measured variables potentially confounding the experimental results. A roofed control allows the quantification of potential artifacts, which is lacking in most experiments.

2: We conducted a drought experiment in experimental grasslands to study artifacts of transparent roofs and the resulting effects of artifacts on ecosystems relative to drought on three response variables (aboveground biomass, litter decomposition and plant metabolite profiles). We established three drought treatments, using (1) transparent roofs to exclude rainfall, (2) an unroofed control treatment receiving natural rainfall and (3) a roofed control, nested in the drought treatment but with rain water reapplied according to ambient conditions.

3: Roofs had a slight impact on air (+0.14°C during night) and soil temperatures (−0.45°C on warm days, +0.25°C on cold nights), while photosynthetically active radiation was decreased significantly (−16%). Aboveground plant community biomass was reduced in the drought treatment (−41%), but there was no significant difference between the roofed and unroofed control, i.e., there were no measurable roof artifact effects.

4: Compared to the unroofed control, litter decomposition was decreased significantly both in the drought treatment (−26%) and in the roofed control treatment (−18%), suggesting artifact effects of the transparent roofs. Moreover, aboveground metabolite profiles in the model plant species *Medicago x varia* were different from the unroofed control in both the drought and roofed control treatments, and roof artifact effects were of comparable magnitude as drought effects.

5: Our results stress the need for roofed control treatments when using transparent roofs for studying drought effects, because roofs can cause significant side effects.

Editor: Kurt O. Reinhart, USDA-ARS, United States of America

Funding: The drought experiment was funded by the University of Jena (grant to Alexandra Weigelt) with additional support of the Universities of Zurich, Göttingen and Freiburg. The Jena Experiment was funded by the Deutsche Forschungsgemeinschaft (DFG; FOR 456). Nico Eisenhauer gratefully acknowledges funding by the DFG (Ei 862/2). The funders had no role in study design, data collection and analysis, decision to publish, or preparation of the manuscript.

* E-mail: anja.vogel@uni-jena.de

Introduction

Precipitation change is an important driver of global change affecting the functioning of ecosystems [1] and is predicted to increase in the future [2]. Consequently, many experiments have investigated the effects of precipitation change on ecosystem functioning. However, the experimentally applied setup of drought experiments may have side effects, hereafter called artifacts, in addition to the intended manipulation of precipitation patterns (summarized in [3]). The implications of such artifacts are particularly serious, when a second important driver of global environmental change with effects on ecosystem functioning, like plant diversity loss (e.g., [1,4]), is manipulated, as it is not known how potential artifacts would interact with a second treatment. Given the importance of multifactor experiments for quantifying the effects of global change on ecosystem functioning, it is crucial to critically evaluate the results and potential conclusions drawn from drought experiments.

Roofs or rain-out shelters are a common tool to induce drought in field experiments. Their design varies among experiments, e.g. they vary in size, shape and transparent material [5]. All roofs are constructed in a way to minimize possible artifacts, which is difficult and rarely tested. The most obvious unwanted artifacts are shading (e.g., [6,7]), and passive warming [8,9], although some authors report only slight increases in air and soil temperature due to roofs [6,10,11] or even no influence [12,13]. In contrast to experiments in forests, where rain can be intercepted below the canopy, roofs in grassland studies must cover the entire plant stand, and thus artifacts might be particularly significant in these ecosystems.

Warming and changes in irradiance may be part of the predicted climate change in certain scenarios [14,15], and one could argue that such roof effects might help to simulate future climate more realistically. Nevertheless, artifacts are not controlled in roof experiments and may not mirror regional projections [2,15]. For example, with increasing frequency of heat waves and drought periods, irradiance is more likely to increase than decrease. Furthermore, roof artifacts may themselves affect ecosystems and therefore confound the results of drought experiments. Warming for example has an effect on several ecosystem functions and increases productivity and decomposition [16,17,18]. Both, drought and roof artifacts might differ in various aspects, as, e.g., different plant communities, and therefore confound the results on interactions.

Shifts in plant diversity are another important driver of global environmental change in addition to climate change. And as both drivers operate simultaneously in real ecosystems, their interacting effects are of particular importance for future studies. However, plant diversity changes the density and aboveground productivity of the plant communities and therefore community structure. Thus, if the interaction of drought with another treatment (i.e. plant diversity) is investigated, one should be aware of the interacting effects of roof artifacts with this second treatment to avoid misinterpretations. For example, species richness was found to decrease resistance of biomass production to drought ([19,20], but see [21]), but it is unknown, so far, whether this relationship was partly confounded by roof artifacts.

As there is currently no way to prevent unwanted roof effects, additional control treatments are needed to separate drought effects from roof artifacts. An obvious control involves roofs under which the collected rain is redistributed to the experimental plots [13,22,23], named "roofed control" hereafter. To our knowledge, there is no published study, investigating whether roof artifacts affect ecosystem processes and therefore confound the conclusions of drought experiments. Furthermore, studies using roofed controls in highly replicated experiments manipulating a second treatment, such as plant diversity experiments, do not exist.

We performed a roof experiment to separate pure drought and roof artifact effects and their influence on three ecosystem functions, aboveground biomass production, litter decomposition and the production of plant metabolites. This experiment was part of a large grassland biodiversity experiment and therefore allowed us to study, whether the effects of roof artifacts varied with a second treatment causing confounding interaction effects between drought and the second treatment. We compared measurements from roofed and unroofed control treatments to assess roof artifacts and to quantify their relative strength compared to the drought effects themselves. We tested 1) whether the roofs caused potential artifacts such as shading or passive warming, 2) the extent of artifact effects on aboveground biomass production, litter decomposition and the production of plant metabolites compared to drought and 3) whether any of these effects interacted with a second treatment, i.e. plant diversity.

Methods

Experimental Design and Drought Manipulations

The Jena Experiment field site in the floodplain of the river Saale in Jena (Thuringia, Germany, 50°55′N, 11°35′E, 130 m a.s.l.) served as platform for our experiment. In 2002, 80 plant communities (experimental plots) were established and assembled out of a pool of 60 mesophilic grassland species typical for the regional Molinio-Arrhenateretea meadows. The plots were arranged in four blocks perpendicular to a gradient in soil texture and moisture from the river Saale. The plant communities varied in species richness (1, 2, 4, 8, 16 and 60 species) and functional group richness (1, 2, 3 and 4 functional groups: grasses, small herbs, tall herbs, legumes). This experimental design was maintained by two to three annual weeding campaigns to eliminate non-target plants. The field site was managed by mowing twice a year (beginning of June and beginning of September). For details of the experimental design see Roscher et al. [24].

Since summer droughts are forecasted to increase in the future [2], our roof treatments excluded all rainfall from July through August every year since 2008. Starting in 2009 (16-Jul to 01-Sep-2009 and from 25-Jul to 03-Sep-2010 and from 11-Jul to 29-Aug-2011), we changed the roof constructions and applied three roof treatments to each of the 80 plots (which means 80 replicates per roof treatment): a roofed treatment without water addition served as "**drought**" treatment, a roofed treatment with water addition served as "**roofed control**" treatment and an ambient treatment served as "**unroofed control**" (Fig. 1). The roof, which covered "drought" and "roofed control" subplots, consisted of a wooden frame of 3×2.5 m and was covered with corrugated PVC sheets of 1.0 mm thickness (www.paruschke-kunststoffe.de, product code: PVCSPK7018K10). It had a height of 1.3 m to 1.5 m for good ventilation with a roof inclination of 4.6° for precipitation runoff. The large size of the roofs enabled us to let the removed water drop on the nearby ground, because we could perform our measurements in an area (1×1 m^2), which was 1 m apart from the lowest edge and about 0.5 m from the other edges of the roofs to minimize edge effects. With this design, the shelters prevented direct precipitation on the subplots, when rainfall was straight or slightly skewed and water only entered on very windy days. However, lateral movement of water by overland flow or lateral diffusion in the soil was possible since the plots were not trenched and water was not collected. Given that the field site is located in a plane area, we expect overland water flow to only occur during extreme rain events. For watering the "roofed control" we collected rain water of a small subset of roofs equipped with guttering and plastic barrels. To prevent the development of algae and to mimic the ambient precipitation pattern as closely as possible, we always watered the subplots the day after each rain event. We added the amounts of precipitation to the "roofed control", which were recorded in 10-minutes intervals at a weather station located on the field site apart from the experimental plots.

Data Sampling

We used a subset of plots for the measurements of potential artifacts and soil moisture to be able to record data simultaneously across a given time interval. We measured potential roof artifacts concerning shading and passive warming as well as litter decomposition in the first year of the experimental manipulations (2009). Soil moisture and biomass data were collected in every year. Measurements of air and soil temperatures were carried out in four plots, one in each block, to cover the edaphic variability

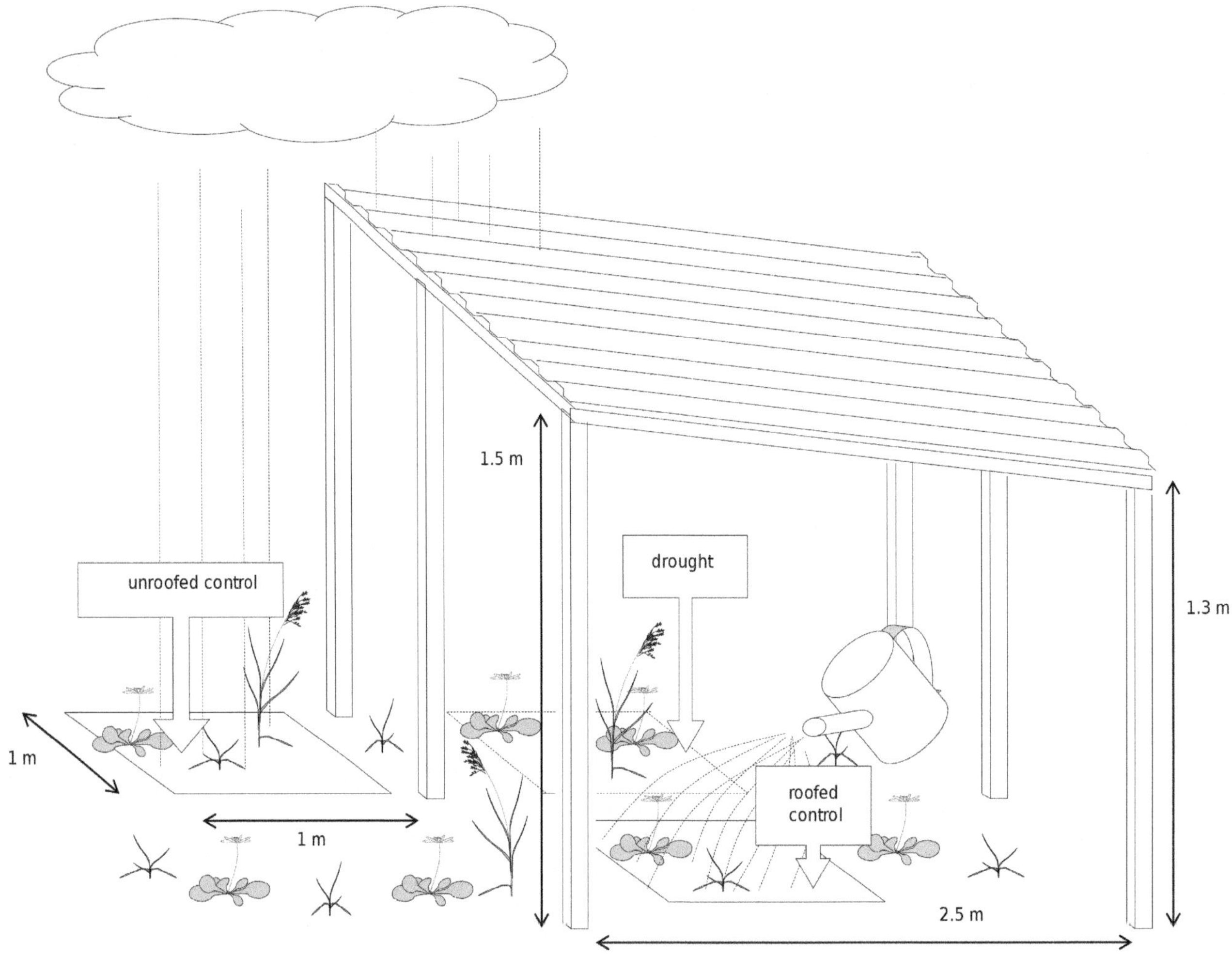

Figure 1. Roof construction and arrangement of the subplots in the field site. Given are the different subplots and the size of subplots and the roof construction. For more details see main text.

across the field site, which means four replicates per roof treatment. The selected plots had similar species richness and high grass cover (two plots with one and one plot with two grass species, one plot with a grass and a small herb species). Soil moisture was measured as volumetric water content every 15 minutes by EC-5 sensors and recorded with Em50 data loggers (Decagon devices, Pullman, USA) continuously since 12-May-2009 in about 7 cm depth of drought, roofed and unroofed control subplots of three of the above mentioned plots, which means three replicates per roof treatment. Soil temperature (in 7 cm depth) and air temperature (in 25 cm height) was recorded by testostor 175 data logger (Testo AG, Lenzkirch, Germany) every 15 minutes between 12-May and 05-Nov-2009.

PAR was measured above the vegetation and in a lower height compared to the roofs in all subplots of four plots, giving again four replicates per roof treatment. In contrast to the selected plots for soil parameters described above, these plots were close to each other for practical reasons. We recorded PAR every 30 minutes from sunrise to sunset on 19-Aug-2009 using the portable sunscan system SS1 (Delta-T, Cambridge, UK).

Aboveground biomass and litter decomposition was measured in all subplots of all 80 plots. The aboveground plant material was cut at a height of 3 cm above soil surface within one frame of 20×50 cm per subplot at the end of the drought periods in 2009 and 2010 (28- to 31-Aug-2009 and 25- to 26-Aug-2010). Plant material was sorted into sown species, unsown species (weeds) and dead material, dried (70°C, 48 h) and weighed separately. Aboveground biomass presented here represents standing biomass (dry mass) of sown species.

Litter decomposition was measured using plastic containers (9×9 cm in size). which were constructed using pots with 4 mm mesh at the bottom. Mesh and lateral cuttings of the pots allowed access of micro-, meso- and macrofauna to litter material. We used ~3 g of dried senescent wheat shoot material (chopped into pieces of ~3 cm, N = 0.4%, C = 45.2%, C:N ratio = 111.5) as standard litter in one container per subplot from 17-Jun to 24-Aug-2009. At the end of the experiment, containers were collected, and the remaining litter material was dried (70°C, 48 h) and weighed.

For the analysis of metabolites, we used the legume *Medicago x varia*. Flower buds (sampled from the apical nodes of leading shoots and showing no signs of petal pigmentation), sink leaves (sampled from the apical nodes of leading shoots) and source leaves (sampled from the 3rd or 4th node below the apical meristem from leading shoots) were harvested from identical plant individuals. We harvested plant material (organs from one plant per subplot) in the fourth year of consecutive summer drought on 22-Aug-2011 from

11 a.m. to 3 p.m. in every subplot of eight plots. In addition, we sampled sink leaves and source leaves from at least seven different individuals from each subplot of a plot of high species cover, to get an estimate of the within-subplot variation. Samples were weighed, frozen immediately in liquid nitrogen and stored at −80°C. Metabolite extraction, derivatization and gas chromatography were done as described in Fester *et al.* [25] based on methods described by Sanchez *et al.* [26] and Desbrosses *et al.* [27]. In short, the frozen material was extracted with methanol and chloroform with the addition of ribitol (0.2 mg/ml dissolved in methanol) as internal standard. Derivatization involved treatments with methoxamin hydrochloride (20 mg/ml in pyridine) and *N*-methyl-*N*-(trimethylsilyl)-trifluoroacetamide. Samples of 1 µl were analyzed in splitless mode using an Agilent GC 6890 equipped with a Rtx-5Sil MS capillary column (30 m×0.25 mm inner diameter; Restek GmbH, Bad Homburg, Germany) and a MSD 5973 (Agilent, Böblingen, Germany). Data evaluation was performed using the programs Metalign [28] for baseline correction and TagFinder [29] for chromatographic deconvolution and quantification of compounds. Retention time indices (RI) calculated by TagFinder and mass fragmentation data were compared with data from the Golm metabolome database [30,31] for metabolite identification.

Statistical Data Analysis

Data on soil moisture, temperature, PAR, aboveground biomass and litter decomposition were analyzed with linear mixed models using the REML algorithm to estimate variance components; and F statistics to test fixed effects. Depending on the dataset and the underlying experimental setup, different sets of fixed and random terms had to be used for different response variables (Table S1). For most analyses the treatment term was split into two contrasts. The first contrast distinguished between either roofed (mean of drought and roofed control treatment) and unroofed or between "moist" (mean of roofed and unroofed controls) and "dry" (drought) treatments. As a second contrast we tested for the **pure drought** effect and compared the (roofed) drought treatment and the roofed control, or we tested for **roof artifacts** and compared the roofed and unroofed control. Temperature was analyzed independently for day and night with the official time of sunrise and sunset as separator. To test whether temperature changed due to roof effects or by local climate, we fitted a contrast, which distinguished between cold and warm days (air: ≥24°C; soil ≥20°C) or cold and warm nights (air: ≥15°C; soil ≥19°C) using data from the field weather station. For PAR we fitted two models, one with all data and one with only noon data (11 am –3 pm) to separate whole day effects from effects of high radiation conditions. Models of time series data (soil moisture, air and soil temperature) were fit using an autoregressive correlation structure of order one (AR1) for the residuals to account for serial correlations. In the case of PAR this model adjustment was dispensable due to a sinus contrast for time which describes the sinus shaped curve of PAR over the day. For aboveground biomass and litter decomposition we fitted several models and included species richness (log-linear), functional group richness, the presence of single functional groups, the treatment contrasts and the two-way interactions of the diversity variables with the treatment contrasts in the fixed-terms part of the model. Because functional group richness and the presence of single functional groups were rarely significant, we excluded them from the models. Mixed effects models were performed using GenStat Release 15.1 (VSN International Ltd.).

Multivariate analysis of metabolite levels of all three plant organs in combination (data from source leaves, sink leaves and flowers) was performed by partial least square discriminant analysis (PLS-DA) using the function plsda() from the R-package *caret* [32] in R 2.14.0 [33]. PLS-DA accounts for the group structure (in our case the roof treatments) of the dataset while still being capable for dimension reduction [34,35] and tests how well the treatments can be separated. We did not use unsupervised methods which can identify the gross variability in a multivariate dataset, because we were not interested in other factors, which might also determine metabolite levels and cannot be tested due to the low numbers of replicates (e.g. species richness or community composition). Metabolite data were fitted to PLS-DAs using the function envfit() from the R-package *vegan* [36]. This function assessed significance of correlations by a correlation test using Monte Carlo permutations (N = 999) of the fitted vectors [37]. The goodness of fit statistics used was squared correlation coefficient (Pearson's product moment correlation coefficient). Significant differences in individual metabolites were analyzed with the same mixed model approach (Table S1) as described for the ecosystem processes excluding, however, the species richness term due to the low numbers of replicates. Data were log- (soil moisture, biomass, metabolites) or square root-transformed (PAR) in case residuals were not normally distributed.

Results

Soil Moisture

In 2009, 53.7 mm rainfall was excluded during the drought period. and precipitation patterns approximated the long-term seasonal trend with the exception of an unusually dry winter (January to March) and a wet autumn (October to December, Table 1). During spring and summer (April to September), April and July were wetter whereas June and especially August were dryer than the long-term average. In 2010 we observed a higher annual precipitation compared to the long-term mean and a higher intra-annual variability (Table 1). As in 2009, winter was dryer and autumn wetter compared to the long-term mean. The high precipitation occurred during summer (August), whereas spring (April and June) was dry. During the drought period in 2010, 196.7 mm rainfall was excluded. In 2009 soil moisture did not differ significantly between the treatments ($F_{1,6.1} = 0.77$, $p = 0.414$) but we found a significant reduction of soil moisture in response to drought over time in 2010 ($F_{1,5.1} = 186.79$, $p<0.001$), whereas soil moisture under ambient conditions and in the roofed control treatment increased over time (Fig. 2).

Air and Soil Temperature

Air temperature during the day was not significantly affected by the roofs (Fig. 3A). In contrast, air temperature during the night was significantly increased by 0.14°C due to the roofs ($F_{1,6} = 32.96$, $p<0.001$, Fig. 3A, B), irrespective of whether nights were cold or warm. Soil temperature during the day was significantly decreased by ~0.45°C under roofs in comparison to ambient, but only on warm days (significant two-way interaction of roofed/unroofed-contrast × cold/warm contrast: $F_{1,5.7} = 15.89$, $p<0.01$; Fig. 3C). In contrast, soil temperature during the night was increased by the roofs by 0.25°C, but only on cold nights (significant two-way interaction of roofed/unroofed-contrast × cold/hot contrast: $F_{1,5.1} = 25.6$, $p<0.01$).

Photosynthetically Active Radiation (PAR)

The roof ($F_{1,9} = 19.23$, $p = 0.002$) and time of day ($F_{1,25.5} = 200.49$, $p<0.001$) showed significant effects on PAR (Fig. 3D). Roofs reduced PAR by around 15% (corrected mean

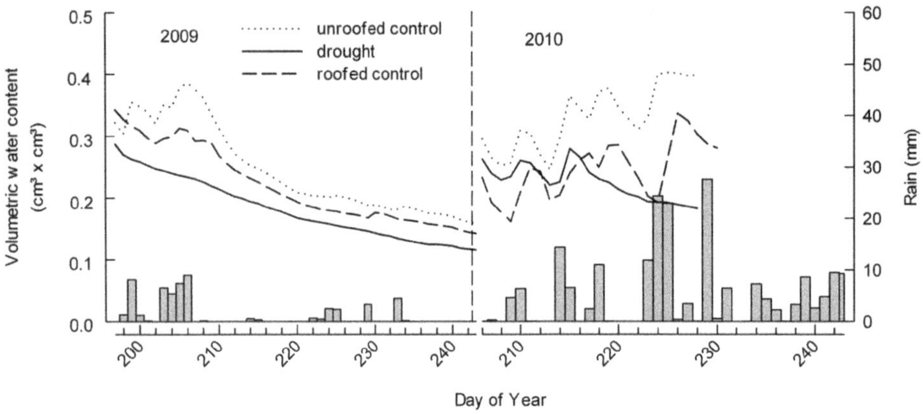

Figure 2. Soil moisture and daily precipitation patterns during the period of induced drought. In summer of 2009 (left) and 2010 (right). Soil moisture data are shown for all three roof treatments (lines, average of N = 3 plots). Daily precipitation patterns (grey bars) were measured on the field site of the Jena Experiment.

over the whole day), and by around 16% during noon ($F_{1,9} = 5.39$, $p<0.045$).

Aboveground Biomass

Aboveground biomass after drought was significantly affected by the treatments only in 2010, but not in 2009 (Table 2, Fig. 4A and B). In 2010 mean biomass was similar in unroofed (mean ± standard error: 132.0±10.5 g * m^{-2}) and roofed control plots (129.9±10.7 g * m^{-2}) and significantly lower in drought plots (76.9±8.1 g * m^{-2}). Although biomass was not significantly different between roof treatments in 2009, the pattern was the same (ambient: 168.4±23.2 g * m^{-2}; roofed control: 170.8±17.5 g * m^{-2}; drought: 151.3±15.5 g * m^{-2}). We did not find a significant interaction between species richness and roof artifacts in either year (Table 2). The relationship between plant diversity and biomass was positive in all treatments.

Litter Decomposition

Litter decomposition was affected by the roof construction (Table 2, Fig. 4C). Litter decomposition was highest in the unroofed control (3.9±0.1 mg * g^{-1} * d^{-1}), moderate in the roofed control treatment (3.2±0.1 mg * g^{-1} * d^{-1}) and lowest in the drought treatment (2.9±0.1 mg * g^{-1} * d^{-1}). The pure drought effect was small compared to the roof artifact (Table 2, Fig. 4C). There was no interaction between species richness and the pure drought or roof artifact (Table 2).

Metabolites

In total, 227 different analytes could be detected in each plant organ, of which 34% could be specified. Multivariate analysis (PLS-DA) of the metabolite profiles from all organs per plot and subplot clearly separated the different roof treatments (Fig. 5). PLS-DA-component 1 separated between the roofed (drought, roofed control) and unroofed treatments. PLS-DA-component 2 separated between the "moist" (roofed and unroofed controls) and

Table 1. Climatic parameters measured on the field site of the Jena Experiment during the study years 2009 and 2010 with the reference period 1961–1990 measured by the German Weather Service DWD in the city center of Jena.

	Air temperature (°C)			Precipitation (mm)			Soil moisture (Vol%)		Soil temperature (°C)	
	1961–90	2009	2010	1961–90	2009	2010	2009	2010	2009	2010
January	0.40	−3.09	−3.92	37.0	9.0	11.0	22.11	35.56	−0.20	1.02
February	1.40	1.15	0.12	34.0	33.7	23.6	33.31	35.26	1.34	0.61
March	4.80	5.04	4.89	43.0	42.5	29.1	37.01	36.65	5.25	4.82
April	8.60	11.58	8.74	57.0	73.7	19.8	31.39	33.04	12.13	9.39
May	13.40	13.89	11.13	62.0	62.6	93.0	31.05	31.49	15.37	12.43
June	16.70	15.01	16.89	75.0	52.9	20.4	28.58	26.70	16.50	17.62
July	18.20	18.34	20.66	52.0	85.1	88.6	31.29	22.95	19.75	21.91
August	17.40	18.59	16.76	63.0	14.6	184.2	22.34	33.15	19.13	18.36
September	14.20	14.56	12.67	42.0	53.6	64.7	23.67	29.98	15.85	14.15
October	9.80	8.42	8.26	39.0	47.3	19.4	28.70	30.91	10.31	9.73
November	5.00	8.06	5.59	41.0	68.3	94.5	34.00	33.95	7.57	7.28
December	1.70	0.64	−4.12	42.0	80.0	56.3	36.22	35.88	3.19	2.20
Year	**9.30**	9.35	8.14	**587.0**	623.3	704.5	29.97	32.13	10.52	9.96

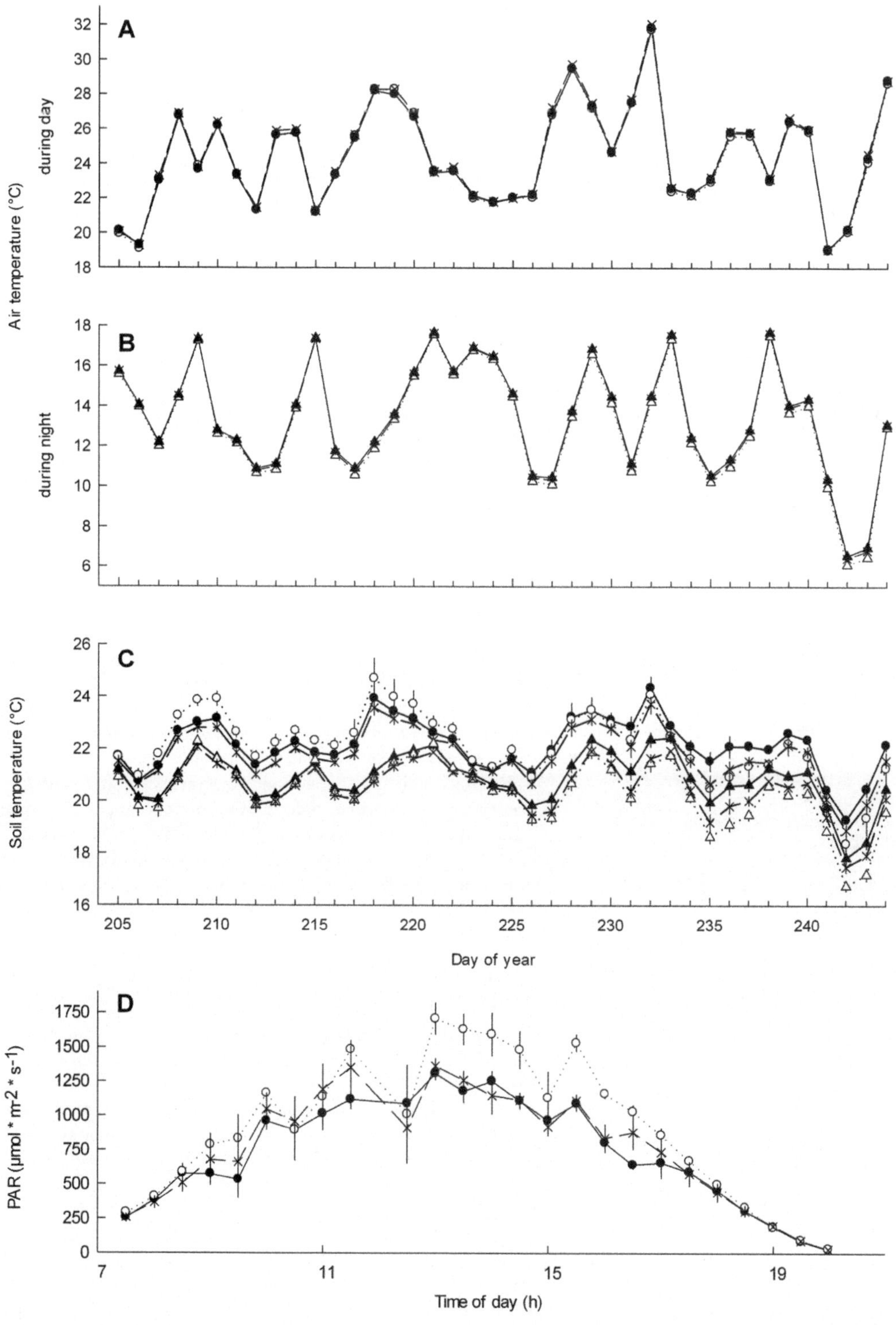

Figure 3. Effects of the presence of roofs on abiotic parameters: air temperature (a, b), soil temperature (c) during day (circles) and night (triangles) and photosynthetically active radiation (d). Given are means and standard errors of the drought treatment (filled symbols, solid lines), unroofed (open symbols, short dashed line) and roofed controls (x symbols, long dashed line) for day (circles) and night (triangles). Data represent mean and standard error of all three treatments in four (respectively three in case of temperature) plots.

the drought treatment. PLS-DA of sink leaves and source leaves obtained from plants from one single plot gave similar results (data not shown). We observed 66 analytes with significant correlations (Monte Carlo-permutations, $p<0.05$), when correlating data of individual metabolites with PLS-DA-components. 22 of these analytes were identified and included in Fig. 5. A subset of these 22 analytes can be grouped into two clusters (cluster I and II), which both positively correlate with the roof effect (component 1). These

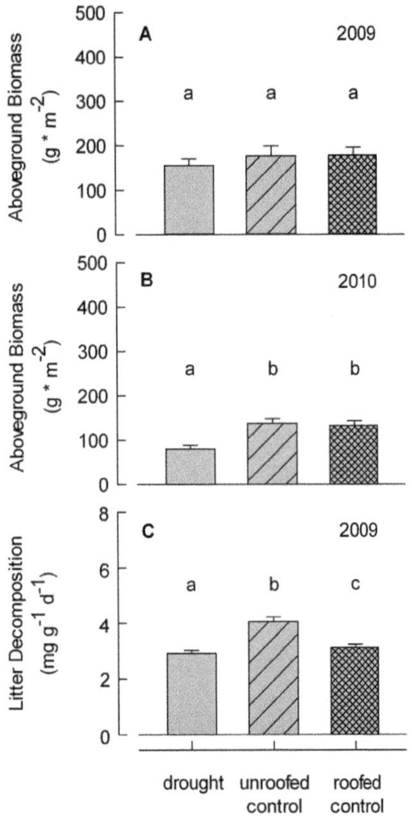

Figure 4. Treatment effects on ecosystem properties. Aboveground biomass production (a, measured in 2009 and b, measured in 2010) and litter decomposition (c). Data represent mean and standard error of all three treatments in 76 plots.

clusters primarily comprise nitrogen-containing compounds, sugars and sugar alcohols, with nitrogen-containing compounds being predominant in cluster I and sugar or sugar alcohols being predominant in cluster II. Interestingly cluster II almost exclusively comprised compounds from source leaves while cluster I comprised mainly compounds from all sink leaves and flowers. There was no drought effect on cluster I (no correlations with component 2), while we found a positive correlation between drought and cluster II. Analysis of significant differences in the levels of individual metabolites revealed drought effects on 20 analytes and effects of the roof artifacts on 27 analytes in at least one plant organ (Table S2). Only three of these analytes were affected both by pure drought and roof artifacts.

Discussion

Many methodological issues are prevalent with drought studies in the field. The problem of comparison arises from the different experimental designs and the size of roof constructions, from missing common metrics to describe the magnitude of the treatment and from the question, how to deal with roof artifacts [3,5,38,39]. The goal of the present study was to investigate the magnitude and ecological relevance of confounding roof artifacts in a drought experiment and their interaction with a second treatment, plant diversity. We found minor effects of passive warming (i.e. soil and air temperature were warmer at night and soil temperature cooler during the day) and strong shading effects of the roofs, confounding both, litter decomposition and plant metabolite profiles, while aboveground biomass remained unaffected. Moreover, roof artifacts did not interact with plant diversity as an additional experimental treatment.

Our roof construction induced a slight increase of air temperature during the night, indicating that roofs retarded cooling by blocking the emission of thermal energy from the soil surface. However, this did not lead to a significant passive warming effect during the day and indicates good ventilation of the roof constructions. Soil temperature differences between roofed and unroofed subplots were low compared to other experiments ([8,22,40]. Concerning shading, our measurements support previous results (e.g. [6,22]) showing that fixed roofs reduce radiation. Automatically closing constructions might minimize or even prevent such roof artifacts, because they only close during rain events and slide back in dry periods [41,42]; although this approach is appropriate, it mostly is not affordable, particularly when having a large number of roof replicates. Similarly, reducing precipitation with portable rain-out shelters is very laborious and only possible when working with a low number of replicates [43].

The ecological relevance of roof artifacts strongly depended on the ecosystem process. Despite the fact that light quantity and temperature have an influence on photosynthesis [44,45] and hence growth, plant community aboveground biomass was not significantly different between the unroofed and the roofed control, indicating roof artifact effects to be small and independent of differences in the plant communities (as roof artifacts did not interact significantly with our second treatment). Drought itself did have the expected strong effect on aboveground biomass production, which highlights the strong dependency of plant biomass production on soil water availability.

For surface litter decomposition, however, we found evidence for significant roof artifact effects. The presence of a roof had stronger effects on litter decomposition than the induced drought, and both effects operated in the same direction by decreasing decomposition rates. Therefore, roof artifacts turned out to be very important for surface litter decomposition, and comparing a drought treatment using a roof with an unroofed control only, would most likely lead to a strong overestimation of drought effects on decomposition. The confounding roof effects on litter decomposition also did not vary across the second treatment, indicating similar confounding effects irrespective of the plant communities. We can only speculate why the roof construction reduced litter decomposition in such a strong way. Austin & Vivanco [46] found a strong impact of photodegradation on litter decomposition in a semi-arid ecosystem. If radiation was completely blocked, decomposition was reduced by 60%, if only UV-B radiation was blocked, there was already a reduction of decomposition by 33%. Photodegradation, especially degradation due to UV radiation, may play a particularly prominent role in arid and semi-arid ecosystems (e.g., [47]), but there is evidence that this mechanism is also important for a wider range of grasslands, including more mesic sites [48,49]. Our experiment was not designed to determine the mechanism behind these effects, but assuming photodegradation plays a significant role in our system, it is conceivable that decomposition was sensitive to the roof artifacts through changes in radiation. Since we have not measured the spectral characteristics of solar radiation below the roofs, we cannot separate the effects originating from changes in light quantity or quality. However, we speculate that changes in light quality, with some blocking of UV radiation, might have had strong effects. Temperature effects of the roof construction seem not to be relevant for the decrease of decomposition under the roofed compared to the unroofed control, because increasing air

Table 2. Summary of mixed effects models for aboveground biomass in August 2009 and 2010 as well as for decomposed wheat litter.

		Biomass 2009			Biomass 2010			Decomposition 2009		
	df_{Num}	df_{Den}	F	*p*	df_{Den}	F	*p*	df_{Den}	F	*p*
Model A										
Species richness (log-scale) = SR	1	69.2	27.9	**<0.001*** **	70.8	21.6	**<0.001*** **	71.3	20.0	**<0.001*** **
Presence of Legumes	1	67.7	8.2	**0.006** **						
Treatment										
roofed vs. unroofed = DRvsU	1	141.2	1.6	0.205	140.8	6.3	**0.014***	146.8	59.8	**<0.001*** **
DvsR	1	141.0	1.8	0.188	141.6	33.3	**<0.001*** **	146.8	4.2	**0.042***
Treatment × SR										
DRvsU × SR	1	143.4	2.9	0.092	141.4	0.6	0.453	146.5	2.3	0.131
DvsR × SR	1	143.0	0.4	0.513	144.0	0.7	0.404	147.4	0.7	0.414
Model B										
Species richness (log-scale) = SR	1	69.2	27.9	**<0.001*** **	70.8	21.6	**<0.001*** **	71.3	20.0	**<0.001*** **
Presence of Legumes	1	67.7	8.2	**0.006** **						
Treatment										
dry vs. wet = DvsRU	1	141.2	0.3	0.611	141.3	39.0	**<0.001*** **	146.6	32.0	**<0.001*** **
RvsU	1	140.6	3.1	0.08	140.8	0.6	0.453	147.3	32.0	**<0.001*** **
Treatment × SR										
DvsRU × SR	1	143.6	0.1	0.787	142.9	1.2	0.271	146.7	0.0	0.948
RvsU × SR	1	141.8	3.2	0.075	142.0	0.04	0.834	147.7	3.0	0.087

For the analysis we used all plots of the Jena experiment, except for the 60-species mixtures. In Model A the treatment contrast was split into a contrast for roofed versus unroofed subplots (DRvsU), which aggregates the drought (D) and roofed control (R) treatments and tested it against the unroofed (U) treatment, and the residual contrast (DvsR), which considers the pure drought effect. In Model B the treatment contrast was split in a contrast of dry versus wet subplots (DvsRU), which aggregates the unroofed and roofed control treatments and tested it against the drought treatment, and the residual contrast (RvsU), which considers the roof artifact effects. Df = degrees of freedom; Num = numerator; Den = Denominator.

temperatures should result in increasing decomposition rates when soil moisture is not limiting [17,50] and therefore should act in an opposite direction compared to our results. Still, drought is also known as a factor that decreases litter decomposition, as demonstrated in laboratory experiments [50,51] or in field experiments with automatically closing rain shelters [52], which do not have strong shading effects. Nevertheless, drought effects on litter decomposition might be largely overestimated, if roof artifacts are not taken into account [53,54]. As litter decomposition is a key process for the global carbon cycle, it is crucial to separate climatic effects from methodic artifacts. However, little is known about the global role of photodegradation and about the influence of different roof types on this process.

Metabolite profiling of above-ground organs from *Medicago x varia* was chosen as a test for more subtle physiological changes induced by roofs, since there had been indications from another rain exclusion experiment that roofs may impact plant metabolite profiles independently from water availability (data unpublished). In accordance with this prior experiment, the clear distinction of metabolite profiles of the roofed control plants in comparison to all other treatments indicates roof artifact effects at the plant physiological level. While analysis using linear mixed models only resulted in a low number of identified significant changes in metabolite levels in our experiment, analysis of metabolites correlating with PLS-DA components showed higher numbers of significant correlations. The large numbers of sugars and sugar alcohols from cluster I and II in Fig. 5 are of particular interest in our context, since they represent compatible solutes typically accumulating under conditions of abiotic stress [55]. It is interesting to note that cluster II, which correlated with both roof and drought effects, comprised compounds almost exclusively from source leaves. Cluster I, in contrast, correlated exclusively with roof effects, comprised compatible solutes from flowers or sink leaves as well as some nitrogen containing compounds. While these observations are difficult to interpret in detail, they nevertheless allow the conclusion that, independently from drought levels, roofs introduced some kind of stress to the plant and that the impact of this stress was stronger than the impact of drought under the conditions prevailing in our experiment. As we did not find confounding roof artifacts on aboveground plant community biomass, the physiological changes induced by the roofs apparently were not strong enough to result in a significant change at the level of plant community biomass. However, they should be taken as a caution that other ecosystem processes that are more sensitive to changes in plant physiology may well be affected, as, e.g., the resistance of plants against herbivores. In addition, our experiment was short-term.We cannot exclude the possibility that persistent maintenance of drought treatments for more than three years might result in artifacts in plant physiological traits and total biomass production.

Conclusions

In field drought experiments it is important to critically test for confounding roof artifacts to be able to correctly predict the consequences of drought. Our analysis shows that, in contrast to aboveground biomass production, roofs had clear confounding effects on litter decomposition and also affected plant physiology.

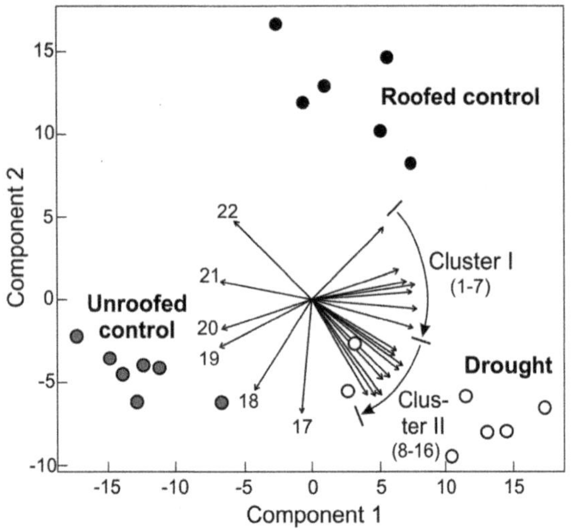

Figure 5. Partial least square discriminant analysis (PLS-DA) of metabolite profiles. (combined data of flowers, sink leaves and source leaves) with one individual (*Medicago x varia*) analyzed for each subplot out of seven plots. Metabolites correlating significantly with PLS-DA data (according to Monte Carlo-permutations, $p<0.05$) are represented as black arrows. Only identified metabolites are shown. These metabolites comprise: asparagine/flower (1), arabitol/flower (2), arabitol/sink leaf (3), citric acid/sink leaf (4), allantoin/flower (5), pinitol/sink leaf (6), asparagine/source leaf (7), arabitol/source leaf (8), 1,6-anhydro-glucose/source leaf (9), sorbitol/source leaf (10), 2,4-diamino-butanoic acid/source leaf (11), glucose/source leaf (12), erythritol/source leaf (13), beta-alanine/sink leaf (14), xylitol/source leaf (15), kestose/sink leaf (16), galactinol/flower (17), mannose/sink leaf (18), glucose-6-phosphate/sink leaf (19), phenylalanine/source leaf (20), phosphoric acid monomethyl ester/sink leaf (21), threonine/sink leaf (22). Retention time index and fragmentation pattern were not sufficient to differentiate between closely related isomers in the case of arabitol, xylitol, 1,6-anhydro-glucose, sorbitol, kestose, galactinol and mannose.

We therefore strongly recommend the use of roofed controls for long-term and short-term studies. As the confounding roof artifacts did not interact with a second treatment, in our case plant diversity, and therefore with changes in vegetation, roofed controls do not necessarily have to be installed in every single replicate, With the use of such roofed controls it is possible to measure pure drought effects, but still under microclimatic conditions, which do not necessarily mimic future climate.

Supporting Information

Figure S1 In the first years of the Jena drought experiment we combined the diversity and drought treatment with an additional management treatment [20] using further roofs of the same construction reported here but with a different roof orientation. Analysis of the extended dataset (including the additional random factor of revealed significant effects on PAR of the roof treatment ($F_{2,8.8}=12.42$, $p=0.003$), time of day ($F_{1,23.0}=298.46$, $p<0.001$) as well as the interaction of the roof treatment and time ($F_{2,281.6}=18.88$, $p<0.001$). Roofs reduced PAR by around 16% (corrected mean). The analysis of the reduced dataset (time span between 11 am and 3 pm) revealed only an effect of the roof treatment ($F_{2,85.4}=34.15$, $p<0.001$), not of the time, indicating that the effect on PAR over the whole day was mainly determined by the difference between noon and the rest of the day. During noon roofs reduced PAR by ~24%.

Table S1 Nested design and statistical model specification for all response variables used in this study. Nested structure gives information on the number of plots were used, whether they were nested in blocks and whether data were time series. All variables and contrasts used in the fixed term and the structure for the random term of the mixed effects models are listed.

Table S2 Fold changes (mean log response ratio, logRR) in metabolite levels due to pure drought (drought/ roofed control) and roof artifacts (roofed control/ ambient). Only metabolites with significant differences due to either drought or roof artifact effects obtained by mixed effects models are presented. Significance of results is indicated by bold lettering.

Acknowledgments

We thank Pete Manning, who gave us important technical advice on the planning of the roof constructions and also very constructive comments to improve the manuscript as well as one anonymous reviewer for comments. We especially want to thank Victor Malakhov, Kyle Siefers, Kymbat Dikambaeva, Yana Simeonova, Carsten Jesch and the gardeners who did the data sampling and maintained the roof installations.

Author Contributions

Conceived and designed the experiments: AV TF NE AW MSL. Performed the experiments: AV AW TF. Analyzed the data: BS AV TF. Contributed reagents/materials/analysis tools: TF. Wrote the paper: AV TF. Commented on and improved earlier versions of the manuscript: NE BS AW MSL WWW.

References

1. Hooper DU, Adair EC, Cardinale BJ, Byrnes JEK, Hungate BA, et al. (2012) A global synthesis reveals biodiversity loss as a major driver of ecosystem change. Nature 486: 105-U129.
2. Christensen JH, Hewitson B, Busuioc A, Chen A, Gao X, et al. (2007) Reginal Climate Projections. In: Solomon S, Qui D, Manning M, Chen Z, Marquis M et al., editors. Climate Change 2007: The Physical Science Basis Contribution of Working Group I to the Fourth Assessment Report of the Intergovernmental Panel on Climate Change. Cambridge, UK and New York, USA: Cambridge University Press.
3. Beier C, Beierkuhnlein C, Wohlgemuth T, Penuelas J, Emmett B, et al. (2012) Precipitation manipulation experiments - challenges and recommendations for the future. Ecology Letters 15: 899–911.
4. Cardinale BJ, Duffy JE, Gonzalez A, Hooper DU, Perrings C, et al. (2012) Biodiversity loss and its impact on humanity. Nature 486: 59–67.
5. Fraser LH, Al Henry H, Carlyle CN, White SR, Beierkuhnlein C, et al. (2013) Coordinated distributed experiments: an emerging tool for testing global hypotheses in ecology and environmental science. Frontiers in Ecology and the Environment 11: 147–155.
6. Gilgen AK, Buchmann N (2009) Response of temperate grasslands at different altitudes to simulated summer drought differed but scaled with annual precipitation. Biogeosciences 6: 2525–2539.
7. Yahdjian L, Sala OE (2002) A rainout shelter design for intercepting different amounts of rainfall. Oecologia 133: 95–101.
8. Svejcar T, Angell R, Miller R (1999) Fixed location rain shelters for studying precipitation effects on rangelands. Journal of Arid Environments 42: 187–193.
9. McGeoch MA, Le Roux PC, Hugo EA, Chown SL (2006) Species and community responses to short-term climate manipulation: Microarthropods in the sub-Antarctic. Austral Ecology 31: 719–731.

10. Bütof A, von Riedmatten LR, Dormann CF, Scherer-Lorenzen M, Welk E, et al. (2012) The responses of grassland plants to experimentally simulated climate change depend on land use and region. Global Change Biology 18: 127–137.
11. Carlyle CN, Fraser LH, Turkington R (2011) Tracking Soil Temperature and Moisture in a Multi-Factor Climate Experiment in Temperate Grassland: Do Climate Manipulation Methods Produce their Intended Effects? Ecosystems 14: 489–502.
12. Kahmen A, Perner J, Buchmann N (2005) Diversity-dependent productivity in semi-natural grasslands following climate perturbations. Functional Ecology 19: 594–601.
13. Jentsch A, Kreyling J, Elmer M, Gellesch E, Glaser B, et al. (2011) Climate extremes initiate ecosystem-regulating functions while maintaining productivity. Journal of Ecology 99: 689–702.
14. Solomon S, Qin D, Manning M, Alley RB, Berntsen T, et al. (2007) Technical Summary. In: Solomon S, D Qin, M Manning, Z Chen, M Marquis, K.B Averyt, M Tignor, H.L Miller, editors. Climate Change 2007: The Physical Science Basis Contribution of Working Group I to the Fourth Assessment Report of the Intergovernmental Panel on Climate Change. Cambridge, United Kingdom and New York, NY, USA: Cambridge University Press.
15. Haywood JM, Bellouin N, Jones A, Boucher O, Wild M, et al. (2011) The roles of aerosol, water vapor and cloud in future global dimming/brightening. Journal of Geophysical Research-Atmospheres 116: 14.
16. Rustad LE (2008) The response of terrestrial ecosystems to global climate change: Towards an integrated approach. Science of the Total Environment 404: 222–235.
17. Aerts R (2006) The freezer defrosting: global warming and litter decomposition rates in cold biomes. Journal of Ecology 94: 713–724.
18. Bontti EE, Decant JP, Munson SM, Gathany MA, Przeszlowska A, et al. (2009) Litter decomposition in grasslands of Central North America (US Great Plains). Global Change Biology 15: 1356–1363.
19. Pfisterer AB, Schmid B (2002) Diversity-dependent production can decrease the stability of ecosystem functioning. Nature 416: 84–86.
20. Vogel A, Scherer-Lorenzen M, Weigelt A (2012) Grassland resistance and resilience after drought depends on management intensity and species richness. Plos One 7: e36992.
21. Wang YF, Yu SX, Wang J (2007) Biomass-dependent susceptibility to drought in experimental grassland communities. Ecology Letters 10: 401–410.
22. Fay PA, Carlisle JD, Knapp AK, Blair JM, Collins SL (2000) Altering rainfall timing and quantity in a mesic grassland ecosystem: Design and performance of rainfall manipulation shelters. Ecosystems 3: 308–319.
23. Fry EL, Manning P, Allen DGP, Hurst A, Everwand G, et al. (2013) Plant Functional Group Composition Modifies the Effects of Precipitation Change on Grassland Ecosystem Function. PLOS One 8: 14.
24. Roscher C, Schumacher J, Baade J, Wilcke W, Gleixner G, et al. (2004) The role of biodiversity for element cycling and trophic interactions: an experimental approach in a grassland community. Basic and Applied Ecology 5: 107–121.
25. Fester T, Fetzer I, Buchert S, Lucas R, Rillig MC, et al. (2011) Towards a systemic metabolic signature of the arbuscular mycorrhizal interaction. Oecologia 167: 913–924.
26. Sanchez DH, Lippold F, Redestig H, Hannah MA, Erban A, et al. (2008) Integrative functional genomics of salt acclimatization in the model legume Lotus japonicus. Plant Journal 53: 973–987.
27. Desbrosses GG, Kopka J, Udvardi MK (2005) Lotus japonicus metabolic profiling. Development of gas chromatography-mass spectrometry resources for the study of plant-microbe interactions. Plant Physiology 137: 1302–1318.
28. Tikunov Y, Lommen A, de Vos CHR, Verhoeven HA, Bino RJ, et al. (2005) A novel approach for nontargeted data analysis for metabolomics. Large-scale profiling of tomato fruit volatiles. Plant Physiology 139: 1125–1137.
29. Luedemann A, Strassburg K, Erban A, Kopka J (2008) TagFinder for the quantitative analysis of gas chromatography - mass spectrometry (GC-MS)-based metabolite profiling experiments. Bioinformatics 24: 732–737.
30. Kopka J, Schauer N, Krueger S, Birkemeyer C, Usadel B, et al. (2005) GMD@CSB.DB: the Golm Metabolome Database. Bioinformatics 21: 1635–1638.
31. Schauer N, Steinhauser D, Strelkov S, Schomburg D, Allison G, et al. (2005) GC-MS libraries for the rapid identification of metabolites in complex biological samples. Febs Letters 579: 1332–1337.
32. Kuhn M, Wing J, Weston S, Williams A, Keefer C, et al. (2012) caret: Classification and Regression Training. R package version 512–04.
33. R development core team (2011) R: A language and environment for statistical computing. Vienna, Austria: R Foundation for Statistical Computing.
34. Barker M, Rayens W (2003) Partial least squares for discrimination. Journal of Chemometrics 17: 166–173.
35. Jansen JJ, Smit S, Hoefsloot HCJ, Smilde AK (2010) The Photographer and the Greenhouse: How to Analyse Plant Metabolomics Data. Phytochemical Analysis 21: 48–60.
36. Oksanen J, Blanchet FG, Kindt R, Legendre P, Minchin PR, et al. (2012) vegan: Community Ecology Package. R package version 2.0–4.
37. Hollander M, Wolfe DA (1973) Nonparametric Statistical Methods. New York: John Wiley & Sons.
38. White SR, Carlyle CN, Fraser LH, Cahill JF (2012) Climate change experiments in temperate grasslands: synthesis and future directions. Biology Letters 8: 484–487.
39. Vicca S, Gilgen AK, Serrano MC, Dreesen FE, Dukes JS, et al. (2012) Urgent need for a common metric to make precipitation manipulation experiments comparable. New Phytologist 195: 518–522.
40. Shinoda M, Nachinshonhor GU, Nemoto M (2010) Impact of drought on vegetation dynamics of the Mongolian steppe: A field experiment. Journal of Arid Environments 74: 63–69.
41. Mikkelsen TN, Beier C, Jonasson S, Holmstrup M, Schmidt IK, et al. (2008) Experimental design of multifactor climate change experiments with elevated CO2, warming and drought: the CLIMAITE project. Functional Ecology 22: 185–195.
42. Grime JP, Brown VK, Thompson K, Masters GJ, Hillier SH, et al. (2000) The response of two contrasting limestone grasslands to simulated climate change. Science 289: 762–765.
43. Eisenhauer N, Cesarz S, Koller R, Worm K, Reich PB (2012) Global change belowground: impacts of elevated CO2, nitrogen, and summer drought on soil food webs and biodiversity. Global Change Biology 18: 435–447.
44. Sage RF, Kubien DS (2007) The temperature response of C-3 and C-4 photosynthesis. Plant Cell and Environment 30: 1086–1106.
45. Pearcy RW (1999) Responses of Plants to Heterogeneous Light Environments. In: F.I.Pugnaire, Valladares F, editors. Handbook of Functional Plant Ecology. New York: Marcel Dekker.
46. Austin AT, Vivanco L (2006) Plant litter decomposition in a semi-arid ecosystem controlled by photodegradation. Nature 442: 555–558.
47. Foereid B, Rivero MJ, Primo O, Ortiz I (2011) Modelling photodegradation in the global carbon cycle. Soil Biology & Biochemistry 43: 1383–1386.
48. Brandt LA, King JY, Hobbie SE, Milchunas DG, Sinsabaugh RL (2010) The Role of Photodegradation in Surface Litter Decomposition Across a Grassland Ecosystem Precipitation Gradient. Ecosystems 13: 765–781.
49. Ballare CL, Caldwell MM, Flint SD, Robinson A, Bornman JF (2011) Effects of solar ultraviolet radiation on terrestrial ecosystems. Patterns, mechanisms, and interactions with climate change. Photochemical & Photobiological Sciences 10: 226–241.
50. Butenschoen O, Scheu S, Eisenhauer N (2011) Interactive effects of warming, soil humidity and plant diversity on litter decomposition and microbial activity. Soil Biology & Biochemistry 43: 1902–1907.
51. Sanaullah M, Blagodatskaya E, Chabbi A, Rumpel C, Kuzyakov Y (2011) Drought effects on microbial biomass and enzyme activities in the rhizosphere of grasses depend on plant community composition. Applied Soil Ecology 48: 38–44.
52. Emmett BA, Beier C, Estiarte M, Tietema A, Kristensen HL, et al. (2004) The response of soil processes to climate change: Results from manipulation studies of shrublands across an environmental gradient. Ecosystems 7: 625–637.
53. Sanaullah M, Rumpel C, Charrier X, Chabbi A (2012) How does drought stress influence the decomposition of plant litter with contrasting quality in a grassland ecosystem? Plant and Soil 352: 277–288.
54. Yahdjian L, Sala O, Austin AT (2006) Differential controls of water input on litter decomposition and nitrogen dynamics in the Patagonian steppe. Ecosystems 9: 128–141.
55. Merchant A, Richter AA (2011) Polyols as biomarkers and bioindicators for 21st century plant breeding. Functional Plant Biology 38: 934–940.

Natural Hazards in a Changing World: A Case for Ecosystem-Based Management

Jeanne L. Nel[1]*, David C. Le Maitre[1], Deon C. Nel[2], Belinda Reyers[1], Sally Archibald[3,4], Brian W. van Wilgen[1,5], Greg G. Forsyth[1], Andre K. Theron[1], Patrick J. O'Farrell[1], Jean-Marc Mwenge Kahinda[4], Francois A. Engelbrecht[4], Evison Kapangaziwiri[4], Lara van Niekerk[1], Laurie Barwell[1]

1 Natural Resources and the Environment, Council for Scientific and Industrial Research (CSIR), Stellenbosch, Western Cape, South Africa, **2** World Wide Fund for Nature (WWF), Cape Town, Western Cape, South Africa, **3** School of Animal, Plant and Environmental Sciences, University of the Witwatersrand, Johannesburg, Gauteng, South Africa, **4** Natural Resources and the Environment, Council for Scientific and Industrial Research (CSIR), Pretoria, Gauteng, South Africa, **5** Centre for Invasion Biology, Department of Botany and Zoology, Stellenbosch University, Stellenbosch, Western Cape, South Africa

Abstract

Communities worldwide are increasingly affected by natural hazards such as floods, droughts, wildfires and storm-waves. However, the causes of these increases remain underexplored, often attributed to climate changes or changes in the patterns of human exposure. This paper aims to quantify the effect of climate change, as well as land cover change, on a suite of natural hazards. Changes to four natural hazards (floods, droughts, wildfires and storm-waves) were investigated through scenario-based models using land cover and climate change drivers as inputs. Findings showed that human-induced land cover changes are likely to increase natural hazards, in some cases quite substantially. Of the drivers explored, the uncontrolled spread of invasive alien trees was estimated to halve the monthly flows experienced during extremely dry periods, and also to double fire intensities. Changes to plantation forestry management shifted the 1:100 year flood event to a 1:80 year return period in the most extreme scenario. Severe 1:100 year storm-waves were estimated to occur on an annual basis with only modest human-induced coastal hardening, predominantly from removal of coastal foredunes and infrastructure development. This study suggests that through appropriate land use management (e.g. clearing invasive alien trees, re-vegetating clear-felled forests, and restoring coastal foredunes), it would be possible to reduce the impacts of natural hazards to a large degree. It also highlights the value of intact and well-managed landscapes and their role in reducing the probabilities and impacts of extreme climate events.

Editor: Vanesa Magar, Centro de Investigacion Cientifica y Educacion Superior de Ensenada, Mexico

Funding: Financial support was provided by the CSIR (www.csir.co.za), Santam Limited (santam.co.za) and the Global Environmental Facility under the auspices of UNEP's Project for Ecosystem Services (www.unep.org). The funders had no role in study design, data collection and analysis, decision to publish, or preparation of the manuscript.

Competing Interests: The authors bring to your attention to their co-funded collaboration in this study with Santam insurance – a commercial enterprise. Santam's role in this study was to help select a suitable study site and then to mainstream the knowledge the authors generated into their organisation and the broader insurance sector. The authors would like to emphasize that, once the study site was chosen, Santam had no role in study design, data collection and analysis, decision to publish, or preparation of the manuscript. Santam's interest in this study was not for direct financial gain, but rather to use as 'proof-of-concept' to inform further in-house business research on how to improve risk assessment techniques in a changing world. Both organizations in the collaboration (Santam and CSIR) committed similar amounts of funding, and the relationship with Santam insurance and the authors can be more correctly described as a 'knowledge partnership' than a 'client-customer relationship'.

* E-mail: jnel@csir.co.za

Introduction

Since the turn of the century, several major natural disasters have attracted international attention, including hurricanes Katrina and Sandy, floods in Thailand and the Indian Ocean tsunami. These disasters, along with countless more frequent disasters of smaller magnitude, have been responsible for the loss of at least a million lives over the last decade, with recovery often taking years and financial losses estimated to be in the trillions of US dollars [1]. Such disasters occur when extreme physical events – or 'natural hazards' – impact adversely on vulnerable and exposed communities and infrastructure, which are the human elements of disaster [2]. As a result, disaster risk is affected by changes in the incidence of natural hazard, as well as alterations in the patterns of societal exposure and vulnerability. This paper deals specifically with changes in the incidence of natural hazards, which are expected to increase into the future. Changes in climate are expected to result in higher sea levels and increased hurricane activity, bringing more frequent and severe storm-waves that flood and erode coastal areas [2,3]. In many regions of the world, climate-induced shifts in the water cycle will result in more frequent and intense periods of flooding and drought [4,5]. Wildfires are also expected to become more widespread and frequent, being closely linked to hot, dry weather conditions and drought [6].

The anticipated increased incidence of natural hazards is not only attributed to climate change. There is a growing concern that rapid and widespread land cover change is leading to the loss of the buffering capacity that healthy ecosystems provide against these natural hazards [7]. For example, healthy mangrove ecosystems, coral reefs and coastal foredunes are able to dissipate wave energy, reducing the impacts of coastal flooding and erosion during storms [8,9]. Areas with intact mangroves were much less affected by the 1999 and 2004 Indian Ocean tsunamis, than areas where mangroves had been removed [10]. Likewise, healthy inland wetland ecosystems and riparian zones help to absorb peak flows and sediment during extreme rainfall events, reducing flooding and sedimentation hazards to downstream areas [11,12]. Invasion of natural vegetation by alien trees that use more water than the indigenous vegetation that they replace, exacerbates the effects of water scarcity in drought-prone regions [13,14]. Invasive alien trees can also increase fuel loads and thus wildfire hazard [15], or can turn an ecosystem that was not prone to burning into a flammable landscape [16].

This growing body of evidence that land cover change influences the frequency and severity of natural hazards presents new opportunities for managing and reducing risks faced by society, and forms the foundation of ecosystem-based approaches to adaptation. These approaches seek to manage, conserve or restore ecosystems and their associated ecosystem services to help people cope with the impacts of climate change [17]. Ecosystem-based adaptation approaches can be used to complement or replace technological or engineering solutions to adaptation, and often present more cost-effective, self-sustaining and flexible alternatives in the long term [18,19,20]. Apart from enhancing the buffering capacity of ecosystems, these approaches often come with multiple co-benefits to humans, such as improved fisheries production [21], timber harvesting [22], biodiversity conservation [23], and recreational value [24].

A key challenge to incorporating ecosystem-based approaches into disaster risk reduction is quantifying the extent to which land cover changes influence the occurrence and consequences of natural hazards. Clear examples are needed to quantify the benefits of ecosystem-based approaches to disaster risk reduction. Such assessments will require, *inter alia*, understanding the multiple land cover and climate drivers of natural hazards, quantifying how ecosystems will respond to changes in these drivers, and examining trade-offs between ecosystem-based management relative to alternative solutions [19]. This paper describes a 'proof of concept' study which aimed to address the first two of these challenges, exploring how land cover and climate change might affect four natural hazards – floods, droughts, wildfires and storm-waves – in the south coastal region of South Africa. These natural hazards frequently affect southern Africa's emerging economies and vulnerable communities. We limit the scope of this paper to natural hazards, but in the discussion reflect on translating natural hazards to disaster risk, which is the product of the likelihood of a natural hazard event occurring and its consequence on society. The intention of this proof of concept was to inform local authorities and businesses about local land cover drivers in their region and their potential for reducing the risk posed by natural hazards, thus contributing to comprehensive strategies for disaster risk reduction and climate change adaptation. But beyond informing local and national stakeholders in South Africa, the methods and lessons developed in this study have broader implications for ecosystem-based approaches and their evidence-base globally [19].

Methods

Study Area

The study area, hereafter 'Eden' (Figure 1), is 3 820 km^2 in size and comprises the local authorities along the southern Cape coast of South Africa within the Eden District Municipality (viz. Mossel Bay, George, Knysna and Bitou municipalities). Eden regularly experiences extremely heavy rainfall events which, combined with the small, steep catchments in the coastal areas, often results in high runoff and flash flooding [25]. The most severe incidences of flooding generally coincide with large storm-waves associated with cut-off low atmospheric pressure systems over southern Africa [26]. Floods are also interspersed with prolonged periods of extremely low rainfall, and Eden is frequently declared a disaster area due to persistent drought conditions [27]. The area also naturally experiences moderately frequent fires (every 10–13 years) due to the co-occurrence of its indigenous, flammable vegetation ('fynbos'), periods of hot, dry weather, and readily available sources of ignition [28]. Intense wildfires pose significant risks when associated with high population densities, and the intensity of wildfires is further exacerbated by invasive alien trees, which increase the fuel loads [29,30]. Natural hazards in Eden coincide with diverse socio-economic contexts, which results in inequalities to prepare for, cope with and adapt to disasters. Direct damage costs between 2003 and 2008 were estimated to be more than 3.5 times higher than the average annual household income in the most vulnerable and exposed communities, providing an indication of the vulnerability of some resident communities [25].

Approximately 68% of Eden's surface area is covered by indigenous vegetation, with agriculture (lucerne, vegetables and hops) and timber plantations respectively comprising 17% and 12% of Eden [31]. Urban areas are concentrated along the coast, and estimated to occupy just over 2% of the area. Eden has a 26% population growth rate [32], which is well above the national growth statistic of 15%, and which places considerable pressure on both natural resources and built infrastructure in the region [27]. Almost all the remaining indigenous fynbos (98%) is invaded by alien plants to some extent, which translates to approximately 12% of Eden being invaded at 100% density [31].

Climatically, Eden is located in a transition zone between a winter and summer rainfall regime [33]. Rainfall occurs throughout the year, peaking in autumn (March) and spring (October) with the lowest monthly rainfall occurring in June. The spring and autumn rainfall peaks coincide with increased frequency of cut-off low pressure systems over southern Africa [26], and about one of every five of these brings flooding and damage to the coastal areas [34]. Being a transition zone makes the area particularly susceptible to climate change, because the climate is influenced by changes in the circulation systems of both climatic regimes. An increase in annual maximum temperatures of about 1.2°C has been observed since 1960 [35]. While trends in projected future annual rainfall are weak relative to temperature [36], the combined effects of temperature, rainfall and evapotranspiration are amplified in the hydrological cycle and can have profound impacts on water resources in southern Africa [37].

Climate Data and Natural Hazard Models

Four models were developed to examine how land cover and climate change influence the natural hazards of floods, droughts, wildfires and storm-waves in Eden. Particular sites and catchments within Eden were selected to facilitate the modelling of each hazard (Figure 1). We identified drivers of each hazard based on literature reviews and expert consultations, highlighting those which were identified as important, relevant, plausible, and for which adequate data were available. The climate change scenario was extracted from the A2 scenario of the IPCC Special Report on Emission Scenarios (SRES) [38], while land cover change scenarios were based on published information and expert knowledge of the region. We then modelled the effect of the scenarios of change on each natural hazard (Table 1; Table S1).

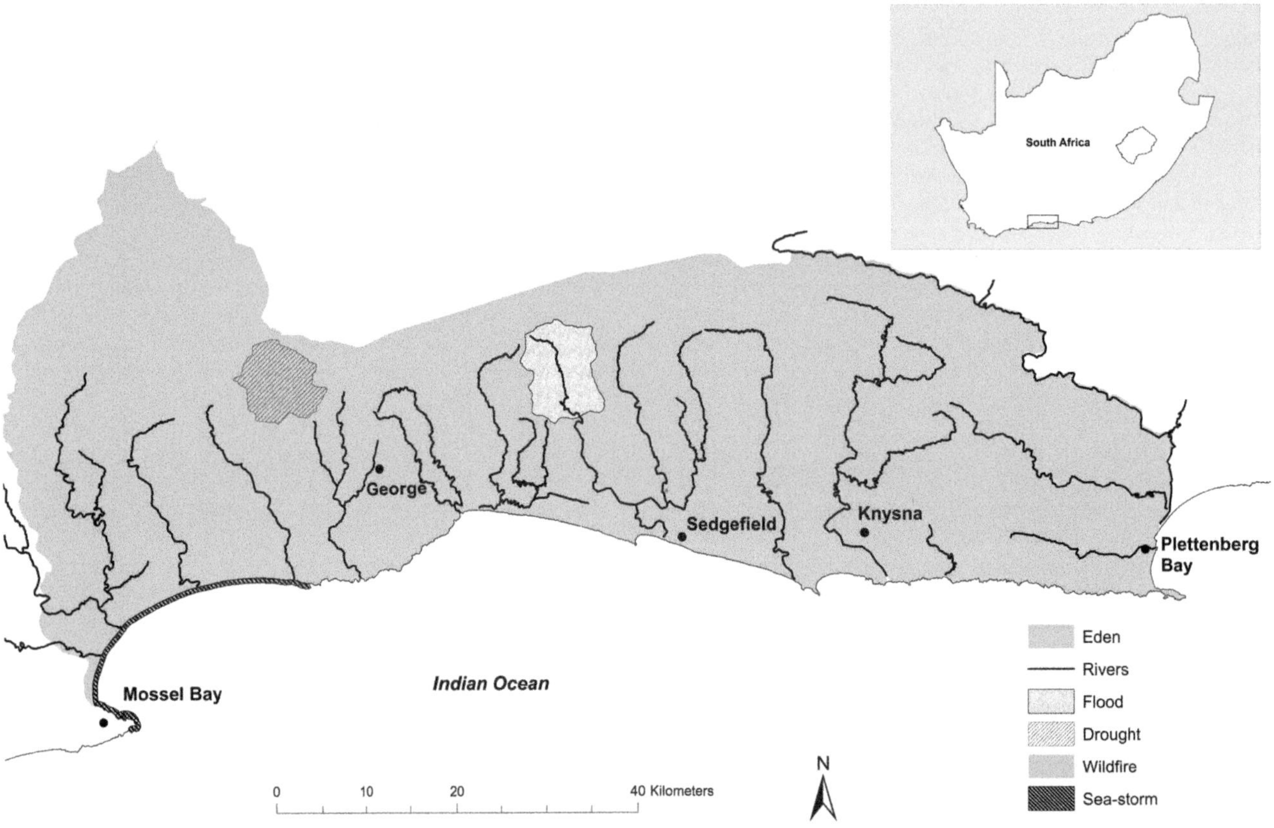

Figure 1. The location of the hazard model study areas within Eden. The inset showing the location of Eden in South Africa.

To explore the changes in flood, drought and storm-wave hazards, we produced simulations that respectively estimated return periods of extreme peak flow events, low flows and wave run-up events. The change in wildfire hazard was examined by calculating the change in fire intensities with future changes in fuel loads and climate.

Climate projections. Projections of future climate, used by the flood, drought and wildfire models, were obtained using a regional climate model, the conformal-cubic atmospheric model (CCAM) [39]. A multiple downscaling procedure was followed to obtain high resolution simulations of future climate change over southern Africa. In the first phase of the downscaling procedure, the sea-ice and bias-corrected sea-surface temperatures of the CSIRO Mark3.5 global climate model [3] was used as lower-boundary forcing in CCAM simulations performed globally at a quasi-uniform resolution of approximately 200 km [3,40]. The simulations used the SRES A2 scenario [38], which was the only detailed downscaled data for the southern African region. The A2 scenario represents a low mitigation, or high emissions, scenario implying rapid continued growth in greenhouse gas concentrations in the atmosphere. It is thus appropriate in examining the impacts of extreme climate events, although we do note throughout the limitations of only having one climate scenario available. CCAM was subsequently applied in a stretched-grid mode, to obtain simulations of approximately 60 km resolution in the horizontal, over southern Africa. This CCAM modelling procedure has been shown to provide satisfactory simulations of annual rainfall and temperature distributions, as well as the intra-annual cycle in rainfall and circulation over the southern African region [40]. It also realistically simulates observed daily climate statistics over southern Africa, such as the frequency of occurrence of extreme precipitation events, and the cut-off low atmospheric pressure systems and tropical cyclones [40,41]. Climate data required for each natural hazard model were extracted for the period 1961–2050, which included daily data for: rainfall, maximum temperature, wind speed, and relative humidity (Dataset S1).

Modelling flood hazard. We developed a hydrological model for an upper catchment within Eden (Figure 1) that drains into a coastal lake near the low-lying town of Sedgefield, which has in the past been highly susceptible to flooding during heavy rains. The model explored how inflows to the lake are affected by catchment land cover and climate change. This upper catchment was chosen because it had reliable gauging weir data for calibrating simulated flows, and included the typical land cover found in Eden.

The agrohydrological modelling system (ACRU) was used to simulate daily flows based on climate data, physical catchment characteristics (particularly soils) and land cover [42]. ACRU contains a soils and vegetation database that was developed specifically for South African conditions, and has been widely used, both locally and elsewhere [43,44]. We used this database, together with 30 m resolution land cover data [45], to identify five 'hydrological response units': pine plantation, clear-felled pine plantation, wattle plantation, indigenous forest and fynbos. We constructed a model for baseline conditions using daily temperature for 1961–1990 from our climate model, and daily rainfall [46] for the same period (Table 1). The latter was considered to better reflect the rainfall gradients within the catchment than the relatively coarse resolution of our climate model. All hydrological response units fed directly into the outflow point at the gauging

Table 1. Scenarios of land cover and climate change used to quantify changes to flood, drought, wildfire and storm-wave hazards.

Hazard	Scenario
Flood	1. BASELINE: current land cover and climate
Flood	2. Clear-felling and non-replanting of the plantation; current climate
Flood	3. Burning down of the plantation by a moderate wildfire with replacement by degraded fynbos; current climate
Flood	4. Burning down of the plantation by a severe wildfire with replacement by degraded fynbos; current climate
Flood	5. Current land cover; future climate
Drought	6. BASELINE: Indigenous natural vegetation with no invasive alien plants or human activities; current climate
Drought	7. Alien trees invade to maximum potential; current climate
Drought	8. Indigenous natural vegetation with no invasive alien plants or human activities; future climate
Wildfire	9. BASELINE: Current levels of invasion by alien trees; current climate
Wildfire	10. Alien trees invade to maximum potential; current climate
Wildfire	11. Current levels of invasion by alien trees; future climate
Wildfire	12. Alien trees invade to maximum potential; future climate
Wildfire	13. Alien trees and shrubs are cleared and maintained at levels below 5% cover; future climate
Storm-wave	14. BASELINE: Current beach slope; current climate
Storm-wave	15. 3° increase in beach slope; current climate
Storm-wave	16. Current beach slope; future climate

More detailed descriptions of each scenario and the associated data used are available in Table S1.

weir, so that simulated flows could be calibrated against observed flows (Table 1). The initial settings in the ACRU model were based on default values in the model for land cover classes and soils [47]. The simulated flows were much greater than the observed flows and too responsive to rainfall events so: (i) soil and effective rooting depths were increased to accommodate regional variations in soils and greater rooting depths for fynbos and pine trees [43]; and (ii) quickflow and baseflow response coefficients were altered according to Royappen et al. [48] for this catchment. These adjustments brought the mean simulated flows to within 10% of the observed daily flows, which were used to describe the baseline for runoff events.

We used this calibrated model to explore the implications of different forestry management practices and climate change on flood events (Table 1). Future daily temperature data for 2021–2050 were extracted from our climate model. To account for the inadequate rainfall resolution, future daily rainfall was calculated using a proportional adjustment to the current daily rainfall data [46], based on regional trends in current and future daily rainfall from the climate model used here. Regional trends were calculated as proportional differences for modelled daily rainfall percentiles and then applied to the corresponding percentiles of the observed data to generate the future rainfall for 1991–2020 and 2021–2050. The exceedance probabilities for extreme flows for both current (1961–1990) and future (2021–2050) climate conditions were calculated from the simulated flow record using the Log Pearson III distribution, which is widely used for calculating extreme values and their return intervals [49]. An extreme rainfall event in the climate model dataset was defined as 25 mm of rain falling within 24 hours over a grid cell of 0.5° latitude $\times$0.5° longitude.

Modelling drought hazard. Many definitions of drought exist – e.g. 'meteorological', 'atmospheric', 'hydrological', 'agricultural' and 'water management' droughts [50] – each differing in their emphasis on describing the characteristics and causes of drought, vulnerability to drought, or impacts of drought. Here, we study the effects of land cover and climate on hydrological drought, which is defined as a persistently low discharge or volume of water in streams or reservoirs, lasting months or years. Although hydrological drought is a natural phenomenon underpinned by dry climate, it is greatly exacerbated by human land use activities, which often affect the magnitude and frequency of the drought [50]. We used flow duration curves to explore changes in hydrological patterns. Flow duration curves plot the percentage of time a flow exceeds a certain threshold, with flows between 70–99% exceedance depicting low flows [51], and flows with >90% exceedance describing extreme low flows and taken here to represent drought.

Flow simulations used to develop flow duration curves were derived for a headwater sub-catchment within Eden (Figure 1) using the Pitman model [52,53,54]. This sub-catchment supplies water for high-value hops farming, and inadequate flow, particularly during the dry season, is frequently problematic [27]. The Pitman model is a conceptual monthly time-step rainfall-runoff model that has been frequently used for water resource assessments in southern Africa for many years and has become a standard method used by many practitioners. It includes explicit routines to simulate interception, infiltration, excess surface runoff, soil moisture (or unsaturated zone) runoff, groundwater recharge and drainage to stream flow, as well evaporative losses from the unsaturated zone and the groundwater storage in the vicinity of the river channel. We used a physically-based parameter estimation procedure described previously [55,56], which uses physical property data at the sub-basin scale (typically 50–1000 km^2). The agricultural land types GIS layer for South Africa [57] provides much of the physical property data required, including soil depths for different parts of the sub-basin (hilltop, valley sides and valley bottoms), soil texture (translated into soil hydraulic properties), topographic slope, and sub-surface geological conditions. Five land types were identified in the study area (Db30, Db32, Fc42, Lb139 and Lb141), based on spatial land cover data [45], and these were used to estimate runoff generating, soil moisture accounting and groundwater parameters for the Pitman model. The model was then run to generate time-series of mean monthly runoff under baseline conditions (Table 1).

Resulting output ensembles were within the 90% confidence intervals of the regionalised mean annual runoff ratios, the gradient of the monthly flow duration curve, and within range of the three recharge estimates of the groundwater resource assessment study [58].

Using the calibrated Pitman model, we then explored changes in flow resulting from maximum potential invasion by alien trees and from climate change (Table 1). An ecological module was used to estimate invasive alien tree water use [59], which links back to the Pitman model to predict changes in streamflow [60]. Alien trees were only allowed to invade untransformed vegetation, which were calculated as those areas not classified as urban, agriculture, forestry plantation or waterbodies in the land cover GIS layer for South Africa [45]. We used monthly temperature and rainfall data from our climate model. The inadequate resolution of the rainfall data was addressed by scaling the projected future rainfall according to the monthly distribution statistics of historical rainfall data as outlined previously [61,62]. Changes in flow for each scenario were compared using flow duration curves, focusing on flows produced at >90% exceedance range.

Modelling wildfire hazard. The damage caused by a wildfire is directly related to Byram's fireline intensity [63], which measures the rate of energy released along the fire front. The relative effort required to control a fire, and the damage it does, are both strongly correlated with fireline intensity [64]. The higher the fireline intensity, the more difficult a fire is to control, which has important implications for human safety and fire-sensitive assets [65]. Wildfires are a regular occurrence in the fire-prone indigenous fynbos in Eden [28]. However, invasion by alien trees have the effect of increasing the above-ground biomass or fuel, and therefore fire intensity [29]. If uncontrolled, invasive alien trees will continue to spread leading to further increases in fire intensity. In addition, climate change is predicted to result in hotter, drier and windier weather, which will further increase the intensity of wildfires.

We used Byram's [63] equation (Equation 1) to estimate changes in fireline intensity across 106 spatial assessment units in Eden under different scenarios. These assessment units are similar-sized irregular areas (mean size = 49 km^2) that are nested within administrative and physiographic boundaries [66].

$$I = Hwr \tag{1}$$

Where I = fireline intensity in kW m^{-1};

H = heat yield (assumed to be constant at 20000 J g^{-1});

w = fuel loads in g m^{-2}; and

r = rate of fire spread in m s^{-1}

For each assessment unit, we assumed that the area covered by untransformed fynbos vegetation (as calculated for the drought model) would be available to burn in wildfires. We considered five scenarios, each with a unique combination of climate and invasive alien trees (Table 1).

We used recent spatial data [31] on the area invaded by pine trees and hakea shrubs (*Pinus* and *Hakea* species) to estimate fuel loads in each assessment unit under different scenarios. Pines have annual spread rates of 3.75–20.6% in the fynbos biome [67,68,69], and hakea can spread at 8% annually [70]. We assumed that pines and hakea would spread at a conservative rate of 4% per year into the available area within each assessment unit (to a maximum of 100% of untransformed vegetation only), and used this to estimate the area that would become invaded by 2050.

Fuel loads in fynbos vegetation are approximately 1 800 g m^{-2} [29]. Invasion by hakea and pines can increase these fuel loads to 3 900, and to 20 000 g m^{-2}, respectively [29,30]. The proportional mix of invasive species in Eden as a whole was 72.5% pine and 27.5% hakea, so we assumed a mean fuel load for assessment units of 15 572 g m^{-2}. Based on these biomass estimates, we used the relative proportions occupied by uninvaded and invaded fynbos respectively to estimate a mean fuel load for each assessment unit under the different scenarios.

We used data on fire weather conditions and associated rates of fire spread in fynbos [71] to establish the relationships between McArthur's Forest Fire Danger Index (FDI) [72] and rates of fire spread (Equation 2). McArthur's FDI is based on observed relationships between the behaviour of fires and the environmental conditions (air temperature, relative humidity and wind speed) under which they burn. The FDI provides an index of the degree of difficulty of suppressing a fire.

$$r = -3.2\left(1.1^{-0.2(FDI-8)}\right) + 3.5 \tag{2}$$

Where r = rate of fire spread in m s^{-1} and FDI = McArthur's Forest Fire Danger Index

Equation 2 was used to convert FDIs to estimates of current (1961–1990) and future (2011–2050) rates of fire spread using daily data from our climate model for temperature, wind speed, rainfall, and relative humidity. We then used the resultant mean rates of spread for current and future conditions (1.1 m s^{-1} and 1.2 m s^{-1} respectively) to estimate the fire intensity under different scenarios (Table 1).

Modelling storm-wave hazard. Storm-waves, in the southern African context, refer to extreme offshore wave events which result in wave impacts that are experienced at the shoreline. The severity of storm-wave impacts is highly dependent on wave run-up height (the maximum point that storm waves can reach on land) and coastal erosion potential [73]. Two models were developed to examine the spatial and temporal variation of wave run-up and coastal erosion potential within the study area and how these are likely to change in the future.

First, we used a numerical wave model, SWAN [74], to translate offshore wave data to inshore wave conditions. Offshore data on wave height, period and direction, from the National Centre for Environmental Prediction (NCEP), were used as input variables [75,76]. Simulations were run to determine inshore wave heights for various return periods. Simulations included offshore wave conditions that result from 1:10 year south-south-westerly swells, as these conditions result in severe inshore conditions in the study area. Second, we used the inshore wave height and period from the SWAN wave model simulations as input into the Nielsen and Hanslow model [77] to calculate wave run-up elevations at 0.5 km points along the coastline, together with corresponding return periods. This model requires information on inshore wave height and period, sea water level and beach slope, and has previously shown an acceptable prediction accuracy ($R^2 = 0.79$) for local South African conditions [73]. We used wave heights from the SWAN model simulations as inputs for the inshore wave height and period, and sea water level was based on spring high tidal level predictions. This is a relatively extreme but realistic scenario as spring high tides occur every 14 days along the coastline, making the chances of storm waves coinciding with spring high tides relatively high. Beach and inshore slopes were calculated using the distance to the 20 m depth contour obtained from the South African Navy's bathymetric charts and available 5 m contour intervals. Using these outputs, baseline conditions of wave run-up elevations and corresponding return periods were plotted for each 0.5 km location along the coast (Table 1; Figure 1).

These models were then used to examine the change in wave run-up elevations resulting from the future potential influence of anthropogenic effects on coastal erosion (resulting in a steeper slope), and the change in future wave climate and sea-level rise (Table 1). Anthropogenic effects on coastal erosion were simulated by assuming a 3° increase in beach slope which was used as input into the Nielsen and Hanslow run-up model [77]. This is a conservative increase in beach slope that local coastal engineers considered to be a realistic effect of coastal erosion, which occurs as a consequence of human activities, e.g. hardening of the coastline through urban and industrial development, removal of coastal foredunes. Future climate conditions were modelled by assuming a 0.5 m rise in sea-level based on reviews of recent publications [78,79], and by applying a 6% increase to offshore extreme waves based on regional projections from metocean climate modelling [80]. We calculated expected changes in wave run-up elevation and return period up to the year 2100 and compared these to baseline conditions.

Results

The flood model for all land cover and future A2 climate scenarios showed an increase in extreme daily flows for equivalent return periods compared to the current baseline condition (Figure 2). Estimated extreme daily flows for a two-year return period showed a 16% increase from the baseline condition for the scenario of clear-felling without replanting. Similarly, the scenarios describing the burning down of the timber plantation under moderate and severe fires show an increase of approximately 24% and 32% respectively. The A2 scenario of future climate change also shows an increased trend in extreme daily flows, estimated to increase by over 24% of the baseline for the 50-year return period. Under the A2 scenario of climate change, flood return intervals are substantially reduced, doubling the frequency of 1:100 year floods to 1:50 years. Land cover changes also resulted in increases in flood frequency, shifting the 1:100 year flood event to about a 1:80 year return period in the most extreme scenario in which timber plantations were burnt down by a severe fire and replaced with degraded fynbos (Figure 2).

The full range of river discharges in the drought study is displayed in Figure 3, from the low flows to the high flows (broadly defined as flows respectively above or below 50% exceedance). Figure 3 shows that both invasion by alien trees and future climate change (under the A2 scenario) will exacerbate extreme low flows (flows >90% exceedance). Flow for alien trees at maximum potential invasion were consistently lower than baseline flows, halving the expected flow at 90% exceedance (0.02 million m^3 under baseline compared to 0.01 million m^3 under full invasion; Figure 3). While the A2 climate change scenario shows an overall trend of higher flows, its flow duration curve drops sharply below the baseline at 90% exceedance, suggesting that under this scenario of climate change much lower flows will be experienced during drought periods.

Mean fireline intensities of around 80 000 kW m^{-1} were estimated under current baseline conditions (Figure 4). This could increase by about 88% (to 150 000 kW m^{-1}) under a scenario in which invasive alien trees continue to spread unhindered, without any climate change, and could potentially more than double (to over 180 000 kW m^{-1}) if the combined effects of alien tree spread and the A2 scenario of climate change are taken into account. Under a hypothetical scenario in which climate change took place, but alien tree invasions remained at current levels, mean fireline intensities would increase by about 9% (to 87 000 kW m^{-1}) under the A2 scenario of climate change compared to current conditions. However, under a scenario in which alien trees are brought under control by reducing and maintaining them at below 5% cover, the estimated mean intensity of fires would be reduced by almost half (to 43 200 kW m^{-1}) when compared to the current situation.

Using a sandy beach location as an example of a typical area prone to storm-waves in Eden, simulations of wave run-up elevation for spring high tide and south-south-westerly swell conditions are currently predicted to range between approximately 5.7 m for a 1:1 year return period to 6.5 m for a relatively extreme event with a 1:50 year return (Figure 5). Under future wave climate and sea-level rise predictions, the 1:50 year wave run-up elevation will be reduced to a 1:3 year return interval at this location. Increasing beach slope by 3° to simulate conservative anthropogenic effects on coastal erosion, produced a substantial increase in wave run-up elevation for respective return periods. In this scenario, the wave heights of a 1:100 year return period are likely to occur on an annual basis.

Discussion

Natural Hazards are Increasing

Our results show that climate change in Eden will increase the frequency of all natural hazard events examined, substantially so in the case of floods, droughts and storm-waves (Figures 2–5), and to a lesser extent for wildfires (Figure 4). When looking beyond climate change impacts, drivers of land cover change appear to have a similar effect on increasing the incidence of natural hazards (Figures 2–5). Allowing the spread of invasive alien trees into untransformed vegetation was estimated to halve the monthly river flows experienced during drought and double fire-line intensities (Figures 3 and 4 respectively). In the case of wildfire (Figure 4), the fireline intensities for invaded fynbos are all orders of magnitude greater than the limits for effective fire control or suppression [81]. This poses significant threats to life and infrastructure [65], and can have significant hydrological impacts [82]. Similarly, severe 1:100 year storm-waves are estimated to occur on an annual basis with only modest human-induced coastal hardening and the removal of coastal foredunes (Figure 5). The impacts of land cover change on floods also attest to the impacts of land management practices on natural hazards by reducing the return time between large flood events by nearly 20% (Figure 2).

In interpreting these findings, we focus on broad trends rather than quantitative measurements because of the inherent uncertainty, gaps in data and models, and limited scenarios explored in predicting land cover and climate change impacts. With only one climate change scenario (focusing on a high emission, high risk scenario), we cannot compare land cover and climate change impacts with great certainty. However, we can point out that even under this extreme climate change scenario, land cover changes were shown to have as great and sometimes greater impacts on natural hazards, highlighting the importance of land cover management in reducing natural hazards. When considering the land cover scenarios, our focus was usually on single drivers of land cover change (e.g. spread of alien trees or forestry management practices); however, even within this small set of land cover drivers, the impacts were substantial on natural hazard incidence. As more climate scenarios become available for southern Africa, as well as data on other land cover and land use drivers relevant to natural hazards (e.g. agricultural practices, ground water abstraction), the additional impacts of these on natural hazards can be incorporated into our models and integrated into other ecosystem-based management options. In moving beyond this study's focus on general trends and evidence of importance of land cover and climate change drivers on natural hazards, future work on new

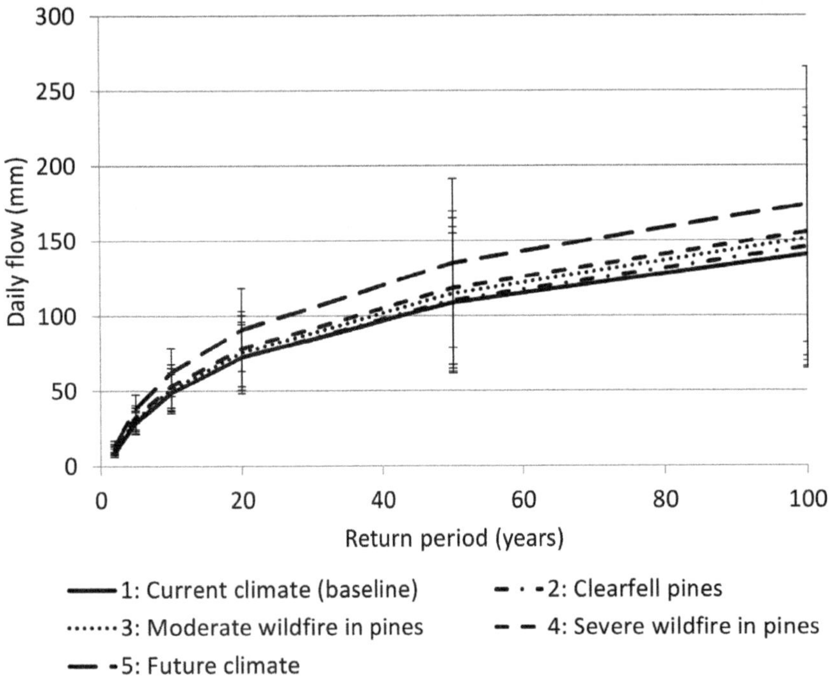

Figure 2. Flood return intervals for different scenarios of land cover and climate change. The numbers prefixing the annotated description of each scenario provides a reference to Table 1, which describes each scenario in more detail. The changes in the values for each return interval illustrate the potential changes in the likelihood of extreme flow events under the different scenarios. For example, the return period of a flood with a daily flow of 150 mm (similar to the May 1981 flood in this area) would decrease from a baseline of more than 100 years to 70 years under future climate (scenario 5).

drivers, data and scenarios would help in better reflecting the ranges of impacts, and the uncertainty associated with the predictions. This in turn will improve the ability to examine trade-offs of alternative solutions to disaster risk reduction and climate change adaptation.

Data and modelling techniques are also continuously evolving, and several improvements can be made to increase the confidence of the hazard models. For example, within specific natural hazard models (e.g. wildfire) the limited availability of fine scale data resulted in the use of a relatively coarse grid scale, possibly underestimating the impacts found. An analysis of extremes (rather than means) at a finer resolution may show higher climate-related change (see Figure S1). Moreover, models that explore interactions between drivers and between natural hazards would be more informative than the separate models produced here. For example, droughts and wildfire are closely linked hazards with similar climate drivers (warmer and drier conditions). Droughts can exacerbate wildfires because they usually increase dry, highly flammable standing biomass [6]. Severe wildfires also exacerbate floods, especially when they precede the onset of rains. This is particularly the case when wildfires occur in densely invaded areas, which burn more intensely than indigenous vegetation, and which result in increased runoff caused by resin-induced water-repellency and associated reductions in infiltration [84].

Natural Hazards can be Reduced by Appropriate Land Management

This study makes a case for the incorporation of ecosystem-based management approaches, in tandem with other approaches (e.g. mitigation or engineering responses), into disaster risk reduction and climate change adaptation. Our findings show that through appropriate and pro-active land use management, it would be possible to reduce the impacts of natural hazards to a large degree. Clear-felling of timber plantations should ideally be associated with rehabilitation and re-vegetation to avoid increasing the flood hazard. Because the timber plantations (mostly *Pinus* sp) are similar to dense stands of invasive alien trees, the flood model also supports the clearing of invasive alien trees to reduce the hazard posed by flood events soon after a wildfire. Clearing invasive alien trees and restoring the natural fynbos vegetation is also an effective tool for reducing wildfire and drought hazards. The wildfire model shows that the impacts of climate change can be substantially reduced by clearing invaded areas and maintaining a healthy cover of indigenous vegetation, lowering the fireline intensity to half that of current levels (Figure 4).

The costs to clear existing invasive alien trees in Eden are much smaller than the estimated losses caused by damaging wildfires to, for example, timber plantations. Estimates of the cost of clearing of invasive trees range from US$ 100 ha^{-1} [85] to US$ 800 ha^{-1} [86], giving estimates of between US$ 2.3–19.2 million to clear in the Eden area. A single wildfire in 2007 caused losses of US$ 200 million to the local timber industry [87]. Similarly, estimated costs of clearing were estimated to be lower than the economic impact invasive alien trees have on the hops industry in Eden (c.a. US$ 250 000 per year for 15 years, compared to US$ 350 000 per year for perpetuity to cover the additional groundwater pumping costs) [88]. Similar invasive alien tree clearing initiatives have been proposed to lower the long term economic impacts of natural hazards to the fruit industry in Eden [27].

Human activities that harden the coastline exacerbate beach erosion, thereby increasing beach slope and wave run-up [8]. Coastal foredunes are the South African equivalent of salt marsh and mangrove wetlands that offer protection from hurricane storm-surge or tsunamis elsewhere in the world [8,21]. Maintain-

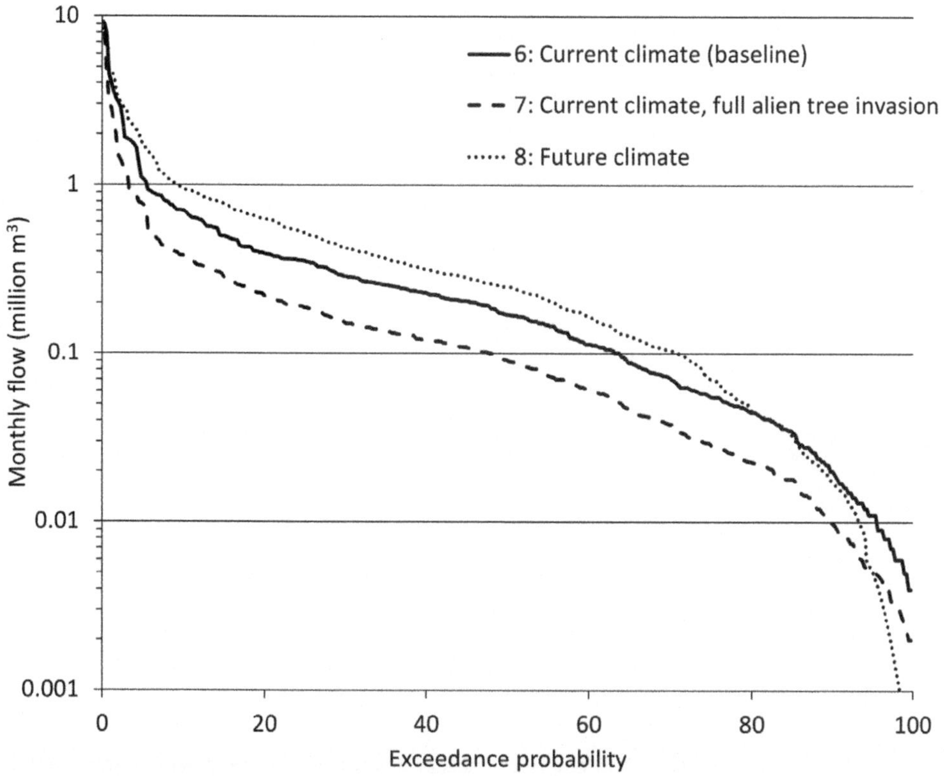

Figure 3. Flow duration curve for different scenarios of land cover and climate change. This shows the cumulative proportion of the months where a flow exceeded a given discharge for the different scenarios. The numbers prefixing the annotated description of each scenario provides a reference to Table 1, which describes each scenario in more detail. Extreme low flows were defined as those with >90% exceedance, which were used in this study to represent severe drought conditions. A log-normal probability curve was used to allow the low and high flow ends of the plot to be more clearly displayed.

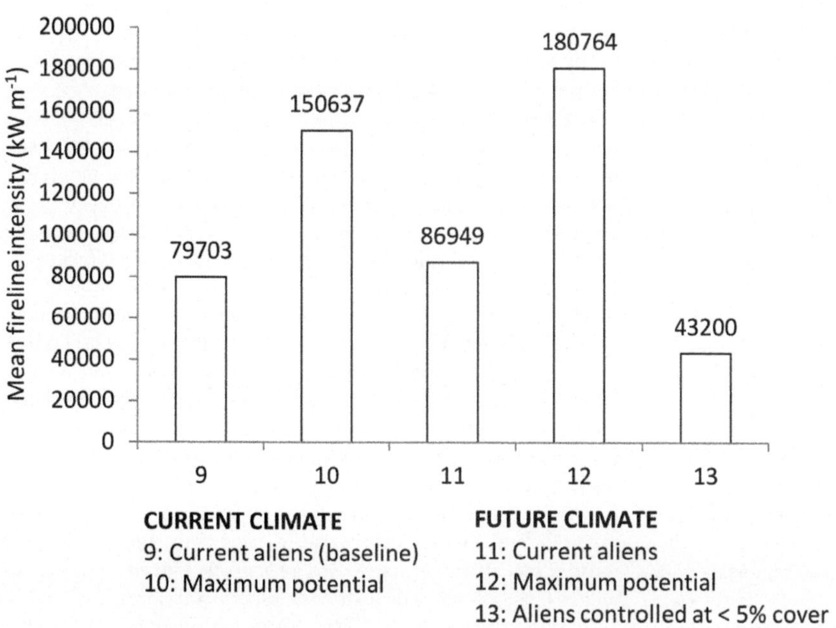

Figure 4. Estimated mean fireline intensity experienced in 106 assessment units for different scenarios of climate change and alien tree management. The numbers prefixing the annotated description of each scenario provides a reference to Table 1, which describes each scenario in more detail.

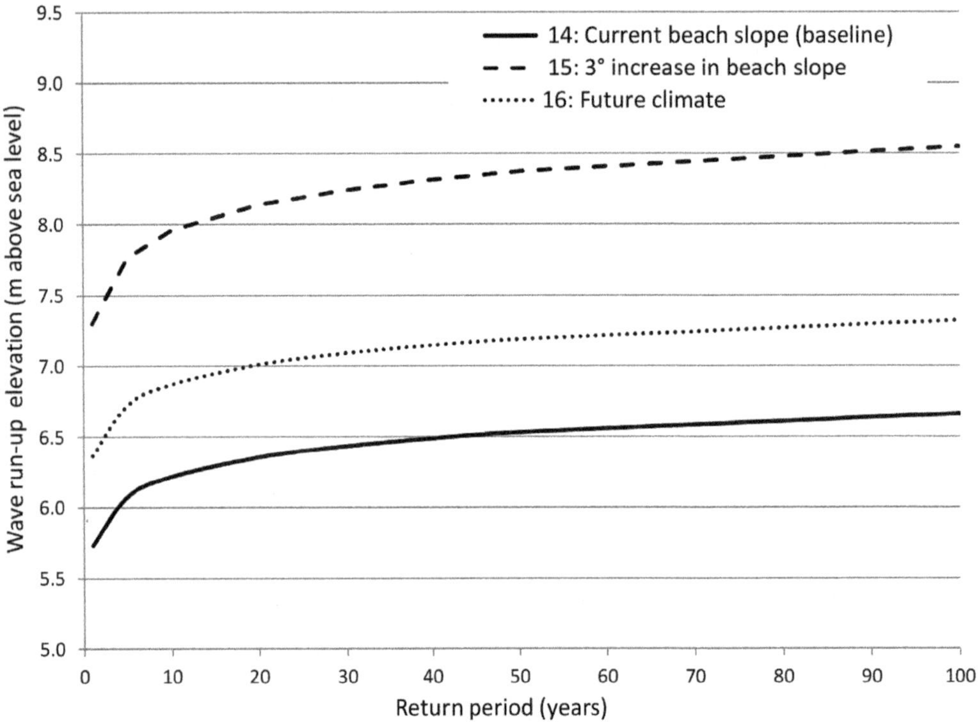

Figure 5. Wave run-up elevations for various storm-wave return intervals for different scenarios of beach slope and climate change. Simulations used here are for a typical sandy beach location in Eden (Tergniet, near Mossel Bay), which is prone to storm-wave damage. Return periods were based on the simulated wave run-up elevations for a south-south westerly swell, and spring high tide levels. The numbers prefixing the annotated description of each scenario provides a reference to Table 1, which describes each scenario in more detail.

ing natural vegetation, sand volume, and natural sediment movement, and restricting developments on foredunes can reduce impacts of wave run-up. Given that human-induced coastal erosion substantially influences the impacts of extreme storm-wave events (Figure 5), rehabilitation of foredunes should also be seriously considered as a means of reducing the impacts of climate change. Indeed, this is the strategy that has seen the US government investing billions of dollars in restoring the coastal salt marshes to protect the Gulf of Mexico from extreme events such as Hurricane Katrina in 2005 [21,89].

Way Forward in Eden and in Ecosystem-based Adaptation

The broad trends from this 'proof of concept' study have provided sufficient evidence to mobilise action within Eden, and several public-private initiatives have been launched to clear invasive alien trees, and restore catchments and foredunes. Implementation mechanisms have also been established to facilitate local action, including the appointment of a catchment manager by the hops farming industry, and the establishment of the 'Business Adopt a Municipality' forum in Eden, which explores how best the insurance sector can support local authorities and communities to manage their natural hazards and environmental risk.

While refining the hazard models is clearly important in identifying on the ground actions, perhaps more important is the need to understand how these natural hazards and associated trends express themselves as risk to communities in Eden, so that they can prepare for, cope with, and recover from disasters [90,91,92]. Disaster risk is widely accepted as the product of a natural hazard and its consequences on society [93]. The latter depends on the exposure and vulnerability of communities to natural hazards [3]. Eden has a naturally high relative exposure, and this coincides with particularly vulnerable communities living in the area [25]. Farmers are highly affected by repeated setbacks from droughts, floods and wildfires, and this has severe ripple effects on rural farm labourers and the entire local economy. This, in turn, increases the vulnerability of communities in Eden, making them more susceptible to subsequent disasters.

Disaster risk reduction is an activity that seeks systemic ways to reduce the severity or occurrence of natural hazards, as well as the consequences that such events have on people [93]. Eden has a relatively well-capacitated Disaster Risk Reduction unit compared to other districts in South Africa. However, much like in many parts of the world, efforts are still very much focussed on recovery from disaster (e.g. through providing disaster relief funding), or short-term disaster preparedness (e.g. through early warning systems, or ensuring adequate supply of fire engines). Longer term efforts to reduce risk are still lacking. These could include efforts to manage drivers of risk through appropriate land use management and restoration, as well as providing opportunities for social learning that promotes individual, collective and institutional capacity to manage risk [91]. Although the disaster risk framework in South Africa (Disaster Management Act No. 57 of 2002), and more widely, acknowledges both short and longer term efforts, there is still a lack of explicit budget allocated to longer term efforts. This expresses itself at the local level where authorities, such as the Eden District Municipality, that are interested in piloting new approaches, have no funding to do so. Interventions at all levels of governance (local, provincial and national) will be required to remove financial implementation barriers.

Conclusions

There is a need to build an evidence-base that addresses the potential of ecosystem-based adaptation approaches [19]. Our models and findings contribute to such a call. Our study shows that land cover change is as important as climate change in influencing the effects of natural hazards. These findings offer the Eden community an empowering message. Through pro-active management of key drivers of land cover change, they will be able to reduce the impacts of floods, droughts, wildfires and storm-waves. In considering the trade-offs of such ecosystem-based approaches with alternative forms of disaster risk reduction, the multiple co-benefits of ecosystem management and restoration need to be considered. For example, clearing invasive alien trees will reduce the impacts of drought, wildfire and flood hazards, create opportunities for employment in rural poor communities, and decrease the vulnerability of agricultural production and thus the overall local economy of the region. Promoting individual, collective and institutional capacity to deal with risk through social learning networks is also increasingly being recognised as an important long term strategy in disaster risk reduction [91,92]. Indeed, a recent study found that socio-institutional interventions tended to offer the most efficient climate change adaptation options in the city of Durban, South Africa [94]. The establishment of the public-private initiatives between land-owners, businesses and government agencies in Eden is a good step in this direction, as it offers the opportunity for cross-sectoral collaboration and social learning around disaster risk reduction.

Supporting Information

Figure S1 Fire Danger Index (FDI) calculated for the current (1960–1990) and future (2010–2050) time periods using the CCAM climate data for temperature, wind speed, rainfall, and relative humidity. Although there is a strong directional change in temperature, this trend is mediated by relatively small changes to wind speed and relative humidity. However, the small change to wind speed and relative humidity may be the result of an averaging effect, because both parameters are highly variable within the relatively coarse grid cells of the CCAM climate model. An analysis of extremes (rather than means) within the grid cell may show higher climate-related change.

Table S1 Scenarios of land cover and climate change used to quantify changes to flood, drought, wildfire and storm-wave hazards, together with the associated data used in each scenario.

Dataset S1 Climate model data used in the natural hazard models for the period 1961–2050, which included daily data for rainfall in mm over a 0.5° latitude×0.5° longitude grid cell (Rain), maximum temperature in °C (MaxT), wind speed in km/h (Wind), and percentage relative humidity (RH).

Acknowledgments

We would like to thank Janis Smith for producing the study area schematic (Figure 1).

Author Contributions

Conceived and designed the experiments: DCN JLN BR SA DCLM BWvW AT JMK LvN FE. Performed the experiments: FE DCLM SA BWvW GF AT LB JMK EK. Analyzed the data: JLN BR PO DCN DCLM BWvW AT JMK EK. Wrote the paper: JLN BR DCLM BWvW PO FE.

References

1. UNISDR (2013) From Shared Risk to Shared Value – the Business Case for Disaster Risk Reduction. Global Assessment Report on Disaster Risk Reduction. Geneva, Switzerland: United Nations Office for Disaster Risk Reduction (UNISDR). 288 p.
2. IPCC (2007) Impacts, adaptation and vulnerability. In: Parry M, Canziani O, Palutikof J, van der Linden P, Hanson C, editors. Cambridge: Cambridge University Press. 982 p.
3. IPCC (2012) Managing the risks of extreme events and disasters to advance climate change adaptation. In: Field CB, Barros V, Stocker TF, Qin D, Dokken DJ, et al., editors. Special report of working groups I and II of the intergovernmental panel on climate change. Cambridge: Cambridge University Press. 582 p.
4. Bates BC, Kundzewicz ZW, Wu S, Palutikof JP (2008) Climate Change and Water. Technical Paper of the Intergovernmental Panel on Climate Change, IPCC Secretariat, Geneva.
5. Milly PCD, Betancourt J, Falkenmark M, Hirsch RM, Kundzewicz ZW, et al. (2008) Stationarity is dead: Whither water management? Science 319: 573–574.
6. Bradstock RA, Cohn JS, Gill AM, Bedward M, Lucas C (2009) Prediction of the probability of large fires in the Sydney region of south-eastern Australia using fire weather. Int J Wildland Fire 18: 932–943.
7. MA 2005. Millennium Ecosystem Assessment: Ecosystems and human well-being: Biodiversity synthesis. Available: http://www.millenniumassessment.org/documents/document.354.aspx.pdf. Accessed August 2013.
8. Shephard CC, Crain CM, Beck MW (2011) The protective role of coastal marshes: a systematic review and meta-analysis. PLoS ONE 6: e27374.
9. Day JW Jr, Boesch DF, Clairain EJ, Kemp GP, Laska SB, et al. (2007) Restoration of the Mississippi Delta: Lessons from Hurricanes Katrina and Rita. Science 315: 1679–1684.
10. Das S, Vincent JR (2009) Mangroves protected villages and reduced death toll during Indian super cyclone. PNAS 106: 7357–7360.
11. Bullock A, Acreman M (2003) The role of wetlands in the hydrological cycle. Hydrol Earth Syst Sc 7: 358–389.
12. Zedler JB, Kercher S (2005) Wetlands resources: status, trends, ecosystem services and restorability. Annu Rev Environ Resour 30: 39–74.
13. Görgens AHM, van Wilgen BW (2004) Invasive alien plants and water resources in South Africa: current understanding, predictive ability and research challenges. S Afr J Sci 100: 27–33.
14. Ehrenfeld JG (2010) Ecosystem consequences of biological invasions. Annu Rev Ecol Syst 41: 59–80.
15. Brooks ML, d'Antonio CM, Richardson DM, Grace JB, Keeley JE, et al. (2004) Effects of invasive alien plants on fire regimes. BioScience 54: 677–688.
16. d'Antonio C, Vitousek P (1992) Biological invasions by exotic grasses, the grass/fire cycle, and global change. Annu Rev Ecol Syst 23: 63–87.
17. Munang R, Thiaw I, Alverson K, Mumba M, Liu J, et al. (2013) Climate change and ecosystem-based adaptation: a new pragmatic approach to buffering climate change impacts. Curr Opin Environ Sustain 5: 67–71.
18. Aronson J, Milton SJ, Blignaut JN (2007) Restoring natural capital. Science, business and practice. Washington: Island Press.
19. Jones HP, Hole DG, Zavaleta ES (2012) Harnessing nature to help people adapt to climate change. Nature Clim Change 2: 504–509.
20. van Slobbe HJ, de Vriend S, Aarninkhof K, Lulofs M, de Vries M, et al. (2013) Building with nature: in search of resilient storm surge protection strategies. Nat Hazards 65: 947–966.
21. Barbier EB, Hacker SD, Kennedy C, Koch EW, Stier AC, et al. (2011) The value of estuarine and coastal ecosystem services. Ecol Monogr 81: 169–193.
22. Moberg F, Rönnbäck P (2003) Ecosystem services of the tropical seascape: interactions, substitutions and restoration. Ocean Coast Manage 46: 27–46.
23. Yu X, Jiang L, Li L, Wang J, Wang L, et al. (2009). Freshwater management and climate change adaptation: experiences from the central Yangtze in China. Clim Dev 1: 241–248.
24. Dubgaard A (2004) Cost-benefit analysis of wetland restoration. J Water Land Dev 8: 87–102.
25. RADAR (2010) RADAR Western Cape 2010: Risk and Development Annual Review. Disaster Mitigation for Sustainable Livelihoods Programme, University of Cape Town, South Africa: PeriPeri Publications. 104 p.
26. Singleton AT, Reason CJC (2007) Variability in the characteristics of cut-off low pressure systems over subtropical southern Africa. Int J Climatol 27: 295–310.
27. Holloway A, Fortune G, Zweig P, Barrett L, Benjamin A, et al. (2012) Eden and Central Karoo drought disaster 2009–2011: The scramble for water. Report

Number LG 10/2011/2012, Disaster Mitigation for Sustainable Livelihoods Programme, Stellenbosch University, South Africa. 132p.
28. van Wilgen BW, Forsyth GG, de Klerk H, Das S, Khuluse S, et al. (2010) Fire management in Mediterranean-climate shrublands: a case study from the Cape fynbos, South Africa. Journal of Appl Ecol 47: 631–636.
29. van Wilgen BW, Richardson DM (1985) The effect of alien shrub invasions on vegetation structure and fire behaviour in South African fynbos shrublands: a simulation study. J Appl Ecol 22: 955–966.
30. Versfeld DB, van Wilgen BW (1986) Impacts of woody aliens on ecosystem properties. In: Macdonald IAW, Kruger FJ, Ferrar AA, editors. The ecology and control of biological invasions in South Africa. Cape Town: Oxford University Press. 239–246 pp.
31. Vromans DC, Maree KS, Holness S, Job N, Brown AE (2010) The Garden Route Biodiversity Sector Plan for the George, Knysna and Bitou Municipalities: Supporting land-use planning and decision-making in Critical Biodiversity Areas and Ecological Support Areas for sustainable development. Garden Route Initiative. South African National Parks, Knysna.
32. Statistics South Africa (2012) Census 2011 Municipal Report: Western Cape Statistics South Africa Report no. 03–01–49. Pretoria. 93 p.
33. Weldon D, Reason CJC (2013) Variability of rainfall characteristics over the South Coast region of South Africa. Theor Appl Climatol DOI 10.1007/s00704-013-0882-4.
34. Taljaard JJ (1985) Cut-off lows in the South African region. South African Weather Bureau Technical Paper No 14. Pretoria. 153 p.
35. Kruger AC, Shongwe S (2004) Temperature trends in South Africa: 1960–2003. International J Climatol 24: 1929–1945.
36. Kruger AC (2006) Observed trends in daily precipitation indices in South Africa: 1910–2004. Int J Climatol 26: 2275–2285.
37. Schulze RE (2000) Modelling hydrological responses to land use and climate change: a Southern African perspective. Ambio 29: 12–22.
38. Nakicenovic N, Alcamo J, Davis G, de Vries B, Fenhann J, et al. (2000) IPCC Special Report on Emissions Scenarios. Cambridge: Cambridge University Press. 599 p.
39. Mcgregor JL,Dix MR (2008) An updated description of the Conformal-Cubic Atmospheric Model. In: Hamilton K, Ohfuchi W, editors. High Resolution Simulation of the Atmosphere and Ocean. Springer Verlag. 51–76.
40. Engelbrecht FA, Landman WA, Engelbrecht CJ, Landman S, Bopape MM, et al. (2011) Multi-scale climate modelling over Southern Africa using a variable-resolution global model. Water SA 37: 647–658.
41. Malherbe J, Engelbrecht FA, Landman WA (2013) Projected changes in tropical cyclone climatology and landfall in the Southwest Indian Ocean region under enhanced anthropogenic forcing. Clim Dynam 40: 2867–2886.
42. Jewitt GPW, Schulze RE (1999) Verification of the ACRU model for forest hydrology applications. Water SA, 25: 483–489.
43. Smithers JC, Schulze RE (1995). ACRU Agrohydrological Modelling System: User Manual Version 3.00. Department of Agricultural Engineering, University of Natal, Pietermaritzburg, South Africa.
44. Warburton ML, Schulze RE, Jewitt GP (2010) Confirmation of ACRU model results for applications in land use and climate change studies 2. Change 2: 3.
45. van den Berg EC, Plarre C, van den Berg HM, Thompson MW (2008) The South African National Land Cover 2000. Agricultural Research Council (ARC) and Council for Scientific and Industrial Research (CSIR), Report No. GW/A/2008/86, Pretoria.
46. Lynch SD (2004) Development of a raster database of annual, monthly and daily rainfall for southern Africa. Report No. 1156/1/04, Water Research Commission, Pretoria.
47. Schulze RE, Maharaj M, Warburton ML, Horan MJC, Kunz RP, et al. (2008) South African Atlas Of Climatology and Agrohydrology. Report No. 1489/1/08, Water Research Commission, Pretoria.
48. Royappen M, Dye PJ, Schulze RE, Gush MB (2002) An analysis of catchment attributes and hydrological response characteristics in a range of small catchments. Report No. 1193/1/02, Water Research Commission, Pretoria.
49. US Interagency Advisory Committee on Water Data (1982) Guidelines for Determining Flood Flow Frequency. Bulletin 17B of the Hydrology Committee, US Geological Survey, Reston, VA, USA.
50. Wilhite DA, Glantz MH (1985) Understanding the drought phenomenon: The role of definitions. Water Int 10: 111–120.
51. Smakhtin VU (2001) Low flow hydrology: a review. J Hydrol 240: 147–186.
52. Pitman WV (1973) A mathematical model for generating monthly flows from meteorological data in South Africa. Report No. 2/73, Hydrological Research Unit, University of the Witwatersrand, South Africa.
53. Midgley DC, Pitman WV, Middleton BJ (1994) Surface Water Resources of South Africa 1990: User's Manual. Report no. 298/1/94, Water Resource Commission, Pretoria.
54. Middleton BJ, Bailey AK (2009) Water Resources of South Africa 2005 Study. Water Research Commission Report Number TT 380/08, Water Research Commission, Pretoria.
55. Kapangaziwiri E, Hughes DA (2008) Towards revised physically-based parameter estimation methods for the Pitman monthly rainfall-runoff model. Water SA 32: 183–191.
56. Hughes DA, Kapangaziwiri E, Sawunyama T (2010) Hydrological model uncertainty assessment in southern Africa. J Hydrol 387: 221–232.
57. AGIS (2007) Agricultural Geo-Referenced Information System. Available: http://www.agis.agric.za. Accessed August 2013.
58. DWAF 2005. Groundwater Resource Assessment II: Recharge Literature Study Report 3A. Report of Department of Water Affairs and Forestry, Pretoria.
59. Le Maitre D, van Wilgen BW, Chapman RA, McKelly DH (1996) Invasive plants in the Western Cape, South Africa: modelling the consequences of a lack of management. J Appl Ecol 33: 161–172.
60. Mallory S. (2011) Extent of Invasive Alien Plants and the Impact of Removal on the Water Resources of Olifants River catchment. Department of Water Affairs Report No. P WMA 04/B50/00/8310/3. Department of Water Affairs, Pretoria.
61. Hughes DA, Mantel SK, Slaughter A (2011) Developing climate change adaptation measures and decision support system for selected South African water boards. Water Research Commission Report K5/2018/2, Water Research Commission, Pretoria.
62. Hughes DA (2004) Problems of estimating hydrological characteristics for small catchments based on information from the South African national surface water resource database. Water SA 30: 393–398.
63. Byram GM (1959) Combustion of forest fuels. In: Davis KP, editor. Forest Fire: Control and Use, New York: McGraw- Hill. Pp. 155–182.
64. Mercer DE, Prestemon JP (2005) Comparing production function models for wildfire risk analysis in the wildland-urban interface. Forest Policy Econ 7: 782–795.
65. Gill AM, Stephens SL, Cary GJ (2013) The worldwide "wildfire" problem. Ecol Appl 23: 438–454.
66. Naude A, Badenhorst AZH, Van Huyssteen WE, Maritz J (2007) Technical Overview Of The Mesoframe Methodology and South African Geospatial Analysis Platform. CSIR Report CSIR/BE/PSS/IR/2007/0104/B, Council for Scientific and Industrial Research, Pretoria.
67. Richardson DM, Brown PJ (1986) Invasion of mesic mountain fynbos by Pinus radiata. South Afr J Bot 52: 529–536.
68. Higgins SI, Richardson DM, Cowling RM (2000) Using a dynamic landscape model for planning the management of alien plant invasions. Ecol Appl 10: 1833–1848.
69. Moeller J (2010) Spatial Analysis of Pine Tree Invasion in the Tsitsikamma Region, Eastern Cape, South Africa: A Pilot Study. Honours dissertation, Department of Geography, Rhodes University, Grahamstown, South Africa.
70. van Wilgen BW, de Wit MP, Anderson HJ, Le Maitre, DC, Kotze IM, et al. (2004). Costs and benefits of biological control of invasive alien plants: case studies from South Africa. S Afr J Sci 100: 113–122.
71. van Wilgen BW, Le Maitre DC, Kruger FJ (1985) Fire behaviour in South African fynbos (macchia) vegetation and predictions from Rothermel's fire model. J Appl Ecol 22: 207–216.
72. Noble IR, Gill AM, Bary GAV (1980) McArthur's fire danger meters expressed as equations. Austral Ecol 5: 201–203.
73. Theron AK, Rossouw M, Barwell L, Maherry A, Diedericks G, et al. (2010) Quantification of Risks to Coastal Areas and Development: Wave Run-up and Erosion. Proceedings CSIR 3rd Biennial Conference 2010, CSIR, Pretoria, South Africa.
74. Booij N, Ris RC, Holthuijsen LH (1999) A third-generation wave model for coastal regions, Part I: Model description and validation. J Geophys Res 104: 7649–7666.
75. Tolman HL, Balasubramaniyan B, Burroughs LD, Chalikov DV, Chao YY, et al. (2002) Development and Implementation of Wind-Generated Ocean Surface Wave Models at NCEP. Weather Forecast 17: 311–333.
76. NCEP (2013) Offshore data on wave height, period and direction. Available: http://www.nco.ncep.noaa.gov/pmb/products/wave/#WW3ENS. Accessed August 2013.
77. Nielsen P, Hanslow DJ (1991) Wave run-up distributions on natural beaches. J Coastal Res 7: 1139–1152.
78. Rossouw M, Theron AK (2012) Investigation of potential climate change impacts on ports and maritime operations around the southern African coast. In: Asariotis R, Benamara H, editors. Maritime Transport and the Climate Change Challenge. Routledge: United Nations Conference on Trade and Development (UNCTAD), Earthscan. 360 p.
79. Parris A, Bromirski P, Burkett V, Cayan D, Culver M, et al. (2012). Global Sea Level Rise Scenarios for the US National Climate Assessment. NOAA Tech Memo OAR CPO-1. 37 p.
80. Mori N, Yasuda T, Mase H, Tom T, Oku Y (2010) Projection of extreme wave climate change under global warming. Hydrological Research Letters 4: 15–19.
81. McCaw L (2013) Western Australia's weather and climate: a synthesis of Indian Ocean climate initiative stage 3 research. Austral For 76: 110–112.
82. Shakesby RA, Doerr SH (2006) Wildfire as a hydrological and geomorphological agent. Earth-Sci Rev 74: 269–307.
83. Hewitson BC, Crane RG (2006) Consensus between GCM climate change projections with empirical downscaling: precipitation downscaling over South Africa. Int J Climatol 26: 1315–1337.
84. Scott DF (1993) The hydrological effects of fire in South African mountain catchments. J Hydrol 150: 409–432.
85. Marais C, van Wilgen BW, Stevens D (2004) The clearing of invasive alien plants in South Africa: a preliminary assessment of costs and progress. S Afr J Sci 100: 97–103.
86. van Wilgen BW, Forsyth GG, Le Maitre DC, Wannenburgh A, Kotzé JDF, et al. (2012) An assessment of the effectiveness of a large, national-scale invasive alien plant control strategy in South Africa. Biol Conserv 148: 28–38.

87. Godsmark R (2007) The impact of the 2007 plantation fires on the SA Forestry and Forest Products Industry. Available: http://www.forestry.co.za/industry-information-protection/. Accessed August 2013.
88. Nel DC, Colvin C, Nel JL, de Lange W, Mwenge Kahinda J, et al. (2010) SAB Hop Farms Water Risk Assessment: Project Report. CSIR Report CSIR/NRE/ECOS/2011/066/B, CSIR, Pretoria.
89. Barbier EB, Georgiou IY, Enchelmeyer B, Reed DJ (2013) The value of wetlands in protecting southeast Louisiana from hurricane storm surges. PLoS ONE 8: e58715.
90. Thomalla F, Downing T, Spanger-Siegfried E, Han G, Rockström J (2006) Reducing hazard vulnerability: towards a common approach between disaster risk reduction and climate adaptation. Disasters 30: 39–48.
91. Cutter S, Barnes L, Berry M, Burton C, Evans E, et al. (2008) A place-based model for understanding community resilience to natural disasters. Glob Environ Change 18: 598–606.
92. Brown K, Westaway E (2011) Agency, capacity, and resilience to environmental change: lessons from human development, well-being, and disasters. Annu Rev Environ Resour 36: 14.1–14.22.
93. UNISDR (2009) The 2009 UNISDR Terminology on Disaster Risk Reduction. United National International Strategy for Disaster Reduction (UNISDR), Geneva, Switzerland.
94. Cartwright A, Blignaut J, De Wit M, Goldberg K, Mander M, et al. (2013) Economics of climate change adaptation at the local scale under conditions of uncertainty and resource constraints: the case of Durban, South Africa. Environ Urban 25: 1–18.

10

3800 Years of Quantitative Precipitation Reconstruction from the Northwest Yucatan Peninsula

Alicia Carrillo-Bastos, Gerald A. Islebe*, Nuria Torrescano-Valle

El Colegio de la Frontera Sur, Unidad Chetumal, Herbario, Chetumal, Quintana Roo, México

Abstract

Precipitation over the last 3800 years has been reconstructed using modern pollen calibration and precipitation data. A transfer function was then performed via the linear method of partial least squares. By calculating precipitation anomalies, it is estimated that precipitation deficits were greater than surpluses, reaching 21% and <9%, respectively. The period from 50 BC to 800 AD was the driest of the record. The drought related to the abandonment of the Maya Preclassic period featured a 21% reduction in precipitation, while the drought of the Maya collapse (800 to 860 AD) featured a reduction of 18%. The Medieval Climatic Anomaly was a period of positive phases (3.8–7.6%). The Little Ice Age was a period of climatic variability, with reductions in precipitation but without deficits.

Editor: Gil Bohrer, The Ohio State University, United States of America

Funding: This work received funding from: Conacyt funding project 127613. www.conacyt.mx. The funders had no role in study design, data collection and analysis, decision to publish, or preparation of the manuscript.

Competing Interests: The authors have declared that no competing interests exist.

* E-mail: gislebe@ecosur.mx

Introduction

The understanding of climate dynamics and the prediction of future changes require the identification of patterns of climatic change at different temporal scales [1]. To find these patterns, it is necessary to take into account continuous readings that go beyond instrumental measurements, which provide a temporally limited perspective.

Fossil pollen is a natural source of climatic information. Vegetation is subject to environmental conditions that favor certain species depending on their climatic preference. By calibrating modern pollen samples with climate variables, it is possible to reconstruct quantitative climatic parameters [2].

Precipitation is one of the most important and sensitive parameters of the tropical climate [3]. The reconstruction of precipitation is of great interest because it permits the validation of climatic change models by examining the patterns of the past, present, and future [4]. On the Yucatan Peninsula, the fossil pollen record has been analyzed in terms of vegetation change. However, the inferences concerning climate have been expressed only in qualitative terms (wetter/drier). In the Yucatan region, the variability in precipitation had an important influence on the cultural history of the Mayan civilization [5]. Previous studies on the peninsula have indicated that the droughts suffered during the Classic period contributed to the collapse of the Mayan civilization [6–9].

A quantitative estimate of the precipitation over the last 1500 years is available for the peninsula. The record is based on measurements of $\delta^{18}O$ values in stalagmites from the Tzabnabh cave [8]. In that study, it was determined that precipitation decreased by 36–52% relative to modern precipitation during the multiple droughts at the end of the Classic period. Recently, based on quantitative analysis of high-resolution climate records, it was concluded that the droughts of the Terminal Classic period were due to the reduction in the frequency and intensity of summer storms [9].

The Yucatan Peninsula is influenced by a number of different components of the climate system, creating a zone of mixed climates. For this reason, it is necessary to investigate whether the changes that occurred in this region were similar in magnitude, frequency and time in order to understand climatic variability. These changes are of particular importance because studying the events of the recent past will permit us to foresee what may occur in the near future, as climatic forcing that operated during the late Holocene did not significantly change [10]. With the purpose of contributing to the knowledge of climatic variability of the Yucatan Peninsula, this study presents a reconstruction of the vegetation from the northwest corner of the peninsula and the first quantitative reconstruction of precipitation for the last 3800 years.

Study Area

The study area is located in the northwest Yucatan Peninsula between the coordinates 21.5661N and 88.0865W in the biosphere reserve of Ria Lagartos (Figure 1). Because of its geographical position and proximity to the sea, the site is influenced by several different patterns of atmospheric circulation, such as the trade winds, polar air masses and the convection currents that provide precipitation [11]. The climate type is BSo(h')w(x)iw, warm-semiarid and evaporation exceeds precipitation. The largest portion of the annual precipitation is received during the summer (62%), when the Intertropical Convergence Zone (ITCZ) and Bermuda Azores high-pressure cell move to the north and the trade winds intensify. The remaining precipitation (38%) occurs during the dry season from November to May. The influence of polar air masses occurs during the months of

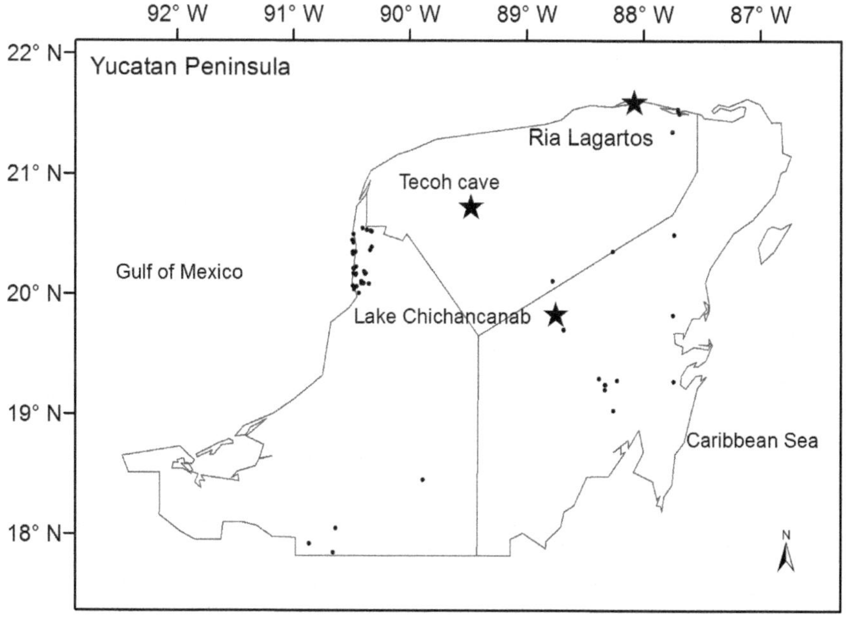

Figure 1. Map of area studied showing the sites mentioned in the discussion and the location of the modern pollen samples.

September to April [12]. The average temperature is 26°C [12], and the annual precipitation is 760 mm.

Several different vegetation regimes are found in the reserve: medium forest, dry forest, low forest, mangroves, coastal dunes and wetlands [11]. Vegetation near the coring site includes *Metopium brownei*, *Bursera simaruba*, *Haematoxylum campechianum*, *Conocarpus erectus*, *Plumeria sp.*, and *Bravaisia sp.*

Methods

Calibration of Pollen Rain Data and Precipitation

Pollen data. Samples of modern pollen rain were collected from different points of the Yucatan Peninsula (Figure 2), from a previous study [13] and present work. All are superficial sediment samples that were taken at a depth between 3 and 5 cm. At each site, 2–3 samples were taken and homogenized into one sample to avoid the effect of over-representation of local vegetation. A total of 57 samples were obtained.

The samples were treated with the standard techniques used for extraction and concentration: KOH, HCl, acetolysis [14], suspended glycerine gelatin (first study [13]) and an ultra-kit at 100% in this study. Tablets of *Lycopodium* were added to all the samples at the beginning of the treatments as an exotic marker. The pollen was identified with the help of the atlas developed by Palacios-Chavez et al. [15] and the palynological collection of Herbario-Ecosur (CIQR).

A minimum of 300 pollen of taxa belonging to pollen sum taxa were counted with an optical microscope with objectives of 400× and 1000×. Aquatic taxa and fungal spores were excluded from the pollen sum.

A total of 78 taxa were identified. To establish criteria of standardization and consistency in the data and to reduce bias, only taxa with percentages ≥1% and present in at least 5% of the samples were included [16]. Following this procedure, 30 species (Figure 2) were selected and the percentages were recalculated accordingly.

Precipitation data. Precipitation values were determined based on 65 meteorological stations on the Yucatan Peninsula from national weather services. At present, there are a total of 157 stations on the peninsula. However, only the 65 selected had at least 25 years of records during the period 1981 to 2010. The precipitation values for each sample of pollen rain were calculated by ordinary Kriging using the ArcGis 10 extension Geostatistical Analyst.

Modern Pollen–precipitation Relationship

The first step in the quantitative reconstruction of precipitation is to determine if the relationship between the plant species and the environmental variable follows a linear or a unimodal model [2]. For this, it is necessary to estimate the gradient length of the environmental variable expressed in standard deviation (SD) units of biological turnover [2]. To estimate this, detrended canonical correspondence analysis (DCCA) [17] was conducted, using the software CANOCO 4.5 [18]. The following parameters were used: precipitation as the only predictor variable, detrending by segment, non-linear rescaling and pollen data transformed to the square root. During this analysis, four samples exhibited extreme values and as a result were discarded when constructing the transfer functions.

Model of the Transfer Function

The length of the gradient of the environmental variable in SD units is an estimate of the behavior of the species along the length of this gradient [2]. If the gradient is small (2 SD units), the taxa behave monotonically through the gradient and the use of linear models of regression and calibration are appropriate [2]. In the DCCA, the length of the gradient obtained was <2 SD. Therefore, the transfer function was estimated using the method of partial least squares (PLS) regression [19], using software C2 1.7.2 [20].

PLS regression is a linear method that, by using inverse regression, calculates the transfer function that relates the modern pollen data to the environmental variable. This function can then be applied to fossil pollen data to reconstruct precipitation [2]. Evaluation of the prediction and selection of the best model with

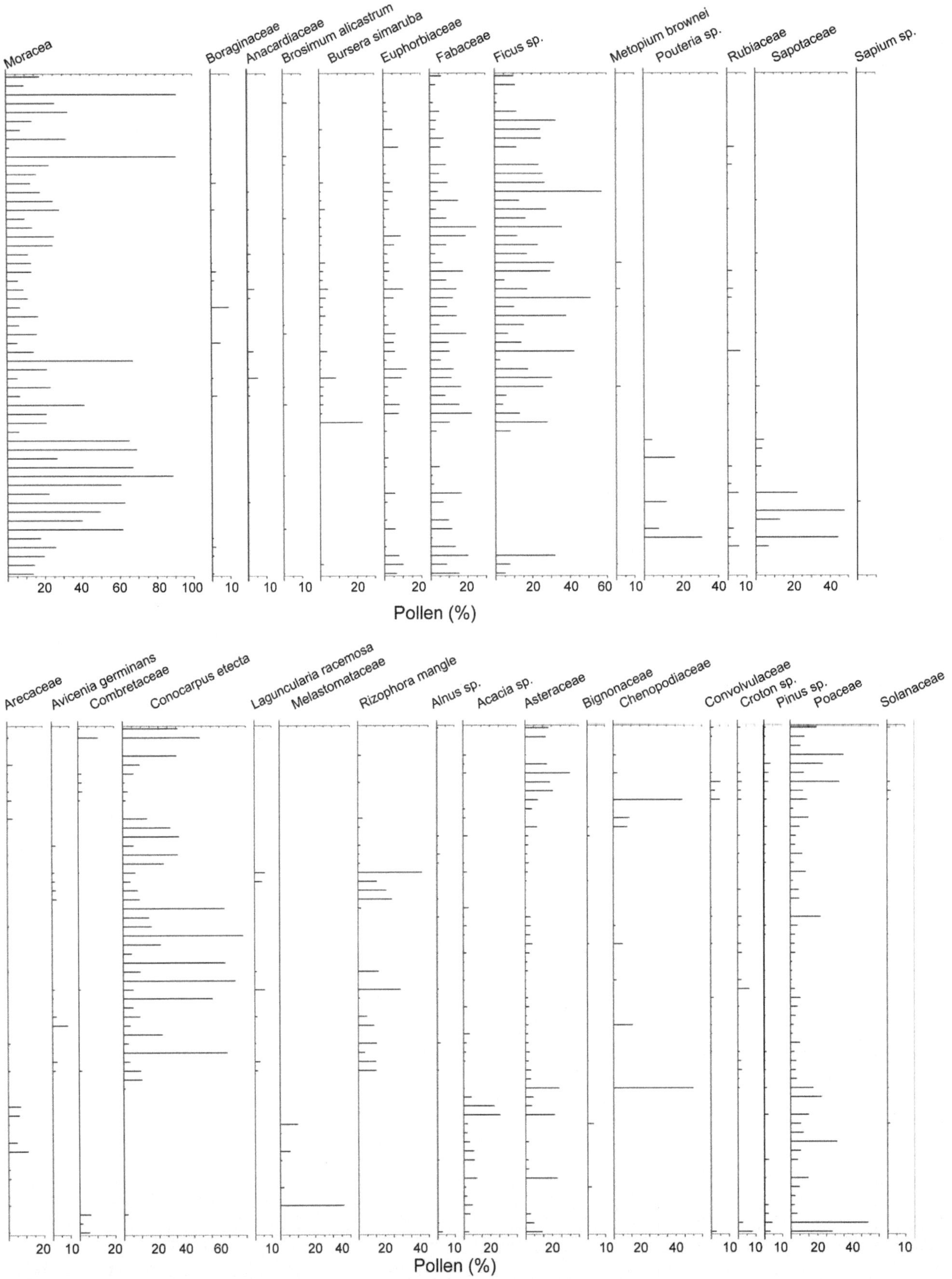

Figure 2. Percentage of taxa used in the calibration. The samples are sorted in ascending order with respect to the precipitation.

the smallest number of components are based on the combination of the values of maximum bias, root mean squared error (RMSEP), the coefficient of determination (r^2) and the residuals, all four of which are based on the method of cross-validation by bootstrapping 1000 interactions. The model selected was a two component PLS regression.

The precipitation anomalies are calculated using the following formula: $A = (R_i - Rm)*100/Rm$, where A is the anomaly of every period, R_i is the estimated precipitation in the same period and R_m is the average of all recorded precipitation.

Fossil Data

With a Russian corer, 50 cm sections of fossil sediment were obtained in a lagoon situated 5 to 7 km from the coastline. The total length of the core was 200 cm. The sediment was packed and transported to the palynology laboratory of Ecosur-Chetumal. The sediment was processed by physical and chemical treatments using the following protocol: treatment with HCl, filtration, digestion by KOH, acetolysis and mounted in Ultra-Kit at 100%. During the treatment with HCl, tablets of *Lycopodium* were added as an exotic marker. A total of 74 samples of fossilized pollen were obtained at an interval of 3 cm.

Taxa were separated by groups of ecological preference: tropical forest, disturbed taxa and mangrove. The pollen data are available in the Neotoma Database (http://www.neotomadb.org/uploads/data_subs/LAGARTOS.zip). The results obtained were graphed in a pollen diagram using the software TILIA 1.7.16 [21]. Pollen zones were determined by classification analysis based on the index of similarity using the tool CONISS, integrated into the same software package. The chronology was estimated by a linear regression applied to three radiocarbon dates (AMS) (Table 1) measured in the organic matter of the samples. The samples were analyzed in the laboratory of Beta Analytic Inc. in Miami, Florida. The radiocarbon dates were calibrated with the program CALIB 6.1.1 [22] using the IntCal09 calibration curve [23], rounded to the nearest decade and converted to calendar ages using 1950 AD as 0 cal yr BP.

Ethics Statement

The samples are from locations where no specific permissions were required, and field studies did not involve endangered or protected species.

Results

Modern Pollen – precipitation Relationship

In the DCCA, the length of the gradient was 1.69 SD. The correlation between the first ordination axis of the taxa and precipitation was 0.803, while the percentage of variance accumulated between taxa and the precipitation was 95.6%. The eigenvalue of the first ordination axis was 0.201. The ordination diagram showed separation of taxa in the first axis, which generated the correlation with the precipitation (Figure 3).

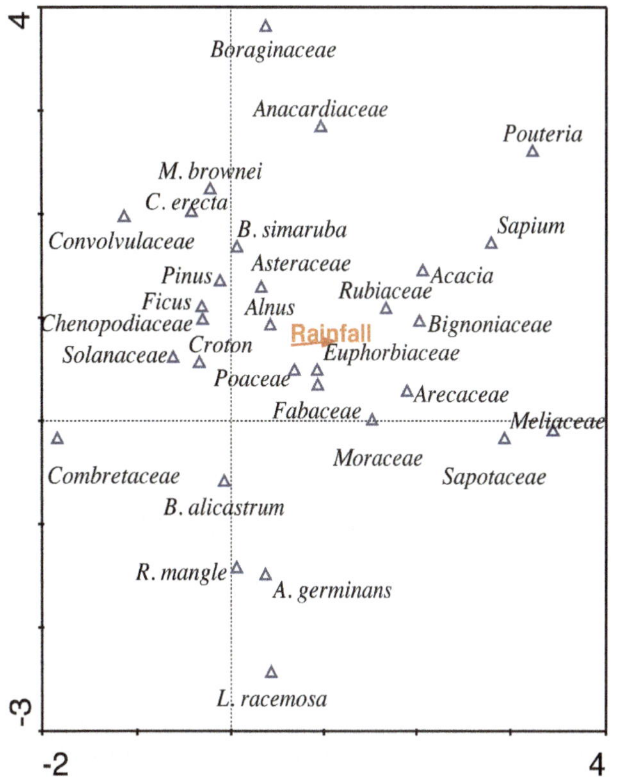

Figure 3. Ordination diagram results of DCCA.

Model of the Transfer Function

In the construction of the transfer function, a model with two components is used (Table 2). This model showed the lowest value of the RMSEP (131 mm) and a medium–high value for the coefficient of determination (0.703). The value of maximum bias (280.612) was not the smallest, but with this number of model components, the percentage of reduction in the RMSEP was 7.041% and the value of significance was 0.04.

Chronology and Reconstruction of the Vegetation

The resulting equation of the linear regression applied in conjunction with the calibrated ages (Table 1) produced, as a result, an age of 1850 BC at the bottom of the record.

Table 1. Radiocarbon ages (AMS) and sedimentation rates.

Sample code	Depth (cm)	Radiocarbon age (yr BP)	Calibrated age (cal yr BP)	Calendar age (BC/AD)	Relative area under probability distribution (2σ)	Sedimentation rate (mm/yr)
Beta-317982	84.1	2160±30	2056–2185	106–235 BC	0.54807	0.40
Beta-325222	171.36	3020±30	3141–3338	1191–1388 BC	0.974839	0.78
Beta-325223	193.8	3130±30	3319–3409	1369–1459 BC	0.847747	1.80

Regression equation for chronology $Age = (depth+9.1809)/0.055$ $r^2 = 0.95987$.

Table 2. Model performance.

PLS component	RMSEP	r^2	Maximum bias	% Change	t-test significance
1	140.97	0.486	348.176	...	...
2	**131.05**	**0.703**	**280.612**	**7.041**	**0.044**
3	133.21	0.735	255.696	−1.653	0.579
4	142.485	0.761	253.917	−6.956	0.984

Performance statistics of the four components of the PLS pollen-rainfall transfer function. Root mean squares error of prediction (RMSEP), coefficient of determination (r^2), and maximum bias and reduction percentage in RMSEP given based on a bootstrapping cross-validation method. The select two-component model is shown in bold.

The average calculated rate of sedimentation is 0.99 mm/yr. In the fossil pollen count, we identified 53 pollen taxa that belonged to 37 families.

Only those taxa composing ≥1% in at least 5% of the samples were used for the quantitative analysis (Figure 4). The curve of *Zea mays* is also represented because it gives valuable information about the presence of human activity. After applying the classification analysis to the percentages of pollen, four pollen zones were established, which represent the principal phases of vegetation change.

Zone I (200-130 cm, 1850-580 BC). In this zone, the main change is between taxa of the tropical forest and the disturbance taxa. The tropical forest oscillates between 50 and 90%, dominated by Moraceae (15–80%), followed by *Ficus sp.*, Fabaceae and Euphorbiaceae. In the disturbed vegetation, the dominant taxa is Chenopodiaceae (40%) and, in lesser proportions, *Croton sp.*, Asteraceae and Poaceae. Mangrove vegetation is scarce and does not reach 5%. The first record of *Z. mays* appeared around 1840 BC.

Zone II (130-97 cm, 580 BC-20 AD). The change during this zone is due to the decrease in disturbance elements (percentages ≤20%) and the increase in forest taxa (70–89%). *Pouteria sp.* and Rubiaceae are present. *Brosimum alicastrum* is present almost continuously, and Moraceae increases (values ≥58%). There is also a slight increase of mangrove (values ≤3.3%), especially *Conocarpus erectus*. *Z. mays* appears discontinuously.

Zone III (97-57 cm, 20–750 AD). This zone is marked by an increase of disturbance vegetation and mangroves and decrease of forest taxa. *B. alicastrum* is not present at the start of this zone, and *Bursera simaruba* appears, while Moraceae drops to 10%. *Z. mays*, Asteraceae and *Pinus sp.* are continuous throughout and in greater percentage than in the previously mentioned zones. Chenopodiaceae and Poaceae predominate in this vegetation zone. The presence of *C. erectus* is continuous and in percentages between 2 and 12%.

Zone IV (57-0 cm, 750 AD - present). Percentages of disturbed taxa are low, such as Chenopodiaceae (≤2%), Poaceae (≤8%), Asteraceae (≤1%), and *Croton sp.* (≤1%). *Z.mays* does not appear, and *Pinus sp.* are found. Moraceae is the dominant element

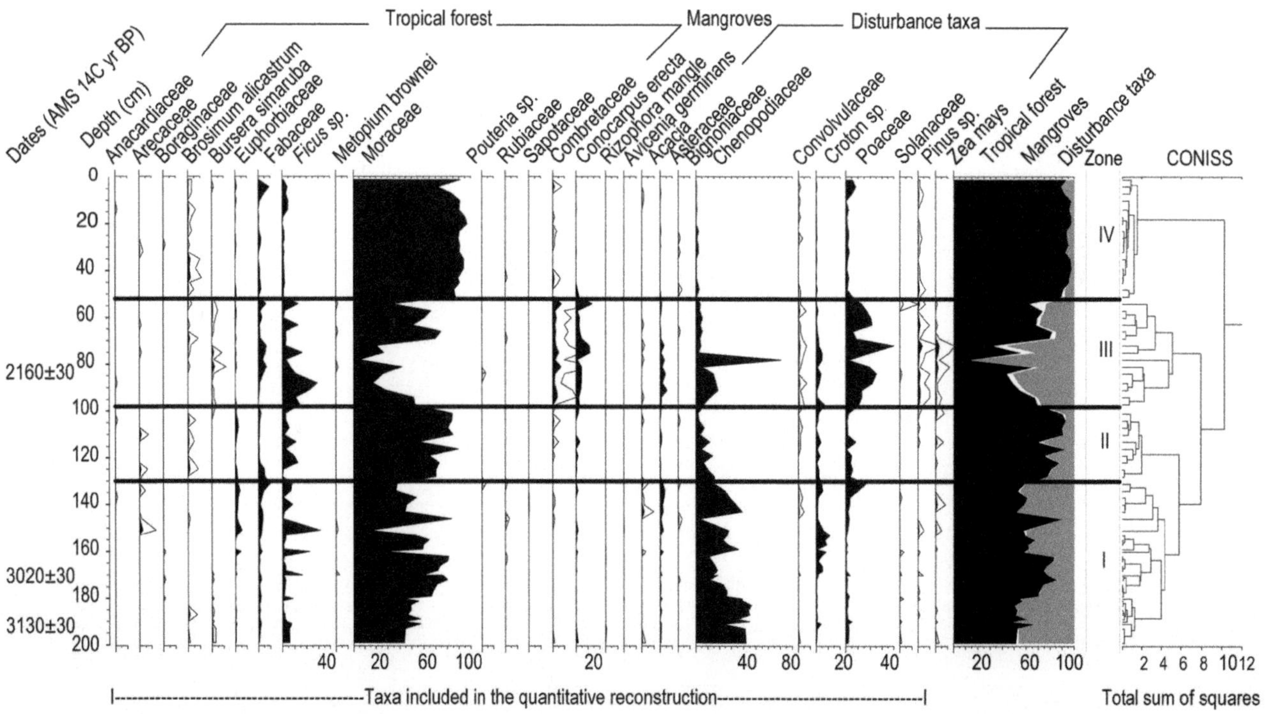

Figure 4. Diagram of fossil pollen.

(≥70%). *Ficus sp.* are lower. *B. simaruba* is not present, while *B. alicastrum* is present continuously. The curve of *C. erectus* is discontinuous.

Quantitative Precipitation Reconstruction

The application of the transfer function to the pollen data of Ria Lagartos permits the reconstruction of precipitation variability during the last 3800 years. The average precipitation for this period is 850 mm with a SD of 66 mm. Four phases of change are detected: the first from 1850-1000 BC with an average of 883 mm, the second from 1000-50 BC with an average of 870 mm, the third from 50 BC-500 AD with an average of 748 mm and the fourth from 500–1770 AD with an average of 898 mm (Figure 5). Upon calculating the anomalies of precipitation, it appears that the greatest deficit occurred during the third phase. The years 180, 420 and 800 AD feature negative anomalies of 21% in the first extreme drought and 18% in the other two droughts.

However, the surplus did not even reach 9%. There were two surpluses in the second phase of change (820 and 540 BC) and two in the fourth phase (1420 and 1600 AD).

Discussion

Modern Pollen – precipitation Relationship

The longitude of the gradient obtained with the DCCA was low (<2 SD). Therefore, the relationship between taxa and precipitation has a monotonic relationship. From ecological studies, it is observed that species abundance exhibits a unimodal behavior with respect to an environmental variable [2]. Each taxon grows best at an optimum value of the environmental variable and is unable to fully develop at higher or lower values [24]. The range captured by the values of precipitation is most likely only a fraction of the unimodal curve. This explains why the taxa in this study showed monotonic and unimodal behavior. For the same reason, it is probable that the ordination diagram will show a short vector for precipitation. However, the diagram can show that the first axis of ordination separates the majority of the groups and their correlation with precipitation was 0.803.

Along this axis, the referred taxa act as indicators of humidity. Conserved vegetation in the paleoecological studies of the region [25–28] are ordered in the direction of the precipitation vector (e.g., Moraceae, Meliaceae, Sapotaceae, *Pouteria sp.*, Rubiaceae and Sapindaceae). Taxa associated with less humid conditions and with disturbance vegetation are grouped on the opposite side of the graph (e.g., Chenopodiaceae, *Croton sp.*, Asteraceae, Solanaceae, Convolvulaceae and Poaceae).

It is important to note that the second axis contributes to the ordering of the taxa. Although it is not possible to determine the responsible variable, it is most likely related to the condition of the soil. Species such as *Rhizophora mangle*, *Laguncularia racemosa* and *Avicennia germinans* are all clearly separated. *C. erectus* is associated with mangroves (but only away from the influence of the tide) and is found grouped together with *Metopium brownie* and *B. simaruba*, which prefer well-drained soils but can withstand periodic floods typical of low areas of the peninsula [29].

Model of the Transfer Function

For the Yucatan Peninsula, no other quantitative reconstructions that use pollen as a proxy exist. This absence rules out a comparison between the statistical values obtained in the construction of the transfer function.

However, the evaluation and selection of a model can be based on a combination of the different parameters. One of these parameters is RMSEP, which in our chosen model took the lowest value (Table 2).

This value indicates the prediction ability of the modern dataset, and therefore, we assert that the calibration model can function as a tool of prediction [2].

It is also necessary to look for a low value for the maximum bias, as it measures systematic differences in the prediction [30]. It is not the lowest value of all the models, but it is much lower in relation to the model of a single component. The medium-high r^2 value indicates a strong relationship between the inferred values and those observed [2]. The maximum bias and r^2 statistics were neither the lowest nor the highest, respectively, but the lower RMSEP was certainly important. For a model to be useful, it is necessary that the RMSEP is reduced to at least 5% in relation to the model with the least number of components [31].

Furthermore, it is important to consider that the p value must be <0.05 to ensure that the estimates of the model are not just due to

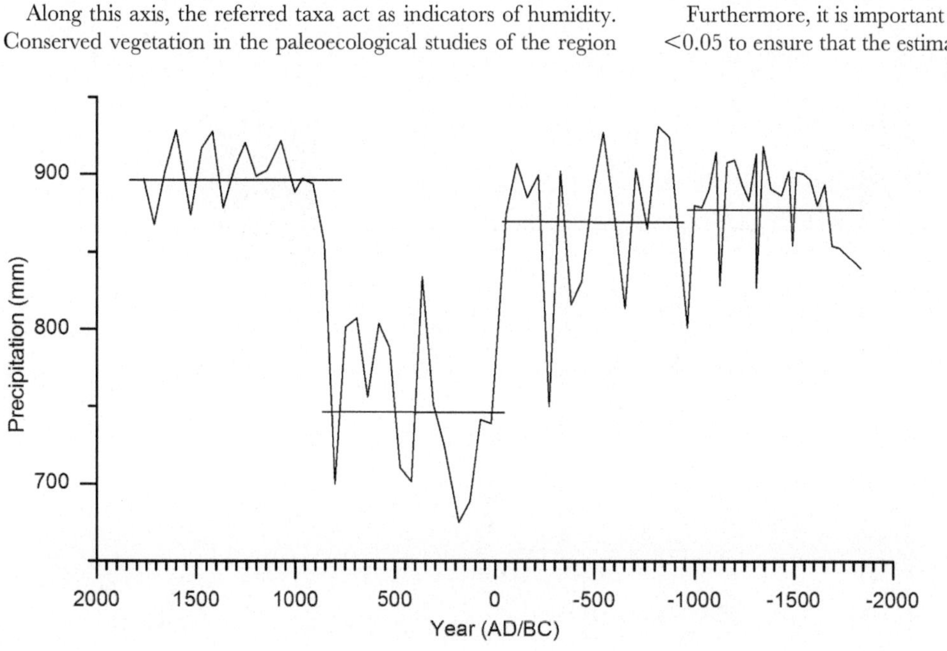

Figure 5. Inferred precipitation from fossil pollen. The horizontal lines indicate the averages for every phase.

chance. In this work, the *p* value of the reconstruction was 0.044. Therefore, we can consider that this value is statistically significant.

The graphs that compare observed values to estimated values (Figure 6a) and the observed values to the residuals (estimated minus observed) (Figure 6b) show that the model predicts values greater than 800 mm relatively well but is less reliable for medium values. However, the low values tend to be overestimated, with differences of at least 160 mm. Overall, the model is generally better with high values than with low values.

Chronology and Reconstruction of the Vegetation

Research has revealed that the vegetation of the region has been subject to the impact of human activities since early times. The first evidence of maize occurs around 1840 BC. The presence of taxa such as Malvaceae, *Acacia sp.*, Burseraceae, *Ficus* sp. and Fabaceae indicates that, during this stage, the vegetation consisted of secondary forest with Chenopodiaceae, Poaceae, and some cultivated plants. Data from other sites on the Yucatan Peninsula show conditions of low humidity during this stage [6], [26]. For this reason, we may infer that the secondary character of the vegetation is a reaction to the climatic conditions. The presence of Rubiaceae and the increase in abundance of Moraceae indicate that the forest recovered in the late phase of the middle Holocene (1550-1130 BC).

At the beginning of the late Holocene, a reduction in Moraceae and an increase of *Ficus sp.* and Chenopodiaceae are recorded, suggesting disturbed conditions in the region. This disturbance was not related to agricultural activity, as neither *Z. mays* nor elements associated with cultivation, such as *Croton sp.*, are present. Circa 950 BC, the vegetation acquired a more open character. This is inferred by the increase in the percentage of *Pinus sp.*, which do not form part of the regional vegetation. The presence of this pollen indicates that the arboreal cover was sparse enough to permit the deposition of allochthonous pollen [32]. During this time, the vegetation was a mosaic of low tropical forest, savannas and crops. From 550 BC to 50 AD, the increase in Moraceae; the almost continuous presence of *B. alicastrum*; and the presence of *Pouteria sp.*, Rubiaceae and Sapotaceae, indicate the recovery of the vegetation.

From 50 to 750 AD, the tropical forest suffered the greatest transformation of the studied period. The pollen record shows that tree cover decreased circa 190 AD. This is demonstrated by the reduction of tropical forest taxa (<50%) such as Moraceae (<20%) and the increase of *Pinus sp.* and *Ficus sp.* This change in the vegetation coincides temporally with the Preclassic drought reported in the Maya region [7], [25–26], [33–35]. In 370 AD, the forest changed to savanna. Pollen from arboreal elements decreased by 15%, while the taxa of disturbance (Poaceae, Chenopodiaceae, Convolvulaceae, *Pinus sp.*, among others) reached a total of 84%. During both transformations, agricultural activity was present, but maize pollen also decreased notably during these two events. This suggests that the climate was one of the triggers of the vegetation transformation.

The highest intensity of agricultural activity occurred in 470 AD based on the increase in pollen from maize as well as an increase in Poaceae and *Pinus sp.*, among other elements of disturbance. These taxa reach values of 56%, while forest taxa contribute only 29%. This suggests an expansion of savannas and crops but with more arboreal vegetation present than the period of land clearing discussed earlier.

The forest later recovered, and taxa suggest a decrease in the expansion of cultivated and grassland areas, although grasslands continued to be an important component of the vegetation (between 14 and 29%).

The vegetation mosaic of this time was composed of low tropical forest, savannas, crops and mangrove. Evidence of maize was last recorded around 800 AD and was absent in the next analyzed date (860 AD). For this reason, we infer a decrease of agricultural activities and the cultural collapse between 800 and 860 AD.

Circa 1779 AD, the youngest estimated age of this study, forest taxa percentages are the highest of the core. *B. alicastrum* is present continuously and, together with the high percentages of Moraceae and the presence of Anacardiaceae, Rubiaceae and Boraginaceae, suggests more conserved vegetation. There are, however, slight variations which coincide with small oscillations in the curve of *Pinus sp.* and Poaceae, which suggest phases of reduced arboreal cover.

Quantitative Precipitation Reconstruction

The statistical values in the construction of the transfer model indicate that the estimated precipitation from the Ria Lagartos core corresponds to changes in vegetation. However, the

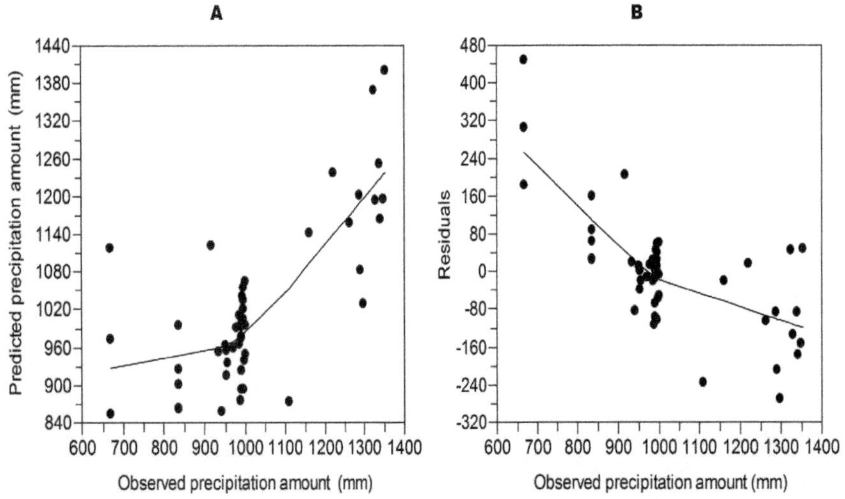

Figure 6. Model diagnostic plots. (A) Relation between estimated (with the transfer function) vs. observed precipitation. (B) Residual relations (observed-estimated) vs. observed precipitation.

vegetation of the region has been impacted since early times by human activity, which may interfere the reliability of the precipitation reconstruction. Nevertheless, as suggested by Leyden et al. [30], even under human influence, the pollen signal maintains a relationship to climatic changes. Therefore, it is indeed useful to compare pollen to the record of precipitation anomalies at Ria Largatos, to the record of *Z. mays* and to the $\delta^{18}O$ record of Lake Chichancanab [6] (Figure 7). With these comparisons, it is clear that precipitation deficits (droughts) and surpluses (wet phases) do not always correlate with rises and falls in the percentage of *Z. mays.*

In the first phase of the precipitation record, surpluses were more frequent than droughts and longer in duration (Figure 5). Precipitation was 3% higher than the average of the last 3800 years. During this time (1850-1000 BC), isotope values of Lake Chichancanab [6] are lower than the mean, suggesting wet conditions. Changes in the maize record do not match changes in the precipitation record, which suggest that variations in precipitation correspond to climate variability. This phase can be divided into two stages. During the first (1850-1250 BC), precipitation is above average and reaches surpluses of 6.7%. This wet stage can also be observed in a previous study of Ria Largatos [33] and in the Cariaco Basin [34]. During the second stage, droughts occur (precipitation 3.4% below average), but the surpluses are of similar magnitudes to those of the first stage. The beginning (950 BC) of the second phase is marked by a drought where the precipitation decreases by 6.4%. This drought coincides with an increase in the $\delta^{18}O$ values of Lake Chichancanab [6], Lake Tzib [26] and Lake Punta Laguna [36], along with the decrease in the percentage of forest taxa in the previous study of Ria Largatos [33]. We detect an increase in the maize record. It is probable then that the decrease in precipitation is less than the reconstruction suggests.

The largest surplus (8.75%) is recorded around 320 BC, which agrees with lower $\delta^{18}O$ values of Lake Chichancanab [6] and with one of the highest percentages of Ti in the last 4000 years in the Cariaco Basin [34]. The recorded increase in precipitation also coincides with an increase in solar radiation [37]. We find that the most intense drought of this phase occurred in 250 BC. The precipitation decreased 12.4%. This drought is apparently a result of the modification of the vegetation by agricultural activities. In Lake Chichancanab, higher $\delta^{18}O$ values are also observed but do

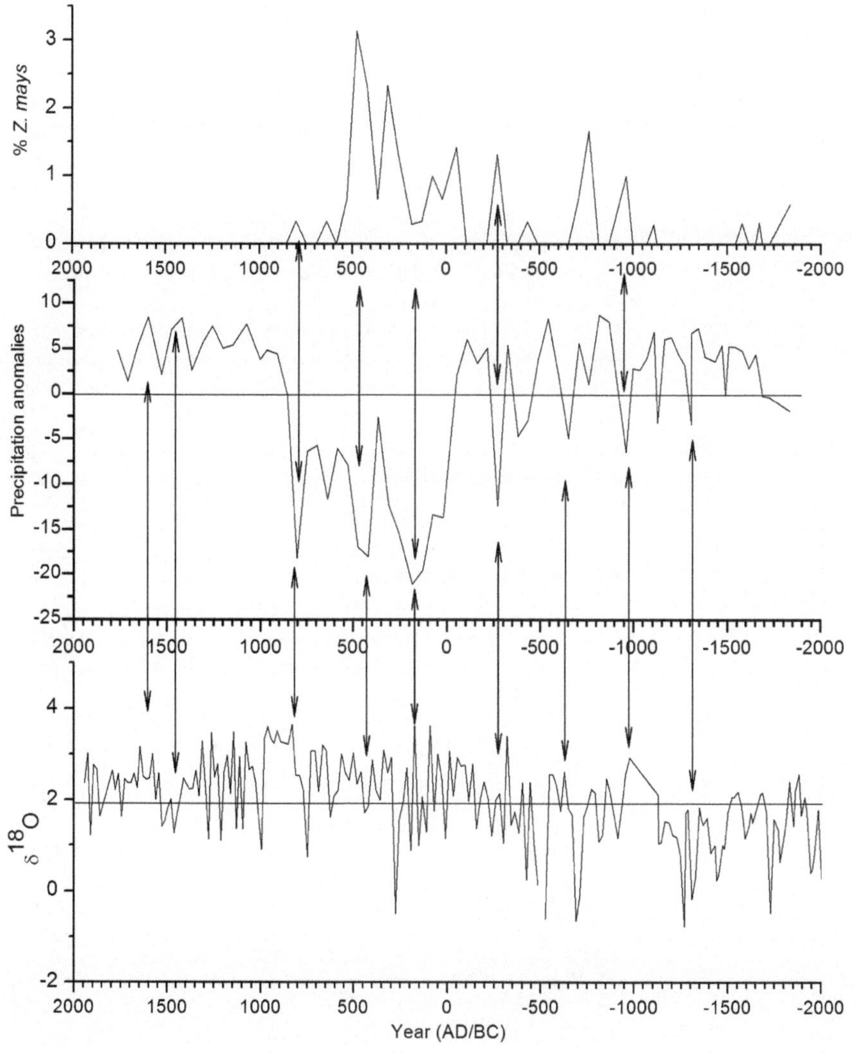

Figure 7. Comparison of records. (A) Record of *Z. mays* of this same study. (B) Anomalies of the inferred precipitation. (C) Record of $\delta^{18}O$ Lake Chichancanab [6].

not hold up a reduction of precipitation, nor do we observe such a reduction in Cariaco. However, changes are detected in the records of maize and other elements associated with cultivation, such as *Croton sp.* The second phase ended with an increase in the precipitation (5–6%).

The third phase is clearly differentiated by repeated droughts. During this time, precipitation was below average by an average of 12.6% but at times more than 18%. The first drought lasted from around 130 to 236 AD. During this drought, precipitation dropped by 21%. This phase coincided with what is known as the Preclassic drought reported in various works in the region [26–27], [33–34]. During this event, important changes in the environment and in Mayan cities occurred, including the collapse of El Mirador (Guatemala), a city that flourished during the Preclassic period [38]. It is important to note that, during this drought, there was no increase in the percentage of maize. The precipitation record confirms that this was a result of the climate and not an artifact of agricultural activity. However, the drought that occurred around 420 AD does coincide with an increase in evidence of maize. In fact, the drought coincides with the highest percentage of maize in the record. Therefore, it is probable that the 18% decrease in precipitation is overestimated. The third drought occurred in 800 AD when precipitation declined 18.2% and, in contrast to the previous drought, no high values of *Z. mays* were observed. This drought coincides with the high $\delta^{18}O$ values in the study of Lake Chichancanab [6] and decreased precipitation in the Cariaco Basin [28]. The drought also coincides temporally with the second phase of the disintegration of Mayan lands, during which various cities were abandoned. Among these cities were Yaxchilan, Calakmul, Piedras Negras, Copan and Naranjo [38]. Following this drought, maize disappeared from the record and is likely related to the collapse around this time. The collapse is estimated to have occurred between 800 and 860 AD.

Medina-Elizalde proposed that the reduction in precipitation during the final Classic Period was 36–52% based on quantitative precipitation reconstruction using isotopes in stalagmites. The values obtained in this study were more conservative. Nevertheless, it is important to consider that the model constructed tended to overestimate the low values (Figure 6). Therefore, it is likely that droughts were more intense than the values suggest here.

The fourth phase of change (800–1760 AD) is characterized as a wet period. The increase in precipitation was rapid, and the anomaly exceeded an average of 5% with no deficits recorded, despite variable precipitation. Medina-Elizalde [8] suggests that the revitalization of the Puuc region in the north of the peninsula was favored by the precipitation increase, which increased the production of food and could subsequently support a larger population. This event coincides temporally (860–900 AD) with the beginning of this fourth phase, during which the precipitation increased considerably. The period from 900 AD to 1300 AD experienced precipitation surpluses of more than 7.5%. This period coincides with the global event known as the Medieval Climatic Anomaly [39].

Between 1300–1760 AD, precipitation was high overall but also more variable. This period falls into the Little Ice Age (LIA), which was drier than the Medieval Climatic Anomaly but still wetter than the rest of the record. The precipitation reconstruction from the Tzabnabh cave [8] suggests that the LIA was relatively more humid, which agrees with the precipitation record of the present study, yet contrary to earlier findings [34]. However, other studies on fossil pollen on the peninsula have found droughts in this phase [27], [33].

Conclusions

The calibration of modern fossil pollen with observed precipitation led to the construction of a predictive and reliable model. The model works best with medium values, while low values tend to be overestimated.

The application of a transfer function permitted the reconstruction of precipitation for the last 3800 years. In the area of Ria Lagartos, precipitation anomalies show that the magnitudes of deficits (21%) exceed the surpluses (<9%). Four phases of change have been detected; the third (50 BC – 800 AD) was the driest. The drought related to the abandonment of the Preclassic Period featured an estimated 21% reduction in precipitation, while the drought of the Mayan collapse (800–860 AD) saw a decrease of 18%. The Medieval Climatic Anomaly was a period of positive anomalies (3.8–7.6%), while the Little Ice Age was a more variable period, with reductions but without droughts. Fossil pollen indicates agricultural activity since the early phases (1840), and those activities provoked important changes in the vegetation during the Classic Period. Moreover, data indicate the vulnerability of the vegetation to the combined effect of agriculture and climate. When both agents operate in conjunction, it is possible to cause major transformations in vegetation, such as occurred in 370 AD.

Acknowledgments

Valuable comments by the reviewers are acknowledged.

Author Contributions

Conceived and designed the experiments: ACB GAI NTV. Performed the experiments: ACB GAI NTV. Analyzed the data: ACB GAI NTV. Contributed reagents/materials/analysis tools: GAI. Wrote the paper: ACB GAI NTV.

References

1. Labeyrie L, Cole J, Alverson K, Stocker T (2003) The History of Climate Dynamics in the Late Quaternary. In Alverson KD, Bradley RS, Pedersen TF, editors. Paleoclimate, global change and the future. Berlin: Springer. 33–61.
2. Birks HJ (1995) Quantitative palaeoenvironmental reconstructions. In Maddy D, Brew JS, editors. Statistical modelling of Quaternary science data. Technical Guide.
3. Riehl H (1979) Climate and Weather in the Tropics. London: Academic Press. 611p.
4. Birks HJ (2005) Quantitative palaeoenvironmental reconstructions from Holocene biological data. In Mackay A, Battarbee R, Birks J, Oldfield F, editors. Global change in the Holocene. Great Britain: Hodder education. 107–123.
5. Hodell DA, Brenner M, Curtis J (2000) Climate change in the northern American tropics and subtropics since the last ice age. In Lentz DL, editor. Imperfect balance, landscape transformations in the pre-Columbian Americas. New York: Columbia University Press. 13–38.
6. Hodell DA, Curtis JH, Brenner M (1995) Possible role of climate in the collapse of classic Maya civilization. Nature 375: 391–394.
7. Curtis JH, Hodell DA, Brenner M (1996) Climate variability on the Yucatan peninsula (Mexico) during the past 3500 years, and implication for Maya cultural evolution. Quatern Res 16: 37–47.
8. Medina-Elizalde M, Burns SJ, Lea DW, Asmerom Y, van Gunten L, et al. (2010) High resolution stalagmite climate record from the Yucatán peninsula spanning the Maya terminal classic period. Earth Planet Sci Lett. 298: 255–262.
9. Medina-Elizalde M, Rohling EJ (2012) Collapse of Classic Maya Civilization related to modest reduction in precipitation. Sci. 335: 956–959.
10. Mann ME (2007) Climate over the past two millennia. Annu. Rev. Earth Planet. Sci. 35: 111–136.
11. CONANP (2007) Programa de conservación y manejo reserva de la biosfera Ría Lagartos. Secretaría de Medio Ambiente y Recursos Naturales. 266p.

12. CNA (2006) Jefatura de Proyecto de Aguas Superficiales. Subgerencia Regional Técnica. Gerencia Regional Península de Yucatán. Comisión Nacional del Agua. México.
13. Torrescano-Valle N (2007) Reconstrucción paleoambiental del Holoceno Medio-Tardio en la parte centro-sur de la península de Yucatán, México. Doctoral Thesis. El Colegio de la Frontera Sur.
14. Erdtman G (1969) Handbook of palynology. An Introduction to the study of Pollen grains and spores. Hafner Publishing Co. 486p.
15. Palacios-Chavez R, Ludlow-Wiechers B, Villanueva-Gutiérrez R (1991) Flora palinológica de la reserva de la biosfera de Sian Ka'an, Quintana Roo, Mexico. CICRO. 321p.
16. Correa-Metrio A, Cabrera KR, Bush MB (2010) Quantifying ecological change through discriminant analysis: a paleoecological example from the Peruvian Amazon. J Veg Sci 21: 695–704.
17. ter Braak CJF (1986) Canonical correspondence analysis: a new eigenvector technique for multivariate direct gradient analysis. Ecology 67: 1167–1179.
18. ter Braak CJF, Šmilauer P (2002) CANOCO Reference Manual and CANODRAW User's Guide: Software for Canonical Community Ordination (Version 4.5). Microcomputer Power (Ithaca, New York). 500 p.
19. Wold S, Ruhe A, Wold H, Dunn WJ (1984) The collinearity problem in linear regression: the partial least squares (PLS) approach to generalized inverses. SIAM J Sci Comput 5: 735–743.
20. Juggins S (2007) C2 version 1.5 user guide. Software for ecological and palaeoecological data analysis and visualization. Newcastle University, Newcastle upon Tyne, UK. 73p.
21. Grimm EC (2011) TILIA software version 1.7.16. Illinois State Museum, Research and Collection Center. Springfield USA. Available: http://intra.museum.state.il.us/pub/grimm/tilia/.
22. Stuiver M, Reimer PJ, Reimer RW (2005) CALIB 5.0 Available: http://calib.qub.ac.uk/calib/.
23. Reimer PJ, Hughen KA, Guilderson TP, McCormac FG, Baillie MGL, et al. (2002) Preliminary Report of the first workshop of the IntCal04 Radiocarbon Calibration/Comparison Working Group. Radiocarbon 44: 653–661.
24. ter Braak CJF (1987) Unimodal models to relate species to environment. DLO-Agricultural Mathematics Group. 266p.
25. Torrescano-Valle N, Islebe GA (2006) Tropical forest and mangrove history from south eastern Mexico: a 5000 yr pollen record and implications for sea level rise. Veg Hist Archaeobot 15: 191–195.
26. Carrillo-Bastos A, Islebe GA, Torrescano-Valle N, González NE (2010) Holocene vegetation and climate history of central Quintana Roo, Yucatan Peninsula, Mexico. Rev Palaeobot Palyno 160: 189–196.
27. Gutiérrez-Ayala LV, Torrescano Valle N, Islebe GA (2012) Reconstrucción paleoambiental del Holoceno Tardío de la reserva Los Petenes, Península de Yucatán, México. Rev Mex Cienc Geol 29: 749–763.
28. Leyden BW (2002) Pollen evidence for climatic variability and cultural disturbance in the Maya Lowlands. Anciente Mesoam 13: 85–101.
29. Pennington DT, Sarukhán J (2005) Árboles tropicales de México: manual para la identificación de las principales especies. Mexico: Universidad Autónoma de México, Fondo de cultura económica. 523 p.
30. ter Braak CJF, Juggins S (1993). Weigthed averaging partial least squares regression (WA-PLS): an improve method for reconstructing environmental variables from species assemblages. Hydrobiologia 269/270: 485–502.
31. Birks HJB (1998) Numerical tools in quantitative palaeolimnology-progress, potentialities, and problems. J. Paleolim 20: 307–332.
32. Islebe GA, Hooghiemstra H, Brenner M, Curtis JH, Hodell DA (1996) A Holocene vegetation history from lowland Guatemala. The Holocene 6: 265–271.
33. Aragón-Moreno AA, Islebe GA, Torrescano-Valle N (2012) A ~3800-yr, high-resolution record of vegetation and climate change on the north coast of the Yucatan Peninsula. Rev Palaeobot Palynol 178: 35–42.
34. Haug GH, Gunther D, Peterson LC, Sigman DM, Hughen KA, et al. (2003) Climate and the Collapse of Maya Civilization. Sci 299: 1731–1735.
35. Hodell DA, Brenner M, Curtis JH, Guilderson T (2001) Solar Forcing of Drought Frequency in the Maya Lowlands. Sci 292: 367–1370.
36. Hodell DA, Brenner M, Curtis JH (2007) Climate and cultural history of the Northeastern Yucatan Peninsula, Quintana Roo, Mexico. Climatic Change 83; 215–240.
37. Steinhilber JAA, Beer J (2008) Solar modulation during the Holocene. Astra 4: 1–6.
38. Gill RB (2008) Las grandes sequías mayas, agua, vida y muerte. México Distrito Federal: Fondo de cultura económica. 561 p.
39. Mann ME (2002) Medieval climatic optimum. In: MacCracken MC, Perry JS, editors. Encyclopedia of global environmental change. Chichester: Jonh Wiley and Sons 514–516.

11

Why Herd Size Matters – Mitigating the Effects of Livestock Crashes

Marius Warg Næss[1]*, Bård-Jørgen Bårdsen[2]

1 Center for International Climate and Environmental Research – Oslo (CICERO), Fram Centre, Tromsø, Norway, **2** Norwegian Institute for Nature Research (NINA), Arctic Ecology Department, Fram Centre, Tromsø, Norway

Abstract

Analysing the effect of pastoral risk management strategies provides insights into a system of subsistence that have persevered in marginal areas for hundreds to thousands of years and may shed light into the future of around 200 million households in the face of climate change. This study investigated the efficiency of herd accumulation as a buffer strategy by analysing changes in livestock holdings during an environmental crisis in the Saami reindeer husbandry in Norway. We found a positive relationship between: (1) pre- and post-collapse herd size; and (2) pre-collapse herd size and the number of animals lost during the collapse, indicating that herd accumulation is an effective but costly strategy. Policies that fail to incorporate the risk-beneficial aspect of herd accumulation will have a limited effect and may indeed fail entirely. In the context of climate change, official policies that incorporate pastoral risk management strategies may be the only solution for ensuring their continued existence.

Editor: Jon Moen, Umea University, Sweden

Funding: The present study is part of the ECOPAST project financed by the Research Council of Norway (the FRISAM program, project number 204174). Webpage: http://www.forskningsradet.no/en/Home_page/1177315753906. The funders had no role in study design, data collection and analysis, decision to publish, or preparation of the manuscript.

Competing Interests: The authors have declared that no competing interests exist.

* E-mail: m.w.nass@cicero.tromso.no

Introduction

More of the Earth's land surface is used for grazing than for any other purpose [1]. Extensive pastoral production occurs in 25% of the global land area from the drylands of Africa and the Arabian Peninsula, to the highlands of Asia and Latin America and the Arctic parts of Fennoscandia and Russia [2]. Specifically, grazing land covers 77% of Australia, 61% of Africa, 49% of Asia and 18% of Europe [1]. It has been estimated that pastoralists produce 10% of the world's meat, and supports some 200 million pastoral households who raise nearly 1 billion head of camel, cattle and smaller livestock [2]. The main livestock species kept by pastoralists are cattle, donkeys, goats and sheep, although they also keep, e.g., alpaca and llamas in the Andes, camels and horses in east-central Asia, the dromedary in Africa and West Asia, reindeer in northern Eurasia, and yak on the Tibetan Plateau [1]. Pastoralism is also economically important, especially in poor regions: compared to settled farmers in Africa, pastoralists produce 50–70% of all the milk, beef and mutton produced on the continent and while comprising only 1.5% of the total population of Iran, pastoralists keep 25% of the national herd [1]. From a global point of view, >1 billion people depend on livestock, and 70% of the 880 million rural poor living on less than USD 1.00 per day are at least partially dependent on livestock for their livelihoods [3]. Accordingly, Dong et al. [4] argue that pastoralism is important from a global point of view because of: (1) the human populations it supports; (2) the food and ecological services it provides; (3) the economic contributions it makes; and (4) the long-standing societies it helps to maintain.

Environmental Hazards and Pastoralism

Environmental hazards, such as drought, floods and icing significantly affect livestock survival and reproduction. For example, in Africa mortality rates for cattle during drought have been estimated to be between 35–75% [5] and 10–25% [6], and 25% on average [7]. Small stock losses have been found to range between 1–35% (mean = 24.2%) for sheep and between 5–30% (mean = 16.6%) for goats [6]. Drought has also been found to increase the number of stockless households from 7 to 12% [7]. As for Mongolia, it has been reported that icing in 1993 resulted in the deaths of three-quarter of a million head of livestock where 110 households lost all animals, and 2 090 households lost >70% of their herds [8]. Between 1999 and 2002, 12 million livestock died in winter disasters, and many thousands of households lost their livelihoods [9]. In Inner Mongolia, about 30% of households have lost nearly all their livestock since 2001 due to continuous drought conditions [10]. On the Tibetan Plateau, six harsh winters with heavy snowfall from 1955 to 1990 resulted in 20–30% livestock losses [11]. Specifically, during the winter of 1996–1997 nomads on the western part of the plateau experienced losses of up to 70% and 25% of juvenile and adult goats respectively, and 20% of their lambs [11]. For Tibet in general, some townships lost up to 70% of their total livestock population, and by April 1998 it was estimated that the region had lost over 3 million head of livestock, which represents an estimated loss of USD 125 million [11]. In northern Norway, the reindeer husbandry utilizes winter pastures characterized by a cold but stable continental climate [12]. Nevertheless, mass starvation due to severe winter conditions, i.e. icing events, have been reported to dramatically reduce reindeer populations:

In 1918 one reindeer population was, for example, reduced by a third [13], and adverse weather events, i.e. too much snow in late winter, also caused substantial reductions in 1958, 1962 and 1968 [14].

The effects of environmental hazards are especially important in the context of climate change because the frequency of extreme weather events are predicted to increase in the future (a trend that has already been observed empirically: e.g. [15,16]), and thus represents a significant challenge for pastoralists [17]. It is therefore important to increase our understanding of both impacts of environmental hazards and the effects of strategies aimed at dampening them to enhance the ability of pastoralists to deal with the negative impacts of climate change.

Predictions

Herd accumulation has been argued, as well as to some extent demonstrated empirically, to be an effective strategy for buffering environmental hazards for short periods of time because it seems that wealthier pastoral households weather calamities better than poorer ones [18]. Nevertheless, few studies have evaluated the long-term effect of herd accumulation (but see [19]) and a case has been made that the effectiveness of herd accumulation should be assessed by analysing changes in livestock holdings during crisis periods, i.e. when pastoralists experience a near collapse in livestock holdings [20].

Consequently, our study assess two predictions: First, while the relationship between pre-collapse herd size and losses due to the crisis should be positive, adding one animal to pre-collapse herd size should result in losing less than one animal during the crisis if herd accumulation is an effective strategy for countering the negative impacts of environmental hazards (this is in line with evidence from Africa, see [21]). Second, we also assess to what extent pre-collapse herd size predicts post-collapse herd size, i.e. the *per se* benefit that herders gain by adding animals to their herds [7,20]. In short, we expect pre-collapse herd size to be a positive predictor of post-collapse herd size, if herd accumulation is an effective strategy for countering the negative impacts of environmental hazards.

Materials and Methods

Ethics Statement

The data utilized in this study were provided by the Norwegian Institute of Nature Research as part of the participation in the project ECOPAST (http://pastoralism-climate-change-policy.com/projects/). The standard of ethics pertaining to the data has been approved by the Norwegian Social Science Data Services in connection with the project 'Beregning av produktivitet i reindrift' ('Calculation of productivity in the reindeer husbandry').

The Saami Reindeer Husbandry in Norway

Saami reindeer husbandry has been said to be the cornerstone of the Saami culture in northern Fennoscandia [22]. Although it is difficult to come up with accurate dating of the origin of reindeer husbanding as a pastoral economy, it developed at least 400 years ago [23] and probably evolved from a hunting culture based on wild reindeer. Traditionally, reindeer pastoralism was based on families, or households, which followed the herds year-round where the pastoral economy was based on reindeer products [24]. The reindeer husbandry has undergone major technological, economic and political changes; most notably the production system has changed from being subsistence based to a motorized and market-oriented industry [25]. During the late 1970s, the Norwegian Government became more directly engaged in the reindeer husbandry through subsidies and regulations. Reforms during the end of the 1970s and early 1980s aimed at increasing both production and co-management [26]. In 1976, negotiation between the Saami Reindeer Herders' Association of Norway and the Norwegian Government resulted in the General Agreement for the Reindeer Industry (GARI). Importantly, this laid the foundation for annual negotiations pertaining to official subsidies and development: an arrangement that continues to this day [26]. The Reindeer Management Act (RMA) from 1978 focused on: (1) the establishment of formal institutions for access to the reindeer husbandry and pasture management; and (2) co-management. Berg [27] argued that the RMA of 1978 and the GARI of 1976 provided the foundation for a change into a corporative reindeer husbandry, i.e. not only production of meat for subsistence and sale but also for official subsidies. Accordingly, in many areas different support and compensatory arrangements have provided around half of the income [27]. The RMA from 2007 broadened the focus on co-management by giving the industry more self-determination, influence and responsibility for its actions [28].

At present, the Saami reindeer husbandry can be distinguished into three different levels of social organization. The *husbandry unit*, lately designated as 'siida shares', is the basic unit of the social organization and consists of a government license that entitles a person to manage a herd of reindeer within a delimited area. The *siida* is a cooperative unit composed of one or more reindeer management families organized on the basis of kinship joined together in social and labour communities for keeping control of herds of reindeer through herding. A *district* is a formal management unit with responsibility to provide the Norwegian Reindeer Husbandry Administration with information as well as ensuring that the reindeer husbandry is managed in accordance with governmental regulations [18,29,30,31].

Data Material – Collapse in the Reindeer Husbandry

To evaluate the efficiency of herd accumulation as a risk reducing strategy the present study utilized data from the reindeer husbandry in Finnmark, Norway (see Text S1 for details). From an historical point of view, the number of reindeer in Finnmark has been characterized by considerable temporal variation: from the early 1900s, there was a decreasing trend that reached a minimum around the Second World War. Afterwards, while fluctuating there has been an upward trend that peaked in the early 1990s (from 90 000 animals in 1976 to 210 000 in 1988, [32]) and decreased until 2000/2001. More recently, reindeer abundance increased by ~40% from 2002 to 2010 [33] and again reached a historical high-level [34]. After the peak in the early 1990s, governmental subsidies resulted in increased harvest rates and a subsequent decline in reindeer abundance [35]. This downward trend was further compounded by the "[...] catastrophic winters with heavy snowfalls in 1997 and 2000 [...]" (Hausner et al. [14], p. 6), which resulted in a low point in reindeer abundance in 2001 (Fig. 1).

By designating the low point in 2001 (Fig. 1) as a 'collapse' in reindeer abundance (but see Text S2) it was possible to shed additional light on the effectiveness of herd accumulation as a risk management strategy by looking at the relationship between: (1) pre-collapse herd size and loss; and (2) pre-collapse and post-collapse herd size.

Study Protocol and Statistical Analyses

As in previous studies, e.g. [36], this empirical study is based on governmental statistics compiled and published annually by the Norwegian Reindeer Husbandry Administration. This dataset contains data on herd size (total number of reindeer in the spring per husbandry unit), covering the period 1998–2008 with data from 20 reindeer husbandry summer districts. Data pertaining to herd size are based on counts made by herders that are regularly checked by the authorities (for more details pertaining to dataset and design, see [18,29,36]). The utilized dataset contains the following variables:

$N_{post-collapse}$ (response).– A continuous (husbandry unit level) variable denoting the total herd size at the end of the period (i.e. 2008).

N_{loss} (response).– A continuous (husbandry unit level) variable denoting the number of reindeer lost from the *pre-collapse* (1998) to the *collapse* (2001) year. The variable was created by subtracting herd size in 2001(collapse) from herd size in 1998 (pre-collapse).

$N_{pre-collapse}$(predictor).– A continuous (husbandry unit level) variable denoting the total herd size at the beginning of the period (i.e. 1998).

To evaluate the efficacy of herd accumulation as a risk reducing strategy, we looked at the relationship between: (1) $N_{pre-collapse}$ and N_{loss}; and (2) $N_{pre-collapse}$ and $N_{post-collapse}$. Statistical analyses and plotting of results were carried out in *R* [37]. All tests were two-tailed and the null-hypothesis was rejected at an α-level of 0.05, and we used Wald statistics to test if estimated parameters were significantly different from zero. Regular linear regression was used to investigate the overall relationships between pre-collapse and post-collapse herd size (grouping effects, e.g. possible differences between districts with respect to natural and/ or social factors, were considered to be negligible, see Text S3 for details). Visual inspection of the data indicated problems related to the homoscedastic assumption [38]. We therefore fitted models with different variance structures in order to assess if violations of the homoscedastic assumption altered our conclusions [38]. As the conclusions from homoscedastic and heteroscedastic models were similar, we present the results from regular linear models in the main text (see Text S4 for the results based on heteroscedastic models).

Results

Pre-collapse herd size had a positive effect on the number of animals lost [effect of$N_{pre-collapse}$: 0.44], which means that husbandry units with more animals in 1998 tended to lose more animals from 1998 to 2001 (Fig. 2A). More specifically, if a herder increased his/her herd by one animal in 1998 his/her losses were expected to increase by 0.44 animals. This indicates that while herd accumulation is effective, it is a costly form of insurance against environmental variability. The heteroscedastic models revealed the same relationship although the estimated effect size was reduced (see Text S4 for details). For a similar analysis pertaining to reported losses, see Text S5.

Pre-collapse herd size had a positive effect on post-collapse herd size [effect of$N_{pre-collapse}$: 0.62], i.e. husbandry units that had more animals in 1998 also tended to have more animals in 2008 (Fig. 2B). While the relationship is not perfect, it provides the underlying rationale for herd accumulation: having a large herd prior to a collapse ensures a large herd after the collapse (a 1% increase in pre-collapse herd size predicts a 0.62% increase in post-collapse herd size, as both variables were $\log_e$-transformed). Herd accumulation thus seems to be an effective strategy for countering the negative impacts of environmental hazards. The heteroscedastic models revealed the same relationship, but again the effect size was reduced (see Text S4 for details).

Discussion

The positive relationship between pre- and post-collapse herd size indicates that herd accumulation is a rational response to environment-induced catastrophes, and provides the rationale for why both reindeer herders [29,31] and pastoralists in general [39] invest labour to increase herd size. Nevertheless, herd accumulation may also stem from cultural values: e.g. prestige or status [40,41,42]; conspicuous display [43]; and provision of bridewealth [44]. However, the fact that pre- and post-collapse herd size was positively correlated clearly shows the economic rationale and whether herd accumulation also results from cultural values is somewhat irrelevant [20]. This point of view is also shared by the herders in Finnmark since 51% of the herders 'agree' or 'strongly agree' that herds size is an important risk reducing strategy, while only 26% 'agree' or 'strongly agree' that herd size is important for social status [45].

A previous study found that herd size one year was positively correlated with herd size the next year [18]. Although the previous study was based on a shorter time series, the present study complements the former in showing the underlying rationale for herd accumulation: (1) a large herd one year result in a larger herd the next year [18]; and (2) a large pre-collapse herd also result in a larger post-collapse herd (this study). In short, since herders with large herds also have comparable larger herds from one year to the next and during crisis periods, herd accumulation maximizes long-term survival for pastoralists (see also [46]). Nevertheless, the positive relationship between pre-collapse herd size and loss indicates that herd accumulation is a costly form of insurance against environmental variability (see [21] for a similar result for Africa; and see [47] p. 36-8 for a discussion pertaining to 'loss' and the effectiveness of herd accumulation).

Herd accumulation may, however, not be the best available risk reducing strategy because of negative density dependence. For the reindeer husbandry it has been demonstrated that reindeer density and climatic conditions have negative effects on individual body mass [12,31,48,49,50] and, consequently, also survival [49,51]. Increased reindeer abundance also increases herders' vulnerability to unfavourable climate as the negative impacts of adverse climatic events increases with increasing reindeer density [12,49,52]. From a risk perspective, the best long-term strategy may thus be to invest in livestock body mass and not herd size. Næss et al. [29] hypothesised that one reason for why this strategy is not utilised by herders is because of competition for access to common winter pastures since access is to a large degree determined by herd size. Herd accumulation, through e.g. cooperative labour investment [29], may thus be a viable strategy for gaining access to winter pastures. More to the point, competition for access to winter pastures may explain why herd accumulation is the dominant and only viable risk reducing strategy in Finnmark. Herd accumulation may thus be taken to support a Tragedy of the Commons (ToC) [53] situation in Finnmark. ToC is a prisoners' dilemma where everybody is better off by coordinating strategies, e.g. by restricting the number of animals on the pastures, but the dominant solution is to always maximize herd size because the cost of high reindeer density is shared between all herders while the benefit of adding additional animals is attached to individual units [18]. Nevertheless, evidence supports both a *presence*, e.g. the aforementioned negative density dependence, and an *absence*, e.g. a positive relationship between slaughter undertaken by neighbouring herders and own slaughter [30], of ToC in Finnmark.

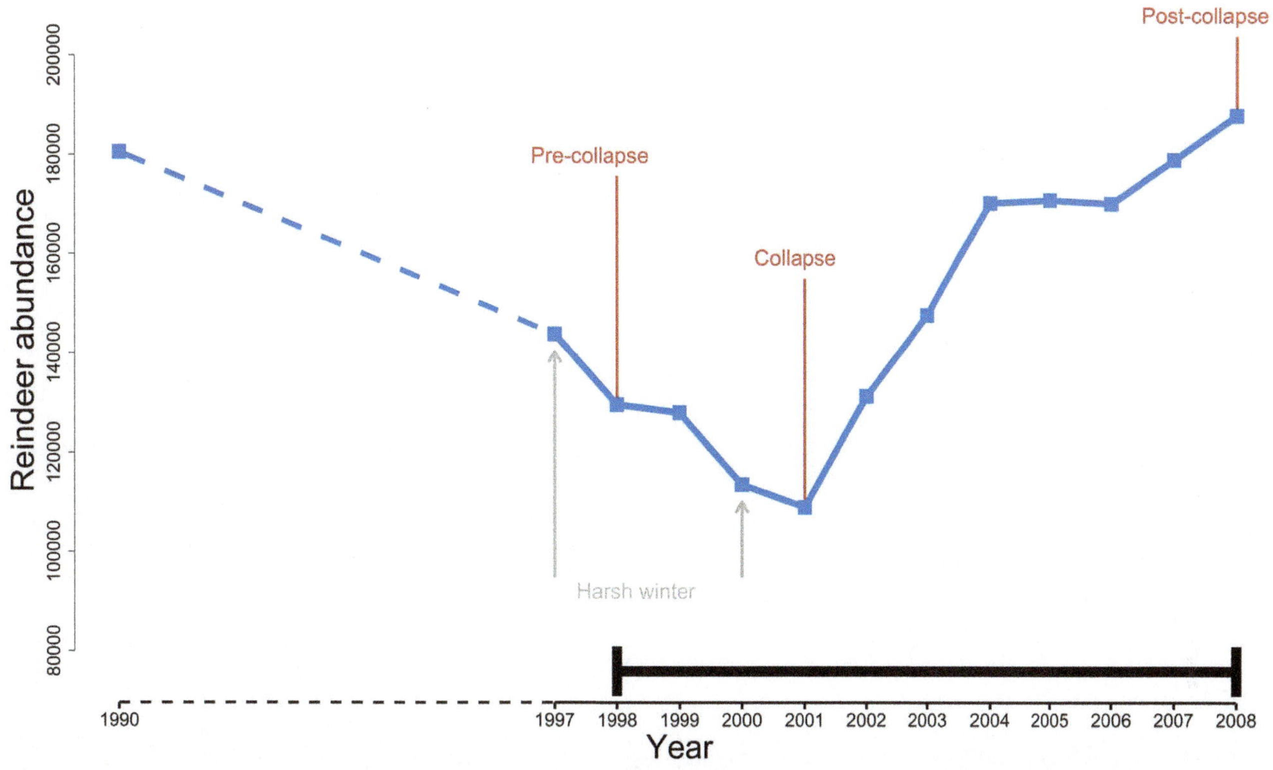

Figure 1. Temporal trend in reindeer abundance in Finnmark, Norway. Thick arrow indicates the period for which there exists official statistics pertaining to herd size for *individual husbandry units* (1998–2008), and thus restricts the period designations 'pre-collapse', 'collapse' and 'post-collapse' used in the analyses. Hatched line indicates missing data (i.e. abundance between 1990 and 1997). Abundance for 1990 from Tømmervik and Riseth [34], while abundance from 1997–2008 are per 31st March for each year from Anonymous [65,66,67].

Future Prospects and Management Implications

Scenarios for future climate change generally predict an increasing average, variance and even a changed distribution of important climatic variables like precipitation and temperature [15,54]. Climate change will most likely result in more frequent extreme weather events and there are indications that extreme events, such as heat waves and precipitation extremes, will increase and already have done so [15,16]. Nevertheless, there is limited information available pertaining to how these changes are going to affect the many pastoral cultures of the world. In Africa, climate change is predicted to increase the variability and frequency of rainfall at the same time as the proportion of arid and semiarid lands is likely to increase by 5–8% by 2080 [55]. Furthermore, in the Greater Horn of Africa droughts have now become the norm rather than the exception [56]. While previously pastoralists experienced one major long-term drought every decade coupled with minor occurrences every 3–4 years, droughts now occur annually [56]. As for the Arctic and Sub-Arctic, scenarios generated by most global climate models predict that the climate is likely to become increasingly unstable during the next half century with concomitant increases in the frequency of extreme weather conditions [57]. For Mongolia, regional climate predictions anticipate an increase in areas affected by droughts and in the frequency of extreme events [58]. Importantly, the frequency of droughts has almost doubled during the last 60 years and the worst droughts on record (covering over 50–70% of the country) have occurred during the last decade [58].

Considering the negative impact that environmental hazards have on livestock survival and reproduction, climate change thus represents a significant threat for the future of pastoral societies on a global scale. Nevertheless, it has been argued that by reinforcing the traditional strategies pastoralists have developed to deal with climate variability, in addition to introducing newer techniques, the economic, social, and cultural well-being of pastoral societies can be supported in the face of climate change [59]. Moreover, a case has been made that pastoralists are in a unique position to tackle climate change due to extensive experience managing environmental variability in marginal areas [2] and it has been argued that the ability to withstand environmental shocks is a *defining* feature of pastoralism [60].

Nevertheless, traditional pastoral risk management strategies, such as herd accumulation, may be insufficient for dealing with climate change [56]. While herd accumulation seems to be an efficient strategy, it is predicated on periods of recuperation when herd growth is possible. In fact, a delay in restocking after environment-induced losses is one of the main problems of pastoral production [61]. Herd accumulation can thus be expected to work less efficiently, if at all, when the frequency of extreme events increases.

As for Finnmark, it could be argued that the number of reindeer in Finnmark is unsustainable high, witnessed by the presence of negative density dependence (see above). Consequently, the Norwegian Government is aiming to reduce the number of reindeer so as to achieve a sustainable balance between pasture resources and number of reindeer [28]. The primary tool utilized are several subsidies that aims to increase slaughter and thus reduce herd size [62,63]. Considering the risk-beneficial aspects of having a large herd, this may, however be viewed as a short-term solution that, if followed, decreases long term viability by reducing the insurance potential of large herds. More to the point, official

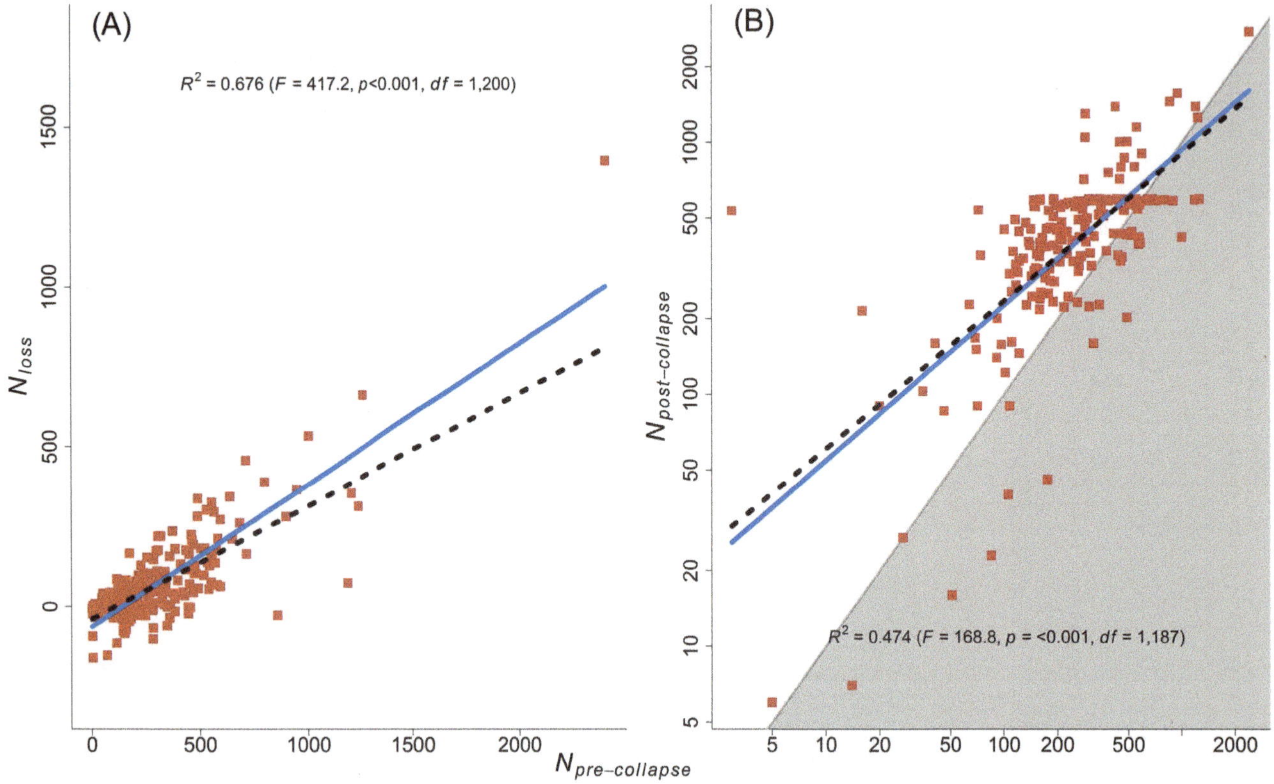

Figure 2. Showing the linear relationship between pre-collapse herd size ($N_{pre-collapse}$) and number of animals lost from pre-collapse to collapse (N_{loss}) (A). Model parameters: Intercept = 75.24 [95% confidence intervals (CI): 63.42, 87.07] and slope ($N_{pre-collapse}$) = 0.44 (95% CI: 0.40, 0.49). The positive relationship indicates that as herd size increases losses also increases: increasing herd size by one animal in 1998 increases the expected losses by 0.44 reindeer. Note that the model parameters are from fitting a model when centring $N_{pre-collapse}$ while the plot shows the relationship on the original scale. Hatched line show the relationship from a Generalized Least Square (GLS) model accounting for potential residual heterogeneity (see Text S4 for details). Showing the linear relationship (on $\log_e$-scale) between pre-collapse ($N_{pre-collapse}$) and post-collapse ($N_{post-collapse}$) herd size (**B**). Points above the shaded area indicate herd increase over the period, while a point on the 45-degree line means that pre- and post-collapse herd size was equal, and points in the shaded region indicate a decrease. The cloud of points above the 45-degree shaded area reflects the overall increase in reindeer abundance for the study area (Fig. 1). Model parameters: Intercept = 2.58 (95% CI: 2.06, 3.09) and slope ($N_{pre-collapse}$) = 0.62 (95% CI: 0.52, 0.71). The positive relationship indicates that as pre-collapse herd size increases so does post-collapse herd size: a 1% increase in $N_{pre-collapse}$ herd size predicts a 0.62% increase in$N_{post-collapse}$ herd size. Hatched line show the relationship from a GLS model accounting for potential residual heterogeneity (see Text S4 for details).

management strategies that fails to incorporate the risk-beneficial aspect of herd accumulation will have a limited effect and may indeed fail entirely [18,36]. Furthermore, production subsidies alone may not properly account for the decision problem facing herders: how to secure a reliable income while at the same time maximizing long-term survival. This decision problem is even more dire in the face of climate change: if herd accumulation works less efficiently when the frequency of extreme events increases, governmental support that incorporates (or preferably increase the effect of) pastoral risk management strategies may be the only solution for ensuring the continued existence of pastoralism. In the end, it may be official policies that disregard the inherent risk reducing logic of pastoral strategies rather than climate change *per se* that represents the greatest challenge for pastoral adaptability [17,64].

Supporting Information

Text S1 Study design and the reindeer husbandry. This text provides a more detailed description of the study design as well as the reindeer husbandry in Norway.

Text S2 Is there really a 'collapse' in reindeer abundance? This text investigates temporal trends in low points of reindeer abundance from 1845–2000.

Text S3 Possible grouping effects. This text investigates possible grouping effects, e.g. differences between summer districts with respect to natural and/or social factors.

Text S4 Finding the correct variance structure. This text explores models with different variance structures in order to assess if violations of the homoscedastic assumption altered the conclusions presented in the main text.

Text S5 Reported loss. This text replicates the analysis pertaining to loss presented in the main text but where 'reported loss' is used as a response.

Acknowledgments

We thank the Reindeer Husbandry Administration for access to data and Torkild Tveraa for preparing the data prior to the statistical analyses. We would also like to thank two anonymous reviewers for improving the manuscript and Matthew Thomas for correcting and improving the language.

Author Contributions

Analyzed the data: MWN BJB. Wrote the paper: MWN BJB.

References

1. Reid RS, Galvin KA, Kruska RS (2008) Global Significance of Extensive Grazing Lands and Pastoral Societies: An Introduction. In: Galvin KA, Reid RS, Behnke JRH, Hobbs NT, editors. Fragmentation in semi-arid and arid landscapes: consequences for human and natural systems. Dordrecht: Springer. 1–24.
2. Nori M, Taylor M, Sensi A (2008) Browsing on fences: Pastoral land rights, livelihoods and adaptation to climate change. Nottingham, UK: International Institute for Environment and Development 29 p.
3. Neely C, Bunning S, Wilkes A (2009) Review of evidence on drylands pastoral systems and climate change - Implications and opportunities for mitigation and adaptation. Rome: Food and Agriculture Organization of the United Nations.
4. Dong SK, Wen L, Liu SL, Zhang XF, Lassoie JP, et al. (2011) Vulnerability of Worldwide Pastoralism to Global Changes and Interdisciplinary Strategies for Sustainable Pastoralism. Ecology and Society 16(2): 10. Available: http://www.ecologyandsociety.org/vol16/iss2/art10/. Accessed 2013 Jun 26.
5. Nkedianye D, de Leeuw J, Ogutu JO, Said MY, Saidimu TL, et al. (2011) Mobility and livestock mortality in communally used pastoral areas: the impact of the 2005–2006 drought on livestock mortality in Maasailand. Pastoralism: Research, Policy and Practice 1: 17 Available: http://www.pastoralismjournal.com/content/1/1/17. Accessed 2013 Jun 24.
6. Little PD, Mahmoud H, Coppock DL (2001) When deserts flood: risk management and climatic processes among East African pastoralists. Climate Research 19: 149–159.
7. McPeak J, Little PD (2005) Cursed If You Do, Cursed If You Don't: The Contradictory Processes of Pastoral Sedentarization in Northern Kenya. In: Fratkin E, Roth EA, editors. As Pastoralists Settle Social, Health, and Economic Consequences of Pastoral Sedentarization in Marsabit District, Kenya. New York and London New York, Boston, Dordrecht, London, Moscow: Kluwer Academic Publishers. 87–104.
8. Templer G, Swift J, Payne P (1993) The changing significance of risk in the Mongolian pastoral economy. Nomadic Peoples 33: 105–122.
9. Janes CR (2010) Failed Development and Vulnerability to Climate Change in Central Asia: Implications for Food Security and Health. Asia-Pacific Journal of Public Health 22: 236S–245S.
10. Xie Y, Li W (2008) Why do herders insist on *otor*? Maintaining mobility in Inner Mongolia. Nomadic Peoples 12: 35–52.
11. Miller DJ (2000) Tough times for Tibetan nomads in Western China: Snowstorms, settling down, fences, and the demise of traditional nomadic pastoralism. Nomadic Peoples 4: 83–109.
12. Bårdsen B-J, Tveraa T (2012) Density dependence vs. density independence – linking reproductive allocation to population abundance and vegetation greenness. Journal of Animal Ecology 81: 364–376.
13. Bjørklund I (1990) Sami Reindeer Pastoralism as an Indigenous Resource Management System in Northern Norway: A Contribution to the Common Property Debate. Development and Change 21: 75–86.
14. Hausner VH, Fauchald P, Tveraa T, Pedersen E, Jernsletten J-LL, et al. (2011) The Ghost of Development Past: the Impact of Economic Security Policies on Saami Pastoral Ecosystems. Ecology and Society 16: 4. Available: http://dx.doi.org/10.5751/ES-04193-160304. Accessed 2013 Jun 24.
15. Sun Y, Solomon S, Dai AG, Portmann RW (2007) How often will it rain? Journal of Climate 20: 4801–4818.
16. Coumou D, Rahmstorf S (2012) A decade of weather extremes Nature Climate Change 2: 491–496.
17. Næss MW (2012) Tibetan Nomads Facing an Uncertain Future: Impacts of Climate Change on the Qinghai-Tibetan Plateau. In: Lamadrid A, Kelman I, editors. Climate Change Modeling for Local Adaptation in the Hindu Kush-Himalayan Region: Emerald Group Publishing Limited. 99–122.
18. Næss MW, Bårdsen B-J (2010) Environmental Stochasticity and Long-Term Livestock Viability-Herd-Accumulation as a Risk Reducing Strategy. Human Ecology 38: 3–17.
19. Fratkin E, Roth EA (1990) Drought And Economic Differentiation Among Ariaal Pastoralists Of Kenya. Human Ecology 18: 385–402.
20. McPeak J (2005) Individual and collective rationality in pastoral production: Evidence from Northern Kenya. Human Ecology 33: 171–197.
21. Lybbert TJ, Barrett CB, Desta S, Coppock DL (2004) Stochastic wealth dynamics and risk management among a poor population. Economic Journal 114: 750–777.
22. Bostedt G (2001) Reindeer husbandry, the Swedish market for reindeer meat, and the Chernobyl effects. Agricultural Economics 26: 217–226.
23. Paine R. (1994). *Herds of the Tundra: a portrait of Saami reindeer pastoralism*, Washington and London: Smithsonian Institution Press.
24. Vorren Ø. (1978). *Bosetning og ressursutnytting under veidekulturen og dens differensiering*, in NOU 18A, Finnmarksvidda-natur-kultur. Oslo: Universitetsforlaget. 145–62.
25. Riseth JÅ (2003) Sami Reindeer Management in Norway: Modernization Challenges and Conflicting Strategies. Reflections Upon the Co-management Alternative. In: Jentoft S, Minde H, Nilsen R, editors. Indigenous Peoples: Resource Management and Global Rights. Delft, Netherlands: Eburon Academic Publishers. 229–247.
26. Riseth JÅ, Vatn A (2009) Modernization and Pasture Degradation: A Comparative Study of Two Sàmi Reindeer Pasture Regions in Norway. Land Economics 85: 87–106.
27. Berg BÅ (2008) Utviklingen av reindriften i nordre Nordland 1750–2000. In: Evjen B, Hansen LI, editors. Nordlands kulturelle mangfold: etniske relasjoner i historisk perspektiv. Oslo: Pax. 151–191.
28. Ulvevadet B (2008) Management of reindeer husbandry in Norway - power-sharing and participation. Rangifer 28: 53–78.
29. Næss MW, Bårdsen B-J, Fauchald P, Tveraa T (2010) Cooperative pastoral production - the importance of kinship. Evolution and Human Behavior 31: 246–258.
30. Næss MW, Bårdsen B-J, Tveraa T (2012) Wealth-dependent and interdependent strategies in the Saami reindeer husbandry, Norway. Evolution and Human Behavior 33: 696–707.
31. Næss MW, Fauchald P, Tveraa T (2009) Scale Dependency and the "Marginal" Value of Labor. Human Ecology 37: 193–211.
32. Riseth JÅ, Johansen B, Vatn A (2004) Aspects of a two-pasture – herbivore model. Rangifer Special Issue 15: 65–81.
33. Anonymous (2012) Riksrevisjonens undersøkelse av bærekraftig bruk av reinbeiteressursene i Finnmark - Dokument 3: 14 (2011–2012). Norway: Riksrevisjonen. 89.
34. Tømmervik H, Riseth JÅ (2011) Historiske tamreintall i Norge fra 1800-tallet fram til i dag. NINA report 672. Tromsø: NINA.
35. Ulvevadet B, Hausner VH (2011) Incentives and regulations to reconcile conservation and development: Thirty years of governance of the Sami pastoral ecosystem in Finnmark, Norway. Journal of Environmental Management 92: 2794–2802.
36. Næss MW, Bårdsen B-J, Pedersen E, Tveraa T (2011) Pastoral Herding Strategies and Governmental Management Objectives: Predation Compensation as a Risk Buffering Strategy in the Saami Reindeer Husbandry. Human Ecology 39: 489–508.
37. R Development Core Team (2011) R: A language and environment for statistical computing. Vienna, Austria: R Foundation for Statistical Computing.
38. Zuur AF, Ieno EN, Walker N, Saveliev AA, Smith GM. (2009). *Mixed effects models and extensions in ecology with R*, New York: Springer.
39. Næss MW (2012) Cooperative Pastoral Production: Reconceptualizing the Relationship between Pastoral Labor and Production. American Anthropologist 114: 309–321.
40. Anonymous (2009) Sluttrapport for prosjektet: "Økosystem Finnmark", Tromsø: University of Tromsø.
41. Herskovits MJ (1926) The Cattle Complex in East Africa. American Anthropologist 28: 633–664.
42. Nilsen R, Mosli JH. (1994). *Inn fra vidda: hushold og økonomisk tilpasning i reindrifta i Guovdageaidnu 1960–1993*, Guovdageaidnu: BAJOS Utviklingsselskap AS/ NORUT Samfunnsforskning AS, Tromsø.
43. Paine R. (2009). *Camps of the tundra: politics through reindeer among Saami pastoralists*, Oslo: Instituttet for sammenlignende kulturforskning.
44. McCabe JT. (2004). Cattle bring us to our enemies: Turkana ecology, politics, and raiding in a disequilibrium system, Ann Arbor: University of Michigan Press.
45. Johannesen AB, Skonhoft A (2011) Livestock as Insurance and Social Status: Evidence from Reindeer Herding in Norway. Environmental & Resource Economics 48: 679–694.
46. Mace R, Houston A (1989) Pastoralist strategies for survival in unpredictable environments: A model of herd composition that maximises household viability. Agricultural Systems 31: 185–204.
47. Næss MW (2009) Pastoral Risk Management - The Importance of Cooperative Production [PhD]. Tromsø: University of Tromsø.
48. Bårdsen B-J (2009) Risk sensitive reproductive strategies: the effect of environmental unpredictability [PhD]. Tromsø: University of Tromsø.
49. Bårdsen B-J, Henden J-A, Fauchald P, Tveraa T, Stien A (2011) Plastic reproductive allocation as a buffer against environmental stochasticity – linking life history and population dynamics to climate. Oikos 20: 245–257.
50. Tveraa T, Fauchald P, Yoccoz NG, Ims RA, Aanes R, et al. (2007) What regulate and limit reindeer populations in Norway? Oikos 116: 706–715.
51. Tveraa T, Fauchald P, Henaug C, Yoccoz NG (2003) An examination of a compensatory relationship between food limitation and predation in semi-domestic reindeer. Oecologia 137: 370–376.

52. Bårdsen B-J, Tveraa T, Fauchald P, Langeland K (2010) Observational evidence of a risk sensitive reproductive allocation in a long-lived mammal. Oecologia 162: 627–639.
53. Hardin GJ (1968) The Tragedy of the Commons. Science 162: 1243–1248.
54. Benestad RE (2011) A New Global Set of Downscaled Temperature Scenarios. Journal of Climate 24: 2080–2098.
55. Galvin KA (2009) Transitions: Pastoralists Living with Change. Annual Review of Anthropology 38: 185–198.
56. Blackwell PJ (2010) East Africa's Pastoralist Emergency: is climate change the straw that breaks the camel's back? Third World Quarterly 31: 1321–1338.
57. Brannlund I, Axelsson P (2011) Reindeer management during the colonization of Sami lands: A long-term perspective of vulnerability and adaptation strategies. Global Environmental Change-Human and Policy Dimensions 21: 1095–1105.
58. Marin A (2010) Riders under storms: Contributions of nomadic herders' observations to analysing climate change in Mongolia. Global Environmental Change-Human and Policy Dimensions 20: 162–176.
59. Secretariat of the Convention on Biological Diversity (2009) Connecting Biodiversity and Climate Change Mitigation and Adaptation: Report of the Second Ad Hoc Technical Expert Group on Biodiversity and Climate Change. Montreal, Technical Series No. 41. 126 p. Available: http://www.cbd.int/doc/publications/cbd-ts-41-en.pdf. Accessed 2013 Jun 26.
60. Hatfield R, Davies J (2006) Global Review of the Economics of Pastoralism. Nairobi: IUCN. 44 p. Available: http://cmsdata.iucn.org/downloads/global_review_ofthe_economicsof_pastoralism_en.pdf. Accessed 2013 Jun 26.
61. Bollig M, Göbel B (1997) Risk, Uncertainty and Pastoralism: An Introduction. Nomadic Peoples 1: 5–21.
62. Anonymous (2008) Ressursregnskap for reindriftsnæringen. Alta: Reindriftsforvaltningen. 164 p.
63. Anonymous (2007) St.prp. nr. 74: Om reindriftsavtalen 2007/2008, om dekning av kostnader vedrørende radioaktivitet i reinkjøtt, og om endringer i statsbudsjettet for 2007 m.m. Oslo: Det Kongelige Landbruks- og Matdepartement. 36.
64. Næss MW (2013) Climate Change, Risk Management and the End of *Nomadic* Pastoralism. International Journal of Sustainable Development and World Ecology 20: 123–133.
65. Anonymous (2001) Totalregnskap for reindriftsnæringen. Alta: Reindriftsforvaltningen. 145 p.
66. Anonymous (2004) Totalregnskap for reindriftsnæringen. Alta: Reindriftsforvaltningen. 143 p.
67. Anonymous (2010) Totalregnskap for reindriftsnæringen. Alta: Reindriftsforvaltningen. 132 p.

Spatial and Temporal Distribution of West Nile Virus in Horses in Israel (1997–2013) - from Endemic to Epidemics

Karin Aharonson-Raz[1], Anat Lichter-Peled[1], Shlomit Tal[2], Boris Gelman[3], Daniel Cohen[2], Eyal Klement[1], Amir Steinman[1]*

1 Koret School of Veterinary Medicine, The Robert H. Smith Faculty of Agriculture, Food and Environment, The Hebrew University of Jerusalem, Rehovot, 76100, Israel, **2** School of Public Health, Sackler Faculty of Medicine, Tel Aviv University, Tel Aviv, 39040, Israel, **3** Kimron Veterinary Institute, Bet-Dagan, 50250, Israel

Abstract

With the rapid global spread of West Nile virus (WNV) and the endemic state it has acquired in new geographical areas, we hereby bring a thorough serological investigation of WNV in horses in a longstanding endemic region, such as Israel. This study evaluates the environmental and demographic risk factors for WNV infection in horses and suggests possible factors associated with the transition from endemic to epidemic state. West Nile virus seroprevalence in horses in Israel was determined throughout a period of more than a decade, before (1997) and after (2002 and 2013) the massive West Nile fever outbreak in humans and horses in 2000. An increase in seroprevalence was observed, from 39% (113/290) in 1997 to 66.1% (547/827) in 2002 and 85.5% (153/179) in 2013, with persistent significantly higher seroprevalence in horses situated along the Great Rift Valley (GRV) area, the major birds' migration route in Israel. Demographic risk factors included age and breed of the horse. Significantly lower spring precipitation was observed during years with increased human incidence rate that occurred between 1997–2007. Hence, we suggest referring to Israel as two WNV distinct epidemiological regions; an endemic region along the birds' migration route (GRV) and the rest of the country which perhaps suffers from cyclic epidemics. In addition, weather conditions, such as periods of spring drought, might be associated with the transition from endemic state to epidemic state of WNV.

Editor: Pierre Roques, CEA, France

Funding: The authors have no support or funding to report.

Competing Interests: The authors have declared that no competing interests exist.

* Email: amirst@savion.huji.ac.il

Introduction

West Nile virus (WNV) is a member of the *Flavivirus* genus in the Flaviviridae family [1]. Since the virus was first isolated from the blood of a febrile woman in Uganda in 1937 [2] it has become recognized as one of the most widely distributed flaviviruses in humans as well as in horses. West Nile virus infection was first reported in the Mediterranean basin, in both Egypt and Israel, in the early 1950s and cases with severe neurological manifestations in humans were reported for the first time in 1957 in Israel [3]. In horses, WNV encephalomyelitis was first recorded in the Middle East in 1959 in Egypt [4]. Antibodies to WNV were detected in horses in Israel already in the 1960s [5], but outbreaks were not reported during the next 3 decades. Until the mid-1990's, WNV was not considered an important differential diagnosis for equine encephalomyelitis disease. However, several large scale, well documented outbreaks among equine populations raised the awareness to this disease. West Nile virus encephalomyelitis outbreaks were reported in Morocco in 1996 [3], in Italy in 1998 [6], in France in 2000 [7] and in Israel in 2000 [8]. The dramatic appearance of WNV in New York City (NYC) area in 1999 and its subsequent spread in North America largely contributed to the overall focus the disease acquired. During the decade since its first discovery in NYC (between 1999 and 2012), a total of 37,008 cases of West Nile fever and 16,196 cases of neuroinvasive West Nile disease were reported in humans in the US, with incidence of neuroinvasive disease between 0.1 and 1.0 per 100,000 [9]. Not only is the incidence of West Nile in horses (~700 per 100,000) substantially higher than in humans [10], the disease in horses seems to be more severe; over 25,000 horses in USA have been affected since 1999, with 33% case-fatality and 40% of survivors with neurological sequelae. In humans, 4–9% case fatality was recorded and 30% of encephalitis survivors with sequelae [11]. Another interesting difference observed was that when data was compared between 904 confirmed human cases gathered from the European Centre of Disease Prevention and Control and 200 confirmed equine cases gathered from the World Organisation for Animal Health during 2010, it was noticed that equine morbidity started three weeks later than humans' [12].

West Nile virus imposes great threat on human and animal health not only by emerging in new geographical areas but also as it causes epidemics in endemic areas. Despite the presence of WNV in most of the United States already a decade ago and its spread to all regions of the continental US throughout this decade [13], WNV epidemic activity was demonstrated in 2012, in which more human disease cases were reported nationally than any year since 2003. This outbreak resulted in 2,873 cases of neuroinvasive disease and 286 human deaths [14,15]. Available data suggest that the increased incidence of WNV disease in 2012 was not likely caused by genotypic changes in the circulating virus strains and it was suggested that a number of inter-related factors, including weather, abundance of birds that maintain the virus, abundance of mosquitoes that spread the virus, and human behaviour are all contributors to the disease epidemics [16]. WNV has caused sporadic outbreaks in humans, horses and birds throughout many of the warmer regions of Europe for at least 20 years. However, currently, WNV appears to be expanding its geographical range in Europe and causing increasing numbers of outbreaks associated with human morbidity and mortality. The increase in the number of WNV cases in the last few years in central and southern Europe is likely due to the introduction of a lineage 2 WNV to central Europe (Hungary) in 2004 [17]. After a few years of adaptation, this virus strain has been heavily spreading first within Hungary and to the eastern part of Austria [18], and thereafter through the Balkan states to Greece [19], causing a significant outbreak of human West Nile neuroinvasive disease. In the last 3 years, in addition to the Balkan area, large human epidemics and WNV human involvement was also detected in Italy, Hungry, the Russian Federation and Ukraine [20,21]. Israel serves as a crossroad for bird migration between Africa and Euroasia, and therefore a major focus of attention during the global spread of WN fever (WNF).

Israel has been considered endemic to WNV since the first recognized human WNF outbreak in 1951. Thereafter, decades of small, intermittent outbreaks were reported in 1952, 1953, 1957 and 1980. However, in 2000 Israel experienced its largest recorded outbreak affecting hundreds of people [22], dozens of horses [8] and several flocks of geese [23]. During the following decade, between 2000 and 2012, nearly 1,400 cases of human WNF were reported. Identified cases were fewer between 2001 and 2004 but greatly increased in 2005 and stayed consistently high since then [24]. Previous studies that raised assumptions for the causes of re-emergence of WNV in endemic areas focused mainly on human data. Biotic factors raised include past immunity and increased in awareness among physicians and more frequent request for laboratory testing, while abiotic factors discussed include climatic parameters (i.e. temperature, relative humidity and precipitation) and environmental drives (i.e. bird migration) [20,25]. Since horses serve as incidental hosts to WNV, similar to humans, characterizing the spatial and temporal distribution of disease occurrence in horses may shed light on the overall prevalence of WNV in an endemic region such as Israel. Therefore, the objectives of this study were to identify the seroprevalence and force of infection of WNV in horses in Israel for a period of more than a decade, between 1997 until 2013, and to characterize demographic and environmental risk factors associated with WNV occurrence. In addition, factors associated with transition from endemic to epidemic state were also examined.

Materials and Methods

Blood samples

Blood samples were collected from clinically healthy horses distributed throughout Israel, in order to determine seroprevalence of WNV in the years 1997, 2002 and 2013. In 1997, 290 horses, located in 12 farms were sampled. In 2002, 827 horses located in 52 farms were sampled and in 2013, 179 horses, located in 16 farms were sampled. All of these horses have not been vaccinated to WNV (vaccination of horses began only after 2002 and in 2013 vaccination status was gathered from the owners). Blood collections were performed under owners' consent and all surveys involving blood collection were approved by the Internal Ethics Review Committee of the Koret School of Veterinary Medicine, The Hebrew University. Farms sampled were located throughout Israel according to the estimated geographical distribution of horse farms in the county, with higher farms density in the center, followed by the north and finally in the south of Israel. The horse population in Israel is estimated to be between 20,000–30,000 horses. Most horses live for many years in the same farm, but traveling between farms occurs occasionally, mainly for competition purposes or for trail rides.

Serum preparation, serum neutralization test (SN) and competitive enzyme-linked immunosorbent assay (cELISA)

Blood samples were collected from the jugular vein of each horse into sterile serum-separating tubes and centrifuged at 3000 rpm for 8 minutes. Sera were collected and stored at − 80°C until further analysis. Seroprevalence of WNV was detected using the SN method for the 1997, 2002, and 2013 serum samples and the ELISA method for the 2013 serum samples.

For the SN test, Vero cells were grown on Eagle's medium with 10% fetal calf serum (Bio-logical Industries Ltd. Beit Haemek, Israel) and 1% antibiotic solution, including penicillin, streptomycin, and amphotericin (Bio-logical Industries Ltd. Beit Haemek, Israel). The cells were maintained in a humidified atmosphere at 37°C in 5% CO_2 and were used for growing virus stocks and virus neutralization assay. Double dilutions from 1/10 to 1/1280 of 10 µl of each serum sample were made in 96 well plates (F 96 cell culture plates, Nunc A/S, Roskilde, Denmark) in Eagle's medium containing 50 µl of the challenge virus WN98 (genbank AY033388) [26] at a concentration of $10^{3.3}$ TCID50 [27]. The dilutions were allowed to stand for 1 h at room temperature. Vero cells were then added to each well, and the plates were placed in a 37°C incubator (5% CO_2) for 4 days, when they were inspected microscopically for cytopathic effect (CPE). The neutralizing antibody titer of the serum was expressed as the reciprocal of the highest dilution at which complete neutralization of the CPE was observed. Each titer was then calculated as the percentage of the titer of a known positive control that was checked on each assay.

Immunoglobulin G (IgG) levels in the 2013 serum samples were also determined using a commercial ELISA kit (ID Screen West Nile Competition Multi-species, IDvet Innovative Diagnostics, France), according to the manufacturer's instructions. Samples presenting an S/N percentage (OD sample divided by OD negative control) ≤ 40 were considered positive. Optical density values were obtained using an automatic plate reader. Seroprevalence analyses relied on the SN results for the 1997 and 2002 samples and on the ELISA results for the 2013 samples.

Collection of demographic and environmental data

During blood collections information on age, gender, colour, breed and geographical location was recorded for every subject. Age was analysed both as a continuous variable and as a categorical one (age groups: 0–3 yr, 3–6 yr, 6–9 yr, 9–12 yr 12–15 yr and >15 yr). The most complete data was gathered for 2002, soon after the large WNV outbreak, and therefore this year was chosen to represent the demographic risk factors analysis. The demographic variable of the breed of the horse was recorded for 808 out of the 827 horses that participated in 2002; with 134 Quarter horses (16.6%), 15 Appaloosa (1.9%), 104 (12.9%) Cross breeds, 27 (33.5%) Local breeds, 40 (5%) Ponies, 18 (2.2%) Thoroughbreds, 139 (17.2%) Warmbloods and 87 (10.8%) Arab horses. For the purpose of geographical/environmental analyses Israel was divided to two zones: (1) Great Rift Valley (GRV) zone (including the Golan Heights, the Steppe/Arava zone, parts of the

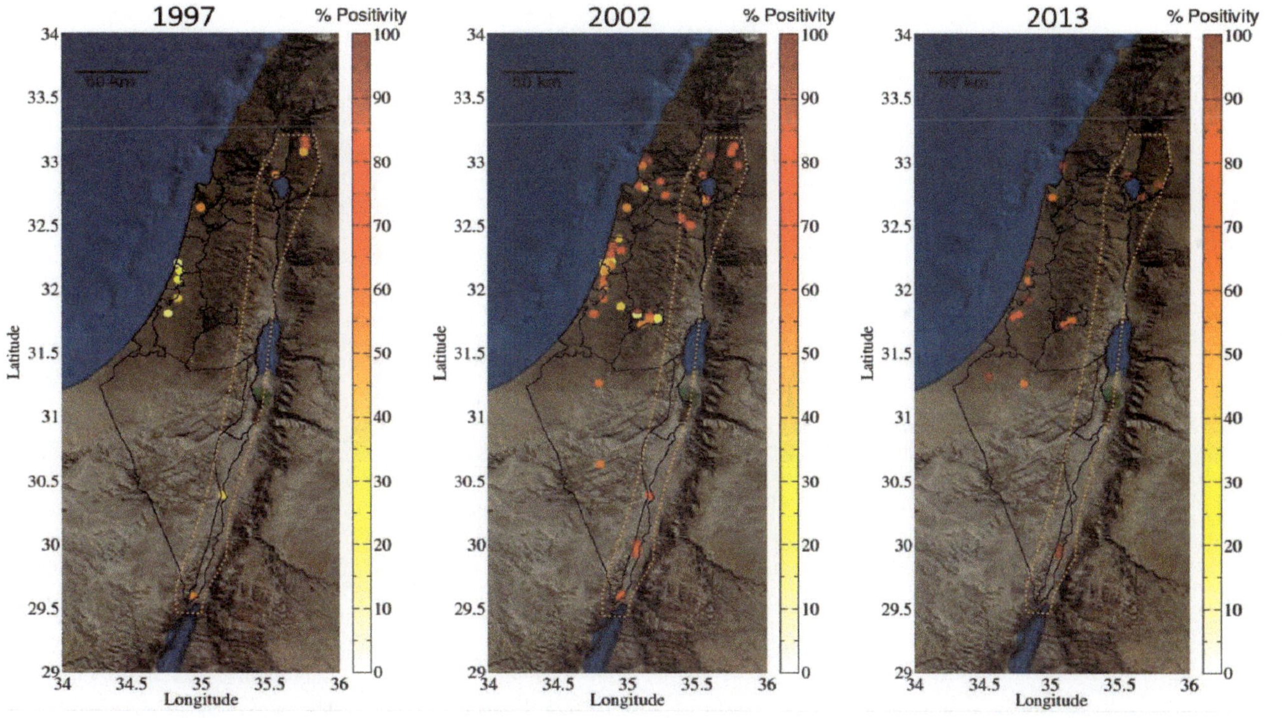

Figure 1. Spatial distribution of WNV seroprevalence in horses in Israel during 1997, 2002 and 2013. Circles inside the dashed orange marking represent farms located in the Great Rift Valley (GRV) area.

Galilee; north/valley/sea of Galilee and the Jordan Valley) representing the dominant route of birds migration in Israel (Figure 1) and (2) the rest of the country.

Calculation of force of infection using age specific data

Force of infection (FOI), a measure of the incidence of infection in a susceptible population, specifying the rate at which susceptible individuals acquire an infection, was estimated using a likelihood method, by maximizing the following likelihood function [28]:

$$\sum_{j=1}^{5} Fj^{Mj} + (1 - Fj)^{Nj - Mj}$$

$$Fj = 1 - \exp(-\lambda * Xj)$$

Where j is the index for the 5 age groups (0–3 yr, 3–6 yr, 6–9 yr, 9–12 yr, >12 yr), Nj is the total number of horses sampled in age group j, Mj is the number of horses that are WNV seropositive in age group j, Xj is the average age in age group j, and λ is the assumed FOI. Maximizing the likelihood function was performed using the solver add-in of Microsoft EXCEL 2010.

Precipitation levels in epidemic vs. non-epidemic years

Information on monthly precipitation (mm) was gathered from the Israel Meteorological Service. Data were collected from all listed weather stations distributed throughout Israel in the years 1997–2009. Spring and autumn/winter precipitation levels were determined by averaging the amount of precipitation for the months of March–May and October–December at all stations, respectively.

In a period of a decade, between 1997–2007, two years demonstrated a prominent increase in incidence rate of WNV in humans: 2000 and 2005 [24]. Precipitation data from these years (2000 and 2005) as well as one year prior to each of these years (1999 and 2004) ("increased WNV incidence") was compared with the precipitation data from the rest of the years; 1997, 1998, 2001, 2002, 2003, 2006, 2007 ("steady WNV incidence").

Statistical analyses

An interrater reliability analysis using the Kappa statistic was performed to determine consistency among the ELISA and the Serum Neutralization results of the 2013 samples.

Seroprevalence and its 95% confidence interval were calculated for each category of each demographic variable (breed, gender, age, and colour) for 2002, as well as for the different geographic zones for 1997, 2002, 2013. Statistical significance of the differences in prevalence between these various groups was assessed by the two-sided Chi-square test. A logistic regression model was constructed with variable inclusion in the model performed by the forward likelihood ratio method. Demographic and geographical variables were included in the model according to statistical significance, and p value for change in the model likelihood ratio after insertion of each additional variable was tested. In each such iteration, a p value of <0.05 was considered as a criterion for inclusion in the model. After each inclusion, variables that were already in the model and showed a p value $>$ 0.1 were excluded.

Independent-samples t-test analyses were used to compare average precipitation in years with increased incidence of human WNV in Israel (1999, 2000, 2004 and 2005) to the rest of the years throughout the decade of 1997–2007. These analyses were performed for the spring and autumn/winter precipitation volumes. Differences were considered significant at p value <

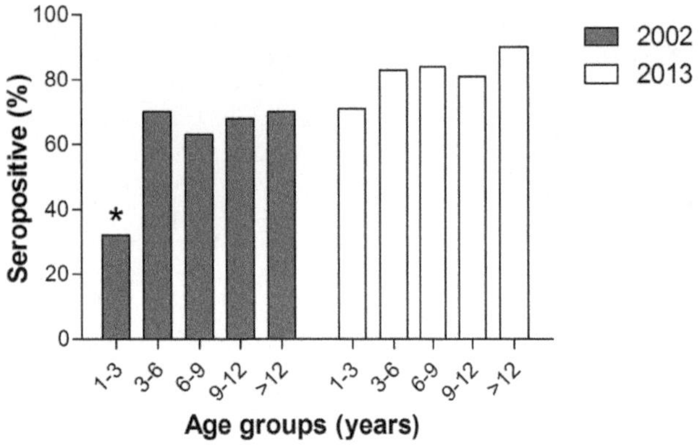

Figure 2. The association of WNV seroprevalence with age in 2002 and 2013. No significant differences were observed in the seroprevalence between the different age groups except for the youngest age group (1–3 yr) in 2002, in which a significant lower seroprevalence was observed. *P<0.05.

0.05. All statistical analyses were performed using SPSS 17.0 software.

Results

Seroprevalence and age-specific force of infection of WNV during 1997, 2002 and 2013

Out of the 290 horses tested in 1997, 113 (39%) were seropositive for WNV, thereafter in 2002, 547/827 (66.1%) horses were seropositive and finally, in 2013 out of 179 horses, 153 (85.5%) were found seropositive for WNV. The above seroprevalence was determined using the SN method for the 1997 and 2002 samples. Kappa coefficient showed an almost perfect agreement between the ELISA results and the SN results of the 2013 samples, with 86% agreement (p value <0.001), and hence further analyses for the 2013 samples were based on the ELISA results.

Seventy one horses were tested both in 1997 and in 2002 and all of them remained in the same geographical location during this time period. Among these 71 horses, 31 (44%) had neutralizing antibodies to WNV already in 1997 and all 31 remained positive when sampled again during 2002. Of the 40 horses that were negative in 1997, 16 (40%) seroconverted during these 5 years and 24 (60%) remained negative during this period. The force of infection using age specific data was 12.3% and 22.3% in 2002 and 2013, respectively. In 2013, a high seroprevalence of 71%–90% was observed in all age groups; 1–3 yr, 3–6 yr, 6–9 yr, 9–12 yr and >12 yr, with no significant differences in seroprevalence between the different age groups. In 2002, no significant differences were observed in the seroprevalence between the different age groups except for the youngest age group (1–3 yr), in which a significant lower seroprevalence was observed (Chi square: odd ratio = 3.62–4.96, p value <0.01) (Figure 2). Not enough data on the age of the horses was collected in 1997 in order to perform adequate statistical analyses.

Geographical/Environmental risk factors for spatial distribution of WNV during 1997, 2002 and 2013

Univariate analysis demonstrated significantly higher seroprevalence in horses located in the GRV area than in those located out of the GRV area in all three years (66.1% vs. 31.6% in 1997, 84.8%vs. 59.7% in 2002 and 100% vs. 78.9% in 2013) (Table 1). This was also the case when logistic regression analysis was performed on the 2002 results (Table 2) as well as on the 1997 and 2013 results (data not shown).

Calculation of FOI using age-specific data for the years 2002 and 2013 revealed higher rate of infection in horses located in the GRV area as compared to those located outside of the GRV area (22.8% in GRV vs. 9.8% outside the GRV and 50% in the GRV and 16.6% outside the GRV in 2002 and 2013, respectively).

Demographic risk factors for WNV in 2002

In the univariate analysis, significant association was detected between WNV seroprevalence and age, colour, gender and breed of the horses. A significantly lower seroprevalence was observed in the youngest age group (0–3 yr), in the dark colour group (black/dark brown/dark grey) as compared to the light colour group (grey/light appaloosa), in males, and in the Arabian, Warmblood, Pony and Appaloosa breeds as compared to the cross breeds (Table 3). Gender showed close association with GRV area as significantly more female horses were sampled in the GRV area (Pearson Chi-Square: 15.891, p value <0.01). Entering all above demographic and geographical location factors into a logistic regression model, age, breed and GRV were found to significantly associate with seroprevalence to WNV; age showed positive association, Warmblood breeds and Ponies were significantly lower in seroprevalence than the Quarter horse breed and horses located in the GRV had significantly higher seroprevalence to WNV than horses located outside this area (Table 2).

Average precipitation in spring and autumn/winter of 1997–2007

During the decade from 1997–2007, the average spring precipitation (mm) was significantly lower in the group of years belonging to the "increased WNV incidence" which include the years 1999/2000 and 2004/2005, as compared to the rest of the years (belonging to the "steady WNV incidence") (15.08 mm versus 36.18 mm, respectively, p value = 0.03). However, no differences in average autumn/winter precipitations were noted between these two groups (51.45 versus 60.06, respectively) (Figure 3).

Discussion

Seroprevalence, force of infection and factors associated with transition from endemic to epidemic state

Although WNV was shown to circulate in Israel since the 1960's, it has been receiving growing attention only in the last decade since the 2000 WNF outbreak among humans and horses. In a serological survey performed during 1959–1960, a 35% seroprevalence to WNV was detected in horses in Israel [29]. Following three decades, as well as a documented WNV outbreak in humans in 1980 [30], a 39% seroprevalence in horses was detected in our study for the year 1997. The findings in the current study indicate that the growing attention to the disease in the equine population in Israel in the last decade is not only due to the increase in the number of horses in Israel and improved veterinary care and services which enabled the identification of the virus, but also to a true surge in virus occurrence among horses in this region. An increase in seroprevalence was detected from 39% seropositivity in 1997, just prior to the WNF outbreak, to 66.1% in 2002, following the large outbreak, and remained high in 2013 with 85.5% seroprevalence. In addition, of the horses that were negative in 1997, 40% seroconverted during a 5-years period. If the infection rate throughout the period between 1997 and 2002 was constant, this demonstrates a rate of infection of 8% per year.

Table 1. Univariate analyses for the association of WNV seroprevalence in horses in Israel and the geographical location of the horses along the Great Rift Valley - GRV (1997, 2002, 2013).

		Positive/total (%)	OR	95% CI	P
1997 GRV	**Rift valley area**	41/62 (66.1)	4.23	2.25–8.08	<0.01
	No Rift valley area	72/228 (31.6)	REF		
2002 GRV	**Rift valley area**	179/211 (84.8)	3.77	2.48–5.86	<0.01
	No Rift valley area	368/616 (59.7)	REF		
2013 GRV	**Rift valley area**	56/56 (100)	1.268*	1.16–1.39	<0.01
	No Rift valley area	97/123 (78.9)	REF		

OR: odds ratio; CI: confidence interval; P: p value; REF: reference variable; GRV: Great Rift Valley.
*Prevalence ratio.

Nevertheless, the true infection rate cannot be estimated due to the large scale outbreak in 2000. The high occurrence of WNV in horses in Israel in the last decade is also demonstrated by the high FOI detected in this study, with 12.3% and 22.3% FOI in 2002 and 2013, respectively. It could be argued that the increase in seroprevalence demonstrated in this study between 2002 and 2013 could be a result of different antibody detection methods, as the ELISA method used for the 2013 samples is considered more sensitive than the SN method used for the 2002 samples [31]; however, it has previously been shown that WNV-neutralizing test was able to confirm 42 out of 49 ELISA-positive horse sera (86%) [32]. Furthermore, in addition to the ELISA method, serum neutralization (SN) test was also employed on the 2013 samples, as was previously performed to evaluate WNV seroprevalence in horses in Germany [33], and excellent agreement was demonstrated between the two detection methods with a kappa of 86.4% (p value <0.001), further supporting the validity of the ELISA results and the conclusions drawn from them.

The increase in seroprevalence of WNV in horses in the last decade is in accordance with human reported cases of WNV in Israel [24]. Although following the 2000 outbreak a decrease in incidence rate was documented, this lasted only until 2005 in which a substantial increase in incidence occurred during that year and remained constant until 2012 [24]. Cyclic epidemics of endemic arboviral diseases are recognized phenomena and have previously been addressed. For example, Eastern equine encephalitis (EEE) virus epidemics occur in the northern regions of the United Stated and tend to cluster over 2 or more consecutive years, separated by periods of dormancy. Several factors were suggested to be responsible for re-emergence of EEE and it has been suggested that although historical risk factors have included above-average rainfall, and mild winters, records show a trend toward milder winters and hotter summers, marked by extremes in both precipitation and drought in the northeastern United States [25,34].

Ambient temperature was constantly shown to play an important role in WNV transmission by affecting the growth rates of vector populations, the interval between blood meals, and viral replication rates [20]. Furthermore, during the WNV outbreak in Israel in 2000, the minimum temperature was found to be the most important climatic factor that encouraged earlier appearance of disease [35]. Nonetheless, the contribution of precipitation to the transmission and spread of WNV is more complex and controversial. Heavy rainfall during spring may increase standing water resources at the beginning of the hot season and therefore increase the vectors natural habitats. In contrast, during drought conditions, the reduced water flow creates stagnant water pools which become richer in the organic material ideal for breeding mosquitoes. This may encourage birds to circulate around small water holes and thus increase the interactions with mosquitoes. These discrepancies regarding the role of precipitation in WNV transmission raised the assumption

Table 2. Logistic regression model for the association of WNV seroprevalence in horses in Israel with demographic (age, gender, breed, colour) and geographical/environmental (Great Rift Valley - GRV) risk factors in 2002 (n = 802).

		Exp(B)	S.E.	P value	95% CI
AGE		1.044	0.016	<0.007	1.044–1.012
BREED	Quarterhorse	REF			
	Arabian	0.692	0.307	0.231	0.379–1.264
	Warmblood	0.304	0.279	<0.001	0.176–0.525
	Thoroughbred	1.123	0.575	0.84	0.364–3.466
	Pony	0.359	0.385	0.008	0.169–0.764
	Local	1.502	0.251	0.105	0.919–2.454
	Cross	1.481	0.318	0.217	0.794–2.764
	Appaloosa	0.470	0.562	0.179	0.156–1.413
GRV		2.773	0.224	<0.001	1.789–4.298

Exp(B): odds ratio (the exponentiation of the B coefficient); S.E.: standard error; CI: confidence interval; GRV: Great Rift Valley; REF: reference variable.

Table 3. Univariate analysis for demographic risk factors associated with WNV seroprevalence in horses in Israel in 2002 (n = 827).

		Positive/total (%)	OR	95% CI	P
AGE (YEARS)	0–3	12/38 (31.6)	0.20	0.09–0.43	<0.01
	3–6	128/184 (69.6)	1.02	0.66–1.57	1.00
	6–9	120/191 (62.8)	0.73	0.48–1.10	0.11
	9–12	89/131 (67.9)	0.91	0.57–1.47	0.73
	12+	193/276 (69.9)	REF		
COLOUR	Paint	16/25 (64)	0.75	0.29–2.04	0.498
	Black/Dark brown/Dark grey	54/102 (52.9)	0.47	0.28–0.80	<0.01
	Bay/Chesnut	313/469 (66.7)	0.85	0.58–1.23	0.415
	Palamino/Buckskin	24/32 (75)	1.26	0.51–3.45	0.678
	Grey/Light appalloosa	140/199 (70.4)	REF		
GENDER	Female	280/396 (70.7)	1.48	1.10–2.01	<0.01
	Male	267/431 (61.9)	REF		
BREED	Arabian	51/87 (58.6)	0.36	0.18–0.71	<0.01
	Warmblood	55/139 (39.6)	0.17	0.09–0.31	<0.01
	Thoroughbred	13/18 (72.2)	0.66	0.19–2.63	0.534
	Pony	19/40 (47.5)	0.23	0.10–0.54	<0.01
	Appaloosa	7/15 (46.7)	0.22	0.06–0.80	<0.01
	Quarterhorse	94/134 (70.1)	0.59	0.31–1.13	0.101
	Local	212/271 (78.2)	0.91	0.49–1.63	0.780
	Cross	83/104 (79.8)	REF		

OR: odds ratio; CI: confidence interval; P: p value; REF: reference variable.
Age was not recorded for 7/827 horses. Breed was not recorded for 19/827 horses.

that the patterns of disease incidence may change over large geographic regions (i.e. tropical vs. Mediterranean regions), depending on differences in the ecology of mosquito vectors [12]. Numerous studies describe the association of drought conditions with WNV human infection in the United States [36]. Some identified inverse relationship between total annual rainfall levels of the previous year and the relative risk of human WNV [37], as well as negative correlation between WNV infection in mosquitoes and the previous year's precipitation [38]. Others report dry spring and summer conditions during the same year an increased risk of human WNV infection was observed [25,39]. Alternatively, widespread drought conditions in the spring, followed by a wetter summer was also linked with increased probability of human WNV cases [40]. Similar correlation between drought periods and outbreaks of WNV has been documented in Europe. In 1996, a significant WNV outbreak occurred in Romania coinciding with prolonged drought conditions (May through October) and excessive heat (May through July) [41]. In the summer of 1999, a large WNV outbreak occurred in Russia following a drought period [42]. In the current study, precipitation data gathered from the Israel Meteorological Services revealed significant lower spring precipitation level during the years with increased WNV human cases (2000 and 2005) and one year prior to them (1999 and 2004) as compared to the rest of the years from 1997 to 2007, supporting the notion of association between WNV epidemics and a prior dry spring season. Similar to the periodic increased incidence of WNV cases in Israel, the United States experienced a resurgence of WN neurologic disease in 2012 with the highest level of disease in the last decade [43]. Interestingly, the resurgence of disease followed a period of nearly 4 years of extremely low incidence rate similar to the 4 years that preceded the increased human incidence in 2005 in Israel. The low incidence rate prior to a resurgence of the disease, observed both in Israel and the US, may suggest the involvement of

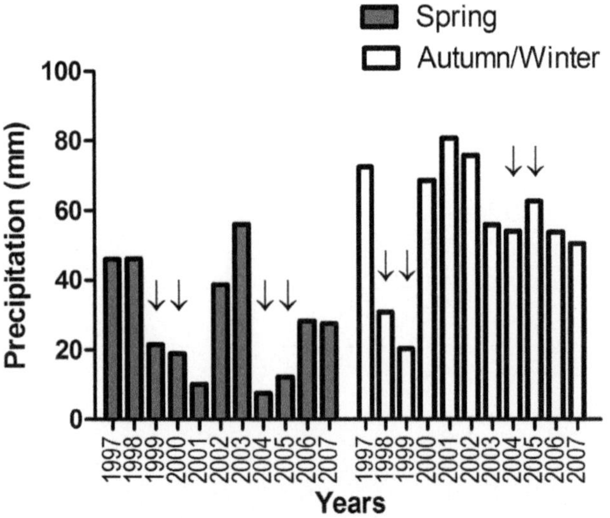

Figure 3. Average spring and autumn/winter precipitation (mm) in the years 1997–2007. Arrows represent years with documented increased human WNV incidence. Between 1997–2007, the average spring precipitation (mm) was significantly lower (P<0.05) in the group of years belonging to the "increased WNV incidence" (arrows), as compared to the rest of the years. No differences in average autumn/winter precipitations were noted.

decreased herd immunity in the incidental host (humans, horses) as well as in the main amplifying host (birds) [44]. Perhaps both the dry springs as well as the decreased immunity in the years prior to the outbreaks contributed, together with many other interrelated biotic and abiotic drivers (i.e. temperature, relative humidity and bird migration), to the WNV epidemics that occurred in humans in Israel during 2000 and 2005.

Spatial and temporal characterization of geographical/ environmental and demographic risk factors for WNV seroprevalence

Locations where large outbreaks as well as sporadic cases of WNV occurred in the United States have varied from year to year, demonstrating their focal nature, and is probably in part a result of variation in the immune state of the susceptible population. In contrast, via spatial and temporal investigation of WNV prevalence, the current study demonstrates a consistent geographical disparity of seroprevalence to WNV among horses in Israel, with a persistent higher seroprevalence in horses located throughout the Great Rift Valley (Syrian-African Rift Valley) area than in horses located outside of this area (in all three years; 1997, 2002 and 2013), as well as higher FOI in horses located in the GRV area in 2002 and 2013 as compared to those located outside of this area. The association between bird migration routes and WNV dispersion worldwide and specifically in Israel has been well documented [20,23,45–47]. Although the whole country of Israel is considered the crossroad for bird migration between Africa and Euroasia, and therefore a major focus of attention during the global spread of WNF, three principal routes exist inside the country, the dominant one along the Great Rift Valley [48]. Based on our findings, we suggest being more cautious with the WNV endemic terminology for Israel, and perhaps a more precise description of the region ought to consider only the Great Rift Valley regions as WNV endemic regions (probably due to the migrating birds passing along, introducing the virus to the region), while all other regions of Israel perhaps suffer from cyclic epidemics of WNV activity every few years. Work et al. (1955) pioneered in dividing the distribution of WNV prevalence around the Nile delta region into "endemic", "transition" and "non-endemic" areas [49]. This division was shown both for wild birds as well as human seroprevalence. Only few children and adults were shown to have antibodies for WNV in the northern portion of the Nile delta along the Mediterranean coast ("non-endemic area", 2% and 30%, respectively) as compared to higher percentage of seropositive children and adults in villages located in the southern part of the delta ("endemic area", 70% and 95%, respectively), and intermediate level in the transitional area, located further inland [50]. Great variation in seroprevalence was also observed in horses in different areas of sub-Saharan Africa, with low seroprevalence in the east of the Sahelian area (9%) and in sub-Sahelian area (3–30%) and highest seroprevalence in the western and central parts of the Sahelian area (92%–97%) [51].

Demographic variables that were shown to associate with seroprevalence to WNV were age and breed of the horses. Previous studies demonstrated the association of age with seroprevalence to WNV in endemic areas. Prior to the 2000 WNF human epidemic in Israel, seroprevalence was shown to increase with age, as in healthy soldiers aged 18–20 years a lower seroprevalence was noted (7.0%) compared to the older age group (40–55 years) with 41.9% seropositivity to the virus [52]. This increase in seroprevalence with age was also observed in Iran, another WNV endemic region, in which a difference of 5 years in horses' age corresponded to a change of seropositivity odds by a factor of 1.3 [53]. Surprisingly, in the current study, when we divided the horses into age groups in the two sampling periods in this study (i.e. 2002 and 2013), seroprevalence to WNV was not shown to differ much between the different age groups. In 2013, the very high overall seroprevlance and FOI could have masked differences between groups as from a very young age horses were largely exposed to the virus. In addition, the great outbreak that occurred in 2000 hit all age groups equally and therefore in 2002 no differences were found between the different age groups, except for the youngest age group (1–3 yr), which include foals that were not yet born during the outbreak and therefore the seroprevalence of that group was significantly lower. The reason for the lower WNV seroprevalence found in our study among Ponies and Warmblood horses, as compared to other breeds, is unknown and might indicate different preference of mosquitoes to different breeds of horses, due to variables such as density of hair and differences in sweat composition. In addition, grooming and anti-pesticides regimes are most likely stricter in the Warmblood breeds since they are mostly used for competitions in Israel while the Quarter horses and cross breeds are used mostly for pleasure riding. The higher seroprevalence detected in the Quarter horses in our study is in accordance with a study performed in Florida, USA, in which Quarter horses were the most commonly affected breed [13], and with a study performed in Saskatchewan, Canada, which indicated that most of the WNV clinical cases were light horse breeds and most were used for pleasure riding [54]. However, as opposed to the above study, in which dark coloured horses were most commonly affected with WNV [54], univariate analysis in the current study demonstrated higher seroprevalence in the lighter colours (grey/light appaloosa; 70.4%) as compared to the darker colours (black/dark brown/dark grey; 52.9%). Though, since these results were not repeated in the logistic regression model, the above findings might be due to association between colour and breed.

Conclusions

The extensive serological survey in horses, throughout duration of more than a decade, covering periods before (1997) and after (2002, 2013) the largest documented WNF outbreak in Israel (2000), permitted the investigation of the spatio-temporal dynamics of WNV in a longstanding endemic region such as Israel. We hereby demonstrate that the 3 serene decades since the 1960's, in which constant seroprevalence to the virus in horses was observed, were interrupted in the last decade when an increase in exposure to the virus was detected from 39% seroprevalence just prior to the 2000 WNF outbreak to 66.1% and 85.5% seroprevalence in 2002 and 2013, respectively. Demographic variables such as age and breed were found to significantly associate with seroprevalence to WNV. But more prominently, environmental factors (geographical location and precipitation amounts), and more specifically the geographical location of the horse along the Great Rift Valley, the main route of birds migration in Israel, was found to be of great impact on the likelihood of WNV exposure throughout the years investigated. Humans, similar to horses, serve as incidental hosts to WNV and therefore may share common epidemiological aspects regarding disease occurrence and spread. Interestingly, weather conditions, and more specifically rainfall conditions, such as periods of spring drought, together with lower incidence rate in the population, were shown to precede transition from endemic state to epidemic state of WNV observed in humans in Israel during the years 2000 and 2005.

Author Contributions

Conceived and designed the experiments: KA AL ST BG DC AS. Performed the experiments: KA AL ST BG. Analyzed the data: KA EK. Contributed reagents/materials/analysis tools: KA AL ST BG EK AS. Wrote the paper: KA AS.

References

1. Beck C, Jimenez-Clavero MA, Leblond A, Durand B, Nowotny N, et al. (2013) Flaviviruses in europe: complex circulation patterns and their consequences for the diagnosis and control of west nile disease. Int J Environ Res Public Health 10: 6049–6083.
2. Smithburn KC, Hughes TP, Burke AW, Paul JH (1940) A Neurotropic Virus Isolated from the Blood of a Native of Uganda. Am J Trop Med 20: 471–492.
3. Murgue B, Murri S, Triki H, Deubel V, Zeller HG (2001) West Nile in the Mediterranean basin: 1950–2000. Ann N Y Acad Sci 951: 117–126.
4. Schmidt JR, Elmansoury HK (1963) Natural and Experimental Infection of Egyptian Equines with West Nile Virus. Ann Trop Med Parasitol 57: 415–427.
5. Akov Y, Goldwasser R (1966) Prevalence of antibodies to arboviruses in various animals in Israel. Bull World Health Organ 34: 901–909.
6. Cantile C, Di Guardo G, Eleni C, Arispici M (2000) Clinical and neuropathological features of West Nile virus equine encephalomyelitis in Italy. Equine Vet J 32: 31–35.
7. Murgue B, Murri S, Zientara S, Durand B, Durand JP, et al. (2001) West Nile outbreak in horses in southern France, 2000: the return after 35 years. Emerg Infect Dis 7: 692–696.
8. Steinman A, Banet C, Sutton GA, Yadin H, Hadar S, et al. (2002) Clinical signs of West Nile virus encephalomyelitis in horses during the outbreak in Israel in 2000. Vet Rec 151: 47–49.
9. Centers for Disease Control, Atlanta, GA. Available: http://www.cdc.gov/westnile/resources/pdfs/cummulative/99_2012_NeuroInvasivebyYear.pdf.
10. Monath TP (2013) Vaccines against diseases transmitted from animals to humans: a one health paradigm. Vaccine 31: 5321–5338.
11. Klee AL, Maidin B, Edwin B, Poshni I, Mostashari F, et al. (2004) Long-term prognosis for clinical West Nile virus infection. Emerg Infect Dis 10: 1405–1411.
12. Paz S, Malkinson D, Green MS, Tsioni G, Papa A, et al. (2013) Permissive summer temperatures of the 2010 European West Nile fever upsurge. PLoS One 8: e56398.
13. Rios LM, Sheu JJ, Day JF, Maruniak JE, Seino K, et al. (2009) Environmental risk factors associated with West Nile virus clinical disease in Florida horses. Med Vet Entomol 23: 357–366.
14. Jansen CC, Ritchie SA, van den Hurk AF (2013) The role of Australian mosquito species in the transmission of endemic and exotic West Nile virus strains. Int J Environ Res Public Health 10: 3735–3752.
15. Kramer LD, Styer LM, Ebel GD (2008) A global perspective on the epidemiology of West Nile virus. Annu Rev Entomol 53: 61–81.
16. Lindsey NP, Staples JE, Delorey MJ, Fischer M (2013) Lack of Evidence of Increased West Nile Virus Disease Severity in the United States in 2012. Am J Trop Med Hyg.
17. Bakonyi T, Ivanics E, Erdelyi K, Ursu K, Ferenczi E, et al. (2006) Lineage 1 and 2 strains of encephalitic West Nile virus, central Europe. Emerg Infect Dis 12: 618–623.
18. Bakonyi T, Ferenczi E, Erdelyi K, Kutasi O, Csorgo T, et al. (2013) Explosive spread of a neuroinvasive lineage 2 West Nile virus in Central Europe, 2008/2009. Vet Microbiol 165: 61–70.
19. Papa A, Bakonyi T, Xanthopoulou K, Vazquez A, Tenorio A, et al. (2011) Genetic characterization of West Nile virus lineage 2, Greece, 2010. Emerg Infect Dis 17: 920–922.
20. Paz S, Semenza JC (2013) Environmental drivers of West Nile fever epidemiology in Europe and Western Asia–a review. Int J Environ Res Public Health 10: 3543–3562.
21. Sambri V, Capobianchi M, Charrel R, Fyodorova M, Gaibani P, et al. (2013) West Nile virus in Europe: emergence, epidemiology, diagnosis, treatment, and prevention. Clin Microbiol Infect 19: 699–704.
22. Weinberger M, Pitlik SD, Gandacu D, Lang R, Nassar F, et al. (2001) West Nile fever outbreak, Israel, 2000: epidemiologic aspects. Emerg Infect Dis 7: 686–691.
23. Malkinson M, Banet C, Weisman Y, Pokamunski S, King R, et al. (2002) Introduction of West Nile virus in the Middle East by migrating white storks. Emerg Infect Dis 8: 392–397.
24. Anis E, Grotto I, Mendelson E, Bin H, Orshan L, et al. (2013) West Nile fever in Israel: The reemergence of an endemic disease. J Infect.
25. Reiter P (2010) West Nile virus in Europe: understanding the present to gauge the future. Euro Surveill 15: 19508.
26. Banet-Noach C, Malkinson M, Brill A, Samina I, Yadin H, et al. (2003) Phylogenetic relationships of West Nile viruses isolated from birds and horses in Israel from 1997 to 2001. Virus Genes 26: 135–141.
27. Banet-Noach C, Simanov L, Malkinson M (2003) Direct (non-vector) transmission of West Nile virus in geese. Avian Pathol 32: 489–494.
28. Cohen DI, Davidovici BB, Smetana Z, Balicer RD, Klement E, et al. (2006) Seroepidemiology of Varicella zoster in Israel prior to large-scale use of varicella vaccines. Infection 34: 208–213.
29. Nir Y, Lasowski Y, Avivi A, Cgoldwasser R (1969) Survey for antibodies to arboviruses in the serum of various animals in Israel during 1965–1966. Am J Trop Med Hyg 18: 416–422.
30. Chowers MY, Green MS, Bin H, Weinberger M, Schlaeffer F, et al. (2005) Post-epidemic serosurvey of West Nile fever in Israel. Eur J Clin Microbiol Infect Dis 24: 851–853.
31. Sanchini A, Donoso-Mantke O, Papa A, Sambri V, Teichmann A, et al. (2013) Second international diagnostic accuracy study for the serological detection of West Nile virus infection. PLoS Negl Trop Dis 7: e2184.
32. Ziegler U, Skrypnyk A, Keller M, Staubach C, Bezymennyi M, et al. (2013) West nile virus antibody prevalence in horses of Ukraine. Viruses 5: 2469–2482.
33. Ziegler U, Angenvoort J, Klaus C, Nagel-Kohl U, Sauerwald C, et al. (2013) Use of competition ELISA for monitoring of West Nile virus infections in horses in Germany. Int J Environ Res Public Health 10: 3112–3120.
34. Armstrong PM, Andreadis TG (2013) Eastern equine encephalitis virus–old enemy, new threat. N Engl J Med 368: 1670–1673.
35. Paz S (2006) The West Nile Virus outbreak in Israel (2000) from a new perspective: the regional impact of climate change. Int J Environ Health Res 16: 1–13.
36. Stanke C, Kerac M, Prudhomme C, Medlock J, Murray V (2013) Health effects of drought: a systematic review of the evidence. PLoS Curr 5.
37. Wang G, Minnis RB, Belant JL, Wax CL (2010) Dry weather induces outbreaks of human West Nile virus infections. BMC Infect Dis 10: 38.
38. Ruiz MO, Chaves LF, Hamer GL, Sun T, Brown WM, et al. (2010) Local impact of temperature and precipitation on West Nile virus infection in Culex species mosquitoes in northeast Illinois, USA. Parasit Vectors 3: 19.
39. Shaman J, Day JF, Komar N (2010) Hydrologic conditions describe West Nile virus risk in Colorado. Int J Environ Res Public Health 7: 494–508.
40. Shaman J, Day JF, Stieglitz M (2005) Drought-induced amplification and epidemic transmission of West Nile virus in southern Florida. J Med Entomol 42: 134–141.
41. Savage HM, Ceianu C, Nicolescu G, Karabatsos N, Lanciotti R, et al. (1999) Entomologic and avian investigations of an epidemic of West Nile fever in Romania in 1996, with serologic and molecular characterization of a virus isolate from mosquitoes. Am J Trop Med Hyg 61: 600–611.
42. Platonov AE, Shipulin GA, Shipulina OY, Tyutyunnik EN, Frolochkina TI, et al. (2001) Outbreak of West Nile virus infection, Volgograd Region, Russia, 1999. Emerg Infect Dis 7: 128–132.
43. Beasley DW, Barrett AD, Tesh RB (2013) Resurgence of West Nile neurologic disease in the United States in 2012: what happened? What needs to be done? Antiviral Res 99: 1–5.
44. Barzon L, Pacenti M, Franchin E, Squarzon L, Lavezzo E, et al. (2013) The complex epidemiological scenario of West Nile virus in Italy. Int J Environ Res Public Health 10: 4669–4689.
45. Jourdain E, Toussaint Y, Leblond A, Bicout DJ, Sabatier P, et al. (2007) Bird species potentially involved in introduction, amplification, and spread of West Nile virus in a Mediterranean wetland, the Camargue (Southern France). Vector Borne Zoonotic Dis 7: 15–33.
46. Pradier S, Lecollinet S, Leblond A (2012) West Nile virus epidemiology and factors triggering change in its distribution in Europe. Rev Sci Tech 31: 829–844.
47. Jourdain E, Gauthier-Clerc M, Bicout DJ, Sabatier P (2007) Bird migration routes and risk for pathogen dispersion into western Mediterranean wetlands. Emerg Infect Dis 13: 365–372.
48. Leshem Y, Yom-Tov Y (1998) Routes of migrating soaring birds. Ibis 140: 41–52.
49. Work TH, Hurlbut HS, Taylor RM (1955) Indigenous wild birds of the Nile Delta as potential West Nile virus circulating reservoirs. Am J Trop Med Hyg 4: 872–888.
50. Murgue B, Zeller H, Deubel V (2002) The ecology and epidemiology of West Nile virus in Africa, Europe and Asia. Curr Top Microbiol Immunol 267: 195–221.
51. Cabre O, Grandadam M, Marie JL, Gravier P, Prange A, et al. (2006) West Nile Virus in horses, sub-Saharan Africa. Emerg Infect Dis 12: 1958–1960.
52. Cohen D, Zaide Y, Karasenty E, Schwarz M, LeDuc JW, et al. (1999) Prevalence of antibodies to West Nile fever, sandfly fever Sicilian, and sandfly fever Naples viruses in healthy adults in Israel. Public Health Rev 27: 217–230.
53. Ahmadnejad F, Otarod V, Fallah MH, Lowenski S, Sedighi-Moghaddam R, et al. (2011) Spread of West Nile virus in Iran: a cross-sectional serosurvey in equines, 2008–2009. Epidemiol Infect 139: 1587–1593.
54. Epp T, Waldner C, West K, Townsend H (2007) Factors associated with West Nile virus disease fatalities in horses. Can Vet J 48: 1137–1145.

13

Effects of Extreme Climate Events on Tea (*Camellia sinensis*) Functional Quality Validate Indigenous Farmer Knowledge and Sensory Preferences in Tropical China

Selena Ahmed[1,2,3]*, John Richard Stepp[3,4], Colin Orians[2], Timothy Griffin[5], Corene Matyas[6], Albert Robbat[7], Sean Cash[5], Dayuan Xue[3], Chunlin Long[3], Uchenna Unachukwu[8], Sarabeth Buckley[2], David Small[9], Edward Kennelly[3,8]

1 Sustainable Food and Bioenergy Systems Program, Department of Health and Human Development, Montana State University, Bozeman, Montana, United States of America, **2** Department of Biology, Tufts University, Medford, Massachusetts, United States of America, **3** College of Life and Environmental Sciences, Minzu University of China, Beijing, China, **4** Department of Anthropology, University of Gainesville, Gainesville, Florida, United States of America, **5** Friedman School of Nutrition Science and Policy, Tufts University, Boston, Massachusetts, United States of America, **6** Department of Geography, University of Gainesville, Gainesville, Florida, United States of America, **7** Department of Chemistry, Tufts University, Medford, Massachusetts, United States of America, **8** Department of Biochemistry, The Graduate Center of the City University of New York, New York, New York, United States of America, **9** School of Engineering, Tufts University, Medford, Massachusetts, United States of America

Abstract

Climate change is impacting agro-ecosystems, crops, and farmer livelihoods in communities worldwide. While it is well understood that more frequent and intense climate events in many areas are resulting in a decline in crop yields, the impact on crop quality is less acknowledged, yet it is critical for food systems that benefit both farmers and consumers through high-quality products. This study examines tea (*Camellia sinensis*; Theaceae), the world's most widely consumed beverage after water, as a study system to measure effects of seasonal precipitation variability on crop functional quality and associated farmer knowledge, preferences, and livelihoods. Sampling was conducted in a major tea producing area of China during an extreme drought through the onset of the East Asian Monsoon in order to capture effects of extreme climate events that are likely to become more frequent with climate change. Compared to the spring drought, tea growth during the monsoon period was up to 50% higher. Concurrently, concentrations of catechin and methylxanthine secondary metabolites, major compounds that determine tea functional quality, were up to 50% lower during the monsoon while total phenolic concentrations and antioxidant activity increased. The inverse relationship between tea growth and concentrations of individual secondary metabolites suggests a dilution effect of precipitation on tea quality. The decrease in concentrations of tea secondary metabolites was accompanied by reduced farmer preference on the basis of sensory characteristics as well as a decline of up to 50% in household income from tea sales. Farmer surveys indicate a high degree of agreement regarding climate patterns and the effects of precipitation on tea yields and quality. Extrapolating findings from this seasonal study to long-term climate scenario projections suggests that farmers and consumers face variable implications with forecasted precipitation scenarios and calls for research on management practices to facilitate climate adaptation for sustainable crop production.

Editor: Darren John Kriticos, CSIRO, Australia

Funding: This research was supported by the TEACRS Program (NIGMS IRACDA- K12GM074869) at Tufts University, Program 111 in Ethnobiology of the Chinese Ministry in Education/Minzu University of China, the Tufts Institute of the Environment, the Tropical Conservation and Development Program at the University of Florida, Tufts University Provost Office (Tufts Collaborates!), NSF REU Program at Tufts University (NSF DBI 1005082), and the NSF Coupled Natural Human Systems (NSF grant #BCS-1313775). The funders had no role in study design, data collection and analysis, decision to publish, or preparation of the manuscript.

Competing Interests: The authors have declared that no competing interests exist.

* Email: selena.ahmed@montana.edu

Introduction

Climate change is impacting agro-ecosystems, livelihoods, and human wellbeing in communities worldwide. In some cases, crop production is at risk due to droughts, wildfires, floods, severe temperature fluctuations, and other extreme weather conditions reducing yields and shifting the geographic range where crops can be cultivated [1,2]. In contrast, increasing temperatures in the higher latitudes in both the northern and southern hemisphere are enhancing the yield of some crops [3]. While the impact of climate change on crop yield has been well documented [1,2,3,4,5], it is also important to understand potential changes in crop quality on farmer livelihoods and food security [6]. This is particularly important for many specialty crops such as fruits, vegetables, herbs, coffee, chocolate, and tea where functional quality is determined by phytonutrients and secondary metabolites, or bioactive food components, that benefit consumers. Bioactive food components can potentially enhance flavor profiles and other sensory experiences [7] and also have the potential to mitigate micronutrient deficiencies and associated risks of chronic disease [6,8]. In addition, producer income can benefit from increased consumer preference for high-quality crops generated through premiums reflected in higher prices. This study uses tea (*Camellia sinensis*; Theaceae), the world's most widely consumed beverage after water, as a case study to explore effects of extreme climate events on crop quality and associated farmer knowledge, preferences, and income.

China is the world's largest supplier of tea, where its production is vulnerable with many of the country's tea agro-ecosystems located in high-risk regions for climate change [9]. In addition, China has an arable land shortage with land suitable for cultivation comprising only 7% of the world's total to support 22% of the global population [10]. Farmer livelihoods in China are at risk with shifts in climate conditions that support their crops. Precipitation and temperature changes have been recognized to impact the yield and geographic range of many crops including tea plants [11,12]. Climate change may also influence tea functional quality that is largely determined by the concentrations of the methylxanthine caffeine and various polyphenolic catechin compounds that are respectively responsible for tea's stimulant, antioxidant, anti-inflammatory, and cardioprotective properties [13].

The secondary metabolites that contribute to crop quality serve as defense compounds in plants and vary in concentration depending on genetic, environmental, and management conditions. For example, the concentrations of methylxanthine and polyphenolic catechin compounds in tea plants vary with geographic location, cultivar, herbivory, season, shade, soil, slope, water availability, and management [13,14,15]. Consumers can perceive changes in the concentrations of tea functional compounds by their sensory characteristics, including astringency, bitterness, and sweetness [7]. Ultimately, crop quality influences consumer-purchasing decisions, farmer livelihoods, and functional benefits derived from crops.

The impacts of climate change are likely to have a notable impact on tea farmers with tea production supporting the livelihoods of millions of farmers and their communities - a growing global market of over $20 billion USD annually. Farmers can mitigate climate effects on crop quality through adaptive management. However, in order for adaptation to occur, farmers must first have knowledge of *how* the climate has changed [16] and impacted their crops. Previous research evaluating individual perceptions of climate change based on seasonal patterns has proven useful for the study of climate change [17].

The present paper integrates plant-defense and socio-ecological theoretical foundations through a coupled natural and human systems framework to examine impacts of extreme climate events on crop quality and socio-economic responses using tea production in China as a study system. Field studies during extreme climate events such as droughts and flooding offer the opportunity to examine effects of climate variability in natural settings under unusual circumstances that are expected to become more frequent and severe with forecasted climate change. Previous work that highlights the underestimation of manipulative studies on plant responses to climate change compared to observational seasonal studies points to the need for research in natural settings [18]. For the present study, we utilized one of the most severe droughts that occurred in recorded history at our study site in China's Yunnan Province during the spring tea harvest along with the onset of the East Asian monsoon that marks the start of the monsoon tea harvest as opportunities to examine the effects of extreme climate events that are expected to become more prevalent in tea producing areas.

Specifically, our study objectives are to provide quantitative data on how climate effects the following aspects of the coupled tea production and consumption system (Figure 1): (1) the impact of seasonal precipitation variability that occurs between the spring drought and monsoon on tea growth and functional quality, (2) variation of tea prices between the spring drought and monsoon tea harvests, (3) farmer knowledge of climate effects on tea quality and associated management, sensory preferences, and livelihoods and, (4) precipitation trends over the past 40 years as well as future climate scenario projections at the study site. We hypothesize the following: (1) the increase in precipitation that occurs with the seasonal transition from the spring drought to the monsoon tea harvests results in an increase in tea yields and a decrease in functional quality, (2) tea prices decrease with a drop in quality resulting from increased precipitation between the spring drought and monsoon tea harvests, (3) farmers at the study site perceive climate variation, have knowledge of climate effects on tea quality, and have sensory preferences on the basis of personal experiences and interactions with the tea market and, (4) precipitation patterns at the study site have varied over the past 40 years and are expected to continue to vary with drought and heavy monsoon rains becoming more extreme and frequent with climate scenario projections.

Materials and Methods

Ethics Statement

Household surveys were conducted according to a human subjects protocol approved by the ethics committee at Minzu University of China. Participants provided their written informed consent to participate in this study. No unique identifiers were attached to the survey data. Permission to conduct interviews and sample tea leaves at the study site was granted by the Secretariat of the Communist Party in Menghai County, Xishuangbanna Dai Autonomous Prefecture, Yunnan Province, China. Sampling of tea leaves was carried out on a private commercialized tea plantation with the permission of the smallholders managing the plantation. No sampling was conducted in a protected area and no wildlife was sampled for this study.

Study site

This project was carried out in an indigenous Akha small-landholder farming community located in evergreen broadleaf montane forest at an altitude between 1 600–1 850 m in Menghai County, Xishuangbanna Dai Autonomous Prefecture in Yunnan Province of southwest China. The study area is situated on the northern edge of the tropical zone in Southeast Asia [19] within the Indo-Burma biodiversity hotspot. The Akha are a Tibeto-Burmese socio-linguistic group that migrated to the region several hundred years ago from the Tibetan plateau. The population at the study site is approximately 400 inhabitants living in family units of a total of 100 households. Each household manages 10 to 30 land-use types including tea agro-forests and monoculture terraces [20]. The study community has been cultivating tea for hundreds of years and produces a premium product sought out by urban connoisseurs in China and elsewhere. Income at the study site is primarily derived from the cultivation, processing, and sale of premium tea to traders and vendors. The study region is characterized by an annual mean air temperature of approximately 22°C, annual precipitation between 1 400–1 500 mm and, a pronounced dry season of six months [19].

Tea harvest seasons

Tea is a perennial plant that is harvested during three distinct seasons in the study area. The three harvest seasons at the study site are dependent on the onset and retreat of the East Asian Monsoon including: (1) a dry spring tea harvest (around mid-March until end of April to mid-May), (2) a monsoon tea harvest marked by the onset of the East Asian Monsoon when warm-wet

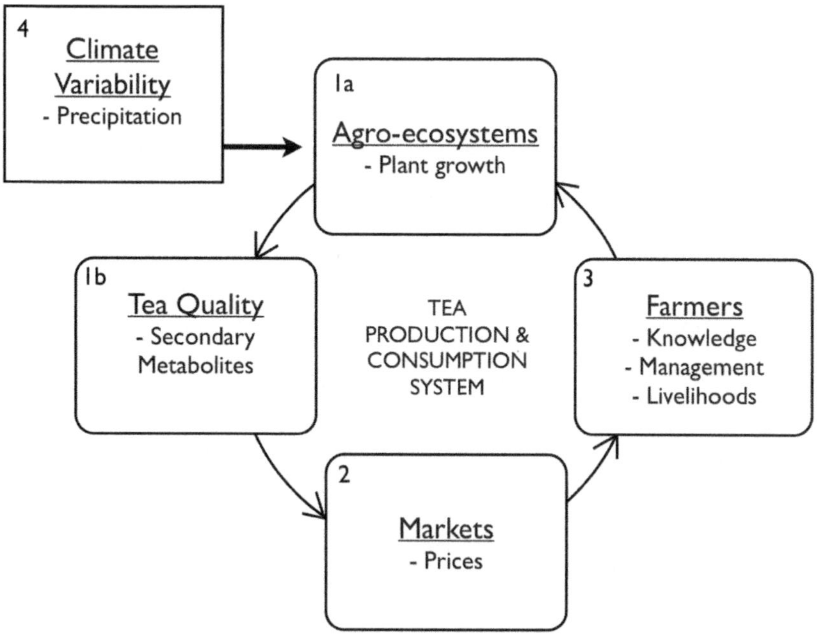

Figure 1. Tea Production and Consumption System. Our study objectives are to examine the tea production and consumption system on the basis of the following dimensions: (1) effects of precipitation variability between an extreme drought that occurred during the dry spring tea harvest season and monsoon tea harvest season on growth and functional quality of tea leaves, (2) tea prices during the dry spring tea harvest season and monsoon tea harvest season, (3) farmer knowledge of climate effects on tea quality and associated management, sensory preferences, and livelihoods and, (4) precipitation trends over the past 40 years as well as forecasted precipitation at the study site.

air masses arrive from the Indian Ocean (around mid-May until end of September) and, (3) a dry autumn tea harvest marked by the retreat of the East Asian Monsoon and arrival of Continental air masses from subtropical origins (around October to mid-November).

Tea growth and functional quality

Stratified time-series sampling was conducted by harvesting tea plants on a daily basis during three sub-seasonal harvest periods starting from the dry spring tea harvest to the monsoon tea harvest. The three harvests are dependent on precipitation with no significant differences in temperature. Daily precipitation and temperature was gathered at the study site during the study period using a weather station to delineate tea harvest seasons that are dependent on precipitation. Each sampling period included five days of sampling corresponding to a tea harvest season or sub-season as follows: (1) spring drought tea harvest from 8–12 May 2012, (2) monsoon onset tea harvest from 13–18 May 2012 and, (3) monsoon tea harvest from 20–25 May 2012.

Sampling was carried out in terrace monoculture teagardens at the study site managed by local smallholders. Three 15 m×15 m plots each consisting of 100–120 tea plants were marked in terraced monoculture teagardens on the same mountain slope at altitudes ranging from 1 787–1 810 m. Tea plants were randomly labeled within each plot. Five tea plants per plot were harvested on each sampling day. No tea plants were harvested more than once to maintain sample independence and to limit the possibility of induction of secondary metabolite defense compounds by leaf removal. For each tea plant, we collected twenty harvest units each consisting of one bud and the adjacent two leaves from various positions on tea plants using sharp shearing scissors. The length of the first fully expanded leaf was measured on ten of the twenty leaves of each sample. All 20 harvest unit samples from an individual plant were combined for secondary metabolite analyses. Leaf samples were processed within two hours of collection to deactivate oxidative enzymes and dried in a microwave oven (700 watts; Haier Electronics Group, Hong Kong) on the basis of an adapted green tea processing protocol [21]. Processed tea leaves were stored in sealed plastic bags in dark conditions at room temperature in the field. Samples were brought to the laboratory, weighed, and sequentially frozen after 20 days from harvest so that each sample was stored at room temperature for the same time period.

Samples were ground to a fine powder in a ball mill and twenty mg of dried leaves were extracted in 1.5 mL 80% aqueous HPLC-grade methanol (Fisher Scientific, Fair Lawn, NJ). The resulting mixture was vortexed for 30 sec (Genie 2, Fisher Scientific, Fair Lawn, NJ) and sonicated for 30 min at 20°C (Quantrex 280, L&R Ultrasonics) and then centrifuged for 15 min at 15 000 rpm (Marathin Micro A, Fisher Scientific).

We measured the concentration of eight antioxidant polyphenol compounds and three methylxanthine compounds that are linked to tea functional quality including its sensory characteristics, health claims, and stimulant properties. The analytes measured for tea functional quality in this study are those recognized by the International Organization for Standardization as well as by the FAO Intergovernmental Panel on Tea's Standards for Tea. Target antioxidant polyphenols included catechin (C), catechin gallate (CG), epicatechin 3-gallate (ECG), epigallocatechin (EGC), epigallocatechin 3-gallate (EGCG), gallic acid (GA), gallo catechin (GC), and gallocatechin 3-gallate (GCG). Target methylxanthines included caffeine (CAF), theobromine (TB) and theophylline (TP). Total methylxanthine content (TMC) was reported from HPLC data by the addition of individual amounts of methylxanthine compounds. All standards were purchased from ChromaDex (Santa Ana, CA).

High Performance Liquid Chromatography (HPLC) was performed as described by some of the study's authors in a previous paper [22]. The method was validated with respect to accuracy, precision, sensitivity, and selectivity. A Waters 2695 (Milford, MA) module HPLC equipped with a 996 photodiode array detector and a 4 µm, 250×4.6 mm ID, C-18 Synergi Fusion, reversed-phase column (Phenomenex, Torrance, CA) was used. Column and autosampler temperatures were maintained at 38°C and 4°C, respectively. Five µL were injected for each sample. The mobile phase was 0.05% (v/v) trifluoroacetic acid in distilled water (Solvent A) and 0.05% (v/v) trifluoroacetic acid in acetonitrile (Solvent B). The following gradient was used at a flow rate of 1 mL/min: 12–21% Solvent B from 0–25 min; 21–25% Solvent B from 25–30 min. The column was flushed with 100% Solvent B for 10 min at the end of each run and was re-equilibrated for 5 min to starting conditions. Spectra were recorded from 254 to 400 nm and relevant peaks were detected at 280 nm on the basis of characteristic absorbance spectra and retention time. Analyte concentrations were determined using peak areas and the linearity determined by plotting signal versus concentration standard curve equations with the limit of detection (LOD) and the limit of quantification (LOQ) in the ranges of 0.05–1 and 0.1–5 g ml^{-1}, respectively.

Total phenolic concentration (TPC) was determined spectrophotometrically using Folin–Ciocalteau reagent as described in a previous paper [22]. Samples were analyzed in triplicates. Absorbance values were measured at 765 nm using a Benchmark Plus Microplate Spectrometer (Bio-Rad, Hercules, CA) and results expressed as gallic acid equivalents (GAE) in mg g^{-1} dry plant material. The concentration of polyphenols in tea samples was derived from a standard curve of gallic acid concentration vs. absorbance between 31.25–500 g ml^{-1}.

Anti-oxidant activity of tea extracts was assessed with the 1-1-diphenyl-2-picrylhydrazyl (DPPH) free radical scavenging capacity assay as described in a previous paper [22]. Gallic acid (0.03–0.25 mg ml^{-1}) and ascorbic acid (0.03–0.50 mg ml^{-1}) were used as positive controls. Each mixture was prepared in quadruplicates of five concentrations. Absorbance values were measured at 517 nm using a Benchmark Plus Microplate Spectrometer (Bio-Rad, Hercules, CA). Radical scavenging capacity of each sample was calculated as the percentage of DPPH free radicals inhibited by the sample in comparison to radical inhibition in the negative water control. Values obtained were plotted against concentration (µg ml^{-1}). Free radical scavenging capacity was expressed as IC_{50} values (concentration of sample required to scavenge 50% of DPPH radicals; the lower the IC_{50} value, the higher the anti-oxidant activity).

Farmer knowledge, sensory preferences, and tea prices

Participant observation and open-ended surveys were carried out in order to develop a structured survey instrument. Structured household interviews using the developed survey instrument were conducted to assess cultural consensus of farmer perceptions, experiences, and management adaptations associated with impacts of climate change and extreme weather events on tea agro-ecosystems and quality. Additional structured household interviews were conducted with smallholder tea farmers at the study site to determine tea prices associated with the spring tea harvest and the monsoon tea harvest over the past decade.

Cultural consensus analysis is a mathematical model that uses factor analysis to test informant agreement regarding a cultural domain on the basis of informant responses to systematic interview questions [23]. It is based on the finding that agreement indicates knowledge within a particular cultural domain. Within the model, consensus of knowledge is met if the ratio of the eigenvalues for the first and second factor is >3:1. The utility of the method lies in its ability to quantify variation, illuminate instances of disagreement, and find the culturally "correct" answer to a particular question. Cultural consensus analysis also allows for a relatively small sample size to provide a statistically high level of confidence in responses. For example, 29 informants with a 0.5 average level of cultural competence provide a 0.99 confidence level [24].

We developed interview questionnaires from pilot testing of sample questions at the study site and surrounding communities. The survey instrument consists of 23 questions that elicit structured responses that are each followed by prompts for open-ended responses. Survey questions are organized into four sections on perceptions and experiences related to: (1) season, precipitation, and temperature, (2) impacts of precipitation and temperature on crop quality, yields, pests, and weeds in tea agro-ecosystems, (3) impacts of climate on income and, (4) management adaptations to climate variability. Surveys were carried out with the assistance of translators trained in household survey methods.

We surveyed 32% of the 100 households at the study site on the basis of suitable sample sizes for cultural consensus analysis as reported in the literature [24]. One adult informant from each household was interviewed. We employed a stratified random design to ensure equal gender and age classes of informants. Age classes are defined here as young to middle-aged informants, whose age is less than 55 years old, and elder informants whose age is 55 years and older.

Data from the structured component of household surveys were analyzed in a binary matrix to determine cultural consensus between households in the community using UCInet 6 [25]. Responses from the open-ended component of the survey were coded individually by two analysts and agreed upon to determine thematic prevalence of responses within the community.

Precipitation variability: historical trends and projected scenarios

Precipitation variability at the study site was examined at three scales: (1) short-term seasonal variability between the spring drought through the onset of the East Asian Monsoon during the sampling period of this study, (2) historical trends over a thirty-year range and, (3) precipitation scenario projections in the 2050-2079 period. Historical trends were calculated using long-term precipitation satellite data for the study site from the pentad (5-day) mean CMAP product over a thirty-year range [26]. The pentad mean precipitation rate is an estimate compiled from satellite, rain gauges, and the output of numerical weather prediction on a 2.5° grid. Satellite data was used because the nearest rain gauge is over 500 km from the study site and is less than 60 percent complete. The precipitation time series used to delineate the wet season is an average of the pentad mean precipitation grids in the 5° by 5° region surrounding the study site. Here, we define the wet season as the period in which 90% of the annual precipitation occurs. The wet is assumed to begin on the pentad when the cumulative precipitation exceeds five percent of the annual total in each year and ends when it exceeds 95%. To validate the models used, we plotted the CMAP data and the modeled data to calculate a measure of temporal overlap of different threshold values.

The significant regional differences in climate change scenario projections produced by different climate models make it difficult to interpret results of individual experiments. Instead of using small ensembles of a single model, we present maps of simulated changes in precipitation for the 21st Century, estimated from an ensemble of single runs from 19 models used in the IPCC AR4 report [27]. The ensemble is the mean of run number one from all of the models. We estimated the mean precipitation rate from the ensemble mean of the 20th Century runs and compared the mean precipitation rain rate of the dry spring season and the monsoon season over the following five-year periods: (1) 1979–1984, (2) 1985–1989, (3) 1990–1994, (4) 1995–1999, (5) 2000–2004 and, (6) 2005–2010. We then employed the ensemble mean of the SRES A1B scenario runs to create future climate scenarios for 2050–2079 to understand possible climate change scenarios and their implications for tea quality and farmer livelihoods at the study site.

Statistical analysis

A Fit Model using a Standard Least Square Means personality function and Analysis of Variance was performed using JMP 10.0 (SAS Institute Inc.) to determine how leaf growth and tea functional quality parameters vary with increasing precipitation from the spring drought through the monsoon onset and monsoon tea harvests. Least Square Means of tea growth values, functional quality values, and farmer perceptions of tea quality (categorized as high, low or medium quality) were plotted against the three tea harvest sampling periods. A multiple comparison using the Least Square Means Tukey HSD method was applied to examine if tea growth and quality significantly increased or decreased with increasing precipitation from the spring drought through the monsoon onset and monsoon tea harvests. Plot and sampling day were used as random variables. Finally, we applied regression analysis to evaluate the relationship between leaf growth measurements and secondary metabolite chemistry with plot as the unit of replication.

Results

Tea growth and functional quality

Tea leaves harvested during the spring drought appeared drier compared to the leaves sampled during the monsoon onset and monsoon tea harvests that appeared tender. Leaf length (n = 2,250) and leaf weight (n = 225) increased significantly ($P<0.01$) with the increase in precipitation over the sampling periods from the spring drought to the monsoon onset and monsoon onset to monsoon tea harvests (Figure 2).

The concentration of six major tea polyphenolic metabolites (EGC, EGCG, ECG, GCG, C and GA) that determime tea functional quality decreased ($P<0.05$) with the onset of the East Asian Monsoon from the spring drought to the monsoon tea harvest (Figures 3). The compounds with the largest change in concentration during the transition from the spring drought to the monsoon tea harvest were EGC, EGCG, and ECG. These metabolites were also found to have the highest concentrations during all three seasonal tea harvest periods. In addition to changes in these polyphenolic metabolites, total methylxanthine concentrations (TMC) significantly decreased from the spring drought to the monsoon onset as well as to the monsoon tea harvest ($P<0.05$; Figure 4). Concurrently, total phenolic concentration and antioxidant activity of tea leaves increased from the spring drought to monsoon onset and monsoon tea harvests (Figure 5). No difference was found between the monsoon onset and monsoon for total phenolic concentration but a significant difference was found for antioxidant activity. Overall, seasonal tea harvest period and samples categorized as 'high quality' by informants were both found to be significant ($P<0.02$) for secondary metabolites analyzed, total methylxanthine concentration, total phenolic concentration, and antioxidant activity independent of plot effects (soil characteristics and minor elevation differences).

The increase in leaf weight was correlated with a decrease in the concentrations of individual phenolic compounds (CAT, ECG, EGC, EGCG, GA and GCG), TMC, and antioxidant activity ($P\leq 0.05$). Total phenolic concentration did not significantly change with leaf weight.

Farmer knowledge, sensory preferences, and tea prices

We found a high degree of agreement between farmer responses to structured survey questions on observed climate patterns and the effects of precipitation variability on tea quality. The eigenvalue ratio between the first and second factors was 10.83:1 and the average competency (or agreement) score was 0.82±0.13. The first eigenvalue of the informant response matrix accounted for 85% of the total response variance. There was complete consensus for 16 out of 23 questions including agreement that precipitation changes during informants' lifetimes have impacted tea yields, quality, and associated livelihoods.

Coded interview responses found that informants perceive that precipitation patterns have changed during their lifetimes including observations of increased duration (78%), strength (78%), and unpredictability (78%) of rains. Additionally, informants perceive that temperatures have increased (78%) during their lifetimes resulting in fewer cold days (78%), warmer winters (78%), warmer summers (72%), and less frequent occurrence of frost (72%). Informants noted these changes in climate patterns have impacted their tea agro-ecosystems with earlier spring tea harvests (94%) and monsoon tea harvests (78%). Field observations with farmers in December 2012 found the unusual occurrence of a fourth tea

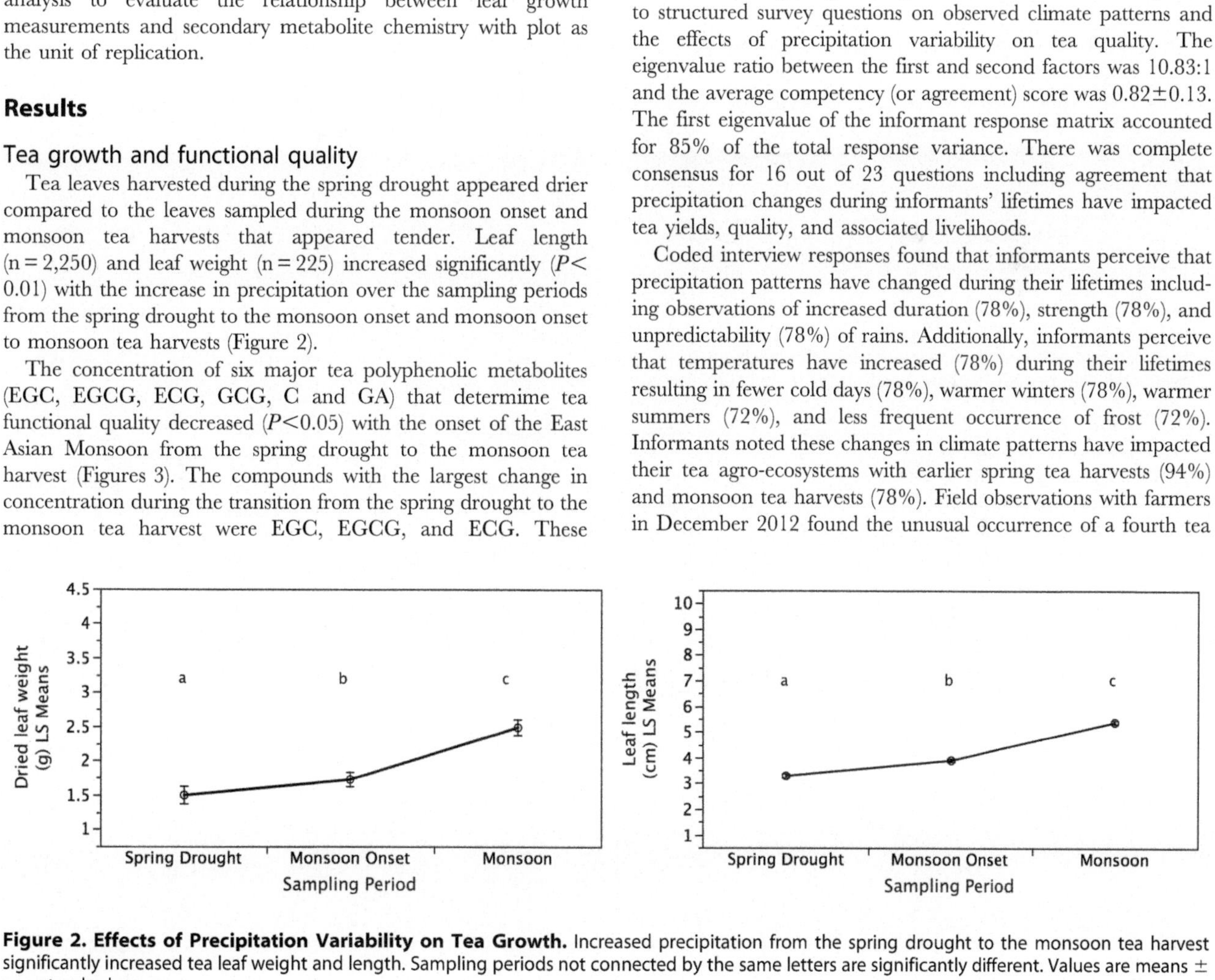

Figure 2. Effects of Precipitation Variability on Tea Growth. Increased precipitation from the spring drought to the monsoon tea harvest significantly increased tea leaf weight and length. Sampling periods not connected by the same letters are significantly different. Values are means ± one standard error.

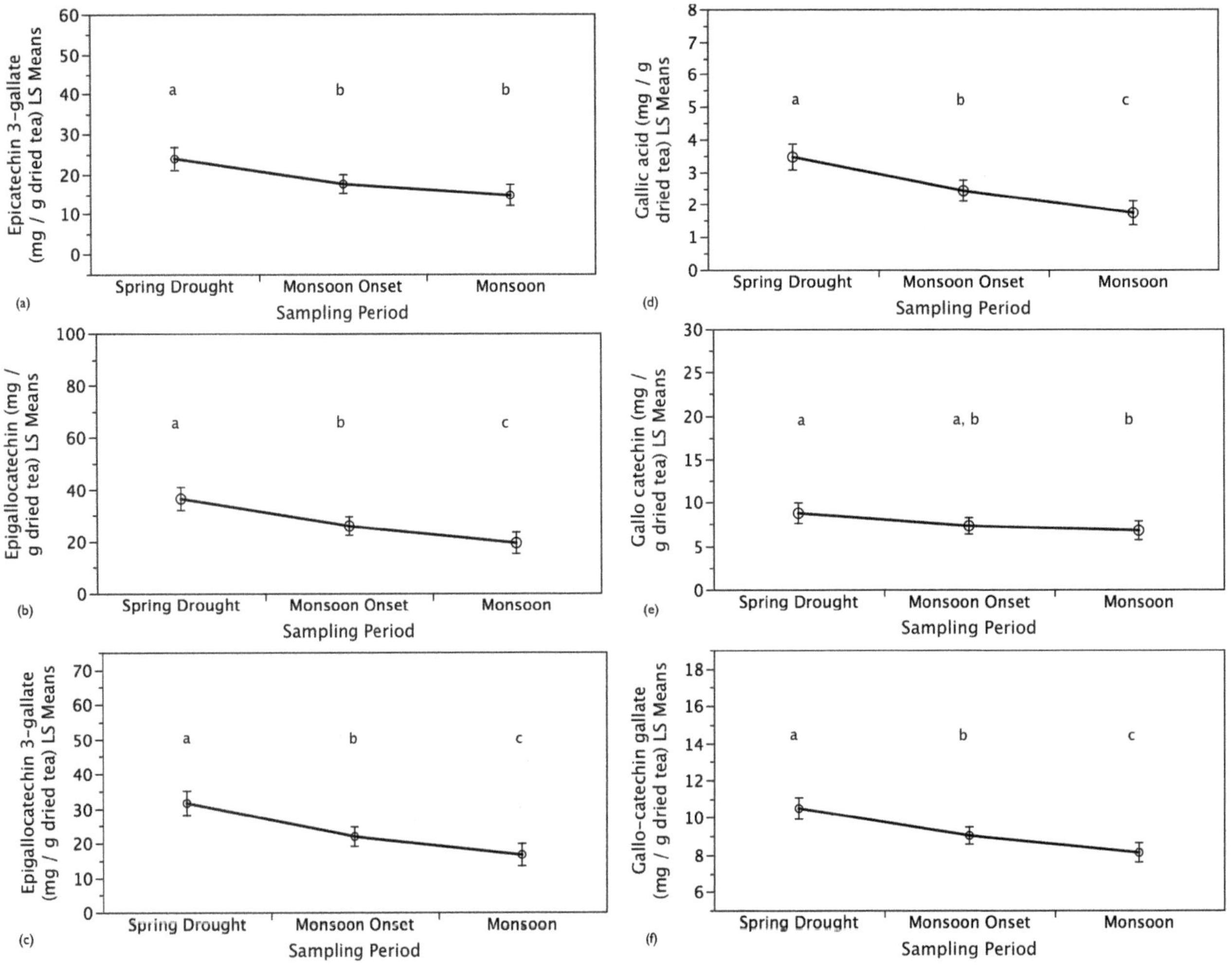

Figure 3. Effects of Precipitation Variability on Tea Polyphenolic Catechins. Increased precipitation from the spring drought to the monsoon tea harvest resulted in significantly lower concentrations of (a) epicatechin 3-gallate (ECG), (b) epigallocatechin 3-gallate (EGCG), (c) epigallocatechin (EGC), (d) gallic acid (GA), (e) gallocatechin (GC), (f) gallocatechin gallate (GCG) as well as catechin (C) and catechin gallate (CG; not shown). Sampling periods not connected by the same letters are significantly different. Values are means ± one standard error.

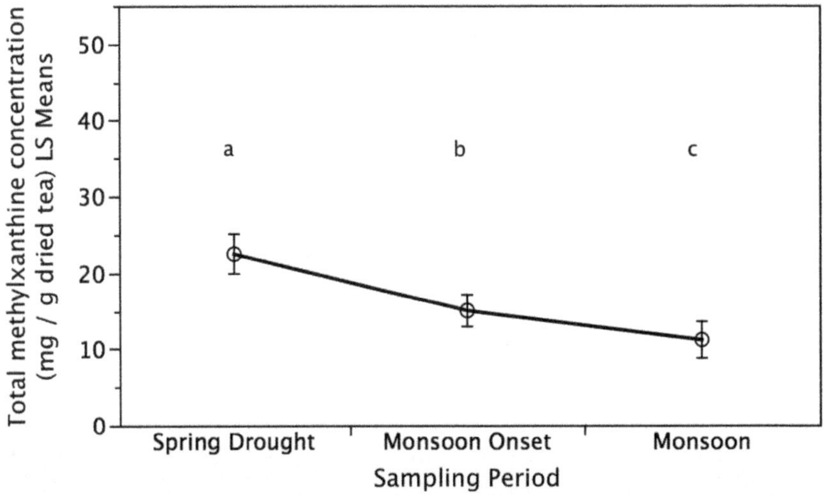

Figure 4. Effects of Precipitation Variability on Total Methylxanthine Concentration. Increased precipitation from the spring drought to the monsoon tea harvest resulted in significantly lower total methylxanthine concentration (TMC) of tea leaves. Sampling periods not connected by the same letters are significantly different. Values are means ± one standard error.

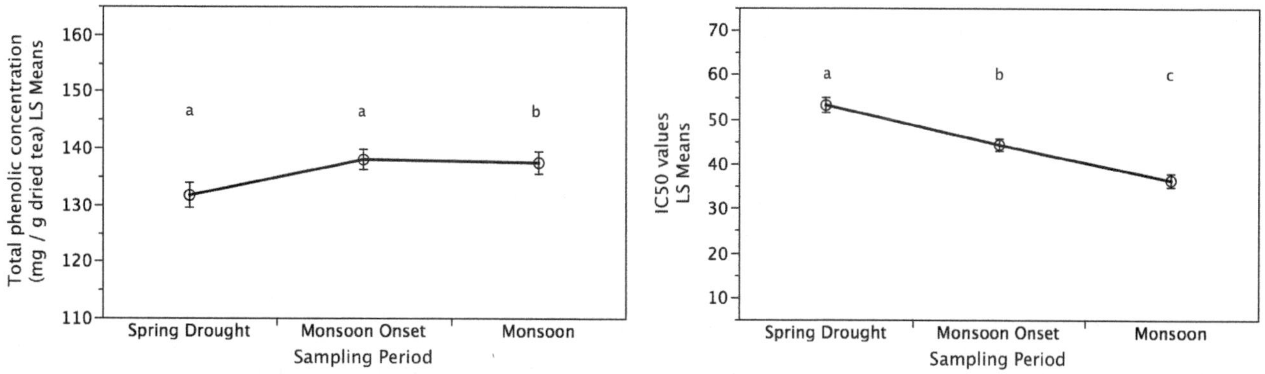

Figure 5. Effects of Precipitation Variability on Total Phenolic Concentration and Antioxidant Activity. Increased precipitation from the spring drought to the monsoon tea harvest resulted in significantly higher total phenolic concentration (TMC) and antioxidant activity of tea leaves. Sampling periods not connected by the same letters are significantly different. Values are means ± one standard error.

harvest with warmer temperatures; tea plants are usually dormant during this period due to colder temperatures.

The majority of informants perceive that precipitation variability impacts tea quality (94%) and yields (81%), although fewer than half agree that temperature variability impacts tea quality (44%) and yields (38%). All informants agreed that quality varies during the three seasonal tea harvests and is inversely related to leaf size. Tea harvested during the dry spring harvest is rated with the highest quality and best sensory preference. Farmers perceive this high quality deteriorates with the arrival of the East Asian Monsoon, when quality and sensory preference is ranked the lowest. Specifically, informants reported that tea harvested during the dry spring harvest has stronger aroma (81%), physiological characteristics (69%), and taste (66%) that many describe as more bitter and bitter-sweet with a sweet lingering aftertaste at the back of the throat referred to as *gaan* (50%). In addition to quality, farmers noted that precipitation levels are directly related to tea growth. Specifically, informants reported that when tea quality is highest during the spring harvest, there is less leaf budding (81%) and tea leaves are drier (88%) and smaller (75%) compared to the monsoon tea harvest season. In addition to direct effects of precipitation variability on tea quality and yields, almost 70% of informants reported that changes in climate patterns have impacted tea quality and yields by altering the quantities and types of pests and weeds in and around tea agro-ecosystems.

Informants reported that tea traders that buy and sell tea from the study site have the greatest demand for tea harvested during the spring and recognize monsoons as deleterious to tea quality and prices while regarding droughts as positive for tea quality but harmful to yields. Tea samples categorized as 'high quality' by informants were significantly correlated ($P<0.006$) with the spring drought tea harvest independent of plot effects. For the past decade, the perceived drop in tea quality and increased leaf growth that occurs with the onset of the East Asian Monsoon has been accompanied with a decrease of 30–50% in livelihoods derived from tea sales. Specifically, for the 2002–2012 period, farmers experienced an average decrease of 51% in on-farm tea prices during the monsoon tea harvests compared to the spring drought tea harvests (Figure 6). Informants (81%) attributed the highest historical tea price that occurred during the dry spring tea harvest of 2012 to an especially high quality crop resulting from the severe drought and not to decreased yields. Only a few farmers (16%) attributed the high spring tea prices to the 25–33% decline in yields resulting from the drought compared to previous dry spring harvests. Regardless of precipitation effects on tea, farmers expressed that climate factors have not negatively impacted their tea sales to date (97%) but have negatively impacted food supply and income from other crops (72%). However, farmers reported concern regarding the drought threshold that is beneficial for tea quality before tea plants loose viability.

All farmers noted having knowledge of management practices to reduce vulnerability of tea crops to climate variability. Management practices employed by farmers at the study site to mitigate climate risk in their agro-ecosystems include cultivating tea from seed rather than from clonal propagules (81%) and growing tea plants as trees rather than as shrubs (50%). Lastly, farmers noted the importance of maintaining canopy coverage (81%) and forested buffer zones around agro-ecosystems (72%) in order to reduce risk in their tea production systems.

Precipitation variability: historical trends and projected scenarios

Significant seasonal variation was found in precipitation during the three sampling periods of the spring, monsoon onset, and monsoon tea harvests while no variation was found in temperature. Historical trends of observed precipitation data show inter-annual variability at the study site from 1979 through 2010. The wet season has increased in length since 1990 by starting earlier and/or ending later (Figure 7). Over the past 30 years, mean daily precipitation has increased during the dry spring tea harvest and decreased during the monsoon tea harvest (Figure 8). Climate scenario projections for the 2050–2079 period suggest that mean precipitation for this period is expected to decrease 4–6% during the dry spring tea harvest and increase by 6–8% during the monsoon harvest compared to the historical 1970–1999 period.

Discussion and Conclusion

This study highlights links between precipitation variability, crop quality, farmer knowledge, sensory preferences, and product prices using tea production and consumption as a model specialty crop system. Compared to the monsoon harvest, the extreme drought sampled during this study was significantly correlated with heightened tea quality on the basis of secondary metabolite concentrations and increased farmer sensory preferences while these quality parameters had an inverse relationship to tea yields. Specifically, increased precipitation from the spring drought to the monsoon rains resulted in a significant increase of up to 50% in tea leaf growth as measured by leaf length and weight. Concurrently, increased precipitation resulted in a decrease of up to 50% in tea

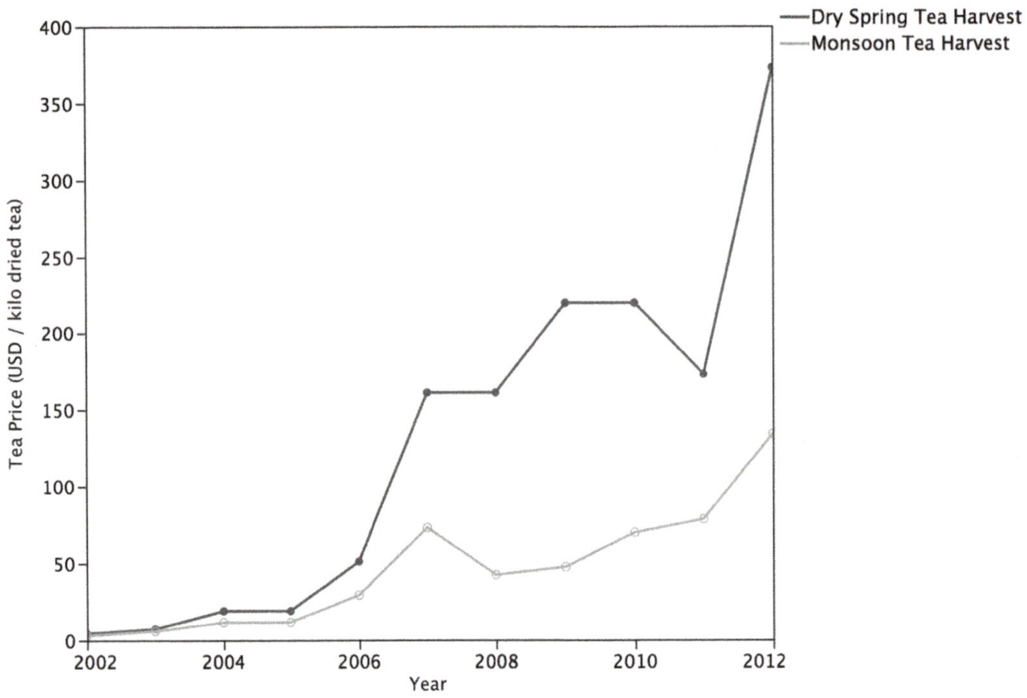

Figure 6. Effects of Precipitation Variability on Tea Prices. Farmers at the study site have experienced an average decrease of 51% in on-farm tea prices received during the Monsoon tea harvests compared to the dry spring tea harvests.

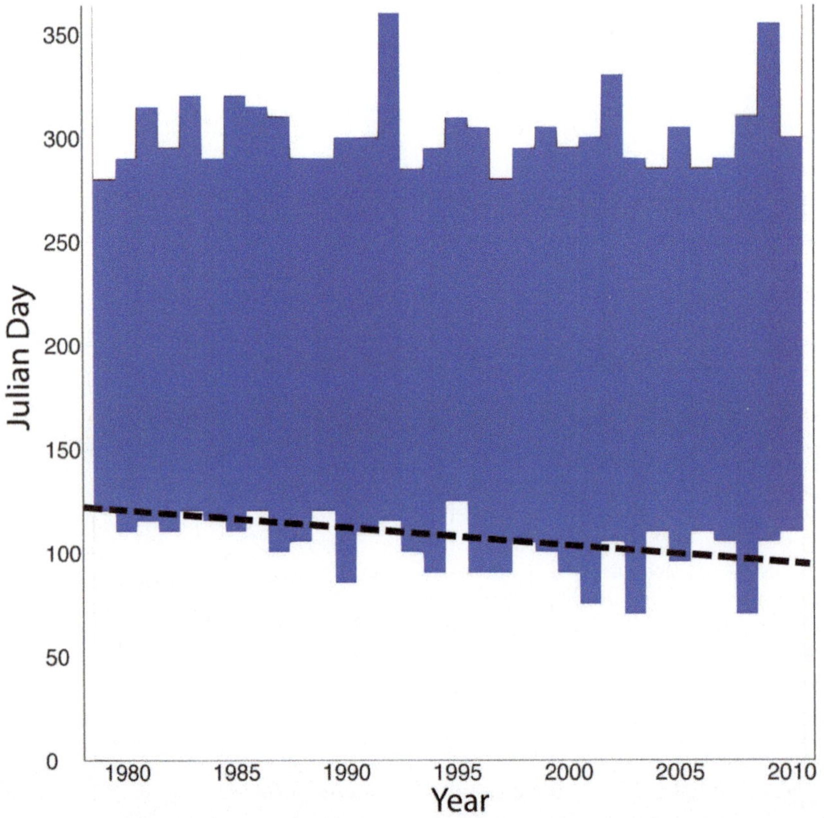

Figure 7. Increased Inter-annual Variability and Total Precipitation. Precipitation data at the study site shows inter-annual variability from 1979 through 2010. The shaded section of the figure represents the period comprising 90% of annual rainfall. In addition to increasing variability since 1990, the monsoon season is arriving earlier as indicated by the downward shift of the trend line.

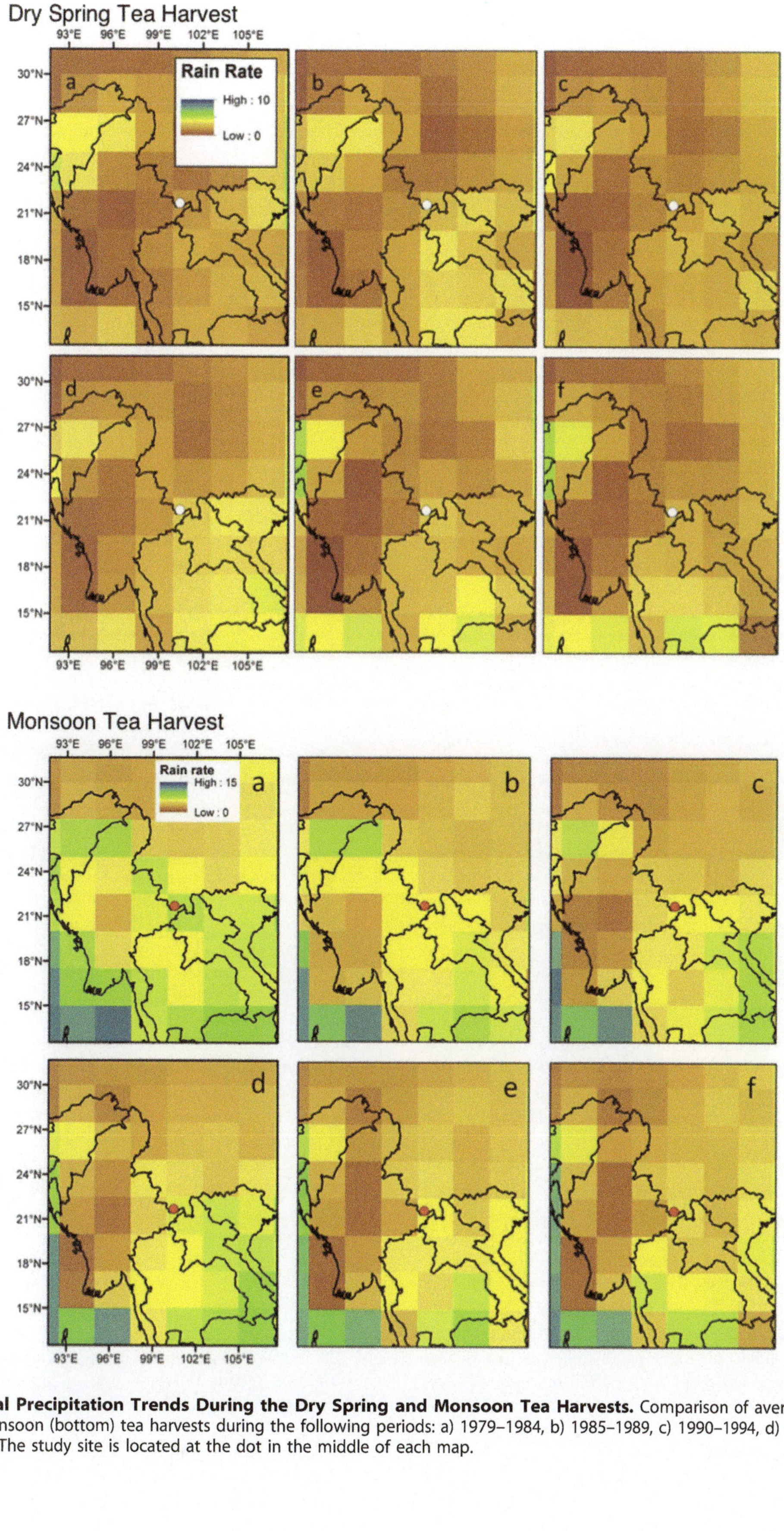

Figure 8. Historical Precipitation Trends During the Dry Spring and Monsoon Tea Harvests. Comparison of average rain rate for dry spring (top) and monsoon (bottom) tea harvests during the following periods: a) 1979–1984, b) 1985–1989, c) 1990–1994, d) 1995–1999, e) 2000–2004, f) 2005–2010. The study site is located at the dot in the middle of each map.

functional quality as determined by concentrations of individual catechin secondary metabolites and total methylxanthine concentrations while total phenolic concentrations and antioxidant activity increased. The decrease in measured tea quality parameters during the monsoon matched farmer perceptions and the decrease in tea prices that were up to 50% lower compared to the spring tea harvest. These findings validate farmer knowledge of climate effects on tea quality and rationalize sensory preferences for spring tea as well at its' higher demand and associated higher prices compared to tea from the monsoon harvest. Extrapolating findings from this seasonal study of extreme precipitation variability coupled with historical precipitation trends and projected climate scenarios in order to understand broader climate change suggests that consumers and producers are likely to experience both positive and negative effects as extreme droughts and heavy rains are becoming more frequent with forecasted drier spring seasons and wetter monsoon seasons.

Results of this study concur with previous studies in other geographic areas that altered precipitation has a notable effect on plant performance [18,28] and significantly impacts both growth and secondary metabolite concentrations of tea plants [29,30,31]. The smaller length and weight of tea leaves harvested during the spring drought compared to the monsoon tea harvest in this study supports the widely accepted recognition that lower 3soil moisture content reduces photosynthesis and growth of plants [32,33]. While soil moisture content is linked to survivability of plants, the spring drought leaves in the present study appeared dry but did not have the physical characteristics of tea leaves under severe water stress suggesting that tea plants can tolerate additional water stress before compromising viability of harvest. Further research is needed to determine precipitation and soil moisture levels that compromise the viability of tea production in various geographic areas and altitudes as well as in agro-ecosystems that follow various modes of management such as tea agro-forests, mixed crop tea gardens, irrigated gardens, and organic production.

This study confirms that rainfall levels are of central importance to tea quality and concurs with previous studies in other climatic zones on the effects of season on tea quality including lower total methylxanthine and catechin concentrations during the monsoon tea harvest when leaf growth is higher compared to the dry spring harvest [31,34,35,36,37,38,39]. For example, Gulati and Ravindranath [31] found that methylxanthine concentrations of tea samples harvested in northeastern India were highest during the early harvest of the dry spring season, lowest during the monsoon season, and improved with the autumn harvest. Further in line with this study's findings, Kottur et al. [38] showed that catechin compounds and overall flavor index of tea harvested in south India were highest during the dry season followed by the pre-monsoon season and were lowest during the monsoon. Similarly, Yao et al. [30] found that catechin compounds were lowest in tea produced during the wet season in Australia while Muthumani et al. [39] reported that the south-west monsoon season resulted in tea of poor quality as determined by biochemical composition including catechins and quality improved with the retreat of the monsoon. Jeyaramraja et al's [37] studies on the impacts of soil moisture stress on tea phytochemicals found that there are optimal drought conditions that upregulate phytochemicals linked to tea quality and water stress beyond this point will result in a decline of these compounds. Further research is needed to determine the optimal soil moisture levels for tea functional quality across various spatial scales and management practices.

While this study did not directly measure volatile aroma compounds, the catechin compounds measured in this study are recognized to be involved in the development of favorable aroma compounds in tea [40], which further explains the higher perceived quality of tea harvested during the dry spring when catechins were the highest. Previous studies support that desirable volatiles and other compounds influencing tea flavor profiles decrease with the monsoon [40,41]. However, previous work also suggests that unlike the catechins and methylxanthines quantified in this study, some phytochemicals and plant constituents in tea including certain amino acids, crude fiber, moisture content, and chlorophyll increase with wet weather conditions and variably impact the aroma and taste of the final brew [38,42]. For example, higher moisture content, crude fiber, and chlorophyll of leaves harvested during the monsoon are recognized to decrease the flavor of brewed tea while certain amino acids favorably increase perceived flavor [38]. Further research is needed to assess the effects of climate on tea quality as a function of the approximately 500-1,000 secondary metabolites in tea that both favorably and unfavorably contribute to tea functional quality.

Our results indicate a discrepancy in the trend of changes in individual secondary metabolite concentrations compared to changes in total phenolic concentration and antioxidant activity with increased precipitation of the monsoon tea harvest. Surprisingly, the decrease in individual secondary metabolite concentrations with increased precipitation was not congruent with the trend of increased total phenolic concentration and antioxidant activity. This discrepancy may reflect the fact that larger tissues can produce higher concentrations of phenolic compounds other than the individual phenolic metabolites measured in this study that determine tea functional quality including tannins and other simple phenolics such as glycosylated flavonols, proanthocyanidins, and phenolic acids and their derivatives that also have antioxidant activity [43,44,45].

The inverse relationship between tea growth and key functional quality parameters found in this study suggests a possible dilution effect of precipitation. Previous research has identified that enzyme reactions controlling the dynamic metabolic systems in tea leaves are also responsible for changes in tea quality parameters during different seasons, with low enzyme activity in the rainy season [39]. In addition, studies have indicated that the availability of compounds for the formation of soluble solids in tea plants is greater during periods of restricted growth such as the drought [40]. Irradiation effects of higher sunshine during the dry season and lower sunshine hours during the monsoon season could be another reason for the differential accumulation of particular phytochemicals in tea shoots [43]. Increased leaf temperatures during the drought have further been attributed to changes in the synthesis of tea phytochemicals [37]. In addition, processing of tea leaves with greater moisture content harvested in the monsoon season requires greater time during the processing steps of pan-frying, steaming, wilting, and drying and this additional processing collectively contributes to degradation of tea phytochemicals.

The methylxanthines and individual catechins that declined with increased precipitation in this study are key secondary metabolites associated with flavor, health claims, stimulant properties, and other quality characteristics of tea [46]. The catechins are the most abundant secondary metabolites in tea plants comprising approximately 30% of dried leaf extract and are also the most biologically active constituents of tea [13]. The consumption of catechins in humans is inversely associated with the incidence of chronic disease including cardiovascular diseases, bone and skin health, inflammation, oxidative stress, cancer, neurodegenerative condition, and the metabolic syndrome [47]. The significant decrease in catechin concentrations during the monsoon suggests decreased consumption of these dietary phytonutrients from tea harvested during forecasted periods of increased precipitation and increased consumption of these dietary phytonutrients during periods of decreased precipitation and subsequent implications for human health.

At the market level, consumer perceptions of the decreased functional quality of tea via its' sensory characteristics during the monsoon harvest [7] drive the decline in tea prices [38]. Tea harvested during the dry season is recognized to fetch higher prices than tea harvested during the monsoon season in major tea production countries [38]. Previous reports of Chinese tea production have documented that high quality tea is primarily produced during the dry season and accounts for approximately 80% of total production value while accounting for only 40% of total production quantity [48]. Projected climate scenarios of increased extreme climate events are thus expected to impact the production value and prices of tea and directly impact farmer livelihoods. These dynamics are important to understand towards designing appropriate climate adaptation strategies and policies that best benefit both producers and consumers in the context of global environmental change.

The large degree of agreement found in the cultural consensus analyses regarding farmer knowledge of climate variability and effects on tea highlights the agro-ecological knowledge [49,50,51] of the study-site community and the customary natural resource institutions and social networks that maintain this knowledge. The maintenance of ecological knowledge in regards to global environmental change is a recognized factor to enable resilience and adaptation to environmental variability [52]. Ultimately, the adaptive capacity of a socio-ecological system determines the extent of climate effects on natural resources, livelihoods, and wellbeing. This study supports that smallholder farmers at the study site have knowledge that climate drives a decline in tea quality either through personal sensory experiences and/or interactions with the market. Farmers at the study site further have knowledge through their agreement that tea cultivated from seed and maintained as trees are more resilient to the effects of precipitation variability compared to tea plants cultivated from clonal propagules and maintained as shrubs. In addition, farmers agree that tea plants grown in diverse, multi-storied agro-forests with canopy coverage and forest buffers are more resilient to the effects of precipitation variability compared to tea plants grown in monoculture terraced plantations. Previous studies have validated that clonal tea shrubs are particularly susceptible to drought effects and show severe water stress during the dry season compared to tea cultivated from seed [53]. Despite such findings, agricultural extension programs and enterprises throughout tea-growing areas of China and worldwide continue to promote the cultivation of tea in monoculture plantations using clonal propagules. Further research is needed to examine how variable management modes of production and practices such as agroforestry systems may buffer the negative impacts of climate variability. For example, previous studies have identified agro-forestry as an adaptive strategy against potential microclimate extremes in several specialty crop systems such as coffee production [54,55]. Studies are needed to identify the variables that promote farmer knowledge of climate effects on crop quality and associated management adaptation. Policies and plans may than use this evidence to foster the variables promote farmer knowledge and adaptive management towards increased resilience of tea agro-ecosystems to global environmental change.

Findings from this study are in agreement with previous research showing vulnerability of tea-producing areas to changing precipitation patterns with consequences for environmental, economic, and social factors [56,11,12]. For example, tea production in Sri Lanka is forecasted to experience more intense dry and wet periods with anticipated implications for top-soil erosion, increased agro-chemical run off, and irreparable yield loss that will subsequently have negative socio-economic repercussions for the laborers that compromise the country's main net foreign exchange earner [56]. However, climate deviations in other tea-producing areas such as a rise in mean temperature during the cold season in Northern India have been reported as beneficial for an early tea crop and extended harvest [57]. Taken together, the totality of evidence emphasizes differential responses of tea production systems across geographic scales.

Extrapolating findings from this seasonal study, previous precipitation trends, and projected climate scenarios to understand broader climate change suggests that farmers are facing increased intra-annual livelihood variability with the increased prevalence and intensity of extreme droughts and heavy rains. The increased length of the monsoon season since 1979 as well as increased precipitation during the dry seasons indicates that tea quality and prices at the study site have been adversely impacted. Looking forward at climate scenarios for the 2050–2079 period during when drier spring harvests and wetter monsoon harvests are projected implies both beneficial and adverse impacts to tea quality within a single year. Changes in the concentrations of phytochemicals have the potential to alter sensory experiences, physiological effects, and consumer willingness to pay for tea products. Such shifts are thereby likely to impact tea prices and income received by tea famers in China and the approximately 60 other tea-producing countries worldwide.

The effects of extreme climate events on tea crops are likely to intensify as prolonged seasonal effects over time may exceed thresholds for plant survival. Further, total effects on crops may intensify with additional shifting dynamics within agro-ecosystems including pest pressures and weeds that farmers reported to already experience in this study. For example, temperature and rainfall changes can shift pest geographic ranges and result in new immigration of pests into tea production areas [48]. Alterations of wind, humidity, and temperatures would present additional dynamics. The feedbacks between these abiotic and biotic changes may alter the geographic range of tea plants as well as those classified as high quality. At the same time, it is important to consider and understand how tea crops may evolve with these shifts in climate variability and accompanying abiotic and biotic changes in tea agro-ecosystems.

In conclusion, this study provides evidence for the relevance of examining crop quality as well as yields in research on the potential effects of climate change on production to facilitate climate adaptation for sustainable crop systems. This study validates farmer perceptions of extreme drought effects on tea quality and rationalizes the higher demand and prices for the spring tea harvest compared to the monsoon tea harvest. Farmer perceptions of the multiple climate variables that impact tea agro-ecosystems points to the need to look beyond only the early onset of the spring season and plant budding that dominates the literature on plant responses to climate change. Lastly, further research is warranted in tea-producing areas located in diverse geographic and climatic zones to compare precipitation effects on tea quality over various spatio-temporal scales and management practices to identify variables that foster resilience of tea production and sustain farmer livelihoods and consumer well-being.

Acknowledgments

This research was supported by the TEACRS Program (NIGMS IRACDA- K12GM074869) at Tufts University, Program 111 in Ethnobiology of the Chinese Ministry in Education/Minzu University of

China, the Tufts Institute of the Environment, the Tropical Conservation and Development Program at the University of Florida, Tufts University Provost Office (Tufts Collaborates!), NSF REU Program at Tufts University (NSF DBI 1005082), and the NSF Coupled Natural Human Systems (NSF grant #BCS-1313775). The authors wish to thank the study site community for hosting this research; Su Yongge for support during field surveys; Anne Elise Stratton, Feven Asefaha and Matthew Reynolds for assisting in weighing and pulverizing of tea leaf samples; Durwood Marshall for guidance on statistics and, Amanda Kowalsick for compiling a data table (that was removed from the final manuscript).

Author Contributions

Conceived and designed the experiments: SA JRS CO TG CM AR SC DX CL UU SB DS EK. Performed the experiments: SA UU SB. Analyzed the data: SA JRS CM. Contributed reagents/materials/analysis tools: SA CO EK DX. Wrote the paper: SA. Organized research logistics and permits in China: SA JRS DX CL.

References

1. Porter JR, Semenov MA (2005) Crop responses to climatic variation. Philos Trans R Soc Lond B Biol Sci. 360: 2021–2035.
2. Lobell DB, Schlenker W, Costa-Roberts J (2011) Climate trends and global crop production since 1980. Science 333: 616–620.
3. Ewert F, Rounsevell MDA, Reginster I, Metzger MJ, Leemans R (2005) Future scenarios of European agricultural land use I. Estimating changes in crop productivity. Agr Ecosyst Environ 107: 101–116.
4. Nelson GC, Rosegrant MW, Koo J, Robertson R, Sulser T, et al. (2009) Climate change: Impact on agriculture and costs of adaptation. Food Policy Report, Briefing Series Number 21 International Food Policy Research Institute. Washington, D.C. http://dx.doi.org/10.2499/0896295354
5. Schlenker W, Lobell DB (2010) Robust negative impacts of climate change on African agriculture. Environ Res Lett 5
6. Myers SS, Zanobetti A, Kloog I, Huybers P, Leakey Andrew DB, et al. (2014) Increasing CO_2 threatens human nutrition. Nature doi: 10.1038/nature13179
7. Ahmed S, Unachukwu U, Stepp JR, Peters CM, Long C, et al. (2010) Pu-erh tea tasting in Yunnan, China: correlation of drinkers' perceptions to phytochemistry. J Ethnopharmacol 132: 176–185.
8. Johns T, Sthapit BR (2004) Biocultural diversity in the sustainability of developing-country food systems. Food Nutr Bull. 2004; 25(2): 143–155.
9. Maplecroft website (2011) Country Risk Briefings. Bath, UK. Available: http://maplecroft.com. Accessed 2013 Apr 16.
10. Piao S, Ciais P, Huang Y, Shen Z, Shushi P, et al. (2010) The impacts of climate change on water resources and agriculture in China. Nature 467: 43–51.
11. Decision and Policy Analyses Group (2011) Future climate scenarios for Kenya's tea growing areas. CIAT Report. Consultative Group on International Agricultural Research. International Center for Tropical Agriculture. Cali, Colombia.
12. Schepp K (2009) Strategy to adapt to climate change for Michimikuru tea farmers in Kenya. AdapCC Report. Deutsche Gesellschaft fur Technische Zusammenarbeit (GTZ) GmbH, Eschborn, Germany. Available: http://adapcc.org/en/kenya.htm
13. Lin Y, Tsai Y, Tsay J, Lin J (2003) Factors affecting the levels of tea polyphenols and caffeine in tea leaves. J Agric Food Chem 51: 1864–1873.
14. Ahmed S, Peters CM, Chunlin L, Meyer R, Unachukwu U, et al. (2012) Biodiversity and phytochemical quality in indigenous and state-supported tea management systems of Yunnan, China. Conservation Letters 5 (6): 28–36.
15. Ahmed S, Orians C, Griffin T, Buckley S, Unachukwu U, et al. (2013) Effects of Water Availability and Pest Pressures on Tea (Camellia sinensis) Growth and Functional Quality. AoB Plants DOI 10.1093/aobpla/plt054
16. Maddison D (2006) The perception of and adaptation to climate change in Africa. CEEPA Discussion Paper No. 10. Special Series on Climate Change and Agriculture in Africa Centre for Environmental Economics and Policy in Africa, University of Pretoria, Republic of South Africa.
17. Spence A, Poortinga W, Butler C, Pidgeon NF (2011) Perceptions of climate change and willingness to save energy related to flood experience. Nature Climate Change 1: 46–49.
18. Wolkovich EM, Cook BI, Allen TM, Crimmins JL, Betancourt SE, et al. (2012) Warming experiments under predict plant phenological responses to climate change. Nature 485: 494–497.
19. Dou J, Zhang Y, Yu G, Zhao S, Song Q (2007) Interannual and seasonal variations of energy and water vapor fluxes above a tropical seasonal rain forest in Xishuangbanna, SW China. Acta Ecologica Sinica 27 (8): 3099–3109.
20. Ahmed S, Stepp JR, Toleno R, Peters CM (2010) Increased market integration, value, and ecological knowledge of tea agro-forests in the Akha highlands of Southwest China. Ecol Soc 15 (4): 27.
21. Gulalti A, Rawat R, Singh B, Ravindranath SD (2003) Application of microwave energy in the manufacture of enhanced-quality green tea. J Agr Food Chem 51: 4764–4768.
22. Unachukwu U, Ahmed S, Kavalier A, Lyles J, Kennelly E (2010) Variation of phenolic and methylxanthine composition and anti-oxidant activity among white and green teas (Camellia sinensis var. sinensis (L.) Kuntze Theaceae). J Food Sci. 75 (6): C541–C548.
23. Romney A, Weller S, Batchelder W (1986) Culture as consensus: A theory of culture and informant accuracy. Am Anthropol 88: 313–338.
24. Weller S, Romney A (1988) Systematic Data Collection. Sage Publications, Newbury Park, California.
25. Borgatti SP, Everett MG, Freeman LC (2011) UCINET 6.362. Natick: Analytic Technologies.
26. Xie P, Arkin PA (1997) Global precipitation: a 17-year monthly analysis based on gauge observations, satellite estimates, and numerical model outputs. Bull. Amer. Meteor. Soc., 78, 2539–2558. doi: http://dx.doi.org/10.1175/1520-0477(1997)078<2539:GPAYMA>2.0.CO;2
27. Meehl GA, Stocker TF, Collins WD, Friedlingstein P, Gaye AT, et al. (2007) Climate Change 2007: The Physical Science Basis. Contribution of Working Group I to the Fourth Assessment Report of the Intergovernmental Panel on Climate Change, S. Solomon et al., Eds. Cambridge Univ. Press, Cambridge 749–845.
28. Cias P, Reichstein N, Viovy N, Grainer N, J Ogée, et al. (2005) Europe-wide reduction in primary productivity caused by the heat and drought in 2003. Nature 437: 529–533.
29. Honow R, Gu KR, Hesse A, Siener R (2010) Oxalate content of green tea of different origin, quality, preparation and time of harvest. Urol Res 38: 377–381.
30. Yao L, Caffin N, D'Arcy B, Jiang Y, Shi J, et al. (2005) Seasonal variations of phenolic compounds in Australia-grown tea (Camellia sinensis). J Agric Food Chem. 53: 6477–6483.
31. Gulati A, Ravindranath SD (1996) Seasonal variations in quality of Kangra tea (*Camellia sinensis* (L) 0 Kuntze) in Himachal Pradesh. J Sci Food Agr 71: 231–236.
32. Kozlowski TT, Kramer PJ, Pallardy SG (1991) The Physiological Ecology of Woody Plants. New York: Academic Press.
33. Coley P (1998) Possible Effects of Climate Change on Plant/Herbivore Interactions in Moist Tropical Forests. Climatic Change 39: 445–472.
34. Dev Choudhury MN, Bajaj KL (1986) Seasonal variation of water soluble sugars in tea shoot and their role in tea quality. Two and a Bud 33: 33–38.
35. Owur PO, Chavanji AM (1986) Caffeine contents of clonal tea: seasonal variations and effects of plucking standards under Kenyan conditions. Food Chem 20: 225–233.
36. Kumar N, Chaudhury TC, Ghosh Hajra N (1992) Seasonal and grade variation in ISO specification for black tea – A Darjeeling perspective. J Plantn Crops 20: 323–326.
37. Jeyaramraja PR, Pius PK, Raj Kumar R, Jayakumar D (2003) Soil moisture stress-induced alterations in bioconstituents determining tea quality. J Sci Food Agric 83: 1187–1191.
38. Kottur G, Venkatesan S, Kumar RSS, Murugesan S (2010) Influence of seasons on biochemical parameters of green shoots and quality parameters of made tea under south Indian conditions. J Biosci Res 1(2): 74–82.
39. Muthumani T, Verma DP, Venkatesan S, Senthil Kuman RS (2013) Influence of climatic seasons on quality of south Indian black teas. J Nat Prod Plant Resour 3 (1): 30–39.
40. Saikia P, Mahanta PK (2002) Specific fluctuations in the composition of lipoxygenase- and glycosidase-generated flavors in some cultivated teas of Assam. J Agric Food Chem 18; 50(26): 7691–9.
41. Harbowy ME, Balentine DA (1997) Tea chemistry. Crit. ReV. Plant Sci. 16; 415–480.
42. Briggs MA, Schultz JC (1990) Chemical defense production in Lotus corniculatus L. II. Trade-offs among growth, reproduction and defense. Oecologia 83: 32–37.
43. Lin L, Chen P, Harnly JM (2008) New phenolic components and chromatographic profiles of green and fermented teas. J Agric Food Chem 56: 8130–40.
44. Wickremasinghe RL, Perera KPWC, Perera VH, Kanapathipillai P (1966) Analyes of polyphenols, amino acid and chlorophyll levels in tea flush at different seasons. Tea Quart. 37: 232–235.
45. Herath NL, Amarakoon AMT, De-Silva MJ (1991) Sri Lanka J.Tea Sci., 60, 36–42.
46. Chen ZM, Wang HF, You XQ, Xu N (2002) The Chemistry of Tea Non-Volatiles.In: Tea: bioactivity and therapeutic potential. Medicinal and Aromatic Plants, Industrial Profiles, Taylor and Francis, London.
47. Bolling B, Chen CY, Blumberg JB (2009) Tea and health: preventive and therapeutic usefulness in the elderly? Curr Opin Clin Nutr Metab Care 12: 42–48.
48. Han W (2012) Climate change and its impacts on tea economy and counteract strategies in China. Report for the round table meeting on climate change and its effects on the world tea economy. Intergovernmental Group on Tea of the FAO Colombo, Sri Lanka.
49. Hunn E (1993) What is traditional ecological knowledge? In N. . Williams and G. . Baines, editors. Traditional environmental knowledge: wisdom for

sustainable development. Center for Resource and Environmental Studies, Australian National University, Canberra, Australia.
50. Ellen R, Harris H (2000) Introduction.In R. F. Ellen, P. Parkes, and A. Bicker, editors. Indigenous environmental knowledge and its manifestations: critical anthropological perspectives. Harwood, Amsterdam, The Netherlands.
51. Rocha J (2005) Measuring traditional agro- ecological knowledge: an example from peasants in the Peruvian Andes. Field Methods 4: 356–372.
52. Nautiyal S, Rao KS, Kaechele H, Raju KV, Schaldach R (2013) Knowledge Systems of Societies for Adaptation and Mitigation of Impacts of Climate Change. Springer USA
53. Tobin MF, Lopez OR, Kursar TA (1997) Drought response of tropical understory species with long and short leaf lifespans. Biotropica 31: 570–578.
54. Verchot LV, van Noordwijik M, Kandji S, Tomich T, Ong C, et al. (2007) Climate change: linking adaptation and mitigation through agro-forestry. Mitig Adapt Strat Glob Chn 12: 901–918
55. Lin B (2007) Agroforestry management as an adaptive strategy against potential microclimate extremes in coffee agriculture. Agriculture and Forest Meteorology 144 (1–2): 85–94.
56. Wijeratne MA (1996) Vulnerability of Sri Lanka tea production to global climate change. Water, Air and Soil Pollution 92(1–2): 87–94.
57. Bhagat RM, Deb Baruah R, Safique S (2010) Climate and Tea Production with Special Reference to Northeastern India: A Review. Journal of Environmental Research and Development 4 (4): 1017–1028.

14

The Influence of Precipitation Regimes and Elevated CO_2 on Photosynthesis and Biomass Accumulation and Partitioning in Seedlings of the Rhizomatous Perennial Grass *Leymus chinensis*

Zhuolin Li[1], Yuting Zhang[1], Dafu Yu[1], Na Zhang[1], Jixiang Lin[2], Jinwei Zhang[1], Jiahong Tang[1], Junfeng Wang[1], Chunsheng Mu[1]*

1 Key Laboratory of Vegetation Ecology, Ministry of Education, Institute of Grassland Science, Northeast Normal University, Changchun, China, **2** Key Laboratory of Saline-alkali Vegetation Ecology Restoration in Oil Field Ministry of Education, Alkali Soil Nature Environmental Science Center, Northeast Forestry University, Harbin, China

Abstract

Leymus chinensis is a dominant, rhizomatous perennial C_3 species in the grasslands of Songnen Plain of Northern China, and its productivity has decreased year by year. To determine how productivity of this species responds to different precipitation regimes, elevated CO_2 and their interaction in future, we measured photosynthetic parameters, along with the accumulation and partitioning of biomass. Plants were subjected to combinations of three precipitation gradients (normal precipitation, versus normal ± 40%) and two CO_2 levels (380±20 μmol mol^{-1},760±20 μmol mol^{-1}) in controlled-environment chambers. The net photosynthetic rate, and above-ground and total biomass increased due to both elevated CO_2 and increasing precipitation, but not significantly so when precipitation increased from the normal to high level under CO_2 enrichment. Water use efficiency and the ratio of root: total biomass increased significantly when precipitation was low, but decreased when it was high under CO_2 enrichment. Moreover, high precipitation at the elevated level of CO_2 increased the ratio between stem biomass and total biomass. The effect of elevated CO_2 on photosynthesis and biomass accumulation was higher at the low level of precipitation than with normal or high precipitation. The results suggest that at ambient CO_2 levels, the net photosynthetic rate and biomass of *L. chinensis* increase with precipitation, but those measures are not further affected by additional precipitation when CO_2 is elevated. Furthermore, CO_2 may partly compensate for the negative effect of low precipitation on the growth and development of *L. chinensis.*

Editor: Senjie Lin, University of Connecticut, United States of America

Funding: The paper was supported by the National Natural Science Foundation of China (grant No. 31172259 and 31370432). Moreover, this paper was supported by the National Natural Science Foundation of China (grant No. 31270464). The funders had no role in study design, data collection and analysis, decision to publish, or preparation of the manuscript.

Competing Interests: The authors have declared that no competing interests exist.

* Email: mucs821@gmail.com

Introduction

In the grassland ecosystems of the eastern Eurasian steppes, *Leymus chinensis* is a typical perennial C_3 grass with rhizomatous propagation that is distributed widely in areas including the Russian Baikal, the northern and eastern parts of Mongolia, the North China Plain, and the Inner Mongolian plateau of China [1], [2]. It is a dominant plant species in those relatively dry areas due to its tolerance of drought and saline-alkaline soils [3]. It is also an economically and ecologically important forage grass in Northern China because it is rich in protein, minerals, carbohydrates, and is palatable to many large herbivores. Recently, the area of grassland on the steppes has been decreasing due to the effects of human disturbances, including poor land use management, overgrazing, and climate change [4]. As a consequence, grassland productivity (dominated by the productivity of *L. chinensis*) has been reduced severely.

Atmospheric carbon dioxide (CO_2) concentrations have been increasing globally at an unprecedented rate [5]–[7]. In particular, regional climate models predict that climatic changes induced by CO_2 will exacerbate the dryness of the semiarid region of China [8], [9].

The ecosystems of China's arid and semiarid regions are driven mainly by precipitation, which is limiting and is therefore a key determinant of vegetation productivity [10]. Net primary productivity is positively correlated with precipitation [11] – [13]. In general, mild or moderate water stress will decrease the transpiration rate (E) of plants by reducing stomatal conductance (g_s), and it will also decrease the net photosynthetic rate (P_n) [13]. Severe drought will damage the photosynthetic system and lead to lower productivity. In many regions with low precipitation, plants are smaller and have relatively smaller leaf areas, but they have more roots to absorb nutrients and water in order to maintain normal growth patterns [14]. Therefore, considerable evidence shows that plant will increase the allocation of biomass to their root systems when water or nutrients are limiting [15]. In contrast, high water content in soils or fully submerged conditions typically lead to a decline in biomass allocation to roots [16].

Elevated CO_2 generally decreases g_s and E, stimulates P_n and increases net primary production [17] – [20]. Elevated CO_2 has been shown to enhance production in many grassland ecosystems, such as those in Switzerland and New Zealand, and Kansas tallgrass prairie and Colorado shortgrass steppe in the United States [21]. However, plants with different photosynthetic pathways have responded differently to elevated levels of CO_2. Increased CO_2 enhances the response of C_3 plants but not C_4 plants, because the ambient CO_2 concentration is enough to satisfy the needs of the unique photosynthetic pathway in C_4 plants [22], [23]. An indirect effect of elevated CO_2 is improving water use efficiency (*WUE*), which can sustain plant growth and development to some degree during dry periods [24] – [26]. As most plants respond to elevated CO_2 with an increase in photosynthesis and biomass, at least in the short term, the allocation of biomass to roots will increase in order to enable better access to nutrients [16], [27] – [28]. However, Nowak et al. [29] suggested that the allocation to below-ground biomass may not increase under elevated CO_2 levels, and the impact on stem and leaf biomass allocation is ambiguous. Overall, there is no clear pattern regarding the effect of elevated CO_2 on biomass allocation in plants [16], [30].

In semiarid grassland ecosystems, water is believed to regulate plant responses to elevated levels of CO_2 in the air; this relationship is fundamental, as CO_2 and precipitation are essential factors that determine plant growth, development and function. Therefore, the interaction between precipitation regime and CO_2 level is even more important for driving plant growth and development in semiarid regions. Evidence shows that plant growth and productivity are stimulated more by elevated CO_2 during water stress than under well-watered conditions [31] – [33]. However, some research suggests that plant responses to elevated CO_2, in terms of growth and productivity, are constrained by drought [34] – [36]. The conflicting conclusions depend mainly on the severity and duration of aridness, as well as the plant species under study.

Earlier studies have considered the effect of water stress or elevated CO_2 on growth of *L. chinensis*. However, to our knowledge few studies have addressed that subject by simulating the gradient involved in real precipitation regimes. The objective of this study was to assess changes in the growth of a dominant grassland species in response to such environmental variation; this is important to understand because overall community properties are strongly influenced by the characteristics of dominants. Thus, we measured photosynthetic parameters, and biomass accumulation and partitioning in *L. chinensis,* along a gradient of three precipitation levels (normal precipitation and normal $\pm$40%) and two CO_2 levels (380$\pm$20 μmol mol^{-1} and 760$\pm$20 μmol mol^{-1}), to investigate the strategy that this drought-tolerant grass uses to adapt to current environmental stresses.

Typically, elevated levels of CO_2 and precipitation stimulate the growth of C_3 species, and rising atmospheric CO_2 improves the efficiency of water use by plants, possibly helping to alleviate the impacts of drought via water-saving effects [37]. We therefore predicted that: (1) there would be a synergistic effect of increased precipitation and CO_2 levels on photosynthesis and biomass accumulation of *L. chinensis*; (2) elevated CO_2 would compensate partly for the negative effect of low precipitation on the response variables; (3) biomass allocation would be altered under elevated CO_2 and different precipitation regimes.

Materials and Methods

Ethics Statement

No specific permissions were required to conduct the field research described, because the Songnen Artificial Grassland Ecological Research Station belongs to Northeast Normal University. The field site is not privately owned or protected in any way, and the study did not involve endangered or protected species. Activities followed the research guidelines of the University.

Soil and plant cultures

Soil and seeds of the "yellow-green" ecotype of *L. chinensis* were obtained from the Field Station, which is part of the Institute of Grassland Science, Jilin Province, in northeastern China (123°44′E, 44°44′N, 167 m elevation). This region has a semiarid, continental monsoon climate with a frost-free period of about 140 d. Annual mean temperature is 6.4°C, and annual mean precipitation is 361.6 mm (2000–2011), most of which (70%) occurs during the summer months of Jun-Aug. The ecosystem's main soil type is mollisol.

The soils were sieved through 2-mm mesh to remove roots and other visible debris, mixed well, and then put into plastic pots with an inside diameter of 19 cm, and height of 14 cm. Each pot was filled with 3.3 kg dry soil. The total nitrogen, organic carbon content, EC and pH of the mollisol soil were 6.8%, 0.3%, 180 μs cm^{-1} and 8.63, respectively. To produce enough seedlings of uniform size, on 1 Sep 2012 about 20 seeds were sown in each of 24 plastic pots for a total of 480 seeds, at 15 d after sowing the seedlings were thinned to 10 per pot.

Experimental pots were placed in the phytotron (LT/ACR-2002 Phytotron System, E-Sheng Tech., Beijing, China) at Northeast Normal University in Changchun. In the phytotron, high-stress sodium lamps (Philips) with photosynthetically active radiation provided light at a rate of 350 μmol^{-2} S^{-1} for 14 h per day. The relative humidity was maintained at 40–60%, and the temperature regime was 22°C from 5:30–8:30, 25°C from 8:30–11:30, 28°C from 11:30–14:30, 25°C from 14:30–17:30, 22°C 17:30–19:30 and 18°C from 19:30–5:30. Air temperature in each chamber was monitored and adjusted every 10 s throughout the day and night, and maintained within $\pm$1°C of treatment set points. The pots were irrigated with 240 mL of water every 3 d (equivalent to 8 mm of precipitation, totaling 80 mm per month), with the soil water content maintained at 50–60% of field capacity. The length of diurnal/nocturnal periods was chosen to mimic the typical length of daylight hours in the Songnen grassland during summer. Temperatures matched the minimum, maximum, and average summer temperatures from 2000–2011 in the same region (Meteorological Bureau of Changling County, China, the grown site of *L. chinensis*) (Fig. 1 and Fig. 2*b*).

Precipitation regime and CO_2 treatments

Precipitation and CO_2 treatments commenced one month after sowing, and the experiment spanned three months. The pots of seedlings were placed randomly (and equally) into two controlled-environment growth chambers. One chamber was randomly assigned to ambient CO_2 at 380$\pm$20 μmol mol^{-1}. The other chamber was assigned to an elevated level of 760$\pm$20 μmol mol^{-1}, because atmospheric CO_2 concentrations are predicted to double by the end of this century, which might have an important influence on the productivity of grasslands dominated by *L. chinensis* and their community structure. The CO_2 was supplied from a tank and delivered through 0.64 cm tubing, and the

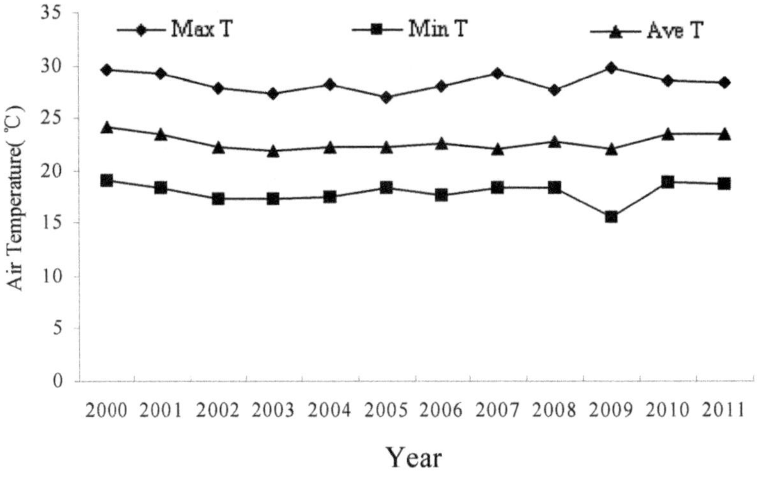

Figure 1. The average maximum, minimum and mean air temperatures on the semi-arid Songnen Grassland during summers (Jun-Aug) from 2000 to 2011. Data were collected by the Meteorological Bureau of Changling County, China.

concentrations were monitored every 5 s and adjusted every 10 s throughout the day and night.

In each CO_2 chamber, pots were assigned to one of three precipitation levels: normal, high (normal +40%), and low (normal –40%). Total monthly precipitation was supplied in ten equal amounts to represent the normal monthly water level (i.e., 8 mm of precipitation every 3d). This was based one the region's average amount and frequency of precipitation during the summers of 2000–2011 (Fig. 2*a* and 2*b*). Based on data for the Songnen grassland over the last 12 years, we found that precipitation in wet and dry years measured 40% higher or lower than the average, so we used that variation to define our experimental water regime.

In order to ensure that each plant experienced similar light conditions, the pot positions were randomly changed every 3 d during the treatment. Further, because there were no chamber replicates in this study, we rotated the treatments between the two chambers every 2 weeks, changing the environmental settings so that all pots were handled as similarly as possible during the experiment.

Leaf gas exchange

Leaf CO_2 exchange parameters were measured with an LI–6400 gas exchange system (LI–6400XT, Li–Cor, Inc., Lincoln, NE, USA) on the youngest available fully expanded leaves (3 leaves per pot, 4 pots per treatment) before sampling.

P_n, g_s, and E were measured with a LiCOR red/blue LED light source in a standard 2×3 cm chamber. The photosynthetically active radiation (PAR) was set at 350 μmol^{-2} S^{-1} to equal the light of the phytotron, and the reference CO_2 concentration was maintained at 380±20 μmol mol^{-1} in the control and 760±20 μmol mol^{-1} in the elevated CO_2 growth chamber using CO_2 control modules. Samples were allowed to acclimate for a few minutes until the P_n stabilized and the coefficient of variation was below 0.5. WUE (defined as mmol of net CO_2 uptake per mol of H_2O lost) was derived from the ratios of P_n to E.

Soil water content and biomass

After three months of treatment, we collected plants from 4 pots per treatment, carefully washed the soil from the roots in running water, and separated the plants into leaves, stems, roots and

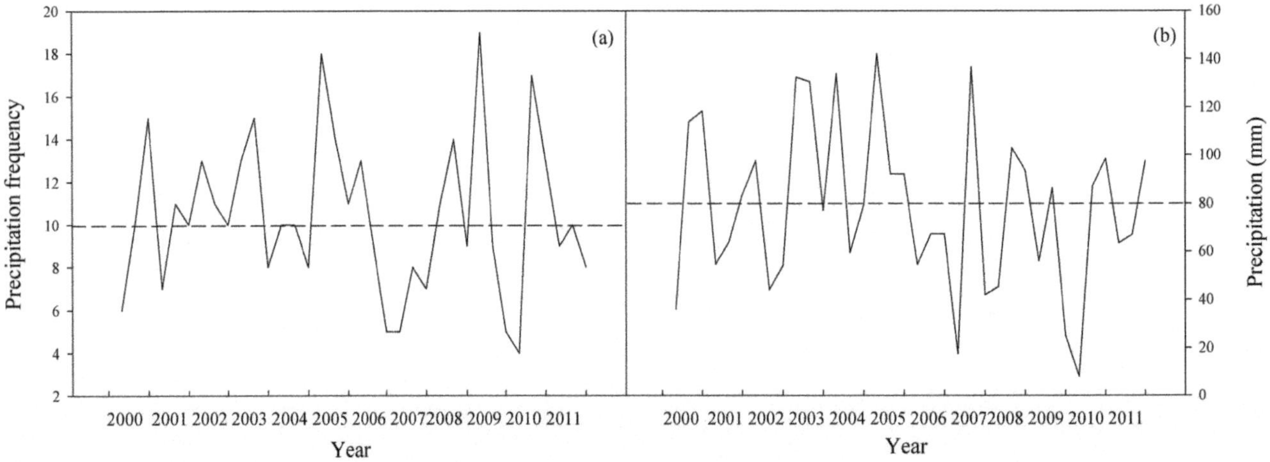

Figure 2. The frequency and amount of rainfall precipitation in summer (Jun-Aug) from 2000 to 2011. The average monthly precipitation was ~80 mm. Data were collected by the Meteorological Bureau of Changling County, China.

rhizomes. A soil sample of about 10 g was collected from each pot and placed in an aluminum can for measurement of soil water content (SWC). Plant parts and soils were dried to a constant mass at 65°C and weighed.

Statistical analysis

Data were analyzed as a split-plot design with CO_2 being the main plot and precipitation regime being the subplot (SPSS Inc, Chicago, IL, USA). Values for the photosynthetic parameters, biomass accumulation and biomass partitioning were tested for normality and homogeneity, and were transformed appropriately if necessary. For each parameter, the difference between the two CO_2 treatments was determined with a t-test, and differences among precipitation treatments were determined with a one-way ANOVA. Levels of $P<0.05$ were considered to be significant.

Results

Soil water content

Soil water content increased significantly with precipitation at both CO_2 levels ($P<0.001$), and showed a weak increase in elevated CO_2 although it was not statistically significant ($P=0.074$). Soil water content was not significantly affected by the interaction of precipitation and elevated CO_2 (Table 1, Fig. 3).

Gas exchange parameters

Both precipitation regime and elevated CO_2 had a significant influence on the P_n, g_s, C_i, E, and *WUE* (Table 1). Ambient CO_2, P_n, g_s and E increased significantly with increasing precipitation (Fig. 4*a, b, d*), whereas *WUE* showed a significant decline under high precipitation, compared to under low and normal precipitation ($P<0.001$) (Fig. 4*e*). Photosynthetic characters of *L. chinensis* under elevated CO_2 responded in a similar way as under ambient CO_2. However, P_n under normal and high precipitation exhibited no significant response, although it was higher in plants grown under high precipitation compared to normal levels (Fig. 4*a*). *WUE* decreased as precipitation increased at the elevated level of CO_2 (Fig. 4*e*).

Elevated CO_2 significantly increased P_n — the greatest change was under low precipitation, at 82.4%, with a change of 36.7% and 14.0% at the normal and high levels of precipitation. There was no difference under high precipitation ($P>0.05$) (Fig. 4*a*). C_i responded in a similar way as P_n (Fig. 4*c*). Compared to in the ambient CO_2 conditions, g_s decreased significantly by 40%, 50% and 58.8% at the low, normal and high levels of precipitation, respectively ($P<0.001$) (Fig. 4*b*), E decreased by 60.5%, 65.6%, and 71.9% ($P<0.001$) (Fig. 4*d*), whereas *WUE* increased by 398.3%, 307.3% and 308% ($P<0.001$) (Fig. 4*e*). These results indicated that elevated CO_2 had a greater effect on the P_n and *WUE* of seedlings, especially under drier conditions, likely to compensate for the negative effects induced by drought.

Biomass accumulation and allocation

Overall, precipitation had a significant influence on above-ground, below-ground, and total biomass. Elevated CO_2 also had a significant influence on above-ground biomass, total biomass and root: shoot ratio. Their interaction, however, affected only below-ground biomass and root: shoot ratio (Table 1). At ambient CO_2, the above-ground, below-ground, and total biomass of individuals increased significantly with increasing precipitation (Table 2; $P<0.05$). Under elevated CO_2, low precipitation significantly decreased individual biomass, while high precipitation caused a slight increase in above-ground biomass. Notably, that enhancement caused by increasing precipitation was not observed under elevated CO_2 when precipitation increased from the normal to high level.

Elevated CO_2 increased above-ground biomass by 29.6%, 32.6% and 22.9%, compared to plants with a similar watering regime (from low to high) at ambient CO_2, although only the effect under normal precipitation was significant ($P<0.05$). Below-ground biomass increased significantly at low precipitation and decreased at high precipitation under CO_2 enrichment compared to ambient CO_2 conditions ($P<0.05$). Therefore, elevated CO_2 significantly increased individual total biomass at low precipitation, but this enhancement was not seen under high precipitation conditions. Furthermore, elevated CO_2 significantly decreased the root: shoot ratio under normal and high precipitation conditions ($P<0.05$). Thus, elevated CO_2 advanced above-ground vegetation growth under favorable water conditions, and increased below-

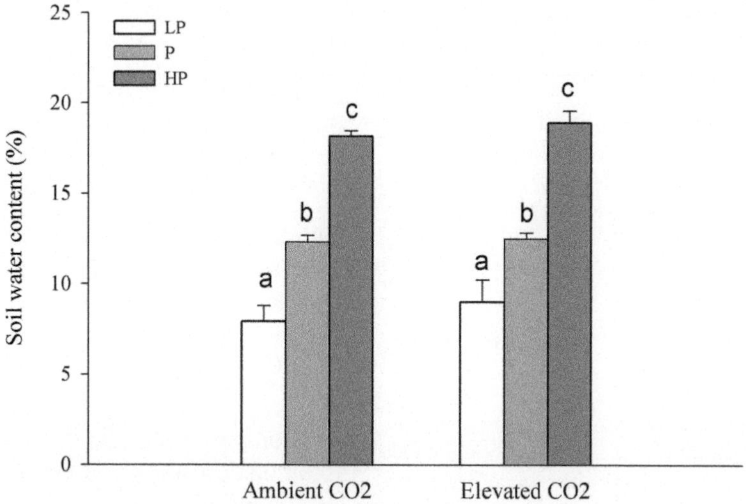

Figure 3. Effect of precipitation on soil water content under two CO_2 concentrations. LP, P, HP represents low precipitation (−40%), normal precipitation and high precipitation (+40%), respectively. Different letters indicate a significance difference among levels of precipitation ($P<0.05$).

Table 1. Analysis of variance to assess the impacts of precipitation, CO_2, and their interaction on soil water content, photosynthetic parameters, biomass accumulation and biomass allocation in the grass species *L. Chinensis.*

	Source of variation					
	CO_2		**Precipitation**		**CO_2×Precipitation**	
	F	**P**	**F**	**P**	**F**	**P**
Soil water content	3.844	0.074	304.408	**<0.001**	0.642	0.543
	Photosynthetic parameters					
Pn	400.232	**<0.001**	12.992	**<0.001**	7.771	**<0.001**
Ci	741.983	**<0.001**	18.223	**<0.001**	4.985	**<0.001**
g_s	290.268	**<0.001**	222.368	**<0.001**	48.112	**<0.001**
E	1116.369	**<0.001**	363.371	**<0.001**	144.112	**<0.001**
WUE	400.232	**<0.001**	12.992	**<0.001**	7.771	**<0.001**
	Biomass					
Above-ground	21.731	**<0.01**	34.501	**<0.001**	0.534	0.599
Below-ground	0.068	0.798	115.944	**<0.001**	20.072	**<0.001**
Total biomass	10.132	**<0.01**	72.149	**<0.001**	2.901	0.094
Root/shoot	27.192	**<0.001**	1.308	0.306	11.576	**<0.01**
	Biomass allocation					
Leaf	1.254	0.285	3.550	0.062	2.089	0.167
Stem	13.809	**0.003**	9.317	**0.004**	1.532	0.255
Root	26.459	**<0.001**	19.799	**<0.001**	1.467	0.269
Rhizome	1.046	0.327	5.690	**0.018**	0.596	0.566

Note:Data are significant at $P<0.05$ level (bolded values).

ground biomass under drought conditions, presumably to facilitate water absorption.

Both precipitation and CO_2 had a significant influence on biomass allocation to stems and roots, whereas no significant effect of interaction was detected between precipitation and CO_2 on biomass allocation (Table 1). Stem biomass allocation increased with precipitation at both CO_2 levels, but only elevated CO_2 caused a significant effect (Fig. 5; $P<0.05$). Biomass allocation to roots responded in opposite directions: it tended to be higher under low precipitation at both CO_2 levels ($P<0.05$), and under normal and high precipitation it was significantly lower at elevated CO_2 ($P<0.05$), but there was no effect with ambient CO_2.

At a given level of precipitation, elevated CO_2 significantly increased biomass allocation to stems under normal and high precipitation compared to ambient CO_2 ($P<0.05$), and decreased biomass allocation to roots under high precipitation conditions ($P<0.05$).

Discussion

In arid and semiarid regions, precipitation is the key variable that affects the growth and development of vegetation, and it can also affect plant metabolism and signal transduction. As elevated levels of atmospheric CO_2 are the main cause of global climate change, we investigated the interaction of precipitation and CO_2 levels to provide a more comprehensive assessment of how current environmental changes may be affecting plant growth in a semiarid region. Substantial research efforts have demonstrated that elevated CO_2 causes an increase in P_n and the accumulation of biomass in plants, while drought has the opposite effect [13], [20], [32] – [33], [38] – [40]. However, results concerning the interaction between precipitation regime and elevated CO_2 have been ambiguous [41] – [43]. The results of our research support our predictions 2 and 3, that increasing precipitation and elevated CO_2 improve biomass and P_n of *L. chinensis*, and moreover, those effects were greater under low precipitation conditions than under high precipitation. Those findings indicated that elevated CO_2 could lead to increased soil water content and compensate for the negative effect of drought on the growth of this grass species. The allocation of biomass among the parts of individual plants had changed too. Furthermore, soil EC decreased with increasing precipitation and elevated CO_2, but pH did not change much (from 8.15 to 8.6; data not shown). Ma and Liang [44] suggested that seed germination and seedling growth of *L. chinensis* are highest when pH is between 8.0 and 8.5. Consequently, elevated CO_2 and precipitation effects mediated through decreased salinity and alkalinity is expected to have minor effects in this study. Intriguingly, we found that P_n did not respond to elevated CO_2 under high precipitation conditions, which, contrary to our first prediction, means that elevated CO_2 and increasing precipitation did not have a synergistic effect on P_n. Therefore, biomass accumulation in *L. chinensis* did not change much in response to elevated CO_2 when precipitation increased by 40%.

Additional precipitation does not enhance biomass accumulation and net photosynthetic rate under CO_2 enrichment

L. chinensis is highly tolerant to drought, and the water holding capacity of soils where it grows site on the Songnen grassland is 50–60% in the summer months (Jun-Aug). In this study, the water holding capacity of soils under high precipitation conditions

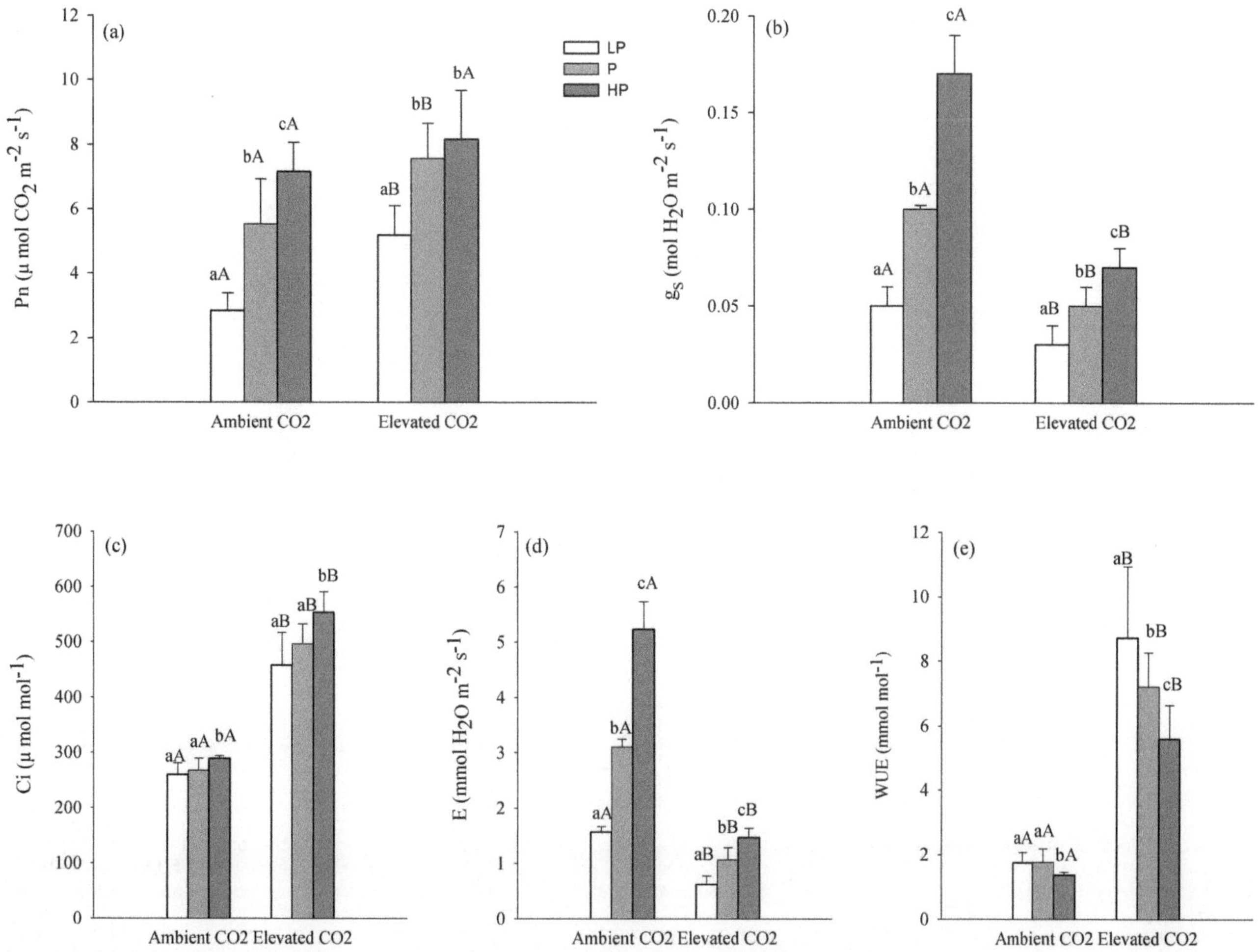

Figure 4. Effect of precipitation on photosynthetic rate (*P_n*), stomatal conductance (*g_s*), intercellular CO_2 concentration (*Ci*), transpiration rate (*E*) and water use efficiency (*WUE*) of *Leymus chinensis* under two CO_2 concentrations. Different lower-case letters indicate a significant difference among different levels of precipitation and different capital letters indicate a significant difference between the two CO_2 levels ($P<0.05$). LP, P, HP represents low precipitation (−40%), normal precipitation and high precipitation (+40%), respectively.

climbed to 80% or more, so above-ground biomass per plant increased weakly when precipitation increased from the normal to high level (Table 2). A similar trend was observed in P_n.

Elevated CO_2 can directly increase the carboxylation efficiency of C_3 species, or induce stomatal closure and then limit the rate of transpiration indirectly, causing an increase in *WUE* [31], [40].

Table 2. Effect of precipitation on biomass of *L. Chinensis* under two CO_2 concentrations.

CO_2 concentration	Index (g plant^{-1})	Precipitation (mm)		
		Low(LP)	Normal(P)	High(HP)
Ambient CO_2 (380 μmol mol^{-1})	Above-ground biomass	0.27±0.02aA	0.43±0.01bA	0.48±0.03bA
	Below-ground biomass	0.36±0.01aA	0.66±0.03bA	0.72±0.01cA
	Total biomass	0.64±0.04aA	1.09±0.03bA	1.21±0.03cA
	Root: shoot ratio	1.34±0.06aA	1.54±0.06aA	1.50±0.06aA
Elevated CO_2 (760 μmol mol^{-1})	Above-ground biomass	0.35±0.03aA	0.57±0.04bB	0.59±0.01bA
	Below-ground biomass	0.49±0.01aB	0.65±0.03bA	0.62±0.02bB
	Total biomass	0.84±0.03aB	1.22±0.07bA	1.21±0.04bA
	Root: shoot ratio	1.42±0.06aA	1.15±0.04bB	1.06±0.06bB

Note: Different lower-case letters indicate a significant difference among precipitation levels and different capital letters indicate a significant difference between the two CO_2 levels ($P<0.05$). Low/High precipitation were defined as normal precipitation +/−40%.

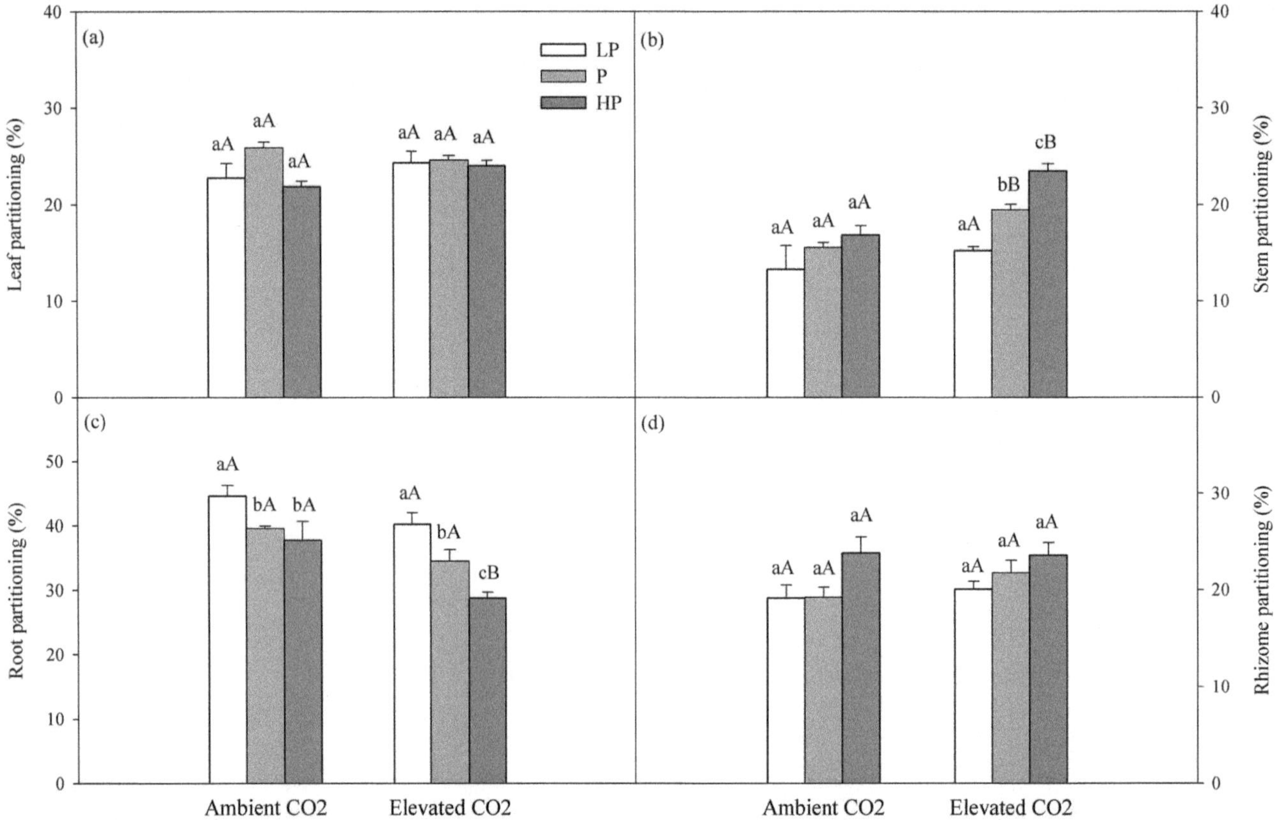

Figure 5. Effect of precipitation on biomass allocation among plant parts under two CO_2 concentrations for *Leymus chinensis*. Different lower-case letters indicate a significant difference among different precipitation levels, and different capital letters indicate a significant difference between the two CO_2 levels ($P<0.05$). LP, P, HP represents low precipitation (−40%), normal precipitation and high precipitation (+40%), respectively.

Elevated CO_2 would not increase above-ground biomass and *Pn* in the grass species that we studied, because decreased transpiration at elevated levels of CO_2 would provide little additional benefit in increased soil moisture [45]. Xu *et al.* [14] also found that mild and moderate drought had no significant influence on the biomass of *L. chinensis*, whereas mild drought (field capacity: 60–65%) stimulates the accumulation of biomass, consistent with the pattern observed in our study. Furthermore, some researchers have found that the below-ground biomass of grasses is suppressed by elevated CO_2 [46]; that trend occurred in our system in the high precipitation treatment, leading to no change in total biomass, which may be because the soil moisture levels were adequate for growth so more roots were not needed to increase water absorption. This may be a typical growth strategy and ecological adaptation of *L. chinensis* in the semiarid Songnen grassland. Therefore, we infer that extra precipitation is essentially redundant, as it does not augment biomass accumulation and P_n under higher CO_2 levels due to the high resistance of this species to soil water stress.

Elevated CO_2 might partly compensate for the negative effect of drought on net photosynthetic rate and biomass

In line with our predictions, elevated CO_2 seemed to partly compensate for the negative effect of drought on *Pn* and biomass of *L. chinensis*. Our results support the notion that some herbaceous species are more stimulated by elevated CO_2 under water stress than under well-watered conditions [23], [24], [31], [33], [45].

Elevated CO_2 reduced g_s, and the effect of elevated CO_2 on the percent change of g_s was smallest at the low precipitation level (Fig. 4*b*). This maybe because both drought and elevated CO_2 caused a decline in g_s to limit E, and elevated CO_2 has a minimum effect on g_s under low precipitation in order to maintain levels of photosynthesis. However, although elevated CO_2 increased the substrate concentration under high precipitation conditions, g_s decreased quickly to restrain absorption of CO_2. Therefore, at the elevated CO_2 level, the increasing percent change of P_n for grass plants grown under low precipitation conditions was considerably higher than for those grown under high precipitation conditions.

In general, in the arid and semiarid grassland, *WUE* is also an important factor that stimulates primary productivity [47], especially in years with low precipitation. We found that elevated CO_2 significantly improved *WUE* (Table 1, Fig. 4*e*), and slightly increased soil water content under low precipitation conditions although the change was not significant (Table 1, Fig. 3*a*). In the Kansas Tallgrass Prairie experiment, volumetric soil water content was generally higher in elevated CO_2 plots than under ambient levels of the gas, mainly during periods when precipitation limited normal plant growth due to drought [21]. This phenomenon occurred in a healthy dry ecosystem. *WUE* improved ~3–4 fold when CO_2 concentration was doubled, which means that the amount of water plants needed to fix one unit of CO_2 decreased by ~3–4 fold [48]. This is important in regions experiencing drought and dry soils due to low precipitation. Therefore, although the effect of elevated CO_2 on plant growth is limited under low precipitation conditions, elevated CO_2 still partly compensated for the negative effect of drought on P_n of *L. chinensis*.

The increase in P_n indicates that the ability of plants to fix carbon can be enhanced. From low to high precipitation levels, elevated CO_2 increased total biomass per plant by 31.3%, 11.9% and 0%, respectively, as compared to at ambient CO_2 levels. The trend for P_n was similar. Furthermore, root: shoot ratios increased at low precipitation, yet decreased at normal and high precipitation, which is consistent with the findings of other studies [23], [34], [35]. Plants still suffer water stress even if elevated CO_2 leads to higher *WUE* and an enhanced ability of soils to maintain water, therefore, plants will increase their below-ground biomass to absorb more water and nutrients. The increase of root: shoot ratio may be the best strategy for plants to adapt to water stress [14]. However, some other studies demonstrate that plants have greater stimulation owing to elevated CO_2 under well-watered conditions than under drought [34] – [36]. The differences among these results may be due to the varying resistance of plants to drought, as well as different drought scales and other variation in the precise conditions studied. Our results indicate that elevated CO_2 could partly compensate for the negative effects of drought on P_n and biomass of *L. chinensis*.

Biomass allocation

Most archetypal vascular plants have leaves that fix carbon, stems that provide mechanical support and a hydraulic pathway, and roots that absorb nutrients and water [16]. Rhizomes function as storage organs in many clonal plants. Allocating biomass differently among these organs enables plants to balance growth and adapt environmental changes [49]. In our study, low precipitation significantly increased the allocation of biomass to roots under two levels of CO_2, effectively enabling plants to absorb more water and nutrients and have more above-ground vegetation growth, which is in line with results of many other studies [14], [16].

We showed that elevating the environmental concentration of CO_2 also altered the allometric relationships of biomass among plant tissues. Under normal and high precipitation conditions, elevated CO_2 significantly increased the biomass allocation to stems, but decreased allocation to roots (Table 1, Fig. 5*b*, *c*). Plants allocate more biomass to stems in order to facilitate competition for light and the acquisition of carbon [17]. However, much evidence shows that elevated CO_2 generally decreases the nitrogen concentration in leaves. Thus, it is possible that a plant allocates more biomass to roots to enable increased uptake of nutrients in order to sustain increases in biomass [16]. In contrast, a plant may regulate its root morphology and physiology rather than allocating more biomass to its roots when soil water and nutrients are abundant [50], [51]. Furthermore, biomass allocation was not affected by elevated CO_2, as has been widely observed in managed grasslands [52], [53]. The differing results are likely due to differences in the conditions being studied, such as other environmental factors, plant species, developmental stages, etc.

In summary, our results suggest that the biomass of *L. chinensis* increases with precipitation levels at ambient levels of CO_2. However, further precipitation is redundant, in that it does not augment biomass accumulation or the net photosynthetic rate of this grass species, and in fact it decreases biomass allocation to roots under conditions of CO_2 enrichment. In effect, our experiments show that elevated CO_2 may partly compensate for the negative effect of low precipitation on the growth and development of *L. chinensis*.

Acknowledgments

The authors sincerely acknowledge the technical assistance of Xindan Ding and Hanqiong He in the laboratory.

Author Contributions

Conceived and designed the experiments: ZLL CSM. Performed the experiments: ZLL YTZ DFY JWZ NZ JT. Analyzed the data: ZLL JXL JFW. Contributed reagents/materials/analysis tools: DFY JWZ CSM ZLL. Wrote the paper: ZLL CSM.

References

1. Wang Y, Zhou G, Wang Y (2007) Modeling responses of the meadow steppe dominated by Leymus chinensis to climate change. Climatic Change 82: 437–452.
2. Wang Z, Li L, Han X, Dong M (2004) Do rhizome severing and shoot defoliation affect clonal growth of *Leymus chinensis* at ramet population level? Acta Oecologica 26: 255–260.
3. Huang ZH, Zhu JM, Mu XJ, Lin JX (2002) Advances on the mechanism of low sexual reproductivity of *Leymus chinensis*. Grassland China 24: 55–60.
4. Jia B, Zhou G, Wang Y, Wang F, Wang X (2006) Effects of temperature and soil water-contention soil respiration of razed and ungrazed Leymus chinensis steppes, Inner Mongolia. J Arid Environ 67: 60–76.
5. IPCC (2012) Summary for policymakers. In: Field CB, Barros V. Stocker TF, et al (eds) Managing the risks of extreme events and disaters to advance climate change adaptation. A special report of working groups I and II of the intergovernmental panel on climate change. Cambridge University Press. Cambridge, New York, pp:1–19.
6. Peter GP, Maland G, Le QC, Boden T, Canadell JG, et al. (2012) Rapid growth in CO_2 emmissions after the 2008-2009 global financial crisis. Nat Clim Change 2: 2–4.
7. Meinshausen M, Smith SJ, Calvin K, Kainuma MLT, Lamarque JF, et al. (2011) The RCP greenhouse gas concentrations and their extensions from 1765 to 2300. Climatic Change 109: 213–241.
8. Ainsworth EA, Rogers A (2007) The response of photosynthesis and stomatal conductance to rising [CO_2]: mechanisms and environmental interactions. Plant Cell Environ 30: 258–270.
9. Li XZ, Liu XD, Ma ZG (2004) Analysis on the drought characteristics in the main arid regions in the world since recent hundred-odd years. Arid Zone Res 21: 97–103.
10. Rao LE, Allen EB (2010) Combined effects of precipitation and nitrogen deposition on native and invasive winter annual production in California deserts. Oecologia 162: 1035–1046.
11. Báez S, Collins SL, Pockman WT, Johnson JE, Small EE (2013) Effects of experimental rainfall manipulations on Chihuahuan Desert grassland and shrubland plant communities. Oecologia 172: 1117–1127.
12. Zhou G, Wang Y, Wang S (2009) Responses of grassland ecosystems to precipitation and land use along the Northeast China Transect. J Veg Sci 13: 361–368.
13. Flexas J, Medrano H (2002) Drought-inhibition of photosynthesis in C3 plants: stomatal and non-stomatal limitations revisited. Ann Bot-London 89: 183–189.
14. Xu ZZ, Zhou GS (2005) Effects of water stress and high nocturnal temperature on photosynthesis and nitrogen level of a perennial grass *Leymus chinensis*. Plant Soil 269: 131–139.
15. Poorter H, Nagel O (2000) The role of biomass allocation in the growth response of plants to different levels of light, CO_2, nutrients and water: a quantitative review. Funct Plant Biol 27: 1191–1191.
16. Poorter H, Niklas KJ, Reich PB, Oleksyn J, Poot P, et al. (2012) Biomass allocation to leaves, stems and roots: meta-analyses of interspecific variation and environmental control. New Phytol 193: 30–50.
17. Kinmonth SH, Kim SH (2011) Carbon gain, allocation and storage in rhizomes in response to elevated atmospheric carbon dioxide and nutrient supply in a perennial C_3 grass, Phalaris arundinacea. Funct Plant Biol 38: 797–807.
18. Cheng W, Sakai H, Yagi K, Hasegawa T (2009) Interactions of elevated [CO_2] and night temperature on rice growth and yield. Agr Forest Meteorol 149: 51–58.
19. Morgan JA, Lecain DR, Mosier AR, Milchunas DG (2001b) Elevated CO_2 enhances water relations and productivity and affects gas exchange in C_3 and C_4 grasses of the Colorado shortgrass steppe. Global Change Biol 7: 451–466.
20. Jackson R, Luo Y, Cardon Z, Sala O, Field C, Mooney H (1995) Photosynthesis, growth and density for the dominant species in a CO_2-enriched grassland. J Biogeogr 22: 1225–1229.

21. Morgan J, Newton P, Nösberger J, Owensby C (2001a) The influence of rising atmospheric CO_2 on grassland ecosystems. Proceedings of the XIX International Grasslands Congress: 973–980.
22. Kakani V, Reddy KR (2007) Temperature response of C_4 species big bluestem (*Andropogon gerardii*) is modified by growing carbon dioxide concentration. Environ Exp Bot 61: 281–290.
23. Mishra R, Abdin M, Uprety D (1999) Interactive effects of elevated CO_2 and moisture stress on the photosynthesis, water relation and growth of Brassica species. J Agron Crop Sci 182: 223–230.
24. Wang H, Zhou GS, Jiang YL, Shi YH, Xu ZZ (2012) Interactive effects of changing precipitation and elevated CO_2 concentration on photosynthetic parameters of Stipa breviflora. Chinese J Plant Ecol 36: 597–606.
25. Körner C (2000) Biosphere responses to CO_2 enrichment. Ecol Appl 10, 1590–1619.
26. Drake BG, Gonzàlez MA, Long SP (1997) More efficient plants: a consequence of rising atmospheric CO_2? Annu Rev Plant Biol 48: 609–639.
27. Arnone JA, Zaller JG, Spehn EM, Niklaus PA, Wells CE, et al. (2000) Dynamics of root systems in native grasslands: effects of elevated atmospheric CO_2. New Phytol 147: 73–85.
28. Milchunas DG, Morgan JA, Mosier AR, Lecain DR (2005) Root dynamics and demography in shrtgrass steppe under elevated CO_2, and comments on minirhizotron methodology. Global Change Biol 11: 1837–1855.
29. Nowak RS, Ellsworth DS, Smith SD (2004) Functional responses of plants to elevated atmospheric CO_2- do photosynthetic and productivity data from FACE experiments support early predictions? New Phytol 162: 253–280.
30. Wand SJ, Midgley G, Jones MH, Curtis PS (1999) Responses of wild C4 and C3 grass (Poaceae) species to elevated atmospheric CO_2 concentration: a meta-analytic test of current theories and perceptions. Global Change Biol 5: 723–741.
31. Morgan JA, LeCain DR, Pendall E, Blumenthal DM, Kimball BA, et al. (2011) C_4 grasses prosper as carbon dioxide eliminates desiccation in warmed semi-arid grassland. Nature 476: 202–205.
32. Atwell BJ, Henery ML, Rogers GS, Seneweera SP, Treadwell M, et al. (2007) Canopy development and hydraulic function in Eucalyptus tereticornis grown in drought in CO_2-enriched atmospheres. Funct Plant Biol 34: 1137–1149.
33. Poorter H, Perez M (2001) The growth response of plants to elevated CO_2 under non-optimal environmental conditions. Oecologia 129: 1–20.
34. Erice G, Irigoyen JJ, Sánchez M, Avice JC, Ourry A (2007) Effect of drought, elevated CO_2 and temperature on accumulation of N and vegetative storage proteins (VSP) in taproot of nodulated alfalfa before and after cutting. Plant Sci 172: 903–912.
35. Xu Z, Zhou G, Wang Y (2007) Combined effects of elevated CO_2 and soil drought on carbon and nitrogen allocation of the desert shrub Caragana intermedia. Plant Soil 301, 87–97.
36. Smith SD, Huxman TE, Zitzer SF, Charlet TN, Housman DC, et al. (2000) Elevated CO_2 increases productivity and invasive species success in an arid ecosystem. Nature 408: 79–82.
37. Volder A, Edwards EJ, Evans JR, Robertson BC, Schortemeyer M. (2004) Does greater night-time, rather than constant warming alter growth of managed pasture under ambient and elevated atmospheric CO_2? New Phytol 162: 397–411.
38. Xu Z, Shimizu H, Yagasaki Y, Ito S, Zheng Y, et al. (2013) Interactive Effects of Elevated CO_2, Drought, and Warming on Plants. J Plant Growth Regul 32: 692–707.
39. Ziska L, Sicher R, George K, Mohan J (2007) Rising atmospheric carbon dioxide and potential impacts on the growth and toxicity of poison ivy (*Toxicodendron radicans*). Weed Science 55: 288–292.
40. Ainsworth EA, Long SP (2005) What have we learned from 15 years of free-air CO_2 enrichment (FACE)? A meta-analytic review of the responses of photosynthesis, canopy properties and plant production to rising CO_2. New Phytol 165: 351–372.
41. Bhattacharya N, Hileman D, Ghosh P, Musser R, Bhattacharya S, et al. (2006) Interaction of enriched CO_2 and water stress on the physiology of and biomass production in sweet potato grown in open-top chambers. Plant Cell Environ 13: 933–940.
42. Ottman M, Kimball B, Pinter P, Wall G, Vanderlip R, et al. (2001) Elevated CO_2 increases sorghum biomass under drought conditions. New Phytol 150: 261–273.
43. Tolley LC, Strain B (1984) Effects of CO_2 enrichment and water stress on growth of Liquidambar styraciflua and Pinus taeda seedlings. Canadian J Bot 62: 2135–2139.
44. Ma HY, Liang ZW (2007) Effects of different soil pH and soil extracts on the germination and seedling growth of *Leymus chinensis*. Chinese Bulletin Bot 24: 181–188.
45. Fay PA, Polley HW, Jin VL, Aspinwall MJ (2012) Productivity of well-watered Panicum virgatum does not increase with CO_2 enrichment. J Plant Ecol 5, 366–375.
46. Shaw MR, Zavaleta ES, Chiariello NR, Cleland EE, Mooney HA, et al. (2002) Grassland responses to global environmental changes suppressed by elevated CO_2. Science 298: 1987–1990.
47. Niu S, Xing X, Zhang Z, Xia J, Zhou X, et al. (2011) Water-use efficiency in response to climate change: from leaf to ecosystem in a temperate steppe. Global Change Biol 17: 1073–1082.
48. Gao SH and Guo JP (2004) Initial study into the CO_2 concentration and soil moisture effects on the photosynthesis impact mechanism of *Leymus chinensis*. Prataculture Sci 21: 23–27.
49. Shipley B, Meziane D (2002) The balanced-growth hypothesis and the allometry of leaf and root biomass allocation. Funct Ecol 16: 326–331.
50. Hirel B, Tetu T, Lea PJ, Dubois F (2011) Improving nitrogen use efficiency in crops for sustainable agriculture. Sustain Sci 3: 1452–1485.
51. Gavito ME, Curtis PS, Mikkelsen TN, Jakobsen I (2001) Interactive effects of soil temperature, atmospheric carbon dioxide and soil N on root development, biomass and nutrient uptake of winter wheat during vegetative growth. J Exp Bot 52: 1913–1923.
52. Ping XY, Zhou GS, Sun JS (2010) Advances in the study of photsynthate allocation and its controls. Chinese J Plant Ecol 34: 100–111.
53. Hill PW, Marshall C, Williams GG, Blum H, Harmens H, et al. (2007) The fate of photosynthetically fixed carbon in Lolium perenne grassland as modified by elevated CO_2 and sward management. New Phytol 173: 766–777.

15

Effects of Climate Variability and Accelerated Forest Thinning on Watershed-Scale Runoff in Southwestern USA Ponderosa Pine Forests

Marcos D. Robles[1]*, Robert M. Marshall[1], Frances O'Donnell[2], Edward B. Smith[3], Jeanmarie A. Haney[1], David F. Gori[4]

1 The Nature Conservancy Center for Science and Public Policy, Tucson, Arizona, United States of America, 2 Northern Arizona University, Flagstaff, Arizona, United States of America, 3 The Nature Conservancy in California, Sacramento, California, United States of America, 4 The Nature Conservancy in New Mexico, Santa Fe, New Mexico, United States of America

Abstract

The recent mortality of up to 20% of forests and woodlands in the southwestern United States, along with declining stream flows and projected future water shortages, heightens the need to understand how management practices can enhance forest resilience and functioning under unprecedented scales of drought and wildfire. To address this challenge, a combination of mechanical thinning and fire treatments are planned for 238,000 hectares (588,000 acres) of ponderosa pine (*Pinus ponderosa*) forests across central Arizona, USA. Mechanical thinning can increase runoff at fine scales, as well as reduce fire risk and tree water stress during drought, but the effects of this practice have not been studied at scales commensurate with recent forest disturbances or under a highly variable climate. Modifying a historical runoff model, we constructed scenarios to estimate increases in runoff from thinning ponderosa pine at the landscape and watershed scales based on driving variables: pace, extent and intensity of forest treatments and variability in winter precipitation. We found that runoff on thinned forests was about 20% greater than unthinned forests, regardless of whether treatments occurred in a drought or pluvial period. The magnitude of this increase is similar to observed declines in snowpack for the region, suggesting that accelerated thinning may lessen runoff losses due to warming effects. Gains in runoff were temporary (six years after treatment) and modest when compared to mean annual runoff from the study watersheds (0–3%). Nonetheless gains observed during drought periods could play a role in augmenting river flows on a seasonal basis, improving conditions for water-dependent natural resources, as well as benefit water supplies for downstream communities. Results of this study and others suggest that accelerated forest thinning at large scales could improve the water balance and resilience of forests and sustain the ecosystem services they provide.

Editor: Julia A. Jones, Oregon State University, United States of America

Funding: The authors have no support or funding to report.

Competing Interests: The authors have declared that no competing interests exist.

* Email: mrobles@tnc.org

Introduction

As we transition into a warmer century, managing forests in the southwestern United States for resilience is more urgent and necessary than ever. Forest conditions across the region have declined significantly. In ponderosa pine (***Pinus ponderosa***) forests of central Arizona, stand densities range from 2 to 44 times greater than during pre-settlement conditions and total basal areas range from 2 to 4 times greater [1–3]. Severe wildfires and drought have caused high tree mortality rates across 14–18% of forests and woodlands in Arizona and New Mexico [4] and these results do not include mortality from two high-severity wildfires in 2011, the Wallow and the Las Conchas wildfires, which burned 217,000 and 63,500 ha (535,000 and 157,000 acres) in Arizona and New Mexico, respectively. Persistent drought conditions have also adversely impacted water-dependent habitats and species within and downstream of forest uplands, as evidenced by heightened mortality of a riparian tree species in central Arizona [5] and declines in number of species and densities of native fish in the Gila River in southwestern New Mexico [6], [7]. Researchers suggest that warming is amplifying the size and severity of wildfire- and drought-mortality events and decreasing snow pack levels [8–11]. Suggested mechanisms involve changes in the water cycle: increased evapotranspiration losses, extended water-stress periods, earlier snowmelt, and lengthened fire seasons. These changes and trends suggest a widespread drying of forests and an increasing risk of uncharacteristic fire and competition-induced water stress and mortality.

At the same time, many communities in the western U.S. are facing critical water shortages and river flows in some basins have diminished or been lost altogether due to unsustainable water management practices [12–15]. Recent studies at the scale of both the Colorado River basin and the state of Arizona concluded that water supplies for many communities will be inadequate to meet future projected demands [16],[17]. Moreover, elevated temperatures are predicted to directly impact river flows. A recent study estimated that for every one-degree Celsius rise in temperature, Colorado river flows will decline by 3–10% [18].

In this study, we explored the ways in which a new era of forest management, including accelerating the pace and scale of forest thinning, could be used to address declining conditions in forests in central and northern Arizona. At about 6,000 hectares per year (15,000 acres/year), the current pace of mechanical thinning in these forests does not begin to match the magnitude of recent forest disturbances. However, a new congressionally funded program called the Four Forest Restoration Initiative (hereafter "4FRI") will accelerate the use of mechanical thinning and prescribed burns across four national forests, treating 238,000 ha (588,000 acres) in the first analysis area over the next 10 years. The objective of the 4FRI project is "...to re-establish forest structure, pattern, and composition, which would lead to increases in forest resiliency and function," including reductions in fire risk and improved watershed function [19]. Mechanical thinning alone or in combination with prescribed fires reduced fire risk, increased runoff, and improved tree water-stress at the plot scale [20–23]. The 4FRI project provided the opportunity to evaluate the effectiveness of forest thinning to improve forest resilience and ecosystem functioning at larger scales.

The objectives of this study were to: (1) investigate how variation in climate and the pace and extent of thinning of ponderosa pine forests affects runoff in the Salt and Verde watersheds of central Arizona; and (2) explore the management implications of using accelerated thinning to improve forest resilience in a time of declining forest and water conditions. We chose the Salt-Verde watersheds because forests from these watersheds provide water for 1.5 million people in Phoenix, which is the 6th largest city in the United States. We focused on ponderosa pine because this forest type produces 50% of the runoff in these watersheds even though it accounts for only 20% of the area [24]. We modified a forest runoff model developed from historical paired watershed experiments in the Beaver Creek sub-watershed within our study area [20] and ran the model in multiple scenarios to estimate additional annual runoff from mechanical thinning. The scenarios simulated forest treatment at two scales (a) landscape-scale thinning planned within the first analysis area of the 4FRI project and (b) watershed-scale thinning in forests across the Salt-Verde watersheds (Figure 1). Scenarios were designed to account for variability in winter precipitation, as well as the pace and extent of forest treatments. We estimated runoff in periods with below-average precipitation, herein referred to as droughts, and in periods with above-average precipitation, herein referred to as pluvials, using a model of 20th century precipitation in ponderosa pine forests in the region [25]. We believe this is the first attempt to estimate the influence of mechanical thinning on runoff over multi-year broad-scale restoration projects that accounts for the effects of climate variability.

Materials and Methods

The subsequent sections provide detailed information on the study area, the development of the modified runoff model, the method we used to simulate variability in winter precipitation, the 4FRI runoff scenarios, and the Salt-Verde runoff scenarios. All statistical analyses and graphics were completed using the statistical software SigmaPlot 11 (Systat software Inc, San Jose, CA), except for regression model fitting which was performed using Matlab 2013 (Mathworks, Natick, MA). Methods and results are reported in SI units, but for accessibility to forest and water managers, we reported English units in parentheses within the text and included all study figures and tables in English units as File S1. Area and water volume amounts were rounded off to three significant figures.

Study Area

Ponderosa pine forests occupy approximately 0.68 million ha (1.68 million acres) of the 3.46 million ha (8.56 million acres) within the Salt-Verde watersheds in central Arizona (Figure 1). A portion of these forests, all within the Verde watershed, will be mechanically thinned in the 4FRI project in the next 10 years. Within the Salt-Verde watersheds, ponderosa pine forests grow at mid-elevations, from 1,800 to 2,600 m (5,900–8,600 feet) [26], [27], on soils derived from basalt (59%) and sedimentary rock (41%) [28]. Mean annual precipitation for ponderosa forests in the southwestern United States ranges from 510 to 760 mm (20 to 30 inches) and is highly variable [29]. Sixty percent of annual precipitation falls primarily as snow in the winter and spring months from October to April, and the melting of the resulting snowpack accounts for 80 to 90% of annual stream flow [29].

Forest Treatment Runoff Model

For the purposes of this study, we were interested in estimating the additional runoff that becomes available due to forest thinning on an annual basis, not absolute values of watershed runoff. We define "additional runoff due to thinning" as the portion of precipitation that appears as surface water at the sub-watershed outlet and that is directly attributable to mechanical thinning treatments. This additional runoff can be considered as "in-place" or "in-situ" as we did not attempt to model runoff accumulation, routing, groundwater outflows or inflows, or channel losses in downstream watersheds.

We modified the Baker-Kovner logistical regression equation from the historical Beaver Creek sub-watershed experiments to calculate annual runoff in our study [20]. We selected this model because the Beaver Creek sub-watershed is actually within our study area (Figure 1), and because the model was based upon measured empirical data. These experiments and others like it demonstrated that reductions in forest basal area of 30–100% in ponderosa forests on basalt-derived soils increased runoff between 15–40% for up to six years after the initial treatment [29]. Runoff increases were negligible for basal area reductions below 30%, six years after treatments, and for years with winter precipitation below 230 mm (9 inches) [20], [29]. The original model predicted annual watershed runoff based on three independent variables: winter precipitation (Oct-Apr), forest basal area, and a solar insolation index, and explained a fair amount of variability in total watershed runoff [20].

To derive increases in runoff directly associated with thinning using the paired watershed method, researchers in the Beaver Creek experiments first established a relationship between stream flows in the control and treated watersheds before treatments [30]. This relationship was used to predict what would have been baseline flows in the treated watershed. The difference between measured flows in the treated watershed and this predicted value were attributed to thinning effects. We compared data calculated in this fashion in the Beaver Creek experiments [30], [31] and one additional nearby site, Castle Creek [32], to output generated by the Baker-Kovner model. The output from the model was calculated as the difference in post-treatment and pre-treatment watershed runoff using post- and pre-treatment basal areas respectively and holding all other variables constant. In this comparison, we found the fit of the model output to the portion of runoff due to thinning to be much poorer ($r^2 = 0.43$, Figure 2a)

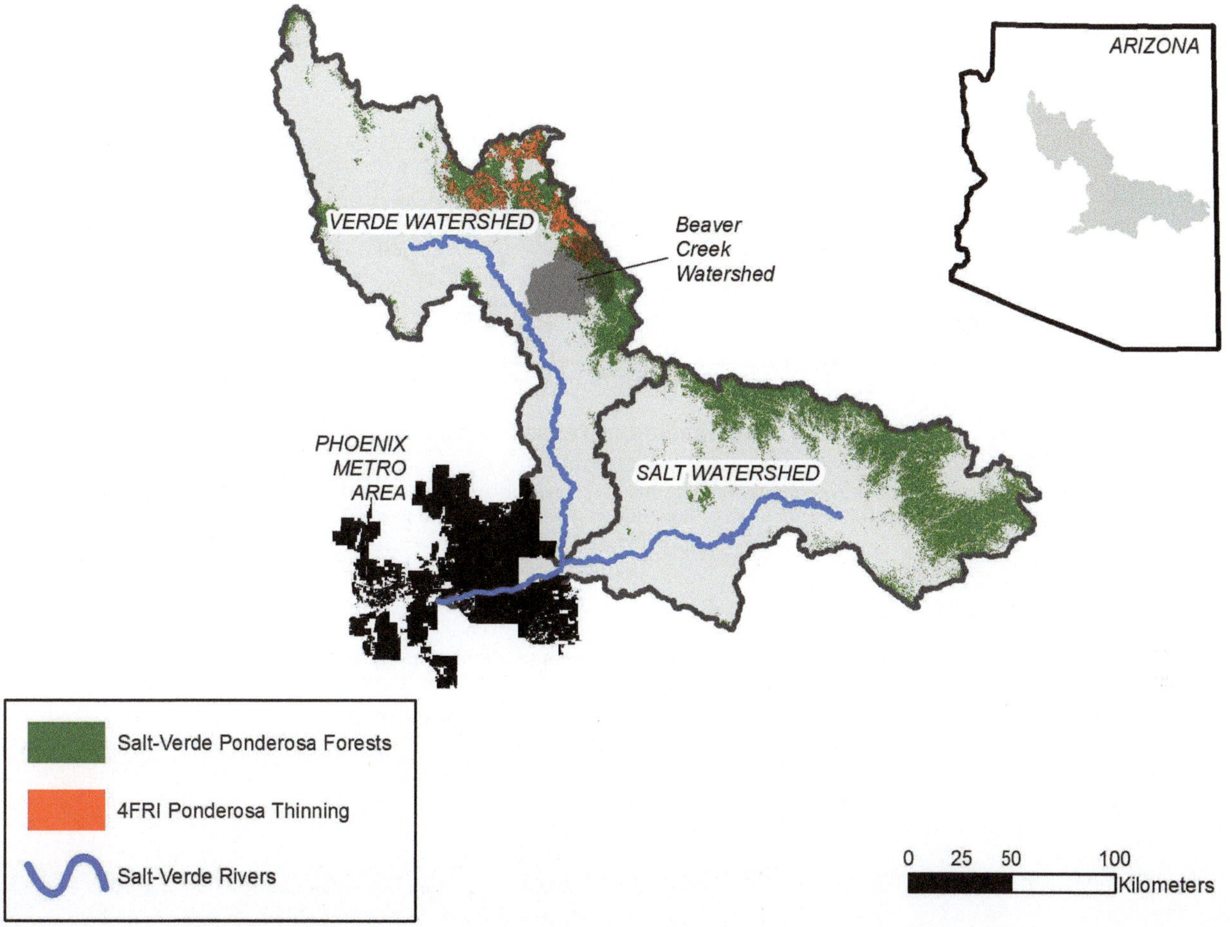

Figure 1. Map of Study Area. Map showing ponderosa pine forests in Salt-Verde watersheds in central Arizona, including those forests that are slated for mechanical thinning within the 4FRI project. Runoff from snowmelt in these forests is primary source of flow to Salt-Verde rivers which in turn are major sources of water for communities in the Phoenix Metro Area. Study used runoff model developed from experimental studies conducted in Beaver Creek watershed [20]. *Inset*: Location of study area in Arizona.

than model fit to total watershed runoff ($r^2 = 0.69$). We concluded that the original regression model was relatively insensitive to the direct effects of forest treatments, including time since treatment, on runoff.

We used a stepwise regression procedure to develop a revised model, beginning with an initial model that included the independent variables (winter precipitation and basal area) that were found to be significant in historical experiments [20]. Input data for all independent variables used to develop the model were derived from empirical data collected during the Beaver Creek and Castle Creek historical experiments and archived in databases or published reports [30–32]. We added independent variables and used an F-test to test the null hypothesis that the coefficient of each term was zero at the 95% significance level. To ensure that the model was not over-fitted, we repeated the stepwise regression procedure with 70% of the original data, selected randomly. We determined that the structure of the model did not change and provided a good fit to the remaining 30% of the data. We examined the residuals from the model and found no discernible patterns of dispersion or bias. Including terms for slope, percent of watershed area with a south-facing aspect, and mean, minimum, and maximum winter temperatures did not significantly improve the model.

The final model included three terms: winter precipitation, an interaction term between winter precipitation and years since treatment, and an interaction term between winter precipitation and basal area. It can be expressed as:

$$R_{treatment} = -28.464 + 0.148 * P - 0.015 * P * Y - 0.092 * P * [\exp(-BA_1/10.33) - \exp(-BA_2/10.33)]$$

$r^2 = 0.67$

where

$R_{treatment}$ = Increase in Annual Runoff attributed to Forest Thinning in mm

P = Total Winter Precipitation (Oct-Apr) in mm

Y = Years since Treatment

BA_1 = Basal Area before Treatment in m^2/ha

BA_2 = Basal Area after Treatment in m^2/ha

Figure 2b shows that the revised model explained more of the variability in predicting additional runoff associated with forest treatments ($r^2 = 0.67$) than did the original model. The model predicted runoff increases in a linear fashion with winter precipitation and basal area reductions, and decreases linearly

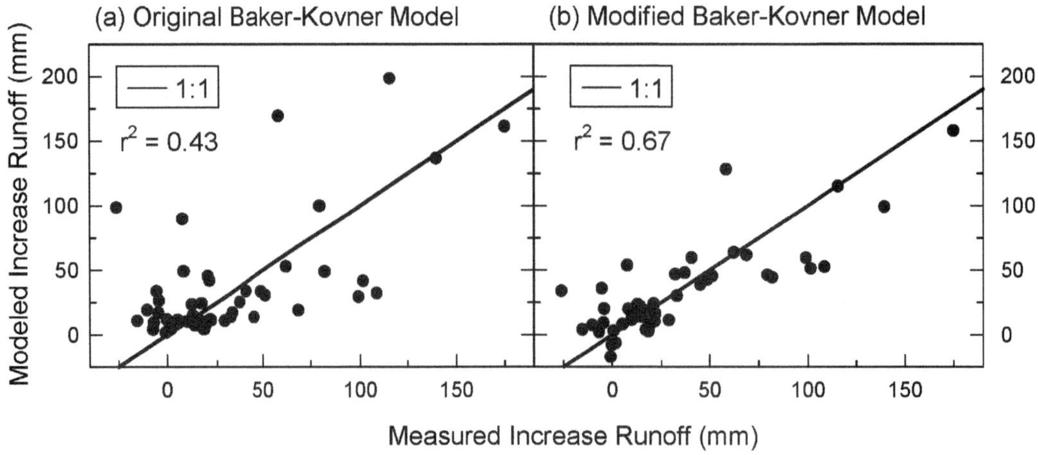

Figure 2. Comparison of models to observed runoff. Fit of (a) original [20] and (b) modified Baker-Kovner regression model output to increases in runoff associated with forests treatments in central Arizona, from Beaver Creek [30],[31] and Castle Creek [32] watersheds.

with years since treatment (Figure 3). To compare the original and modified Baker-Kovner models in English Units, see File S2.

We did use the original Baker-Kovner regression equation in this study to calculate total watershed runoff for forests in their current condition so that we could estimate the percentage increase in runoff associated with treatments. For these calculations, we used the average value for the solar insolation index, $r = 0.7$, reported from the Beaver Creek experiments [20]. In a sensitivity analysis, we found that calculating absolute runoff based on the low ($r = 0.66$) and high values ($r = 0.74$) of this index changed the estimated increase from treatments by only 1–2 percentage points.

Winter Precipitation

In order to account for inter-annual and decadal variability of winter precipitation, we used the PRISM model [25] to extract pluvials and droughts from the 20th century and inserted levels of winter precipitation from these periods into our runoff scenarios. We first extracted total winter precipitation, summed across October to April, for every year from 1900–2012 from the PRISM modeled dataset [25] across 2 scales: (a) ponderosa pine forests within the Verde watershed and (b) these forests within the Salt-Verde watersheds. We selected these geographies to be most representative of conditions for the 4FRI and Salt-Verde runoff scenarios, respectively. From 1900–2012, mean winter precipitation in Verde ponderosa pine forests was 394 mm (15.5 inches) with a range of 99–815 mm (3.9–32.1 inches) (Figure 4). For Salt-Verde forests, the mean was 368 mm (14.5 inches) and the range was 89–747 mm (3.5–29.4 inches) for this same period of time.

To test the accuracy of these modeled data, we compared values from the Verde ponderosa pine PRISM modeled data to measurements of total winter precipitation recorded during the original Beaver Creek experiments [31]. Given that we found the distribution of the data to be non-normal, we ran the non-parametric Spearman's rank order correlation. We found a strong association between modeled and observed data (Rho = 0.9577, $p < 0.001$) although the model underestimated measured winter precipitation over the time period by approximately 25 mm (1 inch) and consistently under-estimated precipitation in wet winters (Figure 4 inset).

We calculated mean winter precipitation across 15-, 25-, and 35-year periods from this modeled data and used this information to select droughts and pluvials. For each scenario, we moved the start and end years that thinning was simulated to occur. For example, time periods for our 15-year scenarios included 1944–1958, 1945–1959, and 1946–1960 (examples illustrated as shaded areas in Figure 4). Using this method, we found two instances of pluvials – early 20th century and late 1970s to late-1990s – and two droughts – during the 1950s and the current drought.

Estimating runoff from 4FRI project

We used the modified regression model to calculate additional annual runoff from individual forests stands that will be thinned in the first analysis area of the 4FRI project, established an annual thinning schedule, and then summed increases in runoff from individual stands across treatment years. We ran a total of 26 scenarios where in each scenario we inserted a different 15-year winter precipitation sequence.

We obtained a Geographic Information System (GIS) dataset of alternative C in the draft environmental impact statement for the first analysis area of the 4FRI project [19]. For each stand to be mechanically thinned, this dataset contained estimates of current basal areas and desired post-treatment basal areas. Forest prescriptions for "group-selection" stands called for two post-treatment basal areas, one for "open" areas that would be evenly thinned and another for "group" areas where thinning would be minimal to enhance wildlife habitats by retaining greater tree densities. For these stands, we calculated a post-treatment basal area that was an average of these two basal areas, weighted by the proportional area of the open and group areas. We selected 4,064 ponderosa pine stands within the Verde watershed slated for mechanical thinning, where the prescription will reduce basal areas by at least 30%. They ranged in size from 0.4 to 217 ha (1 to 536 acres), for a total of 61,900 ha (153,000 acres). The prescriptions within these stands called for reductions of basal area between 30% and 70% with a mean reduction of 48% (Figure 5).

Based on consultation with US Forest Service staff, we found that it was reasonable to assume that an equal number of hectares will be mechanically thinned every year across the 10-year treatment period. So for the purposes of our 4FRI scenarios, we assigned each of these stands to one of ten cohorts until the area of each cohort equaled 1/10th of total area or approximately 6,190 ha (15,300 acres). Figure 6a illustrates the treatment schedule for

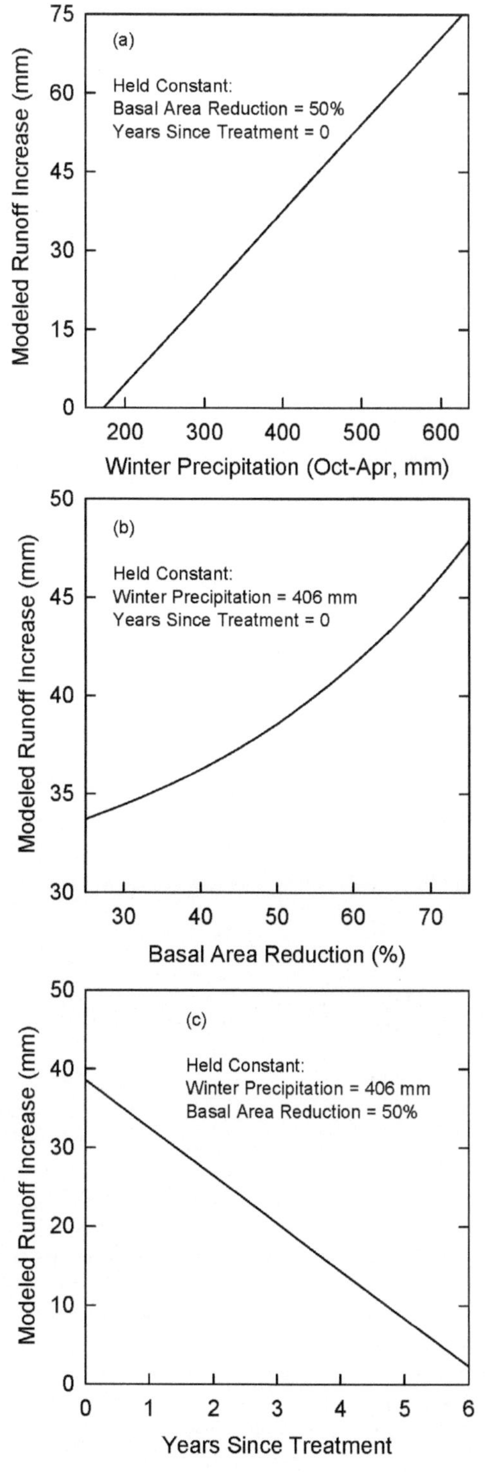

Figure 3. Influence of independent variables on runoff model. Relationships between model output to values for independent variables, including (a) winter precipitation, (b) percent basal area reduction, and (c) years since treatment. In all cases, other independent variables are held constant in order to view relationship of independent variable plotted on X-axis to model output.

the 15-year 4FRI scenarios reported in this study. Cohorts were treated consecutively in the first 10 years and it was assumed that cohorts influenced runoff for 6 years. So, for example, stands treated in cohort 1, contributed to additional runoff in scenario years 1-6 and stands treated in cohort 10 contributed from years 10–15.

We calculated additional runoff associated with thinning and total watershed runoff by inserting values of the independent variables into the revised and original Baker-Kovner regression models, respectively. Pre-treatment and post-treatment basal area values were derived from the 4FRI stand data. Years since treatment ranged from $y=0$ to 5. As described previously, winter precipitation values were droughts and pluvials drawn from the PRISM model for ponderosa forests in the Verde watershed. We set runoff increase to zero in years when winter precipitation was less than 230 mm (9 inches). We transformed the unit of runoff from mm to volume (million m^3) using the area of each stand. We estimated annual increases in runoff at the landscape scale by summing stand level amounts for each scenario year, and calculated summary statistics (mean, median, max, cumulative) that allowed for a comparison of scenarios.

Estimating runoff from Salt-Verde watersheds

Unlike the first analysis area of the 4FRI project, there was not a planned forest restoration project across the larger geography of the Salt-Verde watersheds. We developed a range of estimates for the extent, pace, and intensity of forest thinning that could be conducted over this larger geography, grouped these estimates into runoff scenarios, and ran the scenarios using the revised and original regression models to estimate additional runoff from treatments and total watershed runoff. To estimate the potential areal extent of thinning, we subtracted from the total forested areas those land uses that are typically considered unsuitable for mechanical thinning. We bracketed this initial area estimate with lower values to account for the lack of comprehensive spatial information. Assuming that the pace of treatments at this scale could take longer, we constructed 15-, 25-, and 35-year runoff scenarios. Finally, we assumed that the intensity of forest thinning at this scale would be similar to the range of basal area reductions planned for the first analysis are of the 4FRI project. The remaining parts of this section describe these steps in greater detail.

Based on a methodology developed in a previous study that estimated wood supply on four National Forests in northern Arizona [33], we subtracted from the total area of ponderosa pine in the watersheds – 0.68 million ha (1.68 million acres) – those areas that are typically excluded from mechanical thinning projects due to steepness, restrictions in land management, previous treatments, or considerations of soil, habitat, or wildlife conditions (Table 1). We adopted 6 of the 7 "exclusion" criteria from that study and added one additional factor. We removed one criterion – Northern goshawk nest areas –because the mean basal area reduction for these areas in the 4FRI project was greater than 30% [19] suggesting that thinning in these areas could result in additional runoff based on the results of the Beaver Creek experiments [20]. We added one criterion – high severity burn patches – because a nearby study on the recovery of ponderosa pine forests after wildfire demonstrated that mature forests suitable for thinning would not likely develop in areas that had burned with high severity within the timeframe of this study [34]. While two of the three 4FRI prescriptions for Mexican Spotted Owl Protected Activity Centers (MSO PAC) would result in negligible changes in basal area, one would allow for basal area reductions greater than 30% [19]. However, we were not able to differentiate among these prescriptions from the data we obtained, so we opted for a conservative estimate for this factor by excluding MSO PAC areas from our analysis. Compiling the publicly available geospatial data

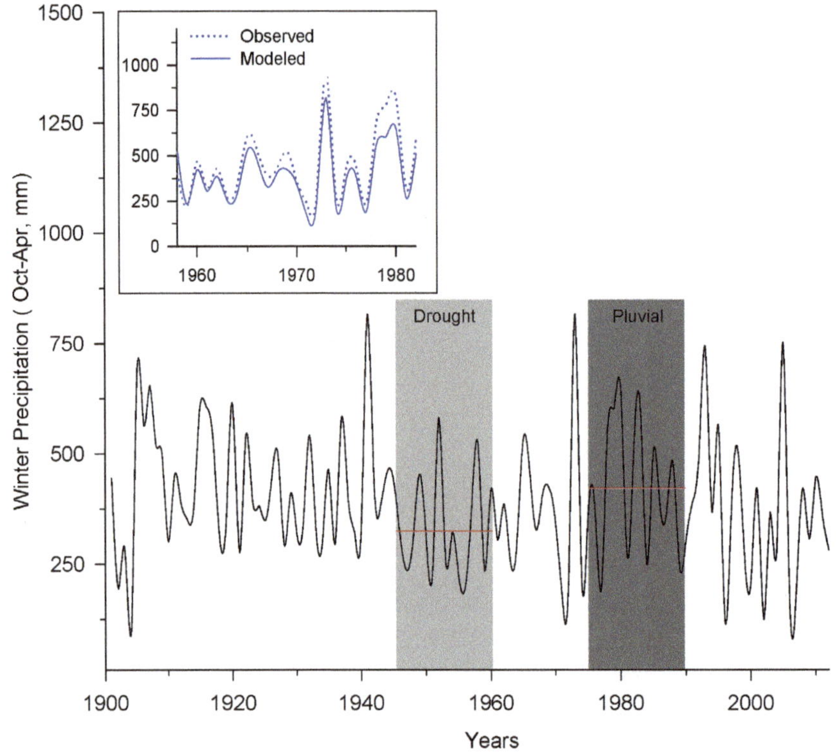

Figure 4. Variability in winter precipitation in ponderosa pine forests. Estimates of historical winter precipitation from 1900–2012 in ponderosa pine forests within Verde watershed from PRISM model [25]. Shaded areas are examples of 15-year droughts and pluvials that were used in study scenarios; horizontal red lines represent mean winter precipitation within these shaded areas. *Inset*: Comparison of measured winter precipitation observed during the historical Beaver Creek watershed experiments [20] from 1958–1982 versus modeled winter precipitation data shown in main figure.

for these factors [26], [27], [33], [35–38], we found that 401,000 ha (992,000 acres), or 59% of ponderosa pine forests, were available for mechanical thinning after accounting for these exclusion areas. The percentage of forests excluded in this estimate – 41% – compared favorably with the average of 37% of forests that were excluded in dozens of restoration projects evaluated in the northern Arizona [33]. We bracketed this high estimate with a moderate estimate – 45% or 301,000 ha (743,000 acres) – and a low estimate – 30% or 204,000 ha (505,000 acres) – because quantitative geo-spatial data on some exclusion areas was missing or incomplete. As illustrated graphically for the moderate estimate in Figure 6b-d, we divided these areas into 10-, 20-, and 30-year treatment cohorts assuming that equal areas would be thinned every year.

We scaled up the forest condition information contained in the 4FRI stand data using a proportion metric, assuming that these stands were representative of stand-level forest conditions across the Salt-Verde watersheds. For example, 4FRI stands that comprised 1,500-ha in a 15,000-ha 4FRI treatment cohort were estimated to be 3,000-ha in a 30,000-ha Salt-Verde cohort. If the thinning prescription for these stands called for a reduction in basal area from 33 to 17 m^2/ha, then the Salt-Verde scenarios calculated additional runoff from 3,000-ha that are thinned from 33 to 17 m^2/ha. After scaling up the area of each stand to the watershed scale, the procedure for estimating runoff in the Salt-Verde scenarios was the same as described for the 4FRI scenarios. We ran 147 scenarios in total, 67 fifteen-year, 52 twenty-five year, and 28 thirty-five year scenarios.

Results

Four Forest Restoration Initiative Runoff Scenarios

Mechanical thinning of ponderosa pine forests in the first analysis area of the 4FRI project – 6,190 ha/year for ten years totaling 61,900 ha (15,300 ac/year; 153,000 acres total) – increased mean annual runoff from 3.13 million m^3 (2,540 acre-feet) in a simulated drought to 7.27 million m^3 (5,890 acre-feet) in a pluvial. Differences in winter precipitation over the 15-year simulation periods significantly influenced runoff gains. A difference in mean winter precipitation of only 130 mm (5 inches), from 330 mm (13 inches) in drought scenario to 460 mm (18 inches) in a pluvial scenario, resulted in a doubling of the annual increase in runoff from treatments (Figure 7).

Inter-annual variability in winter precipitation was also important. Figure 8 illustrates year to year increases in runoff for those scenarios that resulted in the lowest and highest gains in runoff. Years with high winter precipitation played a disproportionate role in additional runoff in both droughts and pluvials. In the majority of scenarios ran (15 out of 26), winter precipitation from only 5 of the 15 years accounted for at least 75% of the increased runoff. Cumulative increases in runoff across 15-year periods resulted in a total increase from 54.3 to 111 million m^3 (44,000–89,800 acre-feet) (Figure 9). Runoff from thinned forests was approximately 20% greater than unthinned forests (as estimated using original Baker-Kovner regression model) in both droughts and pluvials (data not shown).

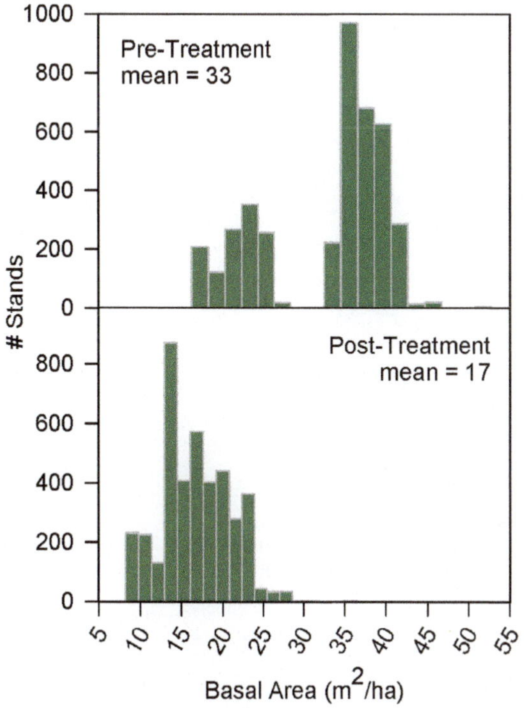

Figure 5. Ponderosa pine basal area reductions in 4FRI project. Histograms showing (top) pre-thinning and (bottom) desired post-thinning basal areas (in m^2/ha) of ponderosa pine stands in the first analysis area of the 4FRI project (excluding stands where basal area reduction <= 30%).

Salt-Verde Runoff Scenarios

Depending on winter precipitation and the forest treatment schedule, mean annual increases in runoff from thinning of ponderosa forests across the Salt-Verde watersheds ranged from 4.76 to 15.0 million m^3 (3,860–12,200 acre-feet) over a 35-year treatment period, 6.18 to 23.4 million m^3 (5,010 to 19,000 acre-feet) over 25 years, and 9.23 to 42.8 million m^3 (7,480 to 34,700 acre-feet) over 15 years (Table 2). Similar to the 4FRI scenarios, additional runoff in the Salt-Verde watersheds was 1.6–2.3 times greater in pluvials than in droughts. Regardless of whether the scenarios occurred in a drought or pluvial, cumulative runoff gains in thinned forests were 20-26% greater than unthinned forests. Cumulative gains ranged from 167 to 525 million m^3 (135,000–426,000 acre-feet) in the 35-year scenarios, 154 to 585 million m^3 (125,000–474,000 acre-feet) in 25-year scenarios, and 138 to 643 million m^3 (112,000–521,000 acre-feet) in 15-year scenarios. In both droughts and pluvials, runoff increased in a positive linear fashion with increases in the pace and the extent of forest thinning (Figure 10). See File S1 to see study figures and tables in English units.

Discussion

This study demonstrated that the pace, extent, and intensity of forest thinning that is planned under the first analysis area of 4FRI and at larger scales could measurably increase runoff in ponderosa pine forests in central Arizona. Modeled runoff from thinned forests was approximately 20% greater than runoff from unthinned forests which is within the range of 10–40% increases in runoff demonstrated in the Beaver Creek experiments [20], [29]. The Salt-Verde runoff scenarios showed that levels of

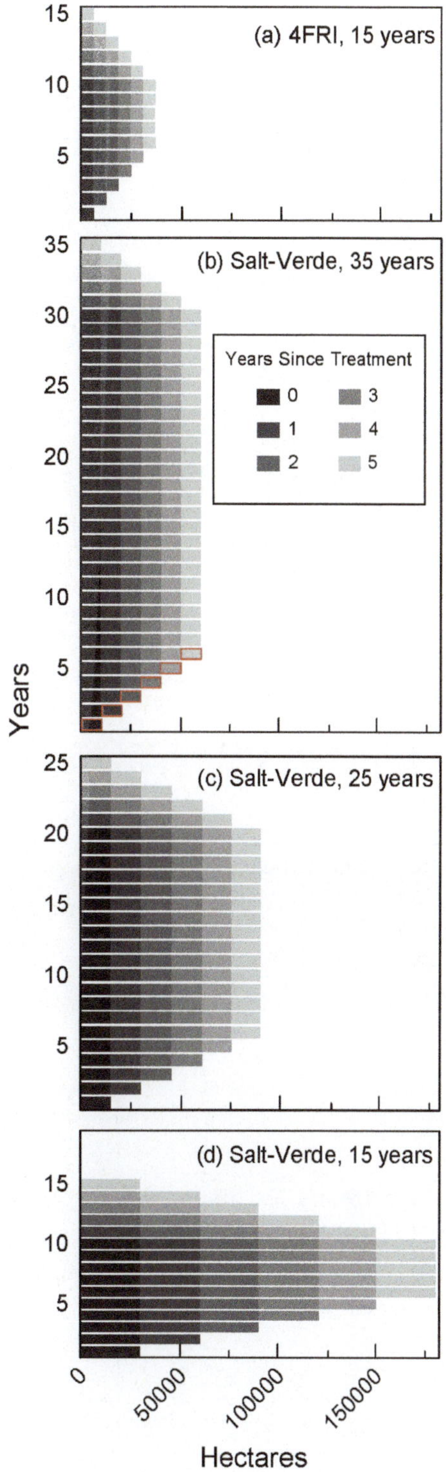

Figure 6. Forest treatment schedules for study scenarios. Graphical depiction of mechanical thinning treatment schedules for (a) 15-year 4FRI scenarios and (b-d) 35-year, 25-year, and 15-year Salt-Verde moderate thinning scenarios (total thinned area was 301,000 ha or 743,000 acres). Scenarios assumed consecutive treatments for 10-, 20-, and 30-year treatment periods shown as black bars in the bottom left portion of each of the figures. Bars outlined in red in (b) show the contribution of one cohort of stands through six years in the scenario.

Table 1. Estimate of hectares of ponderosa pine forests available to mechanical thinning in Salt-Verde watersheds.

Category	**Hectares**	**%**	***Sources***
Salt-Verde Watersheds Ponderosa Pine Forests	*681,000*	-	[27]
Exclusions			
Steep Slopes greater than or equal to 40%	68,000	10%	[26]
Hi-Severity Burn Patches within Wallow and Rodeo-Chedeski Fires	43,700	6%	[35]
Specially Designated Areas including Wilderness Areas, Research Natural Areas, State Wildlife Management Areas[a]	21,800	3%	[36],[37]
Streamside Management Zones within 100 feet perennial reaches in national forests and 200 feet in tribal areas	8,050	1%	[38]
Mexican Spotted Owl Protected Activity Centers where owls have been found to be nesting[b]	57,500	8%	[33]
Erodible Soils where thinning is unsuitable due to erosion risk or rocky conditions[c]	93,100	14%	[33],[37]
Completed Treatments that have been thinned in last 10 years[d]	35,200	5%	[37]
Sub-Total Exclusions (accounting for overlap between layers)	*280,000*	*41%*	
Sub-Total Forest Available Mechanical Thinning	*401,000*	*59%*	

[a] Estimate derived from USDA Forest Service Southwestern Region Wilderness Status layer and Forest Level Special Interest Management Areas layers [37].
[b] Data available for Coconino, Apache-Sitgreaves, Kaibab, and Tonto National Forests only. Added qualitative estimate for this exclusion factor for tribal lands and Prescott National Forest assuming land excluded in these land units would be proportional to lands excluded for this factor on the four national forests above.
[c] List of excluded soils for Coconino, Apache-Sitgreaves, Kaibab, and Tonto National Forests based on Hampton et al. [33]. This list of excluded soils was augmented based on consultation with USDA Forest Service southwestern region soil scientists. Geospatial data for these soils derived from USDA Forest Service Forest Level Terrestrial Ecological Units layers [37]. Added qualitative estimate for this exclusion factor for tribal lands and Prescott National Forest where geospatial data was unavailable, assuming land excluded in these land units would be proportional to lands excluded for this factor on the four national forests above.
[d] Geospatial data for this factor derived from USDA Forest Service Forest Level Activities Layer joined to FACTs table for Coconino, Apache-Sitgreaves, Kaibab, Tonto, and Prescott National Forests [37]. Added qualitative estimate for this exclusion factor for tribal lands based on consultation with Bureau of Indian Affairs forestry personnel.
Summary of hectares and percentages of ponderosa pine forests within Salt-Verde watersheds in relation to potential for thinning treatments, including total hectares, those areas typically excluded from mechanical thinning, and the remaining hectares available for mechanical thinning. Estimates rounded off to three significant digits.

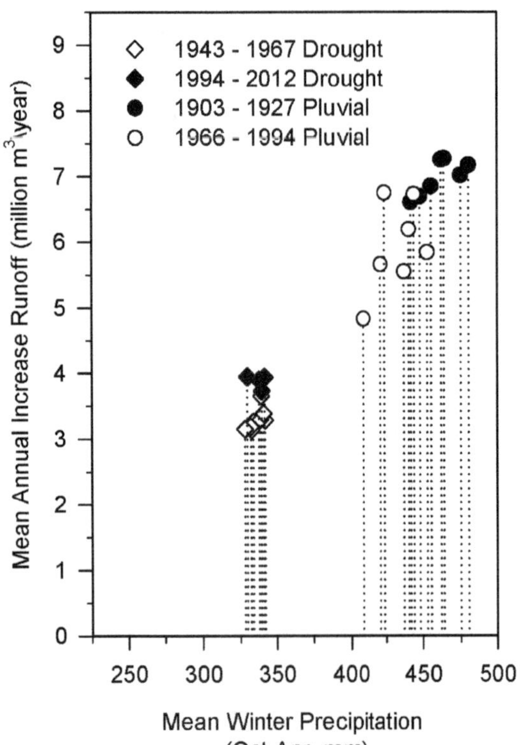

Figure 7. Increases in mean annual runoff from thinning in 4FRI project. Results from 26 scenarios with varying levels of winter precipitation showing increases in mean annual runoff associated with mechanical thinning of ponderosa pine forests in the first analysis area of the 4FRI project. In order to compare scenarios, only increases in *mean* annual runoff are shown. Annual variability in runoff for two of these scenarios is shown in Figure 8.

additional runoff increased proportionally with increases in the pace and extent of thinning. Runoff gains occurred in droughts and pluvials, a surprising outcome that underscored the importance of inter-annual and decadal variability in precipitation. These increases in runoff would likely improve conditions for water-dependent natural resources, such as cienegas, riparian areas, and aquatic habitats, which are vulnerable to low flows that are experienced seasonally, especially in summer months, and also during droughts [5–7]. Runoff gains also could provide incidental benefits to the water supply of downstream users, but the increases were more modest when compared to total runoff from the Salt-Verde watersheds. They would comprise a 0–3% increase from mean annual surface flows from the Salt-Verde rivers of 1.39 billion m^3 (1.13 million acre-feet) and approximately a 1-9% increase of Salt-Verde flows supplied to municipal users in the Phoenix Metro Area on an annual basis [39].

Consistent with the historical experiments, thinning effects on runoff were temporary in this study. Additional runoff would cease six years after the multi-year thinning schedules in our scenarios. Beaver Creek investigators speculated that thinning effects were short-lived because of subsequent regrowth of understory species [29]. We note that prescribed fires are planned as subsequent treatments for all stands that are initially thinned within the first analysis area of 4FRI [19]. These maintenance treatments should provide additional information as to how effective subsequent actions are in terms of removing understory vegetation and sustaining runoff gains.

Our study adds to previous research on the potential for increases in runoff associated with thinning of ponderosa pine forests in the Southwest. Using a simulation model, Brown and Fogel [40] found that thinning approximately 9,300 ha (23,000 acres) of ponderosa pine per year over a 10-year period produced increases in runoff that ranged from 0.28 to 11.8 million m^3/year (230–9,600 acre-feet/year). This range overlapped with but was

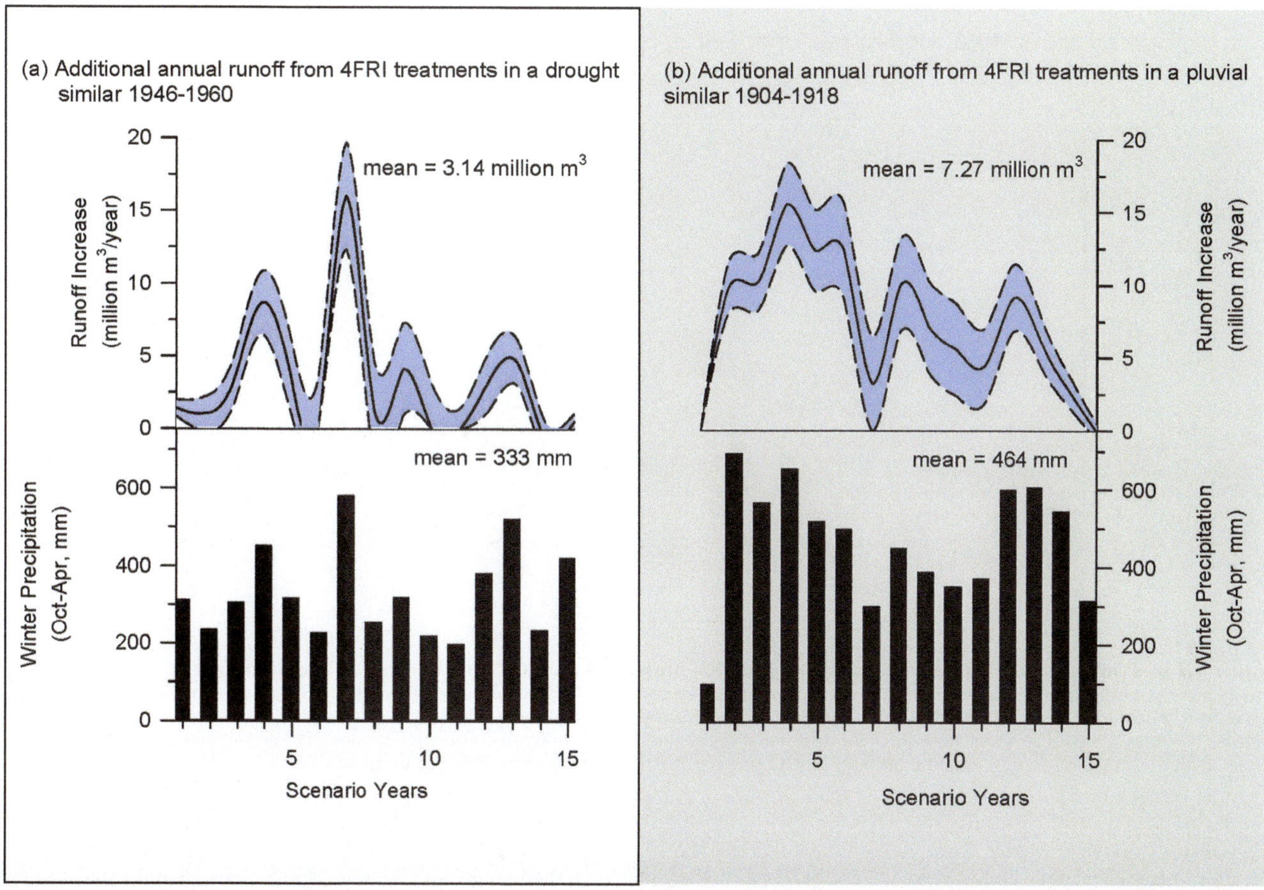

Figure 8. Year to year variability in runoff increases from thinning in 4FRI project. Modeled increases in annual runoff associated with mechanical thinning of ponderosa pine forests in the first analysis area of the 4FRI project during the (a) drought and (b) pluvial that produced lowest and highest gains in runoff respectively. Top panes show increases in annual runoff in million m^3/year. Solid black lines are output values from regression model; dotted lines and blue areas represent 90% confidence intervals. Bottom panes show corresponding winter precipitation values (Oct-Apr, mm) used as one of the independent variables to calculate runoff.

lower than the 6.18 to 11.9 million m^3/year (5,010–9,660 acre-feet/year) increases in runoff for the two Salt-Verde runoff scenarios with annual thinning of approximately 10,000 ha (25,000 acres) (Table 2). One potential reason for this difference was that Brown and Fogel [40] modeled the treatment-runoff relationship to be constant over time, 18 mm (0.71 inches), and did not evaluate variability in precipitation but instead assumed average winter precipitation of 396 mm (15.6 inches). Using similar values for winter precipitation and basal area reduction, our model simulated the runoff response to vary depending on years since treatment, from a high of 38 mm (1.5 inches) in the year immediately following treatments to a low approaching zero mm after six years (Figure 3c).

Study Limitations

In this study, we revised a statistical runoff model from the historical Beaver Creek watershed experiments to explore how variations in climate and forest thinning would affect runoff. To the extent possible, we limited the application of the runoff model to the range of climatic and environmental conditions under which the original data were collected and model developed. However, our modeling approach was constrained by several uncertainties that will require further investigation. The model was limited by the scope of forest practices conducted in the Beaver Creek watershed experiments [20], [30–32]. These experiments used thinning techniques such as strip-thinning and patch clearing that had a similar range of basal area reductions but a different spatial configuration from contemporary thinning prescriptions. They did not measure the effects of maintenance treatments on runoff. They measured "in-place" runoff at sub-watershed outlets and did not measure other aspects of the water budget, such as evapotranspiration, soil moisture, surface routing, and groundwater recharge. The revised model represented statistical relationships between annual runoff and winter precipitation, forest basal area reductions and time since treatment, but these relationships were not necessarily related to physical processes or functions that control water balance [20]. We summarized how important factors that are not directly modeled may increase or decrease the range or runoff due to thinning that is documented in this study in File S3.

Of all the factors where uncertainties in forest hydrology and runoff remain, perhaps the most significant were the effects of soil types, long-term climate variability, and climate change. In terms of soil types, the revised regression model in this study was derived from experimental thinning on basalt-derived soils that support 59% of ponderosa pine forests in the Salt-Verde watersheds. The runoff effects of thinning on the remaining 41% of forests on sedimentary-derived soils are unknown. Two studies demonstrated that runoff from untreated ponderosa forests on sedimentary-derived soils can be

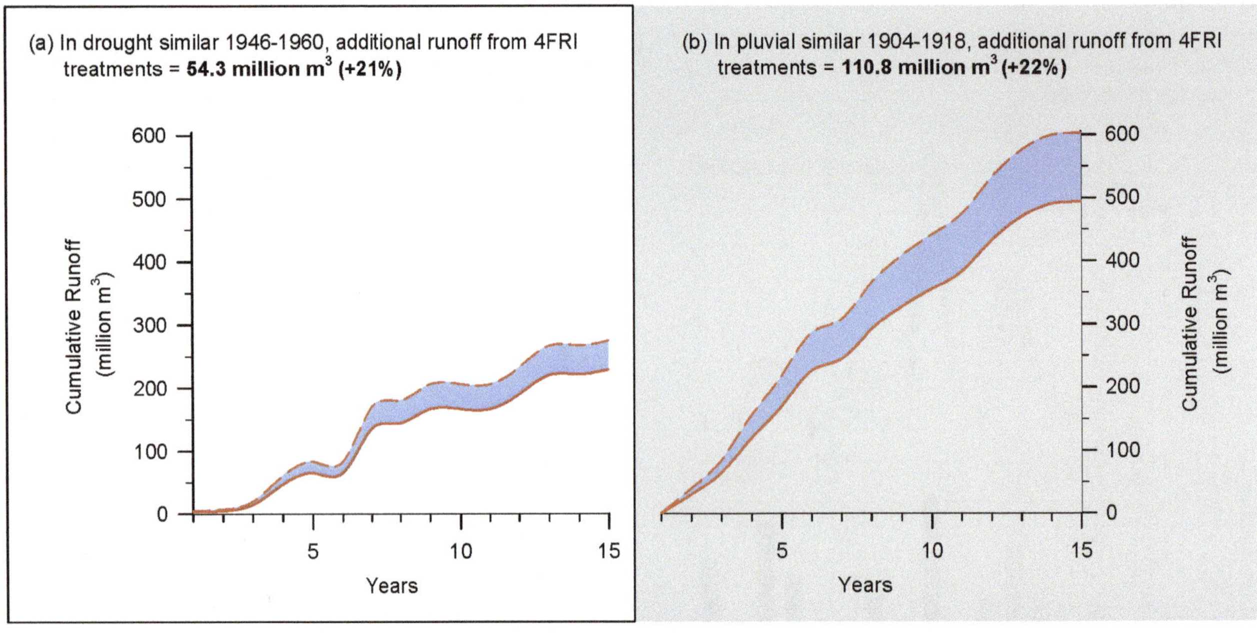

Figure 9. Cumulative runoff increases from thinning in 4FRI project. Estimates of cumulative increases in runoff (million m^3) from planned mechanical thinning of ponderosa pine forests in the first analysis area of the 4FRI project under (a) drought and (b) pluvial conditions. Solid red lines are estimates of cumulative runoff under current forest conditions using original Baker-Kovner regression model [20]. Dotted red lines represent increases in cumulative runoff associated with 4FRI treatments using modified Baker-Kovner regression model. Difference between these two values, shown with blue shading, is additional runoff from forest thinning treatments. Estimated increases in runoff ceased after 15 years.

substantially less than runoff from volcanic soils [41], [42]. Several properties of sedimentary-derived soils relative to basalt-derived soils may lead to greater water permeability and infiltration, including lower clay content, extensive fracturing, and deeper soil horizons [30], [43], [44]. Depending on a number of factors, increased infiltration may lead to increased recharge, which we would expect to result in increased discharge to surface water at some location downstream.

A nearby paired-watershed study in mixed conifer forests on sedimentary-derived soils found that overstory removal of 30-80% led to a slightly higher range of increased runoff, 30–110%, that persisted for more years than the Beaver Creek experiments [45],[46]. Direct comparisons with ponderosa pine forests are not possible because this study was conducted within a higher-elevation mixed-conifer forest that had higher initial basal areas and higher mean annual precipitation.

Relative to climate, this study's scenarios captured the high end of variability associated with pluvials but probably not the low end of variability associated with droughts. Specifically, the early 20th century pluvial may have been the wettest pluvial that has occurred in the last 12 centuries [47], whereas the two occurrences of 20th century drought were eclipsed in terms of duration and severity by several multi-decade "mega-droughts" [48–51]. Output from global circulation models indicates that climate variability will continue to be an important characteristic of the region in the future [52], but that climate change may increase the risk of extreme climatic events such as multi-decade droughts and extreme winter precipitation [53], [54]. Some global circulation models also project that mean winter precipitation in the Southwest will decline by up to 10% [52], but it may take many years to detect effects on stream flows because of precipitation variability [55]. The net effect of changes in precipitation on forest condition and hydrology still needs to be resolved and will require more investigation.

Unlike precipitation, the effects of anthropogenic warming on forest hydrology are clearer. Studies have found that warmer temperatures in recent decades help explain a downward trend in snowpack in the western United States, even after patterns of natural climate variability have been considered [9], [56]. In our revision of the historical runoff model, we attempted to add several parameters of temperature – seasonal means, maximums, and minimums – as explanatory variables to the revised regression model but none were significant. This suggests that the temporal and spatial range of winter temperatures that occurred during the years when the watershed experiments were conducted was not sufficient to add explanatory value to predicting runoff. In a separate trend analysis of winter temperatures (Oct-Apr), we found that temperatures in the months of March and April in the last 25 years, 1988–2012, were significantly warmer by 2 and 1 degrees Celsius (3.5 and 2 degrees Fahrenheit), respectively, than temperatures for the same months in the 25 years when the experiments were conducted, 1958–1982 (see File S4 for more detailed information). The magnitude of observed declines in snowpack in the Southwest, in the range of 20%, is similar to the increases in runoff associated with thinning from this study, suggesting that accelerated thinning may at least offset or ameliorate runoff losses due to climate change.

Management Implications: Accelerated Forest Thinning to Improve Resilience

On balance, we believe there is enough evidence for managers to explore the ways in which accelerated forest thinning can restore forest resilience and functioning to ameliorate the drying of ponderosa pine forests associated with warming and past management activities. Below we summarize the management implications of this and related studies.

Table 2. Annual runoff increases from thinning in Salt-Verde watersheds.

Forest Management: Thinning[a]				Winter Precipitation Regime			Runoff: Annual Increase			Runoff: Cumulative Increase	
Scenario Name	*Total Hectares*	*Hectares per Year*	*Effective Hectares per Year*	*Type*	*Similar to 20th Century Period*	*Mean (mm/yr)*	*Mean (million m^3)*	*Median (million m^3)*	*Maximum (million m^3)*	*Total (million m^3)*	*%*
35-low	204,000	6,800	35,000	Drought	1942–1976	340	4.76	2.99	17.5	167	23%
				Pluvial	1965–1999	396	7.65	5.85	25.7	268	26%
35-mid	301,000	10,000	51,400	Drought	1942–1976	340	7.01	4.39	25.8	245	23%
				Pluvial	1965–1999	396	11.3	8.61	37.7	393	26%
35-high	401,000	13,400	68,800	Drought	1942–1976	340	9.35	5.86	34.4	327	23%
				Pluvial	1965–1999	396	15.0	11.5	50.4	525	26%
25-low	204,000	10,200	49,000	Drought	1942–1966	338	6.18	4.06	26.3	154	20%
				Pluvial	1975–1999	411	11.9	9.65	33.6	299	22%
25-mid	301,000	15,100	72,000	Drought	1942–1966	338	9.09	5.97	38.7	227	20%
				Pluvial	1975–1999	411	17.5	14.2	49.3	438	22%
25-high	401,000	20,100	96,300	Drought	1942–1966	338	12.1	7.98	51.7	303	20%
				Pluvial	1975–1999	411	23.4	19.0	66.0	585	22%
15-low	204,000	20,400	81,700	Drought	1945–1959	315	9.23	5.95	52.3	138	20%
				Pluvial	1975–1989	422	21.8	8.31	60.1	327	22%
15-mid	301,000	30,100	120,000	Drought	1945–1959	315	13.6	8.75	77.0	204	20%
				Pluvial	1975–1989	422	32.1	12.2	88.4	481	22%
15-high	401,000	40,100	161,000	Drought	1945–1959	315	18.1	11.7	103	271	20%
				Pluvial	1975–1989	422	42.8	16.3	118	643	22%

[a] 'Scenario Name' indicates number of years simulated in each scenario (15, 25, 35 years) and extent of forest thinned (low = 30% of forest thinned; mid = 45%; high = 60%). 'Total Hectares' are total number of hectares thinned in each scenario. 'Hectares per Year' are number of hectares thinned every year for 10 consecutive years (15-year scenarios), 20 consecutive years (25-year scenarios), and 30 consecutive years (35-year scenarios). 'Effective Hectares per Year' are the average number of thinned hectares across all years of the scenario that contribute to additional runoff (see figure 6).

Estimates of additional runoff associated with thinning of ponderosa pine forests across Salt-Verde watersheds accounting for variation in treatment periods, total forest extent treated, and drought and pluvial periods of winter precipitation (Oct-Apr). Estimates rounded off to three significant digits.

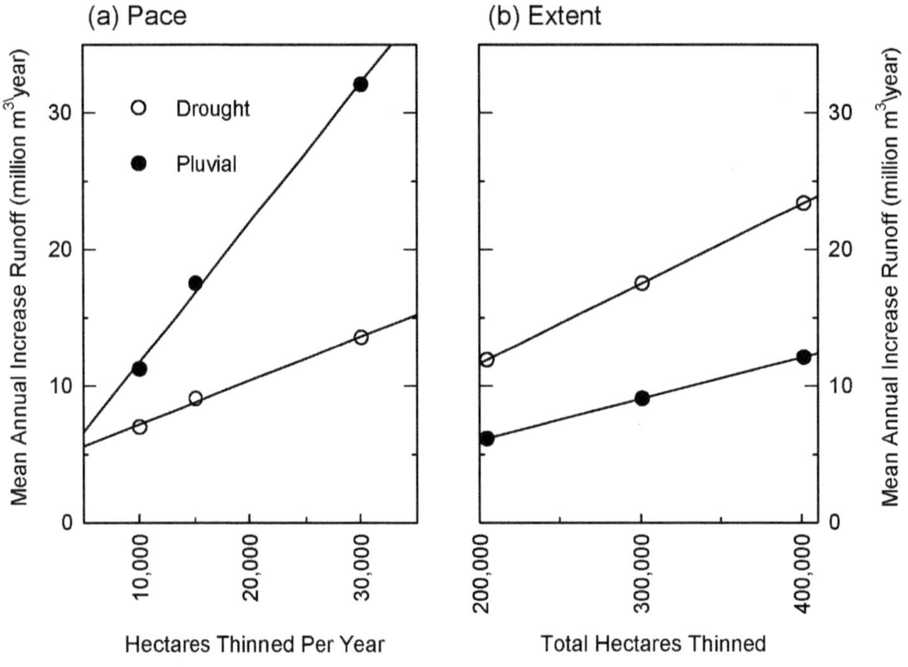

Figure 10. Scale effects of thinning on runoff in Salt-Verde watersheds. Effects of increasing (a) pace and (b) extent of thinning treatments in ponderosa pine forests in Salt-Verde watersheds on increases in mean annual runoff (million m^3/year). In (a) total area thinned is held constant at 301,000 ha (743,000 acres) (scenarios: 35mid, 25mid, 15mid) to show influence of increasing the area thinned per year. In (b) duration of thinning treatments is held constant at 25 years (scenarios: 25low, 25mid, 25high) to show influence of increasing the total area thinned across the scenario. In order to illustrate scale effects, only increases in *mean* annual runoff are shown. Statistics describing annual variability in runoff gains are shown in Table 2 and illustrated graphically for 4FRI scenario in Figure 8.

1. *Maintain & improve ecosystem function* – The current paradigm of forest management has moved away from a debate about managing forests to address one societal need, such as downstream water supply, to recognition that management strategies need to address multiple challenges [29], [57]. The 4FRI restoration project demonstrates this approach with planned benefits that include "…improved vegetation biodiversity, wildlife habitat, soil productivity, and watershed function" [19]. The impaired functioning and heightened vulnerability of southwestern forests to large-scale disturbances increases the urgency to improve ecosystem functioning, and to measure the effectiveness and trade-offs associated with these management activities using robust adaptive management and monitoring protocols. Such a focus may result in more options and flexibility in subsequent years, whereas failure to maintain functioning ecosystems may eliminate future options if soil and hydrologic processes are irreversibly altered.

Our study demonstrated that accelerated thinning can improve surface water runoff, a key ecosystem function, and provided indirect evidence of benefits for soil moisture and productivity. Another study found that average soil water content on low density ponderosa forests (250 trees/ha) was substantially higher than on high density forests (2,710 trees/ha), although these differences in forest density were not a result of mechanical thinning [58]. Researchers in the historical Beaver Creek experiments hypothesized that lower tree densities associated with thinning would reduce evapotranspiration losses, and thereby allow more water to be available for soil moisture, groundwater recharge, and surface water runoff [29]. Although we were unable to evaluate thinning effects on evapotranspiration, Dore et al. [59] found that light thinning (40% basal area reduction) in a ponderosa pine forest reduced stand-level evapotranspiration by 12% over four years, but the effect diminished over time and was not detectable in the fourth post-harvest year. This same study found that removal of all trees in a wildfire-burned site reduced annual evapotranspiration by 20%.

Other studies demonstrated that mechanical thinning can reduce the vulnerability of forests to uncharacteristic crown fires. Using a fire simulation model, Cochrane et al. [60] found that landscape-scale mechanical thinning that preceded actual wildfires could have reduced the average size of six wildfires in ponderosa pine and mixed conifer forests in California and the southwestern United States by an average of 18% (range 0.3 to 65%). A meta-analysis of 54 experimental studies showed that mechanical thinning significantly reduced the susceptibility of western ponderosa and Jeffrey pine forests to crown fires [23]. These studies describe the primary mechanism by which thinning reduces fire risk as a redistribution of fuels, thereby reducing fire spread rates and crowning behavior.

2. *Apply management tools at scale* – Managing forests at a scale that is comparable to recent forest disturbances will be another key factor for improving resilience. At 238,000 ha (588,000 acres) of ponderosa pine to be treated with mechanical thinning and prescribed fire, the 4FRI project is approaching the scale of these amplified disturbances. Our study demonstrated that potential increases in runoff associated with thinning were scale dependent: runoff increased in a linear fashion as the pace and extent of thinning increased in the Salt-Verde runoff scenarios (Figure 10). Recent efforts to reduce fire risk with landscape-scale treatments of ponderosa pine forests have been successful. For example, treatments completed before the largest wildfire in Arizona history, the Wallow Fire in 2011, effectively protected communities and towns by reducing fire severity before it reached key residential areas [61].

Further, accelerated thinning may open up opportunities to use other management tools at an accelerated pace. For example, wildland fire use has been an effective and inexpensive tool for maintaining forest resilience across large areas in remote forests and woodlands – the Gila Wilderness in New Mexico and Grand Canyon National Park in northern Arizona [62], [63] – but has been used less frequently in forests near population centers and towns. Completion of treatments in the first analysis area of the 4FRI project could have a multiplying effect on maintaining forest resilience if the thinning itself allows for increased use of wildland fire or prescribed fire over many more hectares.

3. *Leverage knowledge of climate variability* – The Beaver Creek watershed experiments and others like it demonstrated that winter precipitation was the best predictor of additional runoff from forest thinning [20], [30]. We add to this knowledge by examining how variability in winter precipitation could influence management effectiveness. Our scenarios demonstrated that gains in runoff from thinning could occur even under drought conditions. In simulated droughts, 1 out of 3 years had above-average precipitation, and these years accounted for 75% of the runoff gains in these scenarios. Even though these gains were substantially less than runoff gains during pluvials, additional runoff under drought conditions could help lessen the impact of predicted climate-driven losses in hydrologic connectivity important to native fish in the Salt and Verde watersheds [64].

This study also demonstrated that the same thinning schedule could result in a doubling in additional runoff if conducted in a pluvial typical of 20th century variability as compared to a drought. Scientists have developed decision support tools to help managers understand the likelihood that precipitation regimes will shift from wet to dry conditions or vice-versa [65], [66]. Knowledge of whether management activities are taking place in a pluvial or a drought could improve the ability of managers to predict the likelihood of success for various objectives. For example, thinning in a pluvial might meet substantive objectives to improve runoff, whereas thinning in a drought might be more significant for reducing risks of catastrophic crown fires or drought mortality.

With the notable exception of a mega-drought, the droughts and pluvials examined in this study were characteristic of climate variability in the region. An examination of hydro-climatic trends in the Salt-Verde watersheds revealed that temperatures are non-stationary, increasing significantly in recent decades, but the same is not true for winter precipitation or resulting stream flows [55]. Rather, winter precipitation has fluctuated between droughts and pluvials in these watersheds in the last century with no evidence of directional change, a pattern that has reoccurred in the southwestern United States for the last 1000 years [50]. These precipitation cycles have been linked to persistent anomalies in sea surface temperatures that vary across inter-annual and decadal time steps [67], [68]. If past patterns of precipitation variability remain stable in the near term, then it is probable that precipitation and flows in the Salt-Verde watersheds will shift into wetter conditions within the timeframe examined in this study [66].

4. *Building Knowledge to Reduce Uncertainties* – Landscape-scale restoration projects like 4FRI present the opportunity to learn about the influence of accelerated thinning on forest water budgets and resilience using modern forestry techniques and under a changing and variable climate. Stakeholders in the 4FRI project are developing a robust adaptive management and monitoring program so that progress towards objectives can be measured and timely adjustments to management made. Additionally, a paired-watershed experiment to evaluate the effects of forest treatments, including prescribed fire maintenance treatments, on watershed properties is planned within the first analysis area of 4FRI. It will monitor runoff, precipitation, evapotranspiration, snow water storage, soil moisture, groundwater recharge, and sediment yield, thereby reducing our knowledge gaps regarding the effects of forest management on the water cycle.

Conclusions

The widespread pattern of forest drying, evidenced by ecosystem level moisture deficits and larger and more severe wildfires, indicates we are entering a new era of forest management. The 4FRI project is an example of a restoration effort that is addressing this challenge in two ways: establishing objectives to improve resilience and increasing the scale at which management tools are applied. Our study demonstrated that this type of project can increase runoff by approximately 20% compared to unthinned forests, even under simulated drought conditions. If treatments at this scale are completed and repeated over the next several decades, increases in runoff could help offset the current and projected declines in snowpack and stream flow due to warming while improving the resilience of forest stands. As an incidental benefit in an era of dwindling water supplies and projected water shortages, forest thinning could play a role in augmenting river flows on a seasonal basis, improving conditions for water-dependent ecosystems, and benefitting the water supplies of downstream communities. Accelerated forest thinning to reduce water stress and wildfire risk is one of the only management options under our control, and it is probably the most critical to apply over the short term. Because the effects are temporary, it is important to frame accelerated thinning as a reset of ecosystem resilience, not a permanent cure. Rather, accelerated forest thinning creates opportunities to apply other management tools that restore forest resilience at broader scales, whereas management actions at the current pace will likely not be sufficient to address recent and projected changes in forest conditions.

Acknowledgments

We thank Daniel Majka for his early contributions and input into this study. We thank Dan Huebner and Karen Koestner for providing data from the historical Beaver Creek experiments and Shaula Hedwall and Mark Nigrelli for providing data and expertise from the Four Forest Restoration Initiative and the Northern Arizona Wood Supply Study. We thank Kit MacDonald, Chris Nelson, Estella Smith, Rory Steinke, Joe Ringelberg, and Andrea Thode for providing data and expertise of forest thinning exclusions. We thank Julio Bentancourt, Michael Crimmins, and Gregg Garfin for input on climate variability and climate change, Charlie Ester for input on water management in the Salt-Verde River watersheds, and Robert Steidl for advice on statistical analyses. We thank Pat Graham, Molly Hunter, Dale Turner and Sue Sitko for critical review of this paper.

Author Contributions

Conceived and designed the experiments: MR RM FO JH. Performed the experiments: MR FO. Analyzed the data: MR FO. Contributed reagents/materials/analysis tools: MR FO JH. Wrote the paper: MR RM FO ES JH DG.

References

1. Covington WW, Moore MM (1994) Southwestern ponderosa pine forest structure: Changes since Euro-American settlement. Journal of Forestry 92(1): 39–47.
2. Fule PZ, Heinlein TA, Covington WW, Moore MM (2003) Assessing fire regimes on Grand Canyon landscapes with fire-scar and fire-record data. International Journal of Wildland Fire 12: 129–145.
3. Cocke AE, Fule PZ, Crouse JE (2005) Forest change on a steep mountain gradient after extended fire exclusion: San Francisco Peaks, Arizona, USA. Journal of Applied Ecology 42(5): 814–823.
4. Williams AP, Allen CD, Millar CI, Swetnam TW, Michaelsen J, et al. (2011) Forest responses to increasing aridity and warmth in the southwestern United States. Proceedings of the National Academy of Sciences 107(50): 21289–21294.
5. Gitlin AR, Sthultz CM, Bowker MA, Stumpf S, Paxton KL, et al. (2006) Mortality gradients within and among dominant plant populations as barometers of ecosystem change during extreme drought. Conservation Biology 20(5): 1477–1486.
6. Propst DL, Gido KB, Stefferud JA (2008) Natural flow regimes, nonnative fishes, and native fish persistence in arid-land river systems. Ecological Applications 18(5): 1236–1252.
7. Stefferud JA, Gido KB, Propst DL (2011) Spatially variable response of native fish assemblages to discharge, predators and habitat characteristics in an arid-land river. Freshwater Biology 56: 1403–1416.
8. Breshears DD, Cobb NS, Rich PM, Price KP, Allen CD, et al. (2005) Regional vegetation die-off in response to global-change-type drought. Proceedings of the National Academy of Sciences 102(42): 15144–15148.
9. Mote PW (2006) Climate-driven variability and trends in mountain snowpack in western North America. Journal of Climate 19: 6209–6220.
10. Westerling AL, Hidalgo HG, Cayan DR, Swetnam TW (2006) Warming and earlier spring increase Western US forest wildfire activity. Science 313: 940–943.
11. van Mantgem PJ, Stephenson NL, Byrne JC, Daniels LD, Franklin JF, et al. (2009) Widespread increase of tree mortality rates in the Western United States. Science 323: 521–524.
12. Thomas BE, Pool DR (2006) Trends in Streamflow of the San Pedro River, Southeastern Arizona, and Regional Trends in Precipitation and Streamflow in Southeastern Arizona and Southwestern New Mexico. U.S. Geological Survey Professional Paper 1712. Reston, VA: U.S. Geological Survey. 79 p.
13. Timmons D, Springer A (2006) Prescott AMA groundwater flow model update report. Available: http://www.azwater.gov/AzDWR/Hydrology/Modeling/documents/Prescott_Model_Update_Report.pdf. Accessed 6 May 2014.
14. Webb RH, Turner RM, Leake SA. (2007) The Ribbon of Green: Change in Riparian Condition in the Southwestern United States. Tucson, AZ: The University of Arizona Press. 480 p.
15. Marshall RM, Robles MD, Majka DR, Haney JA (2010) Sustainable Water Management in the Southwestern United States: Reality or Rhetoric? PLoS ONE 5(7): e11687. doi:10.1371/journal.pone.0011687.
16. Water Resources Development Commission (2011) Water Resources Development Commission Final Report: Volume II Committee Reports. Phoenix, AZ: Water Resources Development Commission. 685 p.
17. United States Department of Interior Bureau of Reclamation (2012) Colorado River Basin Water Supply and Demand Study: Study Report. Washington DC: U.S. Department of Interior, Bureau of Reclamation. 95p.
18. Vano JA, Udall B, Cayan DR, Overpeck JT, Brekke LD, et al. (2014) Understanding uncertainties in future Colorado River streamflow. Bulletin of the American Meteorological Society 95(1): 59–78.
19. United States Department of Agriculture Forest Service (2013) Draft environmental impact statement for the Four-Forest Restoration Initiative. Flagstaff, AZ: USDA Forest Service, Southwestern Region, Coconino and Kaibab National Forests. 744 p.
20. Brown HE, Baker MB, Rogers JJ, Clary WP, Kovner JL, et al. (1974) Opportunities for increasing water yields and other multiple uses on ponderosa pine forest lands. USDA Forest Service Research Paper RM-129. Fort Collins, CO: U.S. Department of Agriculture, Forest Service, Rocky Mountain Forest and Range Experiment Station. 36 p.
21. Zausen GL, Kolb TE, Bailey JD, Wagner MR (2005) Long-term impacts of stand management on ponderosa pine physiology and bark beetle abundance in northern Arizona: A replicated landscape study. Forest Ecology and Management 218: 291–305.
22. Stephens SL, Moghaddas JJ, Edminster C, Fiedler CE, Haase S, et al. (2009) Fire treatment effects on vegetation structure, fuels, and potential fire severity in western US forests. Ecological Applications 19(2): 305–320.
23. Fule PZ, Crouse JE, Roccaforte JP, Kalies EL (2012) Do thinning and/or burning treatments in western USA ponderosa pine or Jeffrey pine-dominated forests help restore natural fire behavior? Forest Ecology and Management 269: 68–81.
24. Barr GW (1956) Recovering rainfall. Tucson, AZ: Department of Agricultural Economics, University of Arizona. 33 p.
25. Daly C, Halbleib M, Smith JI, Gibson WP, Doggett MK, et al. (2008) Physiographically-sensitive mapping of climatological temperature and precipitation across the conterminous United States. International Journal of Climatology 28(15): 2031–2064.
26. Gesch D, Oimoen M, Greenlee S, Nelson C, Steuck M, et al. (2002) The National Elevation Dataset. Photogrammetric Engineering and Remote Sensing 68(1): 5–11.
27. US Geological Survey, Gap Analysis Program (2010) National Land Cover, Version 1. US Geological Survey. Available: http://gapanalysis.usgs.gov/gaplandcover/viewer/. Accessed 2012 Dec 23.
28. Arizona Geological Survey (2011) Geologic Map of Arizona. Available: http://www.azgs.az.gov/services_azgeomap.shtml. Accessed 2013 Apr 9.
29. Baker MB (2003) Hydrology. In: P . Friederici, editor. Ecological Restoration of Southwestern Ponderosa Pine Forests. Washington DC: Ecological Restoration Institute, Northern Arizona University, Island Press. pp. 161–174.
30. Baker MB (1986) Effects of ponderosa pine treatments on water yield in Arizona. Water Resources Research 22(1): 67–73.
31. Neary D (2011) Managing Arid and Semi-Arid Watersheds: Beaver Creek Watershed. U.S. Department of Agriculture, Forest Service, Rocky Mountain Research Station. Available: http://www.rmrs.nau.edu/watersheds/highlands/watersheds/beavercreek/. Accessed 2012 Oct 29.
32. Rich LR (1972) Managing a ponderosa pine forest to increase water yield. Water Resources Research 8(2): 422–428.
33. Hampton HM, Sesnie SE, Bailey JD, Snider GB (2011) Estimating regional wood supply based on stakeholder consensus for forest restoration in Northern Arizona. Journal of Forestry: 15–26.
34. Haire SL, McGarigal K (2010) Effects of landscape patterns of fire severity on regenerating ponderosa pine forests (*Pinus ponderosa*) in New Mexico and Arizona, USA. Landscape Ecology 25: 1055–1069.
35. Eidenshink J, Schwind B, Brewer K, Zhu Z, Quayle B, et al. (2007) A project for monitoring trends in burn severity. Journal of Association of Fire Ecology 3(1): 3–21.
36. Conservation Biology Institute (2010) Protected Areas Database of the United States, Version 2. Available: http://consbio.org/products/projects/pad-us-cbi-edition. Accessed 2013 Apr 9.
37. United States Department of Agriculture Forest Service (2013) GIS datasets for the Southwestern Region, United States Department of Agriculture, Forest Service. Available: http://www.fs.usda.gov/detail/r3/landmanagement/gis/?cid=stelprdb5202474. Accessed 2013 Apr 9.
38. Turner DS, List MD. 2007. Habitat mapping and conservation analysis to identify critical streams for Arizona's native fish. Aquatic Conservation: Marine and Freshwater Systems 17: 737–748.
39. Arizona Department of Water Resources (2010) Arizona Water Atlas Volume 8: Active Management Areas Water Atlas. Available: http://www.azwater.gov/azdwr/StatewidePlanning/WaterAtlas/ActiveManagementAreas/default.htm. Accessed 2014 May 5.
40. Brown TC, Fogel MM (1987) Use of streamflow increases from vegetation management in the Verde River Basin. Water Resources Bulletin 23(6): 1149–1160.
41. Campbell RE, Baker MB, Ffolliott PF, Larson FR, Avery CC (1977) Wildfire effects on a ponderosa pine ecosystem: An Arizona case study. USDA Forest Service Research Paper RM-191. Fort Collins, CO: U.S. Department of Agriculture, Forest Service, Rocky Mountain Forest and Range Experiment Station. 12 p.
42. Ffolliott PF, Baker MB (1977) Characteristics of Arizona ponderosa pine stands on sandstone soils. USDA Forest Service General Technical Report RM-44. Fort Collins, CO: U.S. Department of Agriculture, Forest Service, Rocky Mountain Forest and Range Experiment Station. 7 p.
43. Pearson GA (1931) Forest types in the Southwest as determined by climate and soil. Technical Bulletin No. 247. Washington DC: U.S. Department of Agriculture, Forest Service, Southwestern Forest and Range Experiment Station. 144 p.
44. Bosch JM, Hewlett JD (1982) A review of catchment experiments to determine the effect of vegetation changes on water yield and evapotranspiration. Journal of Hydrology 55: 3–23.
45. Rich LR, Gottfried GJ (1976) Water yields resulting from treatments on the Workman Creek experimental watersheds in Central Arizona. Water Resources Research 12(5): 1053–1060.
46. Hibbert AR, Gottfried GJ (1987) Streamflow responses to forest treatments on two Arizona mixed conifer watersheds. In: Troendle CA, Kaufmann MR, Hamre RH, Winokur RP editors. Management of subalpine forest: Building on 50 years of research. USDA Forest Service General Technical Report RM-149. Silver Creek, CO: U.S. Department of Agriculture, Forest Service, Rocky Mountain Forest and Range Experiment Station. pp. 189–194.

47. Woodhouse CA, Kunkel KE, Easterling DR, Cook ER (2005) The twentieth-century pluvial in the western United States. Geophysical Research Letters 32: L07701, doi:10.1029/2005GL022413.
48. Woodhouse CA, Overpeck JT (1998) 2000 years of drought variability in the central United States. Bulletin of the American Metereological Society 79(12): 2693–2714.
49. Stahle DW, Cook ER, Cleaveland MK, Therrell D, Meko DM, et al. (2000) Tree-ring data document 16th century megadrought over North America. Eos 81(12): 121–125.
50. Ni F, Cavazos T, Hughes MK, Comrie AC, Funkhouser G (2002) Cool-season precipitation in the southwestern USA since AD 1000: Comparison of linear and nonlinear techniques for reconstruction. International Journal of Climatology 22: 1645–1662.
51. Cook ER, Seager R, Cane MA, Stahle DW (2007) North American drought: Reconstructions, causes, and consequences. Earth-Science Reviews 81: 93–134.
52. Garfin G, Jardine A, Merideth R, Black M, LeRoy S, editors (2013) Assessment of Climate Change in the Southwest United States: A Report Prepared for the National Climate Assessment. A report by the Southwest Climate Alliance. Washington DC: Island Press. 531 p.
53. Ault TR, Cole JE, Overpeck JT, Pederson GT, Meko DM (2014) Assessing the risk of persistent drought using climate model simulations and paleoclimate data. Journal of Climate e-View doi:http://dx.doi.org/10.1175/JCLI-D-12-00282.1.
54. Dominguez F, Rivera E, Lettenmaier DP, Castro CL (2012). Changes in winter precipitation extremes for the western United States under a warmer climate as simulated by regional climate models. Geophysical Research Letters (39) L05803, doi:10.1029/2011GL050762.
55. Murphy KW, Ellis AW (2014) An assessment of the stationarity of climate and stream flow in watersheds of the Colorado River Basin. Journal of Hydrology 509: 454–473.
56. Pierce DW, Barnett TP, Hidalgo HG, Das T, Bonfils C, et al. (2008) Attribution of declining Western U.S. snowpack to human effects. Journal of Climate 21: 6425–6444.
57. Grant GE, Tague CL, Allen CD (2013) Watering the forest for the trees: An emerging priority for managing water in forest landscapes. Frontiers in Ecology and the Environment 11: 314–321.
58. Zou CB, Breshears DD, Newman BD, Wilcox BP, Gard MO, et al. (2008) Soil water dynamics under low-versus high-ponderosa pine tree density: Ecohydrological functioning and restoration implications. Ecohydrology 1: 309–315.
59. Dore S, Montes-Helu M, Hart SC, Hungate BA, Koch GW, et al. (2012) Recovery of ponderosa pine ecosystem carbon and water fluxes from thinning and stand-replacing fire. Global Change Biology 18: 3171–3185.
60. Cochrane MA, Moran CJ, Wimberly MC, Baer AD, Finney MA, et al. (2012) Estimation of wildfire size and risk changes due to fuel treatments. International Journal of Wildland Fire 21(4): 357–267.
61. Kennedy MC, Johnson MC (2014) Fuel treatment prescriptions alter spatial patterns of fire severity around the wildland-urban interface during the Wallow Fire, Arizona, USA. Forest Ecology and Management 318: 122–132.
62. Fule PZ, Laughlin DC (2007) Wildland fire effects on forest structure over an altitudinal gradient, Grand Canyon National Park, USA. Journal of Applied Ecology 44(1): 136–146.
63. Hunter ME, Iniguez JM, Lentile LB (2011) Short- and long-term effects on fuels, forest structure, and wildfire potential from prescribed fire and resource benefit fire in southwestern forests, USA. Fire Ecology 7(3): 108–121.
64. Jaeger KL, Olden JD, Pelland NA (2014) Climate change poised to threaten hydrologic connectivity and endemic fishes in dryland streams. Proceedings of the National Academy of Sciences. doi: 10.1073/pnas.1320890111.
65. Enfield DB, Cid-Serrano L (2006) Projecting the risk of future climate shifts. International Journal of Climatology 26: 885–895.
66. Gangopadhyay S, McCabe GJ (2010) Predicting regime shifts in flow of the Colorado River. Geophysical Research Letters 37: L20706, doi:10.1029/2010GL044513.
67. Sheppard PR, Comrie AC, Packin GD, Angersbach K, Hughes MK (2002) The climate of the Southwest. Climate Research 21: 219–238.
68. Mantua NJ, Hare SR (2002) The Pacific Decadal Oscillation. Journal of Oceanography 58: 35–44.

16

Dynamics, Patterns and Causes of Fires in Northwestern Amazonia

Dolors Armenteras[1]*, Javier Retana[2]

1 Landscape Ecology and Ecosystem Modelling Laboratory, Department of Biology, Sciences Faculty, Colombia National University, Bogotá, Colombia, **2** CREAF i Unitat d'Ecología, Universitat Autònoma de Barcelona, Barcelona, Spain

Abstract

According to recent studies, two widespread droughts occurred in the Amazon basin, one during 2005 and one during 2010. The drought increased the prevalence of climate-driven fires over most of the basin. Given the importance of human-atmosphere-vegetation interactions in tropical rainforests, these events have generated concerns over the vulnerability of this area to climate change. This paper focuses on one of the wettest areas of the basin, Northwestern Amazonia, where the interactions between the climate and fires are much weaker and where little is known about the anthropogenic drivers of fires. We have assessed the response of fires to climate over a ten-year period, and analysed the socio-economic and demographic determinants of fire occurrence. The patterns of fires and climate and their linkages in Northwestern Amazonia differ from the enhanced fire response to climate variation observed in the rest of Amazonia. The highest number of recorded fires in Northwestern Amazonia occurred in 2004 and 2007, and this did not coincide with the periods of extreme drought experienced in Amazonia in 2005 and 2010. Rather, during those years, Northwestern Amazonia experienced a relatively small numbers of fire hotspots. We have shown that fire occurrence correlated well with deforestation and was determined by anthropogenic drivers, mainly small-scale agriculture, cattle ranching (i.e., pastures) and active agricultural frontiers (including illegal crops). Thus, the particular climatic conditions for air convergence and rainfall created by proximity to the Andes, coupled with the presence of one of the most active colonisation fronts in the region, make this region differently affected by the general drought-induced fire patterns experienced by the rest of the Amazon. Moreover, the results suggest that, even in this wet region, humans are able to modify the frequency of fires and impact these historically well preserved forests.

Editor: Ben Bond-Lamberty, DOE Pacific Northwest National Laboratory, United States of America

Funding: No current external funding sources for this study.

Competing Interests: The authors have declared that no competing interests exist.

* E-mail: darmenterasp@unal.edu.co

Introduction

Long-term decreases in annual rainfall, recent drought episodes related to the Amazon basin and the sensitivity of the Amazon rainforest to these drying trends are being discussed extensively [1,2,3,4]. Both the El Niño Southern Oscillation (ENSO) and other sea-surface temperature events, such as the Atlantic Multidecadal Oscillation, have been associated with drought events in Amazonia [3], and evidence has even been found of extensive droughts that were associated with paleo-ENSO events in south-western Amazonia [5]. The rising temperatures and altered rainfall resulting from these drying trends may decrease the levels of humidity in the region, with the potential to transform the rainforest into savannah vegetation and pastures [2].

Fires for land clearing and management are one of the major sources of carbon emissions from Amazonia and play an important role in the vegetation-atmosphere interactions and the hydrology of the area [6]. The extreme drought conditions described above increase the amount of area burned and the frequency of fires, and they produce longer fire seasons [7]; droughts co-occur with peaks of fire activity [4]. The reason for this relationship is that severe droughts increase the flammability of the forests because hydrological factors, including the amount of floodplain area, are significantly affected by drought [3]. This process affects the sensitivity of the forests to dry conditions, increases leaf litter dryness and produces tinder for wildfires [1,8]. In the Amazon basin, recent fires are also typically associated with drought periods [3,5,9]. Indeed, a reported increase in forest fires in the Amazonian Basin has been associated with extreme climatic events, such as the El Niño Southern Oscillation (ENSO) [10,11] or the warm tropical North Atlantic oscillation [5]. Thus, the potentially critical role of fire has been apparent during the recent droughts in Amazonia, with extensive fires spreading from agricultural zones into flammable forests during the droughts of 1997, 1998, 2005 and 2007 [10,12].

Humans also play a major role in determining fire patterns in the earth system [13]. The use of fire for land clearing and management is one of the major threats to neotropical forests and particularly in the Amazon [14]. Current economic and demographic growth has favored increasing demands for agricultural land and timber products from tropical forests [15] and thus has promoted management practices that increase the risk of forest fires. The land use conversion to crops and thus the expansion of the agricultural frontier has been reported as cause of increased deforestation and fire in the Amazon [6,16]. In some cases this expansion has been strongly influenced by illicit crops, especially in remote areas,frequently related to armed conflict and population displacement [17,18]. Fire frequency is also higher in forests under logging and forests that are being cleared for cattle ranching and converted to pastures [19]. Further fire in the Amazon occurs more frequently in fragmented forests, near roads and close to town centres [19,20].

In 2005, below-normal precipitation occurred across most of South America, except in the western and south-western parts of the continent. During the same year, large areas of the Amazon basin experienced one of the most intense droughts recorded in the past 100 years, with Southwestern Amazonia as its epicentre [3,10,21,22]. Current reports indicate that the 2010 Amazonian drought is likely to have been less severe but more spatially extensive than the 2005 drought [4]. However, these drought periods are not homogeneous but vary spatially and temporally throughout the Amazon basin [3]. Typically, the dry season lasts 4–5 months (with <100 mm of rainfall per month) in much of the eastern and southern portions of the basin [22,23], but Northwestern Amazonia is the least likely part of the basin to experience major drought. The 2005 drought was also characterised by extensive fires in the South-western region, and the cumulative number of fires in Amazonia increased by 33% in 2005 relative to that recorded in previous years [3,24].

Studies on drought periods and their linkages with fire patterns have evaluated the whole Amazon and focused on parts of the basin where droughts have been most extensive [6,10,21]. However, the analysis of active fires previously observed in Northwestern Amazonia suggests that the patterns of fire and climate in this part of the basin do not reflect the dynamics of the rest of the basin [25]. Northwestern Amazonia is one of the wettest tropical rainforest regions, with *ca.* 3000 mm of rain per year. Its dry season is shorter and less intense than the dry season in Southwest, Southeast or Central Amazonia [9]. Moreover, different sources indicate that the drought in 2005 was concentrated in western and southern Amazonia and did not have comparably severe effects in northern or eastern Amazonia [5]. On the other hand, in this area the Andes create conditions for air convergence and rainfall that differ from those found in the rest of the basin [26,27]. Moreover, this region is of a particular nature because it is one of the most active colonization fronts in the region through the expansion of the agricultural frontier and the presence of illicit crops, logging activities and pastures [28], a cluster of factors that determines the pattern of fires in most cases [25].

This study compares the spatio-temporal variability of rainfall with that of fires over the past decade in Colombian Amazonia. We also analyse the role of potential drivers of fire occurrence (i.e., deforestation, population, pastures, crops and illicit crops) on the distribution of fires in the region. Given the statements described above, our main hypothesis is that fire incidence will have stronger relationships with human drivers than the fire response to climate variation observed in other parts of the Amazon

Methods

The Amazon basin has a total area of 6.3 million km^2, with 5 million km^2 in Brazil and the remaining area divided among Bolivia, Colombia, Ecuador, Peru and Venezuela. The study area (Fig. 1) corresponds to the Northwestern part of the Amazon, in particular the Colombian Amazon region. This region includes 7% of the entire Amazon Basin, and 40% of it is south of the equator. The population density of the region is still low, but the area is a very active colonisation front [25,29]. The region is highly diverse, owing mainly to differences in geology and geomorphology that produce environments with a great variety of drainage systems and soil qualities. These conditions produce very important differences in ecosystem composition and structure. Several different types of vegetation occur in the region. Lowland forests (<600 m) constitute the first type and can be classified as either riparian or periodically flooded forest (*Várzea* or *Bosques temporalmente inundables*) or moist or permanently flooded forest (*Igapó* or *Bosques Inundables*). Upland forests are similarly differentiated into riparian upland forest (*Campinarana, Bosques de Tierra Firme or Bosques de Colinas*) and montane upland forest (*Piedemonte, Sierra*). The western Amazon is also home to many isolated summits, which support tepui vegetation and montane savannas characterised by high levels of endemism (Tepuis, Pantepui). Various types of aquatic and swamp vegetation complexes are found along the major rivers.

Rainfall data were obtained from the Tropical Rainfall Measuring Mission dataset (TRMM) from January 1998 to February 2011 at 0.25° (=25 km) spatial resolution. A monthly cumulative rainfall dataset was derived for Colombia from the original daily dataset and from the TRMM V6 dataset [30,31]. The cumulative rainfall per three-month period was calculated by summing the values for December (year *i*-1), January and February (year *i*) (hereon referred to as DJF of year *i*). This trimester was chosen because it is the period during which the highest number of fires occurs in the region and generally corresponds with the dry season [25]. Three-month rainfall standardised anomalies in each year *i* ($TRMM_{anomaly,\,i}$) were also quantified per pixel as the difference between the precipitation of year *i* ($TRMM_i$) and the 2001–2010 mean ($TRMM_{2001-2010}$) normalised by the standard deviation ($\sigma_{2001-2010}$) (eq.1; see [10,25]).

$$TRMM_{anomaly,\,i} = (TRMM_i - TRMM_{2001-2010})/\sigma_{2001-2010} \quad (1)$$

where positive values indicate a wetter DJF period in year *i* than the average rainfall for the 2001–2010 DJF period and negative values indicate a drier-than-average DJF period in year *i*.

We used the fire-hotspots series product MODIS, which is included in the Collection 5.1 temporal thermal analysis active-fire dataset [32]. We downloaded the daily dataset ranging from December 2000 to December 2010; this dataset is available at FIRMS (Fire Information for Resource Management System: Archiving and Distributing MODIS Active Fire Data, Collection 5.1 [32]). Active fire detection was based on detecting the thermal signature of fires using a contextual algorithm [33]. We obtained the number of hotspots in each of the 580 0.25° cells considered in the study area. No specific permits were required to access the datasets for these locations.

Deforestation data from 2000 and 2005 were obtained from the official project "Scientific and institutional capacity building to support Reducing Emissions from Deforestation and Degradation (REDD) projects in Colombia" [34]. Information concerning pasture and crop cover was derived from the official cartography of Colombia's continental, coastal and marine ecosystems, which originally used Landsat 2000–2002 images [35] that were available on a 1:500,000 scale. One of the limitations of the study was that we could only use information matching the start of the study period, but the data described above are the only information available for our study period that covered the whole study area. Information regarding the growth of illicit crops in the region was obtained from the global illicit crop-monitoring program of the United Nations Office on Drugs and Crime [36], which is available in Colombia through the Integrated System for Illicit Crops Monitoring project or SIMCI (Sistema Integral de Monitoreo de Cultivos Ilícitos). This system has been consistently monitoring illicit crops in the region since 2000 and provided us

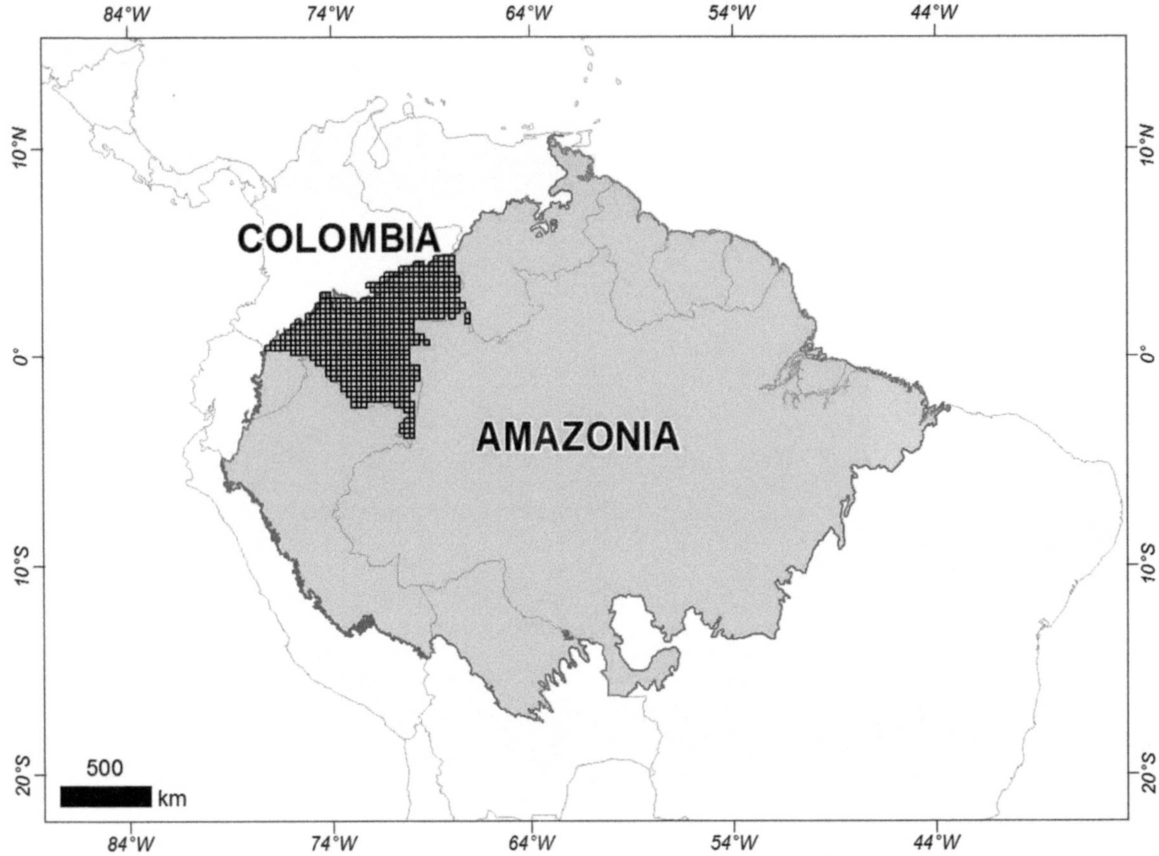

Figure 1. Map of the study area that depicts the limits of Amazonia and the Colombian part of the basin covered in this study.

access to coca survey data for the study area. We used the GIS package ArcGIS (ESRI) to conduct all digital spatial analyses. Finally, we included data on urban population obtained from the 1985, 1993 and 2005 population censuses [37].

The analyses of the variations of numbers of fire hotspots and climatic variables among the different years of study (from 2001 to 2010) were performed with repeated measures ANOVA tests, with the year serving as the repeated measures variable. In these analyses, the sampling units in which we estimated rainfall and fire values were the 580 0.25° cells. Statistical analyses were carried out using STATISTICA 6.0.

The relationships between climate and fire variables were analysed using non-parametric Spearman correlations with two different datasets. First, we analysed the relationship between the number of hotspots, precipitation and anomalies in precipitation in each of the ten years of the study using the values obtained in the 580 cells sampled each year. We also carried out the correlation analyses of fire and climate variables using the mean values per year of the cell values. In this case, the sampling units were the ten years of the study period (from 2001 to 2010).

We also carried out non-parametric Spearman correlations to determine the relationship between the number of hotspots per cell and the following five explanatory variables: % deforestation, % pastures, % crops, surface of coca crops and population. The 580 0.25° cells were again used as the sampling size in these analyses.

Results

Fig. 2 illustrates the spatial distribution of hotspots (a) and precipitation anomalies (b) in the study area for the DJF of each year analysed. In general, these maps show areas with high densities of fires and higher values to the west, towards the Andes. The correlation between the numbers of hotspots per cell in different years was highly significant (correlation values between years ranged between 0.69–0.92), confirming the high recurrence of fires. Slightly different results were found for the precipitation anomalies. A distinct pattern was found for each year. The years 2001, 2003, 2004 and 2010 clearly had more negative precipitation anomalies (i.e., were drier) than 2002, 2005, 2006, 2007, 2008 or 2009. In 2004, the negative precipitation anomalies occurred around the central and the western parts of the study area, whereas in 2010, the negative precipitation anomalies occurred around the northeast of the study area. In 2001, negative anomalies were registered in the central and the western parts of the study area, but in 2003, the drier areas were generally found in the lowland Andes.

The fire hotspots in the study area exhibited strong variability among the years of study (repeated measures ANOVA, $F_{9,5211} = 57.5$, $p<0.001$). The highest number of fire hotspots detected was 18.3±2.1 hotspots per cell (mean±standard error, N = 580 cells) in 2007, followed by 14.5±1.7 in 2004. These two years had exceptionally high values compared with the rest of the period analysed (Fig. 3a).

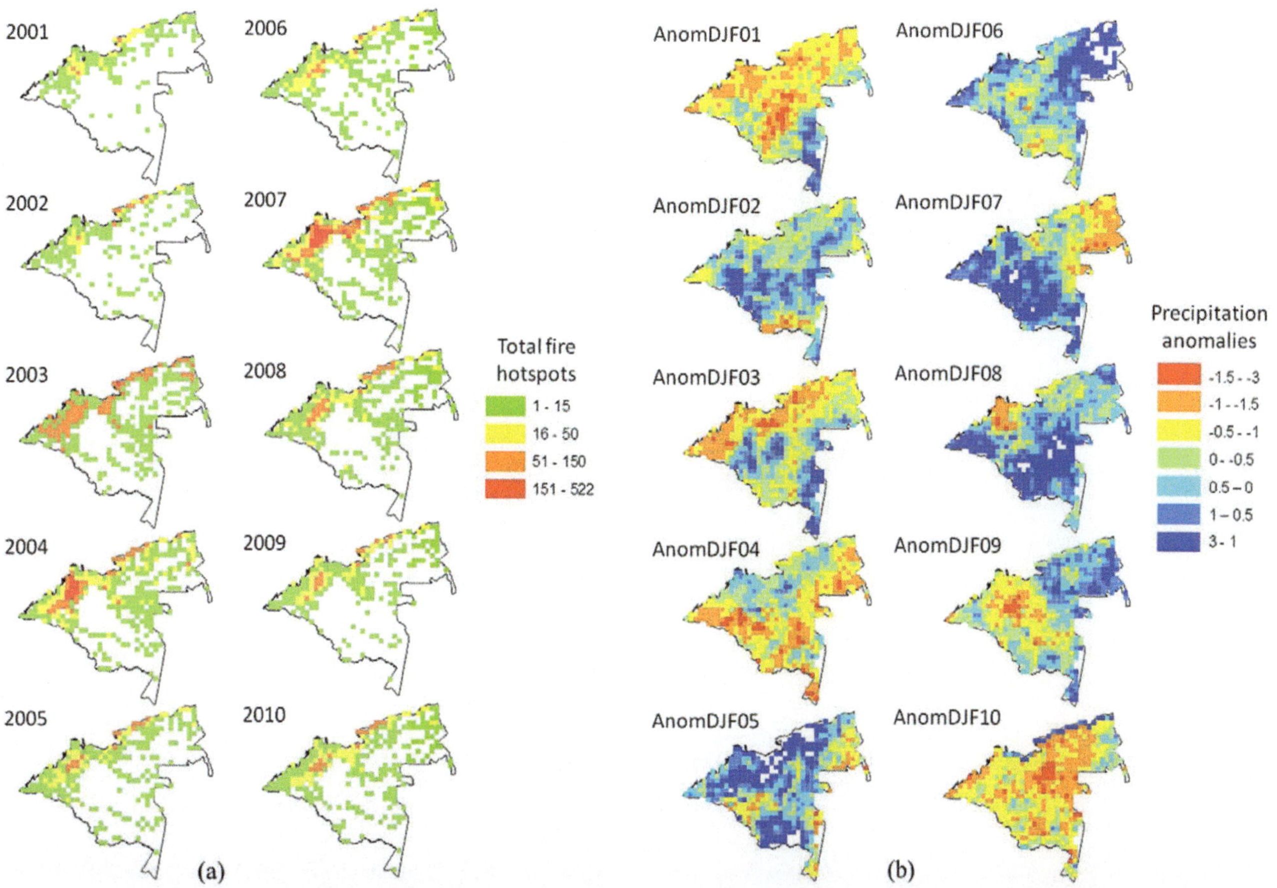

Figure 2. Spatial distribution of the (a) number of fire hotspots and (b) rainfall anomalies in the period including December of the previous year and January and February of the current year (DJF) from 2001 to 2010 in the entire study area.

The temporal patterns of precipitation analysed for the DJF periods between 2001 and 2010 also indicated large interannual variability ($F_{9,5211} = 197.5$, $p<0.001$, Figure 3b). The years 2005, 2006 and 2008 were the wettest years, with 626±11, 639±10 and 634±13 mm of rainfall accumulated in the DJF trimester, respectively. At the other extreme, the years 2001, 2004 and 2010 had only 450±12, 434±8 and 428±9 mm of rainfall, respectively. The same pattern was observed for the rainfall anomalies ($F_{9,5211} = 232.9$, $p<0.001$; Figure 3c). In all cases, the anomalies reported were less than one standard deviation, indicating low deviations from the mean. The most positive anomalies indicated wetter-than-average periods in 2006 (0.55±0.37), 2008 (0.54±0.036) and 2005 (0.59±0.04). In contrast, the driest periods with negative anomalies were found in 2010 (−0.76±0.02), 2004 (−0.61±0.02) and 2001 (−0.61±0.02).

The spatial relationships of the different climate and fire variables in the different years did not show similar patterns. The relationship between the number of fire hotspots per cell and the accumulated rainfall per cell was negative in all ten years considered (Spearman correlation; Rho between −0.15 and −0.55; $p<0.001$ and $n = 580$ in all cases), indicating that the number of fire hotspots was higher in cells with lower precipitation. The relationship between rainfall anomalies and hotspots was generally not consistent among the years of study. Two years showed a positive relationship (2004: Rho = 0.14; 2005: Rho = 0.27; $p<0.05$ and $n = 580$ in both cases), and six years showed a negative relationship (2001: Rho = −0.23; 2002: Rho = −0.17; 2003: Rho = −0.39; 2007: Rho = −0.11; 2008: Rho = −0.20; 2009: Rho = −0.16; $p<0.01$, $n = 580$); in 2006 and 2010, the relationship was not significant. In 9 out of 10 years, the relationship between the accumulated rainfall and the rainfall anomaly was positive (Rho between 0.33 and 0.79; $p<0.01$ and $n = 580$ in all cases), indicating that the anomalies in rainfall were higher in wet cells than in dry cells.

When considering mean annual values of fire and climate variables from all cells, no correlation was found between the number of fire hotspots and either the rainfall or the precipitation anomalies ($p>0.80$ in all cases, $n = 10$ years).

There was strong relationship between the number of hotspots per cell and % deforestation (Rho = 0.72, $p<0.001$, $N = 580$; Figure 4). The other four human-related variables considered showed a high correlation with % deforestation (% pastures: Rho = 0.64, surface of coca crops: Rho = 0.44, %, population: Rho = 0.27, crops: Rho = 0.18; $p<0.001$, $N = 580$ in all four cases), confirming their role on forest loss. The consequence is that the relationship between the number of hotspots per cell and these four explanatory variables was also highly significant and positive: % pastures (Rho = 0.76, $p<0.001$, $N = 580$), surface of coca crops (Rho = 0.64, $p<0.001$, $N = 580$), population (Rho = 0.25, $p<0.001$, $N = 580$) and % crops (Rho = 0.17, $p<0.001$, $N = 580$).

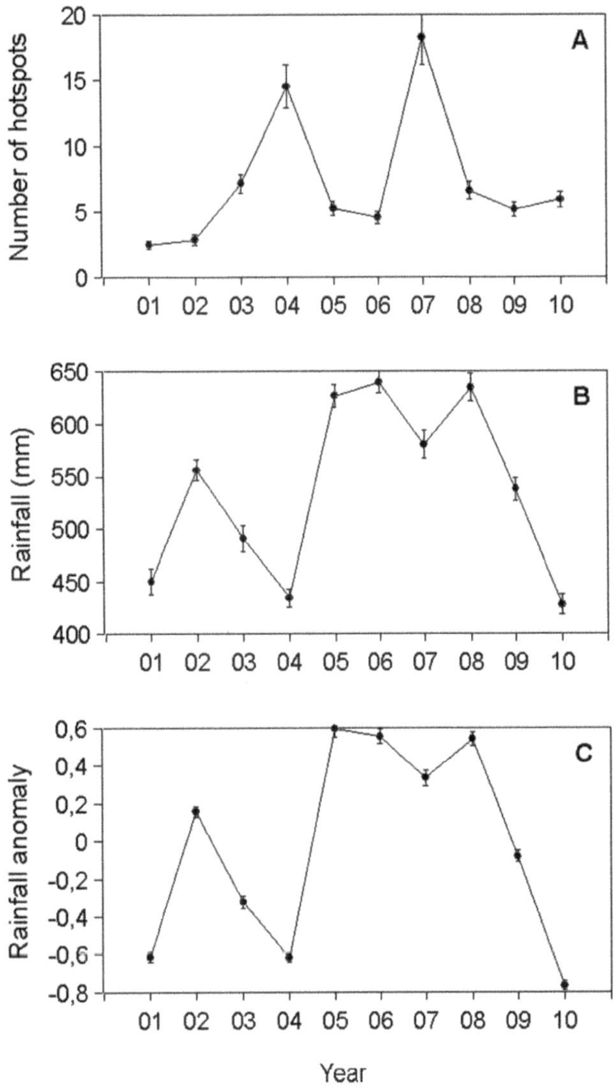

Figure 3. Mean values (±standard error of the mean) of the (a) number of fire hotspots (HP) in the period including December of the previous year and January and February of the current year (DJF); (b) accumulated rainfall in the DJF period (mm); and (c) precipitation anomalies in the DJF period between 2001 and 2010 (without units).

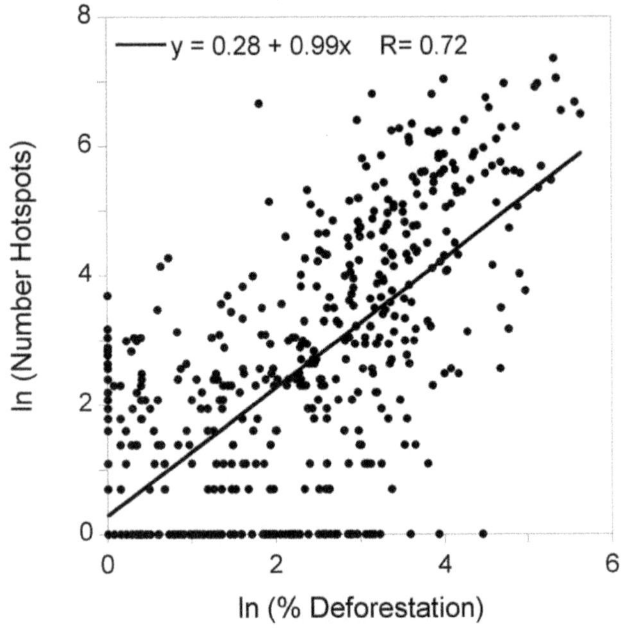

Figure 4. Relationship between the number of hotpots per cel (log-transformed) and % deforestation (log-transformed). N = 580 cells.

Discussion

The spatio-temporal variability in rainfall and fire patterns differs throughout the Amazon basin. Northwestern Amazonia is one of the wettest tropical rainforest regions and does not yet exhibit the strong effects of drought reported for other areas of Amazonia. The Andes, specifically their eastern slope, create positive conditions for air convergence and rainfall in Colombian Amazonia [26]. Our results confirm the findings of Malhi et al. [21] and others [5,10,26] that Northwestern Amazonia has a weaker dry season than southwest, southeast or central Amazonia [9], with very weak (less than 1 standard deviation) anomalies reported for the ten years considered.

The 2005 drought was reported to be more severe in the Southwestern Amazon than in the Central and Eastern region [5]. Precipitation anomalies for the 2005 DJF period in Northwestern Amazon were unlike the drought conditions experienced elsewhere in the basin. On the contrary, the DJF period in this area was quite wet. The 2004 and 2007 years are associated to the ENSO phenomenon, as it has been previously reported for Colombia [25,27]. Precipitation anomalies for these two years were two of the lowest for the period considered (Fig. 3). In 2010, the DJF period was dryer, with more precipitation anomalies. Nevertheless, neither an ENSO dry season as 2004 nor the 2010 DJF period registered rainfall below 400 mm. Dought expansion should be monitored for this region in the future because precipitation anomalies are rarely reported to occur in the north-western part of the Amazon, as this area has been the least affected by historical climate variability [2].

The results of this study reveal that, although there is a relationship between the number of fire hotspots and the accumulated rainfall, no strong interaction exists between climatic events and fire incidence, the latter of which is relatively low when compared to other regions in the Amazon where fire is a major problem and may reach more than 25 counts per 0.25° cell [10]. These results confirm that drought-enhanced fires do not occur equally over the entire Amazon and that Northwestern Amazonia has a pattern particular to its geography. In our analysis (see also [6]), Northwestern Amazonia experienced the highest number of fires in 2004 and 2007, and this did not correspond to the years of highest fire incidence and extreme drought recorded in other parts of Amazonia. These two years were El Niño years and thus ENSO events might be affecting the dynamics of fire more strongly in this part of the basin. In fact during the extreme Amazonian droughts of 2005 and 2010, Northwestern Amazonia exhibited a relatively low number of hotspots and 2005 was one of the wettest periods. Furthermore, there were a large number of fires in 2004, one of the driest periods recorded during our study period but not in other parts of the basin. This finding indicates that fires are more common in Northwestern Amazonia during periods of lower rainfall associated to El Niño events (although we cannot call them strictly drought periods), which is contrary to the association to increased SST in the Atlantic Ocean trends in other subregions of the Amazon (e.g., [3,10]).

There was no relationship between climate and fire occurrence at the annual level, and this was in agreement with the results obtained by Armenteras *et al.* [25]. However, the spatial analysis at the cell level revealed a negative relationship between the number of fire hotspots and rainfall in all ten years tested. This result suggests that fire activity is more severe in the driest areas. In addition, fires show a clear recurrence over the years. The same areas show the greatest fire activity in multiple years (Fig. 3a), probably as a consequence of the impact of social drivers of fire (see below). However, the cell-level relationship between the number of fires and the rainfall anomaly showed contradictory patterns among the years of study. This result confirmed the lack of a clear interaction between fires and climate in the region.

All these results underscore the general notion presented in previous studies [7,25,38] that the pattern of forest loss and fire occurrence in Northwestern Amazonia is not fully limited by the extent of the dry season but rather is influenced much more strongly by socio-economic factors [7]. The process of immigration began recently (in the 1960s) in Amazonia and this region contains currently active colonization and deforestation fronts [28,29]. Many of the active fires detected occurred in areas of transition between the submontane and lowland Andes and Amazonia, where the expansion of the agricultural frontier via the river network is evident [28]. This study provides evidence of fire-driven deforestation, land-use change and colonisation fronts occurring in this part of the Amazon basin.

We have shown that fire occurrence correlates well with anthropogenic drivers. There was a high correlation between fire occurrence and deforestation, indicating that there was a clear association between the two phenomena (see also [4]). Moreover, the other human-related variables considered, which are closely correlated with deforestation, also determined the pattern of fires. In fact, small-scale agriculture, cattle ranching (i.e., pastures) and active agricultural frontiers (including illegal crops) are evident in the region, and fire is the primary tool used to clear forest and maintain areas of pastures and farmlands. Pasture conversion is the primary land use change occurring while human-related fires for the purpose of land clearing and for soil preparation are also common in the area. Recurring fire activity in the process of conversion to pastures has been reported for other parts of the Amazon [8] as one of the strongest land use changes. The conversion of forest to pastures has been recognised as a continuous process across Latin America [19]. Besides the fact that land occupation in the region is based on the clearing of primary forest on pastures, the expansion and intensification of agriculture does not yet exist. This might be partly due to the illicit crop economy, which is a temporary exploitation of a crop strongly linked to armed and illegal groups that frequently leads to land abandonment, resulting in an increase in secondary transformed ecosystems and fire risk.

There is still much to understand regarding the impacts of fires, the exact mechanisms by which fire operates in this part of the Amazon and how activities such as cattle ranching or the selective logging of forests may increase the risks of the fire dispersal into the rainforest. Malhi *et al.* [2] suggest that "rain forests may become seasonally flammable in dry years but without anthropogenic ignition sources fire is a rare occurrence". Thus, obtaining a better understanding of the interaction dynamics between human activities, fire, and climate (as in [20,38] becomes a priority for this area. The results obtained in this study suggest that, even in wet regions such as Northwestern Amazon, where fires were historically rare, humans are able to modify the frequency of fires and impact these well preserved forests (as predicted by [14]. Ultimately (governmental) initiatives to use alternative land management techniques in the colonisation front are urgently required, together with the establishment of consistent and strong fire and land use monitoring programmes and the reinforcement of a sustainable development plan for the region.

Acknowledgments

Thanks are due to Jordi Bosch and Marc Cochrane for their suggestions; to Carol Franco for her continuous help with data handling and the design of the figures; to Graciela Metternich and Silvia Giada (UNEP/ROLAC) and Adolfo Kindgard (Facultad de Agronomía UBA) for providing the Amazonian border datasets; to NASA FIRMS for providing datasets on precipitation and on active fires; to the Instituto de Hidrología, Meteorología y Estudios Ambientales (IDEAM) and its team members Ederson Cabrera, Maria Claudia García, Maria Fernanda Ordoñez, Gustavo Galindo and Diana Vargas for providing the deforestation data maps from the national Colombia REDD (Reducing Emissions from Deforestation and Degradation) project; and to the ever-generous team of the United Nations Office on Drugs and Crime program (UNDOC, 2006), in particular we are very grateful to Leonardo Correa and Orlando González. for providing us with access to data on illicit crops. Finally we want to thank the anonymous reviewer who challenged and pushed us in the revision process and helped us to considerably improve this manuscript.

Author Contributions

Conceived and designed the experiments: DA, JR. Performed the experiments: DA JR. Analyzed the data: DA JR. Contributed reagents/materials/analysis tools: DA JR. Wrote the paper: DA JR.

References

1. Li W, Fu R, Negrón Juárez RI, Fernandes K (2008) Observed change of the standardized precipitation index, its potential cause and implications to future climate change in the Amazon region. Philos T Roy Soc B 363: 1767–1772.
2. Malhi Y, Roberts JT, Betts RA, Killeen TJ, Li W, et al. (2008) Climate change, deforestation and the fate of the Amazon. Science 319: 169–172.
3. Asner GP, Alencar A (2010) Drought impacts on the Amazon forest: the remote sensing perspective. New Phytol 187: 569–578.
4. Lewis SL, Brando PM, Phillips OL, van der Heijden GMF, Nepstad D (2011) The 2010 Amazon Drought. Science 331: 554.
5. Marengo JA, Nobre CA, Tomasella J, Cardoso MF, Oyama MD (2008) Hydro-climatic and ecological behaviour of the drought of Amazonia in 2005. Philos T Roy Soc B 363: 1773–1778.
6. Morton DC, Defries RS, Randerson JT, Giglio L, Schroeder W, et al. (2008) Agricultural intensification increases deforestation fire activity in Amazonia. Glob Change Biol 14: 2262–2275.
7. van der Werf GR, Randerson JT, Giglio L, Gobron N, Dolman AJ (2008) Climate controls on the variability of fires in the tropics and subtropics. Global Biogeochem Cy 22: GB3028. doi:10.1029/2007GB003122.
8. Giles J (2006) Drought of 2005 is a taste of things to come. Nature 442: 726–727.
9. Malhi Y, Wright J (2004) Spatial patterns and recent trends in the climate of tropical rainforests regions. Philos T Roy Soc B 359: 311–329.
10. Aragao LEOC, Malhi Y, Roman-Cuesta RM, Saatchi S, Anderson LO, et al. (2007) Spatial patterns and fire response of recent Amazonian droughts. Geophys Res Lett 34: L07701.
11. Ray D, Nepstad D, Moutinho P (2005) Micrometeorological and canopy controls of fire susceptibility in a forested Amazon landscape. Ecol App 15: 1664–1678.
12. Alencar AAC, Solorzano LA, Nepstad DC (2004) Modeling forest understory fires in an Eastern Amazonian landscape. Ecol Appl 14: S139–S149.
13. Bowman DMJS, Balch JK, Artaxo P, Bond WJ, Carlson JM, et al. (2009) Fire in the Earth System. Science 324: 481–484.
14. Silvestrini RA, Soares-Filho BS, Nepstad D, Coe M, Rodrigues H, et al. (2011) Simulating fire regimes in the Amazon in response to climate change and deforestation. Ecol Appl 21: 1573–1590.
15. Rudel TK, Defries R, Asner GP, Laurance WF (2009) Changing drivers of deforestation and new opportunities for conservation. Conserv Biol 23(6): 1396–1405.

16. Metzger JP (2002) Landscape dynamics and equilibrium in areas of slash-and-burn agriculture with short and long fallow period (Bragantina region, NE Brazilian Amazon). Landscape Ecol 17: 419–431.
17. Dávalos LM, Bejarano AC, Hall MA, Correa HL, Corthals AP, et al. (2011) Forests and drugs: coca-driven deforestation in global biodiversity hotspots. Environ Sci Technol 45: 1219–1227.
18. Etter A, McAlpine C, Pullar D, Possingham H (2005) Modeling the age of tropical moist forest fragments in heavily-cleared lowland landscapes of Colombia. Forest Ecol and Manag 208(1–3): 249–260.
19. Wassenaar T, Gerber P, Verburg PH, Rosales M, Ibrahim M, et al. (2007) Projecting land use changes in the Neotropics: The geography of pasture expansion into forest. Global Environmen Chan 17: 86–104.
20. Cochrane MA (2003) Fire science for rainforests. Nature 421: 913–919 21.
21. Malhi Y, Aragao LEOC, Galbraith D, Huntingford C, Fisher R, et al. (2009) Exploring the likelihood and mechanism of a climate-change-induced diebak of the Amazon rainforest. P Natl Acad Sci USA 106: 20610–20615.
22. Marengo JA (1992) Interannual variability of surface climate in the Amazon Basin. Int J Climatol 12: 853–863.
23. Marengo JA, Tomasella J, Uvo CR (1998) Trends in streamflow and rainfall in tropical South America: Amazonia, eastern Brazil, and Northwestern Peru. J Geophys Res 103: 1775–1784.
24. Brown IF, Schroeder W, Setzer A, de Los Rios Maldonado M, Pantoja N, et al. (2006) Monitoring Fires in Southwestern Amazonia Rain Forests. EOS Transactions 87: 253–264.
25. Armenteras-Pascual DA, Retana-Alumbreros J, Molowny-Horas R, Roman-Cuesta RM, Gonzalez-Alonso F, et al. (2011) Characterising fire spatial pattern interactions with climate and vegetation in Colombia. Agr Forest Meteorol 151: 279–289.
26. Espinoza JC, Ronchail J, Guyot JL, Cochonneau G, Naziano F, et al. (2009) Spatio-temporal rainfall variability in the Amazon basin countries (Brazil, Peru, Bolivia, Colombia, and Ecuador). Int J Climatol 29: 1574–1594.
27. Poveda G, Waylen PR, Pulwarty RS (2006) Annual and inter-annual variability of the present climate in Northern South America and southern Mesoamerica. Palaeogeogr Palaeocl 234: 3–27.
28. Rodríguez N, Armenteras D, Molowny-Horas R, Retana J (2011) Patterns and trends of forest loss in the Colombian Guyana. Biotropica 44: 123–132.
29. Armenteras D, Rudas G, Rodríguez N, Sua S, Romero M (2006) Patterns and causes of deforestation in the Colombian Amazon. Ecol Indic 6: 353–368.
30. Kummerow C, Barnes W, Kozu T, Shiue J, Simpson J (1998) The Tropical Rainfall Measuring Mission (TRMM) sensor package. J Atmos Ocean Tech 15: 809–817.
31. National Aeronautics and Space Administration – NASA (2009) Monthly 0.25°×0.25° TRMM and other sources rainfall. NASA Distributed Active Archive Center, Goddard Space Flight Center Earth Science, Greenbelt, MD. Available: http://disc.sci.gsfc.nasa.gov/precipitation/.
32. Davies DK, Ilavajhala S, Wong MM, Justice CO (2009) Fire Information for Resource Management System: Archiving and Distributing MODIS Active Fire Data. IEEE T Geosci Remote 47: 72–79.
33. Giglio L, Descloitres J, Justice CO, Kaufman YJ (2003) An Enhanced Contextual Fire Detection Algorithm for MODIS. Remote Sens Environ 87: 273–282.
34. Instituto de Hidrología, Meteorología y Estudios Ambientales - IDEAM (2011) Memoria Técnica Cuantificación de la Deforestación histórica para Colombia a nivel nacional, escalas gruesa y fina. Bogotá. 85 p. In press.
35. IDEAM, IGAC, IAvH, Invemar, Sinchi, et al. (2007) Ecosistemas continentales, costeros y marinos de Colombia. Instituto de Hidrología, Meteorología y Estudios Ambientales, Instituto Geográfico Agustín Codazzi, Instituto de Investigación de Recursos Biológicos Alexander von Humboldt, Instituto de Investigaciones Marinas y Costeras José Benito Vives de Andréis, Instituto Amazónico de Investigaciones Científicas Sinchi e Instituto de Investigaciones Ambientales del Pacífico Jhon von Neumann. Bogotá, D.C.
36. United Nations Office on Drugs and Crime (2010) Integrated System for Illicit Crops Monitoring project.
37. Departamento Administrativo Nacional de Estadística (1985,1993,2005) National Census of population for 1985, 1993 y 2005. Colombia.
38. Aragao LEOC, Malhi Y, Barbier N, Lima A, Shimabukuro Y, et al. (2008) Interactions between rainfall, deforestation and fires during recent years in the Brazilian Amazonia. Philos T Roy Soc B 363: 1779–1785.

17

microRNAs Associated with Drought Response in the Bioenergy Crop Sugarcane (*Saccharum* spp.)

Thaís Helena Ferreira[1☯], Agustina Gentile[1☯], Romel Duarte Vilela[2], Gustavo Gilson Lacerda Costa[3], Lara Isys Dias[1], Laurício Endres[2], Marcelo Menossi[1*]

1 Laboratório de Genoma Funcional, Departamento de Genética, Evolução e Bioagentes, Instituto de Biologia, Universidade Estadual de Campinas. Campinas, São Paulo, Brazil, **2** Centro de Ciências Agrárias, Universidade Federal de Alagoas. Rio Largo, Alagoas, Brazil, **3** Laboratório Central de Tecnologias de alto desempenho, Universidades Estadual de Campinas. Campinas, São Paulo, Brazil

Abstract

Sugarcane (*Saccharum* spp.) is one of the most important crops in the world. Drought stress is a major abiotic stress factor that significantly reduces sugarcane yields. However the gene network that mediates plant responses to water stress remains largely unknown in several crop species. Although several microRNAs that mediate post-transcriptional regulation during water stress have been described in other species, the role of the sugarcane microRNAs during drought stress has not been studied. The objective of this work was to identify sugarcane miRNAs that are differentially expressed under drought stress and to correlate this expression with the behavior of two sugarcane cultivars with different drought tolerances. The sugarcane cultivars RB867515 (higher drought tolerance) and RB855536 (lower drought tolerance) were cultivated in a greenhouse for three months and then subjected to drought for 2, 4, 6 or 8 days. By deep sequencing of small RNAs, we were able to identify 18 miRNA families. Among all of the miRNAs thus identified, seven were differentially expressed during drought. Six of these miRNAs were differentially expressed at two days of stress, and five miRNAs were differentially expressed at four days. The expression levels of five miRNAs (ssp-miR164, ssp-miR394, ssp-miR397, ssp-miR399-seq 1 and miR528) were validated by RT-qPCR (quantitative reverse transcriptase PCR). Six precursors and the targets of the differentially expressed miRNA were predicted using an *in silico* approach and validated by RT-qPCR; many of these targets may play important roles in drought tolerance. These findings constitute a significant increase in the number of identified miRNAs in sugarcane and contribute to the elucidation of the complex regulatory network that is activated by drought stress.

Editor: Baohong Zhang, East Carolina University, United States of America

Funding: This study was funded by the following Brazilian funding institutions: FINEP (The Financiadora de Estudos e Projetos), FAPESP (São Paulo Research Foundation), CNPq (The National Council for Scientific and Technological Development) and CAPES (Co-ordination of Improvement of Higher Education Personnel). The funders had no role in study design, data collection and analysis, decision to publish, or preparation of the manuscript.

Competing Interests: The authors have declared that no competing interests exist.

* E-mail: menossi@unicamp.br

☯ These authors contributed equally to this work.

Introduction

miRNAs are a class of small, non-coding RNAs, approximately 21 nucleotides in length, that are endogenous to both plants and animals [1,2] and function to regulate gene expression by sequence-specific interaction with target mRNAs [3,4]. In plants, each primary miRNA is transcribed by the RNA polymerase II enzyme [5] and forms an imperfect foldback structure that is processed into a stem-loop precursor (pre-miRNA) by Dicer-like 1 (DCL1), a nuclear RNase III-like enzyme. The pre-miRNA is then cleaved into a double-stranded RNA duplex, called miRNA/miRNA*, by the same enzyme [3,6,7]. One of the strands, called the mature miRNA, is incorporated into the RNA-induced silencing complex (RISC), whereas the other strand, called miRNA*, is usually degraded. The incorporated mature miRNA guides the RISC to a target mRNA by base pairing, leading to mRNA cleavage or translational repression [3,4,7].

Plants miRNAs were initially described in *Arabidopsis thaliana* [7] and since then, an increasing number of miRNAs have been identified in plants. miRNAs are known to be involved in developmental and physiological processes such as flowering, leaf and flower differentiation and the auxin response [2,8]. Recently, the expression of a number of miRNAs has been found to be sensitive to abiotic and biotic stresses [9,10,11,12,13,14,15], with evidence that several play roles in drought stress [9,16]. To date, 34 sugarcane miRNAs are reported (http://www.mirbase.org). Considering that drought stress significantly reduces sugarcane yields [17], the identification of sugarcane miRNAs that respond to this stress will help to elucidate the molecular basis of drought stress tolerance in this important bioenergy crop.

Deep sequencing strategies have revolutionized the discovery of small RNAs, constituting the most effective method for plant miRNA detection [18]. This strategy has allowed the discovery of species-specific and miRNAs that are expressed at very low levels [19,20]. Because miRNAs are conserved across species, bioinformatics strategies [21,22] based on expressed sequence tags (ESTs) from sugarcane and genomic survey sequences (GSSs) from sorghum were used in this study and homology and structural similarity. In this work, our goal was to identify and characterize miRNAs that may be regulated by water deficit in sugarcane.

Results

Molecular markers for drought in sugarcane plants

Two sugarcane cultivars differing in their tolerance to drought stress, RB867515 (higher tolerance, HT) and RB855536 (lower tolerance, LT), were grown in the greenhouse for three months. Water was withheld for 2, 4, 6 or 8 days, and stress symptoms, such as leaf rolling and senescence, appeared early on the second day. By the sixth day, almost all of the stressed plants were severely affected (Supporting Information, Figure S1 and Figure S2). Based on these results, we decided to evaluate the plants that were stressed at two and four days.

To characterize the degree of stress at the molecular level, we performed RT-qPCR (quantitative reverse transcriptase PCR) amplification of a sugarcane gene encoding a dehydrin (Fig. 1). The sugarcane assembled sequence (SAS) SCQGLR1085F11.g is induced by drought in sugarcane [23], and several homologues have been reported to be modulated by this stress in other species [24,25,26,27]. This gene was induced after two and four days of stress (Fig. 1) in both sugarcane varieties. However, the cultivar with lower tolerance displayed greater induction at both experimental time points.

Transcriptome analysis of small RNAs in sugarcane

To identify the miRNAs involved in drought stress in sugarcane, eight small RNA libraries from mature leaves were sequenced using the Solexa technology. These libraries were representative of plants stressed for two and four days and control plants from both sugarcane cultivars. A total of 90 million reads were obtained, ranging from 8 to 15 million reads per library (Table 1), similar to the available sequencing data for citrus, sorghum and maize [19,28,29]. All Solexa reads were aligned against the GenBank (http://www.ncbi.nlm.nih.gov/) and RFam (http://rfam.sanger.ac.uk/) databases to sort the reads into categories (Table 1). All of the sequences were then regarded as miRNAs for further analysis.

The size distribution of all of the sequences is presented in Fig. 2. The majority of the reads were 21 to 24 nt in length for all libraries, with 21 nt being the most redundant species, followed by 24 nt (Fig. 2). However, sequences of 24 nt were the most represented class of non-redundant species (Fig. 3). In the 21 nt reads, approximately 80% was composed of U or A at the first base position at the 5′end in all libraries (Fig. 4), with equal number of each nucleotide.

Identification of new potential miRNAs in *Saccharum* spp.

To identify sugarcane miRNA candidates within our sequenced set, unique small RNA species were aligned against the *Sorghum bicolor* genome to identify loci corresponding to putative miRNA precursors. A total of 21 miRNAs were detected corresponding to 18 miRNAs families in sugarcane: miR156, miR160, miR164, miR166, miR167, miR169, miR171, miR172, miR319, miR390, miR393, miR394, miR396, miR397, miR399, miR528, miR529 and miR1432 (Table 2). The sugarcane miRNAs were named based on their homology to sorghum miRNAs.

Precursors were also identified among the sugarcane ESTs (Table 3 and Fig. 5) derived from six genomic loci because six different SASs were found to correspond to six different miRNA genes (Table 3) and their RNA sequences have the intramolecular capacity to fold into hairpin structures (Fig. 5). Five precursor sequences were from the SUCEST (http://sucest-fun.org/) database, whereas only one was found in the SoGI (http://compbio.dfci.harvard.edu/) database (Table 3). Two distinct SASs were found for the ssp-miR167 precursor. The MFEs (minimum free energy) ranged from −145 to −338 kcal/mol, and the G/C content ranged from 42 to 49% (Table 3).

Analysis of sugarcane miRNAs modulated by drought stress

Among all of the miRNAs identified, seven were differentially expressed during drought (Fig. 6), of which six were differentially expressed after two days of stress (Table 4) and five were differentially expressed after four days of stress (Table 5). The ssp-miR164 miRNA was differentially expressed only at two days of stress. Additionally, three miRNAs (ssp-miR164, ssp-miR397 and ssp-miR528) were up-regulated in the RB867515 (higher tolerance, HT) cultivar, and none were down-regulated by drought. In the RB855536 (lower tolerance, LT) cultivar, four miRNAs (ssp-miR164, ssp-miR394, ssp-miR399-seq 1 and ssp-miR1432) were down-regulated, and only one (ssp-miR397) was up-regulated by drought. Only ssp-miR397 displayed the same pattern in both cultivars, as it was induced after two days of water stress (Table 4).

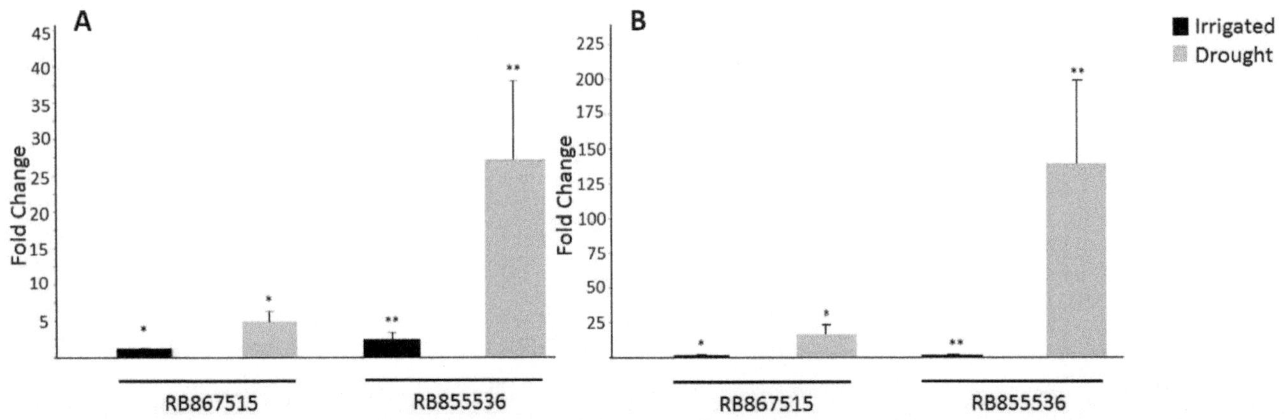

Figure 1. Real-time PCR of a sugarcane gene encoding a dehydrin (SCQGLR1085F11.g). RB867515 (higher drought tolerance) and RB855536 (lower drought tolerance) plants were irrigated (black bars) or subjected to drought stress by withholding irrigation (gray bars) for two (A) and four (B) days. Error bars represent the standard error (n = 4). * $p<0.07$ and ** $p<0.05$. Statistics was calculated between irrigated and drought treatments in each cultivar using the permutation mean test. The expression in irrigated RB867515 plants was considered as 1.

Table 1. Small RNA deep-sequencing data from sugarcane leaves.

Category	HTD2				HTD4			
	Unique RNAs	Percent (%)	Total RNAs	Percent (%)	Unique RNAs	Percent (%)	Total RNAs	Percent (%)
antisense exon	12,762	0.50	108,151	1.22	15,312	0.47	86,674	0.65
sense exon	76,415	2.98	651,383	7.34	139,293	4.24	814,958	6.10
miRNA	23,950	0.94	1,375,593	15.49	26,926	0.82	3,467,791	25.95
rRNA	58,355	2.28	925,787	10.43	63,945	1.95	866,844	6.49
siRNA	59,384	2.32	370,579	4.17	61,333	1.87	552,249	4.13
snRNA	2,094	0.08	7,456	0.08	4,056	0.12	39,774	0.30
snoRNA	780	0.03	1,516	0.02	1,487	0.05	4,854	0.04
tRNA	13,254	0.52	549,361	6.19	19,375	0.59	1,399,409	10.47
unannotated	2,313,810	90.35	4,888,685	55.06	2,951,339	89.90	6,128,310	45.87
total small RNAs	2,560,804		8,878,511		3,283,066		13,360,863	
Category	**HTI2**				**HTI4**			
	Unique RNAs	**Percent (%)**	**Total RNAs**	**Percent (%)**	**Unique RNAs**	**Percent (%)**	**Total RNAs**	**Percent (%)**
antisense exon	10,400	0.49	102,953	1.21	16,899	0.44	131,528	0.87
sense exon	77,102	3.66	831,951	9.78	76,252	2.00	628,631	4.18
miRNA	19,404	0.92	1,310,300	15.40	32,423	0.85	4,305,127	28.62
rRNA	65,322	3.10	1,214,350	14.28	61,239	1.60	861,743	5.73
siRNA	50,116	2.38	286,130	3.36	88,861	2.33	763,503	5.08
snRNA	1,904	0.09	6,942	0.08	2,382	0.06	7,433	0.05
snoRNA	878	0.04	1,897	0.02	809	0.02	1,766	0.01
tRNA	14,825	0.70	788,156	9.27	13,975	0.37	520,672	3.46
unannotated	1,867,650	88.61	3,963,113	46.59	3,528,379	92.34	7,819,918	51.99
total small RNAs	2,107,601		8,505,792		3,821,219		15,040,321	
Category	**LTD2**				**LTD4**			
	Unique RNAs	**Percent (%)**	**Total RNAs**	**Percent (%)**	**Unique RNAs**	**Percent (%)**	**Total RNAs**	**Percent (%)**
antisense exon	13,525	0.52	130,805	1.30	17,226	0.59	154,743	1.17
sense exon	96,298	3.72	1,044,692	10.38	76,303	2.62	735,969	5.58
miRNA	22,271	0.86	1,501,828	14.93	22,744	0.78	3,483,293	26.42
rRNA	77,634	3.00	1,521,502	15.12	60,765	2.09	926,250	7.03
siRNA	58,478	2.26	399,760	3.97	63,817	2.19	587,248	4.45
snRNA	2,755	0.11	11,421	0.11	2,464	0.08	8,252	0.06
snoRNA	1,179	0.05	2,660	0.03	757	0.03	2,020	0.02
tRNA	17,741	0.69	797,948	7.93	14,831	0.51	727,495	5.52
unannotated	2,298,296	88.80	4,649,978	46.22	2,652,688	91.11	6,557,000	49.74
total small RNAs	2,588,177		10,060,594		2,911,595		13,182,270	
Category	**LTI2**				**LTI4**			
	Unique RNAs	**Percent (%)**	**Total RNAs**	**Percent (%)**	**Unique RNAs**	**Percent (%)**	**Total RNAs**	**Percent (%)**
antisense exon	14,157	0.64	131,400	1.40	14,757	0.55	161,302	1.33
sense exon	81,911	3.72	792,014	8.44	67,798	2.52	506,603	4.17
miRNA	20,041	0.91	1,515,240	16.15	23,936	0.89	3,682,698	30.32
rRNA	65,089	2.96	1,150,319	12.26	54,672	2.03	676,834	5.57
siRNA	63,279	2.87	469,568	5.00	68,251	2.53	664,751	5.47
snRNA	2,849	0.13	13,901	0.15	2,362	0.09	8,234	0.07
snoRNA	1,180	0.05	3,190	0.03	973	0.04	2,283	0.02
tRNA	14,087	0.64	736,610	7.85	13,576	0.50	415,579	3.42
unannotated	1,939,448	88.08	4,571,452	48.72	2,447,018	90.85	6,028,355	49.63

Table 1. Cont.

Category	LTI2				LTI4			
	Unique RNAs	Percent (%)	Total RNAs	Percent (%)	Unique RNAs	Percent (%)	Total RNAs	Percent (%)
total small RNAs	2,202,041		9,383,694		2,693,343		12,146,639	

HTD: RB867515 (higher drought tolerance) plants under drought stress (without watering). HTI: RB867515 plants under irrigation. LTD: RB855536 (lower drought tolerance) under drought stress. LTI: RB855536 (lower drought tolerance) plants under irrigation. 2: two days of treatment. 4: four days of treatment.

ssp-miR393 was only differentially expressed at four days of stress. Four miRNAs (ssp-miR394, ssp-miR397, ssp-miR399-seq 1 and ssp-miR528) were down-regulated by drought in the HT cultivar. Two other miRNAs (ssp-miR399-seq 1 and ssp-miR528) were down-regulated, and one (ssp-miR393) was up-regulated in the LT cultivar. Only miRNAs ssp-miR399-seq 1 and ssp-miR528 were down-regulated in both cultivars (Table 5).

Four miRNAs (ssp-miR394, ssp-miR397, ssp-miR399-seq 1 and miR528) were differentially expressed at both time points during the stress period, and only one miRNAs (ssp-miR399-seq 1) presented similar expression profiles in the LT cultivar (Table 4 and 5).

Stem-loop RT-qPCR [30,31] was used to validate the expression of all seven differentially expressed miRNAs during drought according to the sequencing data. The expression patterns of five miRNAs (ssp-miR164, ssp-miR394, ssp-miR397, ssp-miR399-seq 1 and ssp-miR528) were validated by this approach. Among the 15 differentially expressed profiles observed with the sequencing approach (Fig. 6), five (33.33%) displayed the same expression profile using RT-qPCR (Fig. 7). The RT-qPCR analysis confirmed two out of eight (25%) profiles after two days of stress and four out of seven (57%) after four days. In the HT cultivar, five out of seven (71.4%) profiles were confirmed, whereas in the LT cultivar, no profile was confirmed. In all of the RT-qPCR profiles, the miRNA levels of the rehydrated plants demonstrated a tendency to return to the control levels (Fig. 7).

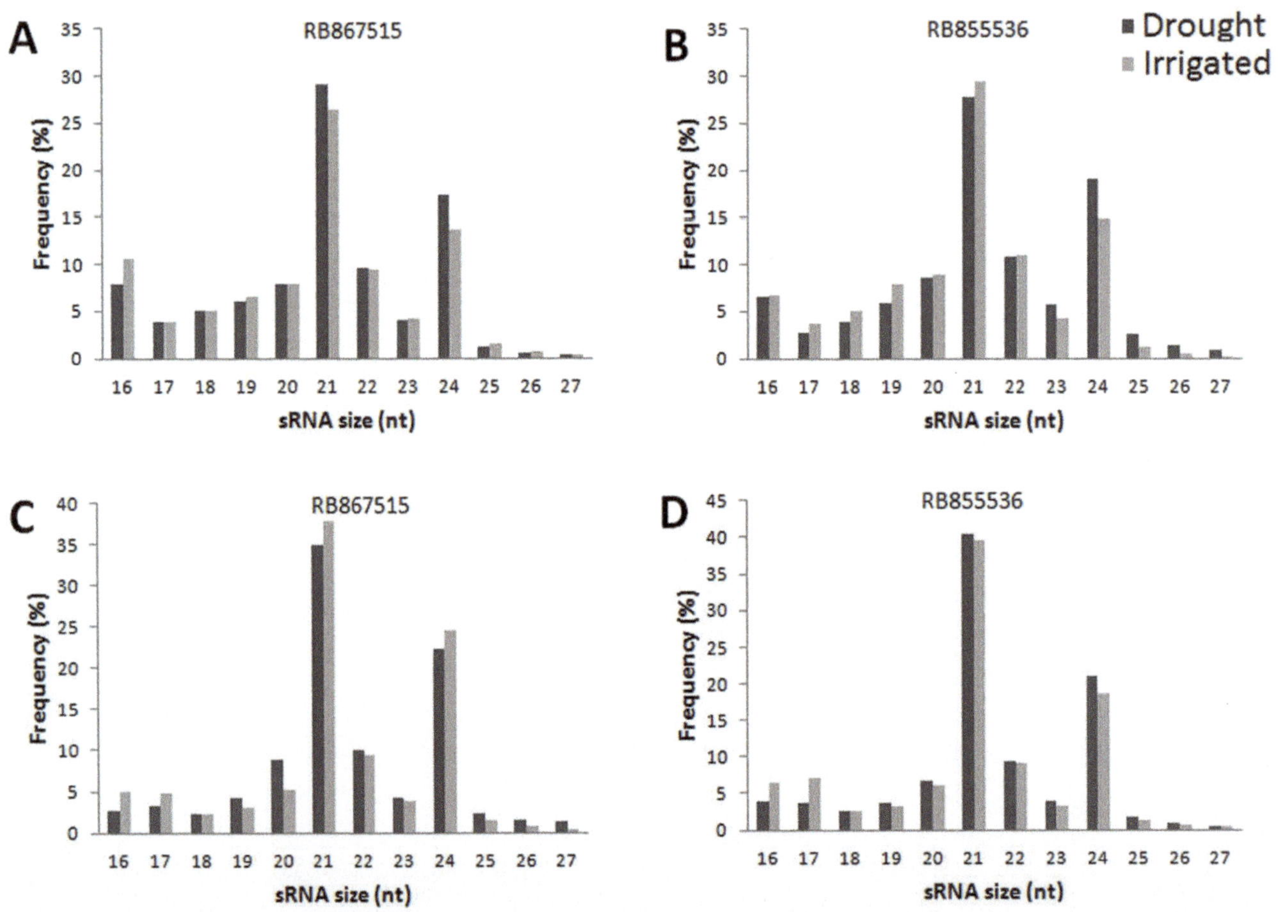

Figure 2. Size distribution of small RNA (sRNA) redundant sequences in two sugarcane cultivars. Cultivars RB867515 (A and C) and RB855536 (B and D) were irrigated (gray bars) or drought-stressed (black bars) for two (A and B) and four (C and D) days. RB867515 is the higher drought tolerant genotype, and RB855536 is the lower drought tolerant genotype.

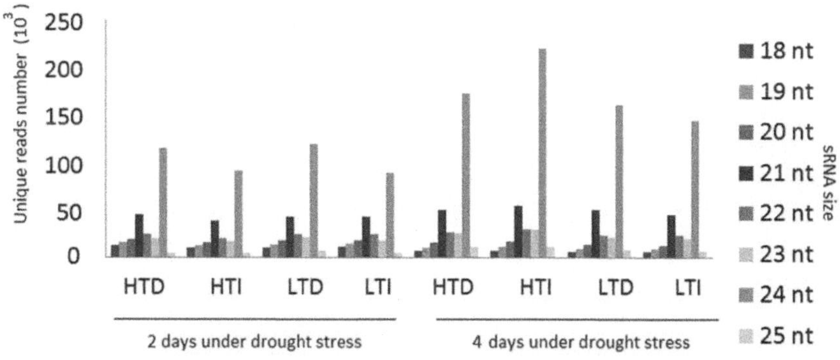

Figure 3. Size distribution of small RNA (sRNA) non-redundant sequences in two sugarcane cultivars. HTD: RB867515 (higher tolerance cultivar) plants without watering. HTI: RB867515 plants under irrigated conditions. LTD: RB855536 (lower tolerance cultivar) plants without watering. LTI: RB855536 plants under irrigated conditions.

Prediction of miRNA targets

The mature sequences of the seven miRNAs modulated by drought were used to search for their targets in sugarcane (Table 6 and Supporting information, Table S1). All of these miRNAs had putative targets in the SUCEST database (Table 6), several of which encode transcription factors (ssp-miR164, ssp-miR394, ssp-miR528 and ssp-miR1432), growth or development regulators (ssp-miR393), proteins associated with floral development (ssp-miR394) and several phosphatases, kinases, and oxidases (ssp-miR394, ssp-miR397, ssp-miR399-seq 1, ssp-miR528 and ssp-miR1432), among others (Supporting Information, Table S1). The targets had homologous proteins in several plant species, especially maize, rice and *Brachypodium*, suggesting that the pathways where these targets act may be conserved.

Target validation by RT-qPCR

RT-qPCR was used to validate the expression of seven putative target genes, one for each of the miRNAs modulated by drought (Supporting Information, Figure S3). To correlate the expression profiles of miRNAs and target genes we calculated the expression ratios between drought-stressed and control plants (Table 7). Considering the 15 treatments where miRNA expression were statistically significant ($P<0.05$, Fig. 6), we found that in seven treatments target genes had the expected profile, i.e., induced expression of miRNA and repressed expression of the target gene and vice-versa (Table 7, underlined ratios). In two out the seven target profiles, RT-qPCR expression was also statistically significant at $P<0.05$. This data suggest that in several cases sugarcane miRNA might regulate target genes by translational repression and not by mRNA degradation.

Discussion

The sugarcane microRNAs

Although the miRNAs of several plant species have been recently studied, neither miRNA sequence identification nor an analysis of differential miRNA expression in response to drought stress has been performed in sugarcane. We identified 21 sugarcane miRNAs comprising 18 families using deep sequencing (Table 2). Because the sugarcane genome has not yet been sequenced, *Sorghum bicolor* was chosen as a reference organism because of the high level of identity between the genes in both species [32,33,34]. One of the strategies for the characterization of miRNAs is to examine the phylogenetic conservation of their sequences [35], which permitted the identification and classification of the miRNAs from eight sugarcane libraries. The identification of these miRNAs was accomplished by precursor sequence folding into genuine hairpin structures of the *S. bicolor* sequences where the reads matched, and their classification was based on homology to *S. bicolor* miRNAs in the miRBase database.

In this study, only six putative precursors were found in the sugarcane EST databases, representing five miRNA families (Fig. 5 and Table 3). Because such precursors have short lifetimes in

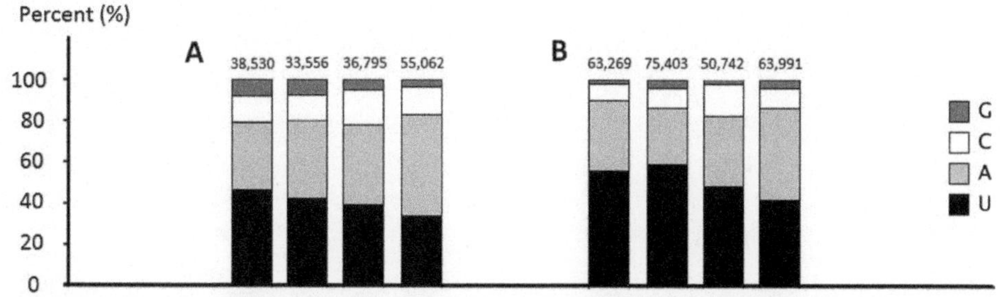

Figure 4. Composition of nucleotides in the first base positions of all 21 nt sequences. A: Control and stressed plants after two days of drought conditions. B: Control and stressed plants after four days of drought conditions. HTD: RB867515 (higher tolerance cultivar) plants without watering. HTI: RB867515 (higher tolerance cultivar) plants under irrigated conditions. LTD: RB855536 (lower tolerance cultivar) plants without watering. LTI: RB855536 (lower tolerance cultivar) plants under irrigated conditions. The total numbers of 21 nt reads in each library are shown at the tops of the bars.

Table 2. Sugarcane miRNAs identified by Solexa sequencing.

miRNA family	miRNA	mature sequence	Sorghum precursor
miR1432	ssp-miR1432	UCAGGAAAGAUGACACCAA	sbi-MIR1432
miR156	ssp-miR156	UUGACAGAAGAGAGUGAGCAC	sbi-MIR156b
			sbi-MIR156c
			sbi-MIR156d
			sbi-MIR156e
			sbi-MIR156g
			sbi-MIR156h
			sbi-MIR156i
miR160	ssp-miR160seq1	UGCCUGGCUCCCUGUAUGCCA	sbi-MIR160c
miR164	ssp-miR164	UGGAGAAGCAGGGCACGUGCA	sbi-MIR164b
miR166	ssp-miR166seq1	UCGGACCAGGCUUCAUUCCCC	sbi-MIR166a
			sbi-MIR166b
			sbi-MIR166c
			sbi-MIR166d
			sbi-MIR166h
			sbi-MIR166j
miR167	ssp-miR167b	UGAAGCUGCCAGCAUGAUCUGA	sbi-MIR167c
			sbi-MIR167d
			sbi-MIR167e
			sbi-MIR167g
			sbi-MIR167h
			sbi-MIR167i
miR169	ssp-miR169seq1	CUAGCCAAGAAUGACUUGCCU	sbi-MIR169f
	ssp-miR169seq2	CAGCCAAGGAUGACUUGCCGA	sbi-MIR169a
miR171	ssp-miR171seq1	UUGAGCCGCGUCAAUAUCUCC	sbi-MIR171h
	ssp-miR171seq2	UGAUUGAGCCGUGCCAAUAUC	sbi-MIR171g
			sbi-MIR171i
			sbi-MIR171k
	ssp-miR171seq3	UGAGCCGAACCAAUAUCACUC	sbi-MIR171c
			sbi-MIR171f
miR172	ssp-miR172	AGAAUCUUGAUGAUGCUGCAU	sbi-MIR172a
			sbi-MIR172d
			sbi-MIR172e
miR319	ssp-miR319	UUUGGAUUGAAGGGUGCU	sbi-MIR319b
miR390	ssp-miR390	AAGCUCAGGAGGGAUAGCGCC	sbi-MIR390
miR393	ssp-miR393	CUCCAAAGGGAUCGCAUUGAU	sbi-MIR393b
miR394	ssp-miR394	UUGGCAUUCUGUCCACCUCC	sbi-MIR394a
miR396	ssp-miR396seq1	UCCACAGGCUUUCUUGAACUG	sbi-MIR396d
miR397	ssp-miR397	UUGACUGCAGCGUUGAUGAGC	sbi-MIR397
miR399	ssp-miR399seq1	UGCCAAAGGAGAGUUGCCCUG	sbi-MIR399a
			sbi-MIR399f
			sbi-MIR399h
			sbi-MIR399i
			sbi-MIR399j
miR528	ssp-miR528	UGGAAGGGGCAUGCAGAGGAG	sbi-MIR528
miR529	ssp-miR529	AGAAGAGAGAGAGUACAGCCU	sbi-MIR529

miRNAs found in the leaves of two sugarcane cultivars (RB867515 and RB855536). Two mismatches were allowed against sorghum mature miRNAs.

Table 3. Bioinformatics prediction of sugarcane miRNAs precursors.

Precursor name	Cluster	MFE (Kcal/mol)	G/C (%)	Mature miRNA sequence (5′ - 3′)
ssp-MIR156	SCSBAD1086B12	−155	43	UUGACAGAAGAGAGTGAGCAC
ssp-MIR166	TC144774	−145.32	45	UCGGACCAGGCUUCAUUCCCC
ssp-MIR167a	SCMCSD2060C04	−156.92	46	UGAAGCUGCCAGCAUGAUCUGA
ssp-MIR167b	SCSFSD1065B12	−225.44	42	UGAAGCUGCCAGCAUGAUCUGA
ssp-MIR169d	SCJLRZ1019E10	−338.87	48	CAGCCAAGGATGACTTGCCGA
ssp-MIR528	SCUTSD1026H02	−204.5	49	UGGAAGGGGCAUGCAGAGGAG

The corresponding precursor clusters in SOGI and SAS in the SUCEST databases are indicated in the Cluster column. MFE: minimum free folding energy. G/C contents and the mature miRNA sequences are also shown.

plants due to rapid processing by Dicer-like 1 in the nucleus [3,36], EST databases do not typically have sufficient coverage to facilitate the discovery of a great number of precursors. For example, in *Brachypodium distachyon*, a model organism for grass species, only 0.05% of the ESTs and 0.012% of the GSSs contain potential miRNAs [37]. The five families found in sugarcane have been previously described, and the secondary structures of the pre-miRNAs have been evaluated using an *in silico* approach [38]. The majority of the small RNAs in sugarcane are 21 or 24 nt in length, as expected in plants (Fig. 2) [4]. The 24 nt species are the most abundant in the non-redundant pool of short reads in sugarcane (Fig. 3), and similar distributions have been reported in studies with several other organisms [18,39,40]. This distribution is likely because 24 nt RNAs are siRNAs and unlike miRNAs, a longer double strand of RNA can result in several different small sequences that act in post-transcriptional gene silencing (PTGS).

It has been reported that uracil appears to be dominant as the first nucleotide at the 5′ ends of mature miRNAs [41]. In sugarcane, the first positions of the mature sequences were equally composed of U and A (Fig. 4). The 5′end nucleotide of a small RNA strand determines the identity of at least several of the recruited AGO proteins [42]. Thus, the 5′end nucleotide has important implications for miRNA function because AGO1 preferentially associates with 21 nt RNAs with a U at the 5′ end, resulting in PTGS activity. In contrast, AGO2 seems to preferentially associate with 21 nt RNAs that have an A at the 5′ end, abolishing their silencing activity and rendering the miRNA functionally inert [43].

miRNAs associated with drought response in sugarcane

To identify miRNAs that are differentially expressed during drought, eight libraries from sugarcane leaf were analyzed: four from RB867515, a cultivar with high tolerance to drought, and four from RB855536, which has a lower drought tolerance. A total of seven miRNAs were differentially expressed during drought (Fig. 6). Six of these were differentially expressed at two days (Table 4), and five miRNAs were differentially expressed at four days of stress (Table 5). For five of these seven miRNAs, the differential expression profiles obtained by sequencing displayed the same expression profiles by RT-qPCR (Fig. 7): ssp-miR164, ssp-miR394, ssp-miR397, ssp-miR399-seq 1 and ssp-miR528. The lack of correlation observed for the other two miRNAs is likely due to cross-amplification of miRNA variants or to the presence of very similar miRNAs in the RT-qPCR samples.

The putative targets of the miRNA modulated by drought provide interesting clues to the drought response in sugarcane. For instance, a NAC transcription factor was found among the targets of ssp-miR164 (Table 6), in agreement with previous studies [44,45]. It is known that miR164 regulates the expression of five NAM/ATAF/CUC (NAC) proteins in *Arabidopsis*. This class of plant-specific transcription factors has important roles in development, growth and stress responses, such as cold, drought and pathogen attach [46,47,48,49,50]. In one study, overexpressing a member of the NAC gene family resulted in enhanced salt tolerance in tobacco [51]. In addition, NAC is responsible for transmitting auxin signals [52]. Another ssp-miR164 target was an MDR-like ABC transporter that also participates in auxin transport [53]. In this study, ssp-miR164 was differentially expressed in response to drought after only two days, suggesting

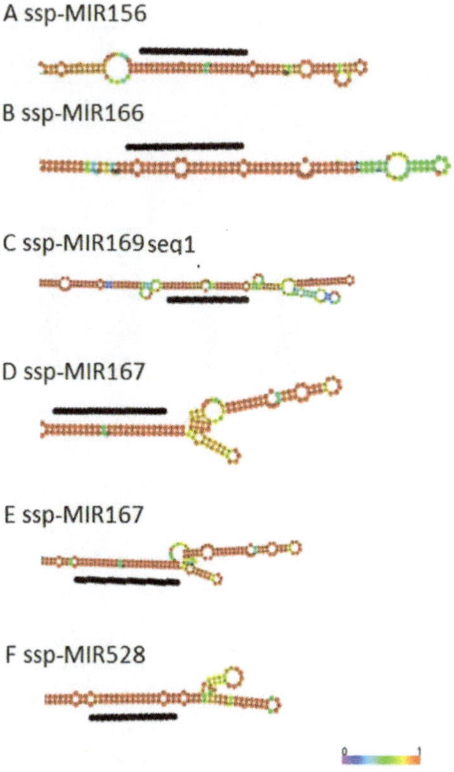

Figure 5. Primary transcripts containing the predicted stem-loop structures of the precursors of the sugarcane miRNAs. The mature miRNAs identified in the sugarcane sRNA library are highlighted in black. The sizes of the precursors may be slightly longer than represented. The colors represent the probabilities for sequence alignment. Red is the highest probability of alignment (1), and purple is the lowest probability of alignment (0).

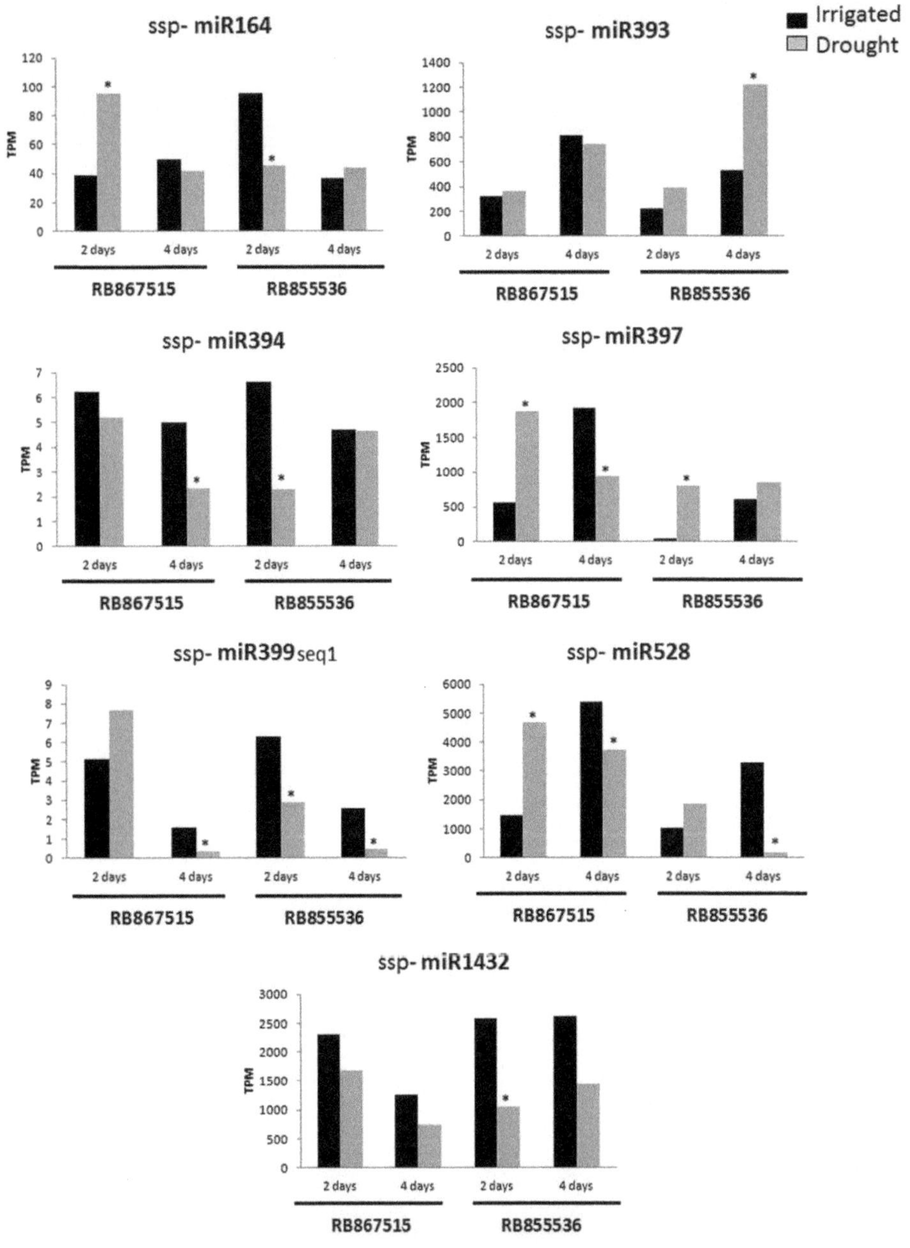

Figure 6. Expression profiles of seven sugarcane miRNAs modulated by drought stress. The values are expressed as the number of transcripts per million (TPM) for irrigated plants (control, black bars) and drought-stressed plants (gray bars). RB867515 (higher drought tolerance) and RB855536 (lower drought tolerance) plants were evaluated after two and four days of stress, as indicated below each graphic. * $p<0.05$ and fold change >2.0. Statistics was calculated between irrigated and drought treatments in each cultivar using the Audic-Claverie method [78].

that this miRNA acts in the early stage of the stress response. Considering that the targets of this miRNA presumably help sugarcane plants to withstand drought stress, we would expect that ssp-miR164 targets are repressed in both cultivars. This was the case for the LT cultivar (Fig. 6 and 7), suggesting that both NAC and the MDR-like ABC transporter, and possibly auxin, play roles in the drought response. However, in the HT cultivar, ssp-miR164 was induced by drought. These data indicate that the two analyzed cultivars differ in their activation of molecular mechanisms in response to drought and that the target genes may play a more relevant role in the LT cultivar.

It is known that miR393 is commonly up-regulated during drought stress in *Arabidopsis*, *Oryza sativa*, *Medicago truncatula* and *Pinguicula vulgaris* [54]. In this work, the ssp-miR393 miRNA was also up-regulated under drought stress in the LT cultivar after four days (Fig. 6). The predicted target of ssp-miR393 encodes a putative protein similar to TIR1, an auxin receptor in *Arabidopsis thaliana*. TIR1 recognized the 3-indol-acetic acid (AIA) and promote the degradation of the Aux/AIA repressor by a protein ubiquitin ligase that binds to a conserved domain of the repressor. This degradation releases the transcription of auxin-regulated genes [55]. Recently, the expression of *TIR1* was related with the response to inorganic phosphate (Pi) in roots of *Arabidopsis thaliana*, [56]. Our data points to an involvement of auxin in sugarcane responses to drought.

Table 4. Sugarcane miRNAs differentially expressed after two days of drought stress.

Family	miR name	Mature sequences	Nt	HTD (TPM)	HTI(TPM)	LTD(TPM)	LTI (TPM)	HTD/HTI	LTD/LTI
miR164	ssp-miR164	UGGAGAAGCAGGGCACGUGCA	21	95.07	38.57	45.43	95.28	2.46*	−2.10*
miR394	ssp-miR394	UUGGCAUUCUGUCCACCUCC	20	5.19	6.24	2.30	6.62	−1.20	−2.88*
miR397	ssp-miR397	UUGACUGCAGCGUUGAUGAGC	21	1873.63	562.10	799.76	42.42	3.33*	18.85*
miR399	ssp-miR399seq1	UGCCAAAGGAGAGUUGCCC	19	7.67	5.18	2.89	6.30	1.48	−2.18*
miR528	ssp-miR528	UGGAAGGGGCAUGCAGAGGAG	21	4674.10	1445.14	1874.15	1025.19	3.23*	1.83
miR1432	ssp-miR1432	UCAGGAAAGAUGACACCAA	19	1687.91	2300.44	1053.92	2583.96	−1.36	−2.45*

Nt: number of nucleotides of the mature miRNA. HTD: RB867515 (higher drought tolerance) plants under drought stress (without watering). HTI: RB867515 plants under irrigation. LTD: RB855536 (lower drought tolerance) under drought stress. LTI: RB855536 (lower drought tolerance) plants under irrigation. In the last two columns, negative values indicate down-regulated miRNAs and positive values indicate miRNAs that are up-regulated during drought. Asterisks indicate minimum fold change >2.00 and $p<0.05$.

ssp-miR394 was down-regulated under drought stress in the two sugarcane cultivars (Fig. 6), indicating that this miRNA is important in the drought stress response independently of the plant genotype. The predicted target of ssp-miR394 is the gene encoding a glyceraldehyde-3-phosphate dehydrogenase (GAPDH); see Table 6. GAPDH catalyzes the oxidation of D-glyceraldehyde-3-P (D-G3P) to 3-phosphoglycerate (3-PGA) with the generation of NADPH in the sixth step of glycolysis [57]. Due to the increased need for available ATP and $NADH_2$ under drought conditions, glycolysis is usually enhanced. This hypothesis is supported by the fact that rice plants exposed to drought display increased levels of *OsGAPDH* transcripts [58]. In this context, a decrease in the ssp-miR394 level would increase the GAPDH level and consequently the ATP content.

We also found that the target of ssp-miR397 is a gene encoding a laccase (Table 6), in agreement with previous studies [59]. Laccases are glycoproteins with roles in lignin synthesis [60] and iron acquisition [61]. It has been proposed that laccases are involved in cell wall modifications, such as lignification, acting to reduce cell wall extensibility and elongation [62,63]. In maize [64] and tomato [65], increases in laccase transcripts were reported at high concentrations of NaCl. Under drought stress, ssp-miR397 was up-regulated in both cultivars on the second day (Fig. 6), suggesting a reduction in laccase expression levels. However, on the fourth day of drought stress, this miRNA was down-regulated in the HT cultivar (Fig. 6), likely leading to increased laccase levels, which would be reflected as a decrease in cell elongation due to lignin accumulation in the cell wall in response to stress. The complex expression pattern observed between the two cultivars indicates that lignification may act in different ways during drought stress.

miR399 has been described as a negative regulator of inorganic phosphate concentration because it targets a pyrophosphatase [66]. In this study, we also found an inorganic pyrophosphatase 2-like protein to be a target of ssp-miR399-seq 1 (Table 6). Overexpression of *AVP1 (Arabidopsis* vacuolar pyrophosphatase gene) results in increased drought and salt tolerance in transgenic *Arabidopsis*, tomato, rice and cotton plants [67,68,69,70]. An increased proton pump activity by the vacuolar pyrophosphatase is the molecular explanation for drought resistance in these plants. This activity leads to a lower water potential in the plant vacuole and increases the activity of secondary transporters, preventing ion accumulation in the cytoplasm [70]. The down-regulation of ssp-miR399-seq 1 under drought stress in both sugarcane cultivars likely leads to increased pyrophosphatase levels. These data suggest that pyrophosphatases also play a role in sugarcane responses to drought stress and that this role is at least partially controlled by the microRNAs.

ssp-miR528 targets a gene encoding a UBX domain-containing protein (Table 6). UBX domains have been identified in several proteins with diverse functions in eukaryotes. Among the 15 UBX-containing proteins encoded by the *Arabidopsis* genome [71], PUX1 was shown to play important roles in plant growth and development. Loss-of-function mutant *pux1* plants display accelerated growth in various plant organs, including roots and inflorescence shoots [72]. In this work, ssp-miR528 was down-regulated under drought stress in both cultivars on the fourth day

Table 5. Sugarcane miRNAs differentially expressed after four days of drought stress.

Family	miR name	Mature sequences	Nt	HTD(TPM)	HTI(TPM)	LTD(TPM)	LTI(TPM)	HTD/HTI	LTD/LTI
miR393	ssp-miR393	CUCCAAAGGGAUCGCAUUGAU	21	743.15	808.97	1220.51	582.06	−1.09	2.10*
miR394	ssp-miR394	UUGGCAUUCUGUCCACCUCC	20	2.33	5.00	4.64	4.70	−2.14*	−1.01
miR397	ssp-miR397	UUGACUGCAGCGUUGAUGAGC	21	940.44	1926.63	858.82	602.07	−2.05*	1.43
miR399	ssp-miR399seq1	UGCCAAAGGAGAGUUGCCC	19	0.38	1.61	0.47	2.56	−4.18*	−5.51*
miR528	ssp-miR528	UGGAAGGGGCAUGCAGAGGAG	21	3717.14	5397.83	182.68	3293.76	−1.45*	−18.03*

Nt: number of nucleotides of the mature miRNA. HTD: RB867515 (higher drought tolerance) plants under drought stress (without watering). HTI: RB867515 plants under irrigation. LTD: RB855536 (lower drought tolerance) under drought stress. LTI: RB855536 (lower drought tolerance) plants under irrigation. In the last two columns, negative values indicate down-regulated miRNAs and positive values indicate miRNAs that are up-regulated during drought. Asterisks indicate minimum fold change >2.00 and $p<0.05$.

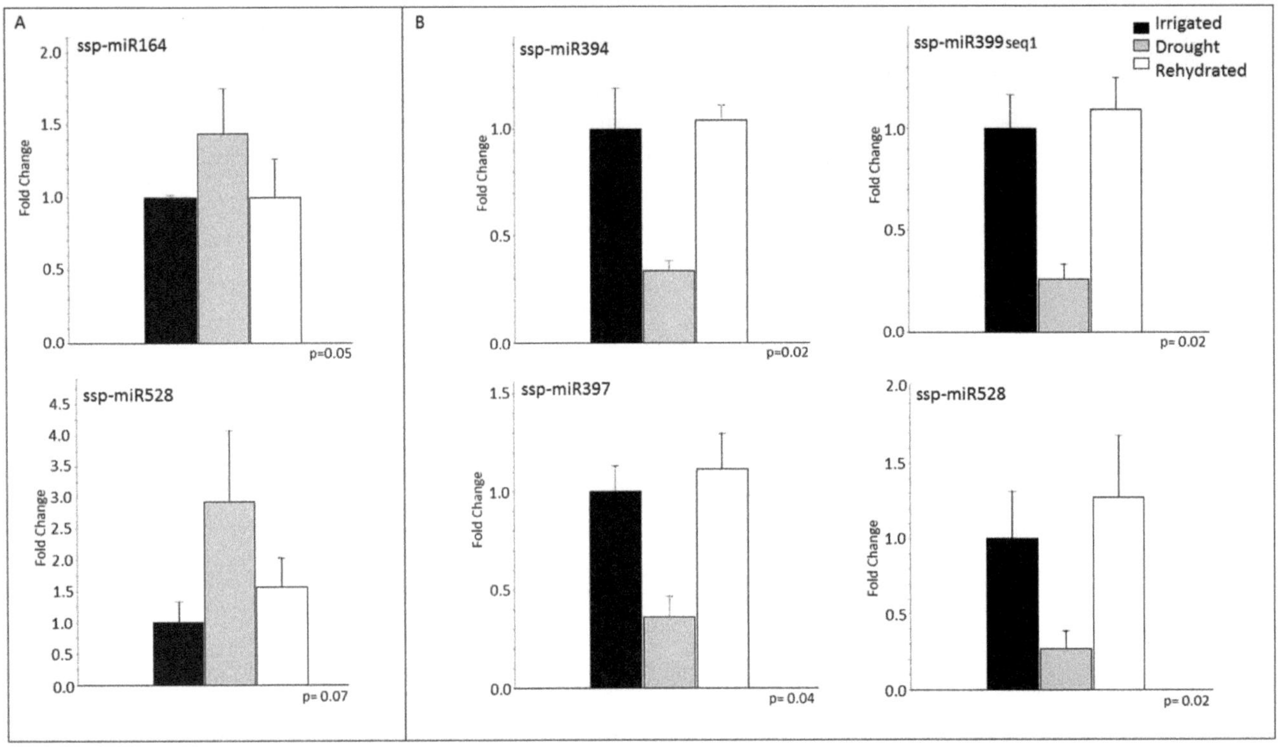

Figure 7. RT-qPCR expression profiles of five sugarcane miRNAs modulated by drought stress. The values are expressed as fold changes relative to the irrigated control for each gene. The bars represent the average of the irrigated plants (control, black bars), drought-stressed plants (gray bars) and rehydrated plants (white bars) for RB867515 (higher drought tolerance) after two (A) and four (B) days of stress. Error bars represent the standard error (n = 4). Means followed by different letters are statistically different ($p<0.05$) using the permutation mean test.

(Fig. 6), suggesting that this miRNA may be involved in the reduction of growth under this stress.

The expression levels of ssp-miR397, ssp-miR399-seq 1 and ssp-miR528 were quite variable in the control plants (Fig. 6), and this variability is likely due to other environmental conditions and/or intrinsic changes, such as developmental influences.

A gene encoding a bZIP transcription factor was determined to be an ssp-miR1432 target (Table 6). Interestingly, several bZIPs are known to play important roles in conferring stress tolerance to plants [73] by regulating the expression of genes that are involved in mechanisms that are essential for stress adaptation, such as cytoplasmic ion homeostasis and osmotic adjustment [74]. A triple knockout in the genes encoding the bZIPs AREB1, AREB2, and ABF3 in *Arabidopsis* displayed increased tolerance to ABA and reduced drought tolerance [75]. ssp-miR1432 was down-regulated by drought at similar levels at all time points in both cultivars (Fig. 6), suggesting that the sugarcane bZIP activates genes that help sugarcane plants to overcome drought stress. However, because the expression patterns observed in both cultivars are similar, this transcription factor alone may not be responsible for the different levels of drought tolerance in RB867515 and RB855536.

It is notable that the expression patterns of the majority of the miRNAs did not display clear correlations with the differences in drought tolerance observed in the two sugarcane cultivars. Nevertheless, it is clear that the microRNAs differ considerably between the two cultivars, suggesting that the miRNAs with differential expression do not fully explain the drought higher tolerance observed in RB867515.

In summary, our results provide insight into the sugarcane microRNAs, highlighting the regulatory network triggered by drought stress in an important bioenergy crop. We found that transcription factors, kinases, phosphatases and chaperones may be targets of the miRNAs modulated by drought in sugarcane. Further work with transgenic sugarcane overexpressing or silencing the miRNAs or their targets will increase our knowledge of the molecular mechanism of drought stress response and tolerance in sugarcane.

Materials and Methods

Plant samples

Sugarcane cultivars RB867515 (high tolerance to drought) and RB855536 (low tolerance to drought) were obtained from RIDESA (Rede Universitária para o Desenvolvimento do Setor Sucroalcooleiro). Plants were grown in a greenhouse at the Federal University of Alagoas (Alagoas, Brazil, 9°45′32″S, 36°13′09″W) under normal irrigation for three months. To develop drought stress, water was withheld from the test plants. The rehydrated plants received normal watering for two days after the period without irrigation. The control plants received normal watering throughout the experiment. Tissue was collected from Leaf+1 (the highest unfolded leaf with a visible dewlap) in quadruplicate after 2, 4, 6 and 8 days from irrigated, non-irrigated (drought-stressed) and rehydrated plants. The samples were quick-frozen and maintained at −80°C until RNA extraction using the miRVana miRNA isolation kit (Life Technologies, USA) according to the manufacturer's protocol. A pool of two replicates was used for Solexa sequencing, whereas four individual plants were used for RT-qPCR validation.

Table 6. Target prediction of the miRNAs that are differentially expressed in drought-stressed sugarcane plants.

miRNA	Target Acc.	Expectation	UPE	mature miRNA	Target start	Target end	miRNA aligned fragment	Target aligned fragment	Inhibition	Target description
ssp-miR164	SCEPRT2048G05.g	1.0	20,052	20	699	718	UGGAGAAGCAGGGCACGUGC	GCAGGUGCCCUGCUUCUCCA	Cleavage	NAC transcription factor [*Hordeum vulgare*]
	SCCCAM1001A03.g	2.5	20,361	19	1774	1792	UGGAGAAGCAGGGCACGUG	CACGUGUCCACCUUCUCCA	Translation	MDR-like ABC transporter [*Oryza sativa Japonica Group*]
ssp-miR394	SCQGAM2027G09.g	3.0	19,627	19	759	777	UUGGCAUUCUGUCCACCUC	UUGGCAUUCUGUCCACCUC	Cleavage	Glyceraldehyde-3-phosphate dehydrogenase [*Triticum aestivum*]
ssp-miR528	SCJFRT2058D11.g	2.5	22,413	20	101	120	UGGAAGGGGCAUGCAGAGGA	UUCUCGGCAUGCCCCUUCUG	Cleavage	UBX domain-containing protein [*Oryza brachyantha*]
ssp-miR397	SCQSAD1056B07.g	2.5	22,911	20	64	83	UUGACUGCAGCGUUGAUGAG	UUCAUCAACGCCGCACUCAA	Translation	laccase-23-like [*Brachypodium distachyon*]
ssp-miR1432	TC134052	3.0	17.32	19	630	648	UCAGGAAAGAUGACACCAA	UUGGUGUUUUCUUCCCUGA	Translation	ABRE-binding factor BZ-1 [*Zea mays*]
	SCSFFL4085D03.g	3.0	15,088	19	624	642	UCAGGAAAGAUGACACCAA	UUGGUGUUUUCUUCCCUGA	Translation	bZIP transcription factor1 [*Zea mays*]
ssp-miR393	TC120009	1.0	20,653	19	302	320	CUCCAAAGGGAUCGCAUUG	CAAUGCGAUCCCUUUGGAU	Cleavage	Auxin-responsive factor TIR1-like protein [*Populus tomentosa*]
ssp-miR399seq1	SCACHR1037A06.g	1.5	19,642	19	463	481	UGCCAAAGGAGAGUUGCCC	GGGUAGUUCUCCUUUGGCA	Cleavage	inorganic pyrophosphatase 2-like [*Brachypodium distachyon*]

Target Acc: accession number in the SUCEST or SoGI databases; Expectation: value assigned to the alignment of the mature miRNA and the target, ranging from 0 (perfect alignment) to 5; UPE: the energy required to open the secondary structure of the target at the recognition site (less energy means better accessibility to the target); Mature miRNA: miRNA mature size (in nucleotides); Target start: the base position where the annealing with the miRNA starts; Target end: the base position where the annealing with the miRNA ends; Inhibition: the type of regulation by the miRNA; and Target description: description of the target according to a BLAST search in the GenBank database, including the name of the organism presenting the best hit.

Table 7. Validation of the target genes found for the miRNAs differentially expressed under drought.

miRNA Target	SAS/SoGI cluster name	Target description	HT2	HT4	LT2	LT4
target miR164	SCEPRT2048G05.g	NAC transcription factor	2.5*/1.2	**0.8/3.4**	**0.5*/3.9**	1.2/2.5*
target miR393	TC120009	Auxin-responsive factor TIR1-like protein	1.1/0.6*	0.9/0.3	1.7/0.9	2.3*/2.0
target miR394	SCQGAM2027G09.g	Glyceraldehyde 3-Phosphate Dehydrogenase	0.8/0.7	0.5*/0.6	0.3*/1.0	1.0/1.7
target miR397	SCQSAD1056B07.g	Lacasse 23-Like	3.3*/1.1	**0.5*/6.9***	**18.9*/0.7**	1.4/11.5
target miR399	SCACHR1037A06.g	Inorganic pyrophosphatase 2-like	1.5/2.3	0.2*/0.1	**0.5*/1.8**	0.2*/0.5
target miR528	SCJFRT2058D11.g	UBX Domain Containing Protein	**3.2*/0.4***	0.7*/0.4*	**1.8/0.5***	**0.1*/2.0**
target miR1432	SCSFFL4085D03.g	B-ZIP transcription factor	**0.7/4.0***	**0.6/4.1**	**0.4*/2.0**	**0.5/11.7***

The expression rations between drought-stressed and control plants are shown. The first number in each pair indicates miRNA levels and the second indicates the target gene expression.
*indicates ratios where differences in the expression levels between irrigated and drought-stressed plants are statistically significant ($P<0.05$). Expression rations that are in bold indicate that miRNA induction or repression correlates with repression or induction of target genes, respectively. SAS: Sugarcane Assembled Sequence; SoGI: Sugarcane Gene Index; HT: higher tolerance cultivar; LT: lower tolerance cultivar; 2: two days of stress; 4: four days of stress.

Small RNA Sequencing

RNAs from 16 to 27 nt long were selected by polyacrylamide gel electrophoresis, ligated with adaptors at both ends and the products used for cDNA synthesis at BGI (Beijing Genomic Institute, Tai Po, Hong Kong). The sequencing was performed using the Solexa platform at BGI.

Bioinformatics analysis

To identify the sequences corresponding to true mature miRNAs, we predicted their precursors (pre-miRNAs). Reads were preprocessed by removing adapters and discarding reads <18 nt. The miRDeep-P program [21] was used to map the reads to the reference sequences (*Sorghum bicolor* genomic sequence and sugarcane ESTs). For a given mapped read, a 250 bp window was used from which reference sequences were extracted to predict RNA secondary structure [76]. The miRDeep core algorithm with a plant-specific scoring system based on known characteristics of plant miRNA genes was used to determine the secondary structure of the extracted reference sequences [35]. RNA sequences were classified as miRNA precursor candidates based on the following parameters: the folding of the RNA sequence into a characteristic stem-loop hairpin secondary structure, the position of the mature miRNA in one arm of the hairpin structure (allowing a maximum of six mismatches with the miRNA* sequence in the opposite arm), negative MFE values for the predicted secondary structures, and a G/C content in the 30–70% range [38], using the RNAfold [77]. After the normalization of the number of reads in each library, the expression of each miRNA was calculated using the Audic-Claverie method [78].

The targets of the miRNA were predicted using psRNATarget (http://plantgrn.noble.org/psRNATarget/) by searching for target genes based on both complementarity scoring and secondary structure analyses [79]. The BlastX algorithm [80] was used to find the hits in the Sugarcane Assembled Sequences (SAS) against the NCBI database to identify the coding strands because several SASs corresponded to the minus strand. Only hits complementary to the coding strand of the sugarcane mRNAs were selected.

Stem-loop reverse transcription and RT-qPCR validation

RT-loop primers (loop-RT), sequence-specific forward PCR primers (loop-FW) and universal reverse primers were designed following Chen et al., 2005 [30] for reverse transcription and PCR amplification of sugarcane miRNAs (Table S2). Reverse transcriptase reactions and cDNA amplification were performed as described by Varkonyi-Gasic et al., 2007 [31]. The miRNA and reference gene reverse transcriptase reactions were conducted under the same reaction condition.

RT-qPCR miRNA and target validation

To validate and quantitate the expression levels of the miRNAs in sugarcane leaf tissues, RT-qPCR was performed using SYBR Green PCR Master Mix (Life Technologies, USA) on a 7500 Real-Time PCR System (Life technologies, USA). Each PCR reaction (18 µL) included 2 µL cDNA, 10 µL SYBR Green Master Mix (1×), 1 µL sequence-specific forward primer (10 µM), 1 µL universal reverse primer (10 µM) and 4 µL water. The sugarcane polyubiquitin gene [81] was used as a reference (Table 7). The reactions were performed at 95°C for 10 minutes followed by 40 cycles of 95°C for 15 seconds and 60°C for 1 minute, with a final dissociation curve analysis. All reactions were run with four biological replicates, each in triplicate. The real-time PCR data analysis was performed by considering the reaction efficiencies to calculate the fold changes in miRNA levels using a web-based QPCR system [82].

For target validation, each reverse transcriptase reaction contained 2.5 µg of total DNA-free RNA, 1 µL oligo $d(T)_{17}VN$ (50 µM) and 1 µL of dNTPs mix (10 µM). The reaction was incubated for 10 minutes at 65°C and then laid on ice for 2 minutes. Subsequently, 5× First Strand Buffer, DTT, RNAse-Out and Superscript II enzyme (Life technologies, USA) were added. This reaction was incubated in a VerityTM Thermal Cycler (Applied Biosystems, USA) for 30 minutes at 42°C, followed by 15 minutes at 70°C. Forward and reverse specific PCR primers (Supporting Information, Table S2) were designed for amplification of the target genes. The reactions were carried out using the SYBRGreen PCR Master Mix (Applied Biosystems, USA) on 7500 Real Time PCR System (Applied Biosystems, USA) as described before.

Supporting Information

Text S1 Supporting information text. Physiological data.

Figure S1 Quantum potential efficiency (Fv/Fm) in sugarcane plants. The sugarcane cultivars RB855536 (LT - lower drought tolerance) and RB867515 (HT- higher drought tolerance) were maintained under irrigation, without irrigation and without irrigation and then rehydrated, as indicated. * indicates differences between irrigated and drought-stressed plants; +indicates differences between irrigated and rehydrated plants.** and ++indicate $p<0.01$, and * and +indicate $p<0.05$ using the t-test. ns - not significant.

Figure S2 Fluorescence quantum yield (ΦPSII) in sugarcane plants. The sugarcane cultivars RB855536 (LT - lower drought tolerance) and RB867515 (HT- higher drought tolerance) were maintained under irrigation, without irrigation and without irrigation and then rehydrated, as indicated. * indicates differences between irrigated and drought-stressed plants; +indicates differences between irrigated and rehydrated plants. ** and ++indicate $p<0.01$, and * and +indicate $p<0.05$ using the t-test. ns - not significant.

Figure S3 Expression profile of the predicted target genes for the seven sugarcane miRNAs modulated by drought. The values are expressed as fold changes relative to the irrigated control for each gene. The bars represent the average of the irrigated plants (control, black bars) and drought-stressed plants (gray bars) for RB867515 (HT) and RB855536 (LT) after two (2) and four (4) days of stress: A) HT2; B) HT4; C) LT2; D) LT4. Error bars represent the standard deviation (n = 3). Statistics was calculated between irrigated and drought treatments using the t- test. * indicates differences between irrigated and drought-stressed plants, with $p \leq 0.05$.

Table S1 Target prediction of the miRNAs differentially expressed in drought-stressed sugarcane plants. (All bioinformatics data without selection.) Target Acc: accession number in the SUCEST or SoGI databases; Expectation: value assigned to the alignment of the mature miRNA and the target, ranging from 0 (perfect alignment) to 5; UPE: the energy required to open the secondary structure of the target at the site recognition (less energy means better accessibility to the target); Mature miRNA: miRNA mature size (in nucleotides); Target start: the base position where the annealing with the miRNA starts; Target end: the base position where the annealing with the miRNA ends; Inhibition: the type of regulation by the miRNA; and Target description: description of the target according to a BLAST search in the GenBank database, including the name of the organism presenting the best hit.

Table S2 Primers used in the reverse transcription and real-time PCR analyses of sugarcane miRNAs and target genes expression. RT: primer loop for reverse transcription; FW: forward primer for real-time PCR; Rv: reverse primer for real-time PCR; reverse universal: reverse primer for miRNA real-time PCR; PUB: polyubiquitin gene primer; tgt: target gene; the complete sequence of each primer is shown and also the number of nucleotides (Nt) and the melting temperature (Tm) in °C.

Author Contributions

Conceived and designed the experiments: THF AG LE MM. Performed the experiments: THF AG RDV GGLC. Analyzed the data: THF AG RDV GGLC LE MM. Contributed reagents/materials/analysis tools: LID. Wrote the paper: THF AG MM.

References

1. Bartel DP (2009) MicroRNAs: Target Recognition and Regulatory Functions. Cell 136: 215–233.
2. Carthew RW, Sontheimer EJ (2009) Origins and Mechanisms of miRNAs and siRNAs. Cell 136: 642–655.
3. Bartel DP (2004) MicroRNAs: Genomics, biogenesis, mechanism, and function. Cell 116: 281–297.
4. Chapman EJ, Carrington JC (2007) Specialization and evolution of endogenous small RNA pathways. Nat Rev Gen 8: 884–896.
5. Lee Y, Kim M, Han JJ, Yeom KH, Lee S, et al. (2004) MicroRNA genes are transcribed by RNA polymerase II. Embo J 23: 4051–4060.
6. Lee RC, Feinbaum RL, Ambros V (1993) The C-Elegans Heterochronic Gene Lin-4 Encodes Small Rnas with Antisense Complementarity to Lin-14. Cell 75: 843–854.
7. Reinhart BJ, Weinstein EG, Rhoades MW, Bartel B, Bartel DP (2002) MicroRNAs in plants. Genes Dev 16: 1616–1626.
8. Sunkar R, Girke T, Jain PK, Zhu JK (2005) Cloning and characterization of microRNAs from rice. Plant Cell 17: 1397–1411.
9. Kantar M, Lucas SJ, Budak H (2011) miRNA expression patterns of Triticum dicoccoides in response to shock drought stress. Planta 233: 471–484.
10. Liu HH, Tian X, Li YJ, Wu CA, Zheng CC (2008) Microarray-based analysis of stress-regulated microRNAs in *Arabidopsis thaliana*. RNA 14: 836–843.
11. Sunkar R, Zhu JK (2004) Novel and stress-regulated microRNAs and other small RNAs from Arabidopsis. Plant Cell 16: 2001–2019.
12. Zhang ZX, Wei LY, Zou XL, Tao YS, Liu ZJ, et al. (2008) Submergence-responsive microRNAs are potentially involved in the regulation of morphological and metabolic adaptations in maize root cells. Ann Bot 102: 509–519.
13. Zhang XH, Zou Z, Gong PJ, Zhang JH, Ziaf K, et al. (2011) Over-expression of microRNA169 confers enhanced drought tolerance to tomato. Biotechnol Lett 33: 403–409.
14. Zhao BT, Liang RQ, Ge LF, Li W, Xiao HS, et al. (2007) Identification of drought-induced microRNAs in rice. Biochem Biophys Res Commun 354: 585–590.
15. Kulcheski FR, de Oliveira LF, Molina LG, Almerao MP, Rodrigues FA, et al. (2011) Identification of novel soybean microRNAs involved in abiotic and biotic stresses. BMC Genomics 12: 307.
16. Zhou L, Liu Y, Liu Z, Kong D, Duan M, et al. (2010) Genome-wide identification and analysis of drought-responsive microRNAs in Oryza sativa. J Exp Bot 61: 4157–4168.
17. Inman-Bamber NG, Smith DM (2005) Water relations in sugarcane and response to water deficits. Field Crop Res 92: 185–202.
18. Fahlgren N, Howell MD, Kasschau KD, Chapman EJ, Sullivan CM, et al. (2007) High-Throughput Sequencing of *Arabidopsis* microRNAs: Evidence for Frequent Birth and Death of MIRNA Genes. Plos One 2: e219.
19. Song C, Wang C, Zhang C, Korir NK, Yu H, et al. (2010) Deep sequencing discovery of novel and conserved microRNAs in trifoliate orange (*Citrus trifoliata*). BMC Genomics 11: 431.
20. Zhu QH, Spriggs A, Matthew L, Fan LJ, Kennedy G, et al. (2008) A diverse set of microRNAs and microRNA-like small RNAs in developing rice grains. Genome Res 18: 1456–1465.
21. Yang XZ, Li L (2011) miRDeep-P: a computational tool for analyzing the microRNA transcriptome in plants. Bioinformatics 27: 2614–2615.
22. Unver T, Budak H (2009) Conserved microRNAs and their targets in model grass species Brachypodium distachyon. Planta 230: 659–669.
23. Rocha FR, Papini-Terzi FS, Nishiyama MY, Vencio RZN, Vicentini R, et al. (2007) Signal transduction-related responses to phytohormones and environmental challenges in sugarcane. BMC Genomics 8.
24. Islam MS, Wang MH (2010) Expression Patterns of an Abiotic Stress-inducible Dehydrin Gene, *LeDhnT14*, in Tomato. Hortic Environ Biotechnol 51: 556–561.
25. Ingram J, Bartels D (1996) The Molecular Basis of Dehydration Tolerance in Plants. Annu Rev Plant Physiol Plant Mol Biol 47: 377–403.
26. Vaseva I, Akiscan Y, Demirevska K, Anders I, Feller U (2011) Drought stress tolerance of red and white clover-comparative analysis of some chaperonins and dehydrins. Scientia Hortic 130: 653–659.
27. Giordani T, Buti M, Natali L, Pugliesi C, Cattonaro F, et al. (2011) An analysis of sequence variability in eight genes putatively involved in drought response in sunflower (*Helianthus annuus* L.). Theor Appl Genet 122: 1039–1049.
28. Wang L, Liu H, Li D, Chen H (2011) Identification and characterization of maize microRNAs involved in the very early stage of seed germination. BMC Genomics 12: 154.
29. Calvino M, Bruggmann R, Messing J (2011) Characterization of the small RNA component of the transcriptome from grain and sweet sorghum stems. BMC Genomics 12: 356.

30. Chen CF, Ridzon DA, Broomer AJ, Zhou ZH, Lee DH, et al. (2005) Real-time quantification of microRNAs by stem-loop RT-PCR. Nucleic Acid Res 33.
31. Varkonyi-Gasic E, Wu RM, Wood M, Walton EF, Hellens RP (2007) Protocol: a highly sensitive RT-PCR method for detection and quantification of microRNAs. Plant Methods 3.
32. Paterson AH, Bowers JE, Bruggmann R, Dubchak I, Grimwood J, et al. (2009) The *Sorghum bicolor* genome and the diversification of grasses. Nature 457: 551–556.
33. Jannoo N, Grivet L, Chantret N, Garsmeur O, Glaszmann JC, et al. (2007) Orthologous comparison in a gene-rich region among grasses reveals stability in the sugarcane polyploid genome. Plant J 50: 574–585.
34. Dillon SL, Shapter FM, Henry RJ, Cordeiro G, Izquierdo L, et al. (2007) Domestication to crop improvement: genetic resources for Sorghum and Saccharum (Andropogoneae). Ann Bot 100: 975–989.
35. Meyers BC, Axtell MJ, Bartel B, Bartel DP, Baulcombe D, et al. (2008) Criteria for Annotation of Plant MicroRNAs. Plant Cell 20: 3186–3190.
36. Kim VN (2005) MicroRNA biogenesis: Coordinated cropping and dicing. Nature Reviews Molecular Cell Biol 6: 376–385.
37. Unver T, Budak H (2009) Conserved microRNAs and their targets in model grass species *Brachypodium distachyon*. Planta 230: 659–669.
38. Zanca AS, Vicentini R, Ortiz-Morea FA, Del Bem LEV, da Silva MJ, et al. (2010) Identification and expression analysis of microRNAs and targets in the biofuel crop sugarcane. BMC Plant Biol 10.
39. Rajagopalan R, Vaucheret H, Trejo J, Bartel DP (2006) A diverse and evolutionarily fluid set of microRNAs in *Arabidopsis thaliana*. Genes Dev 20: 3407–3425.
40. Martinez G, Forment J, Llave C, Pallas V, Gomez G (2011) High-throughput sequencing, characterization and detection of new and conserved cucumber miRNAs. Plos One 6: e19523.
41. Zhang BH, Pan XP, Stellwag EJ (2008) Identification of soybean microRNAs and their targets. Planta 229: 161–182.
42. Mi SJ, Cai T, Hu YG, Chen Y, Hodges E, et al. (2008) Sorting of small RNAs into *Arabidopsis* argonaute complexes is directed by the 5′ terminal nucleotide. Cell 133: 116–127.
43. Voinnet O (2009) Origin, Biogenesis, and Activity of Plant MicroRNAs. Cell 136: 669–687.
44. Raman S, Greb T, Peaucelle A, Blein T, Laufs P, et al. (2008) Interplay of miR164, CUP-SHAPED COTYLEDON genes and LATERAL SUPPRESSOR controls axillary meristem formation in *Arabidopsis thaliana*. Plant J 55: 65–76.
45. Guo HS, Xie Q, Fei JF, Chua NH (2005) MicroRNA directs mRNA cleavage of the transcription factor NAC1 to downregulate auxin signals for Arabidopsis lateral root development. Plant Cell 17: 1376–1386.
46. Ooka H, Satoh K, Doi K, Nagata T, Otomo Y, et al. (2003) Comprehensive Analysis of NAC Family Genes in Oryza sativa and *Arabidopsis thaliana*. DNA Res 10: 239–247.
47. Kikuchi K, Ueguchi-Tanaka M, Yoshida KT, Nagato Y, Matsusoka M, et al. (2000) Molecular analysis of the NAC gene family in rice. Mol Gen Genet 262: 1047–1051.
48. Collinge M, Boller T (2001) Differential induction of two potato genes, Stprx2 and StNAC, in response to infection by *Phytophthora infestans* and to wounding. Plant Mol Biol 46: 521–529.
49. Hegedus D, Yu M, Baldwin D, Gruber M, Sharpe A, et al. (2003) Molecular characterization of Brassicanapus NAC domain transcriptional activators induced in response to biotic and abiotic stress. Plant Mol Biol 53: 383–397.
50. Ditt RF, Gentile A, Tavares RG, Camargo SR, Fernandez JH, et al. (2011) Analysis of the stress-inducible transcription factor *SsNAC23* in sugarcane plants. Sci Agric 68: 454–461.
51. Liu QL, Xu KD, Zhao LJ, Pan YZ, Jiang BB, et al. (2011) Overexpression of a novel chrysanthemum NAC transcription factor gene enhances salt tolerance in tobacco. Biotechnol Lett 33: 2073–2082.
52. Berger Y, Harpaz-Saad S, Brand A, Melnik H, Sirding N, et al. (2009) The NAC-domain transcription factor GOBLET specifies leaflet boundaries in compound tomato leaves. Development 136: 823–832.
53. Wu G, Lewis DR, Spalding EP (2007) Mutations in *Arabidopsis* multidrug resistance-like ABC transporters separate the roles of acropetal and basipetal auxin transport in lateral root development. Plant Cell 19: 1826–1837.
54. Sunkar R (2010) MicroRNAs with macro-effects on plant stress responses. Semin Cell Dev Biol 21: 805–811.
55. Dharmasiri N, Dharmasiri S, Estelle M (2005) The F-box protein TIR1 is an auxin receptor. Nature 435: 441–445.
56. Perez-Torres CA, Lopez-Bucio J, Cruz-Ramirez A, Ibarra-Laclette E, Dharmasiri S, et al. (2008) Phosphate availability alters lateral root development in *Arabidopsis* by modulating auxin sensitivity via a mechanism involving the TIR1 auxin receptor. Plant Cell 20: 3258–3272.
57. Jeong MJ, Park SC, Kwon HB, Byun MO (2000) Isolation and characterization of the gene encoding glyceraldehyde-3-phosphate dehydrogenase. Biochem Bioph Res Co 278: 192–196.
58. Arumugam Pillai MAP, Lihuang ZL, Akiyama TA (2002) Molecular cloning, characterization, expression and chromosomal location of *OsGAPDH*, a submergence responsive gene in rice (Oryza sativa L.). Theor Appl Genet 105: 34–42.
59. Abdel-Ghany SE, Pilon M (2008) MicroRNA-mediated systemic down-regulation of copper protein expression in response to low copper availability in *Arabidopsis*. J Biol Chem 283: 15932–15945.
60. Ranocha P, Chabannes M, Chamayou S, Danoun Sd, Jauneau A, et al. (2002) Laccase down-regulation causes alterations in phenolic metabolism and cell wall structure in poplar. Plant Physiol 129: 145–155.
61. Hoopes JT, Dean JFD (2004) Ferroxidase activity in a laccase-like multicopper oxidase from *Liriodendron tulipifera*. Plant Physiol Biochem 42: 27–33.
62. Neumann PM, Azaizeh H, Leon D (1994) Hardening of root cell walls: a growth inhibitory response to salinity stress. Plant Cell Environ 17: 303–309.
63. Cachorro P, Ortiz A, Barcelo AR, Cerda A (1993) Lignin deposition in vascular tissues of *Phaseolus vulgaris* roots in response to salt stress and Ca^{2+} ions. Phyton-Ann Rei Bot A 33: 33–40.
64. Liang M, Haroldsen V, Cai X, Wu Y (2006) Expression of a putative laccase gene, *ZmLAC1*, in maize primary roots under stress. Plant Cell Environ 29: 746–753.
65. Wei JZ, Tirajoh A, Effendy J, Plant AL (2000) Characterization of salt-induced changes in gene expression in tomato (*Lycopersicon esculentum*) roots and the role played by abscisic acid. Plant Sci 159: 135–148.
66. Wang T, Chen L, Zhao M, Tian Q, Zhang W-H (2011) Identification of drought-responsive microRNAs in *Medicago truncatula* by genome-wide high-throughput sequencing. BMC Genomics 12: 367.
67. Gaxiola RA, Li J, Undurraga S, Dang LM, Allen GJ, et al. (2001) Drought- and salt-tolerant plants result from overexpression of the AVP1 H+-pump. Proc Natl Acad Sci U S A 98: 11444–11449.
68. Park S, Li JS, Pittman JK, Berkowitz GA, Yang HB, et al. (2005) Up-regulation of a H^+-pyrophosphatase (H^+-PPase) as a strategy to engineer drought-resistant crop plants. Proc Natl Acad Sci USA 102: 18830–18835.
69. Zhao FY, Zhang XJ, Li PH, Zhao YX, Zhang H (2006) Co-expression of the *Suaeda salsa SsNHX1* and *Arabidopsis AVP1* confer greater salt tolerance to transgenic rice than the single *SsNHX1*. Mol Breeding 17: 341–353.
70. Zhang H, Shen G, Kuppu S, Gaxiola R, Payton P (2011) Creating drought- and salt-tolerant cotton by overexpressing a vacuolar pyrophosphatase gene. Plant Signal Behav 6: 861–863.
71. Park S, Rancour DM, Bednarek SY (2007) Protein domain-domain interactions and requirements for the negative regulation of *Arabidopsis* CDC48/p97 by the plant ubiquitin regulatory X (UBX) domain-containing protein, PUX1. J Biol Chem 282: 5217–5224.
72. Rancour DM, Park S, Knight SD, Bednarek SY (2004) Plant UBX domain-containing protein 1, PUX1, regulates the oligomeric structure and activity of *Arabidopsis* CDC48. J Biol Chem 279: 54264–54274.
73. Golldack D, Luking I, Yang O (2011) Plant tolerance to drought and salinity: stress regulating transcription factors and their functional significance in the cellular transcriptional network. Plant Cell Rep 30: 1383–1391.
74. Yang O, Popova OV, Suthoff U, Luking I, Dietz KJ, et al. (2009) The *Arabidopsis* basic leucine zipper transcription factor *AtbZIP24* regulates complex transcriptional networks involved in abiotic stress resistance. Gene 436: 45–55.
75. Yoshida T, Fujita Y, Sayama H, Kidokoro S, Maruyama K, et al. (2010) AREB1, AREB2, and ABF3 are master transcription factors that cooperatively regulate ABRE-dependent ABA signaling involved in drought stress tolerance and require ABA for full activation. Plant J 61: 672–685.
76. Yang XZ, Zhang HY, Li L (2011) Global analysis of gene-level microRNA expression in *Arabidopsis* using deep sequencing data. Genomics 98: 40–46.
77. Hofacker IL (2003) Vienna RNA secondary structure server. Nucleic Acids Res 31: 3429–3431.
78. Audic S, Claverie JM (1997) The significance of digital gene expression profiles. Genome Res 7: 986–995.
79. Dai XB, Zhao PX (2011) psRNATarget: a plant small RNA target analysis server. Nucleic Acid Res 39: W155–W159.
80. Altschul SF, Madden TL, Schaffer AA, Zhang J, Zhang Z, et al. (1997) Gapped BLAST and PSI-BLAST: a new generation of protein database search programs. Nucleic Acid Res 25: 3389–3402.
81. Papini-Terzi FS, Rocha FR, Vencio RZN, Oliveira KC, Felix JD, et al. (2005) Transcription profiling of signal transduction-related genes in sugarcane tissues. DNA Res 12: 27–38.
82. Pabinger S, Thallinger GG, Snajder R, Eichhorn H, Rader R, et al. (2009) QPCR: Application for real-time PCR data management and analysis. BMC Bioinformatics 10: 268.

18

Improving Index-Based Drought Insurance in Varying Topography: Evaluating Basis Risk Based on Perceptions of Nicaraguan Hillside Farmers

André Kost[1]*, Peter Läderach[2], Myles Fisher[3], Simon Cook[4], Lorena Gómez[2]

1 University of Bonn, Dept. of Geography, Bonn, Germany, 2 International Center for Tropical Agriculture, Managua, Nicaragua, 3 International Center for Tropical Agriculture, Cali, Colombia, 4 International Water Management Institute, Sunil Mawatha, Pelawatte, Battaramulla, Sri Lanka

Abstract

This paper discusses a methodology to model precipitation indices and premium prices for index-based drought insurance for smallholders. Spatial basis risk, which is borne by the insured, is a problem, especially in variable topography. Also, site-specific drought risk needs to be estimated accurately in order to offer effective insurance cover and ensure financial sustainability of the insurance scheme. We explore farmers' perceptions on drought and spatial climate variability and draw conclusions concerning basis risk with regards to the proposed methodology. There are technically many options to represent natural heterogeneity in index insurance contracts while serving the customer adequately and keeping transaction costs low.

Editor: John P. Hart, New York State Museum, United States of America

Funding: Research was partially financed by Gesellschaft für Internationale Zusammenarbeit. The funders had no role in study design, data collection and analysis, decision to publish, or preparation of the manuscript. No additional external funding received for this study.

Competing Interests: The authors have declared that no competing interests exist.

* E-mail: andre.kost@arcor.de

Introduction

1. Agricultural Risk Management and Index Insurance

Index-based insurance for agriculture [1,2,3] was developed in industrialized countries to avoid adverse selection and moral hazard in traditional agricultural insurance [4,5,6]. The key is the index, which is commonly based on measured rainfall [7,3,5] as a proxy for yield shortfalls due to adverse weather. The index-insurance contract is not about crop loss, but a predefined range of values of the index. In the case of a precipitation index, insurance cover is based on the rainfall measured by a gauge at a predefined site.

Index insurance lowers transaction costs because farm inspections are unnecessary [8], and because the index is used to set the premiums and assess any indemnity. Low-cost index insurance is a possibility for smallholder farmers in developing countries where insurance companies are few and have limited capacity, and where it is difficult to access rural areas because of poor infrastructure.

Although farmers are accustomed to manage risk, poor smallholders invariably use strategies that minimize it, typically by investing as little as possible in their subsistence crops. Index-based insurance could be a supplementary strategy [9] to allow them to reduce risk [10] and give them access to credit by hedging their investment in crop inputs. Index insurance could also partially substitute for emergency aid when there are catastrophic harvest failures [10,11].

There is a 14-year history of testing index insurance in Nicaragua for commercial crops of rice, peanuts, maize, soy, and sorghum in the western lowlands using a statistical approach [5,12]. In 2009, the total sum insured was about US$2 million, but the scheme was not for poor smallholders. It was carried out by the public Nicaraguan insurer INISER and the World Bank [3]. In the same year, the international insurer Lafise opened first the national market for agricultural insurance within an initiative of the Interamerican Federation of Insurance Companies (FIDES), insuring more than 700 ha of irrigated rice and peanuts. In 2010, coverage was expanded to maize, beans, and sorghum. Insurance was supposed to be sold to producers of any type via cooperatives and local financial institutions. By this means, it was expected to reach around 5000 producers, mainly in the central and northern parts of Nicaragua [13,14,15].

However, a statistical approach, which Lafise and INISER currently use, may not be sufficiently precise to determine a farmer's risk of crop loss [16,17]. This is even truer where the topography and microclimates are heterogeneous as it is the case for north-central mountains of Nicaragua. In this study, we explore a more precise method using a weather generator and crop physiological models to determine crop risk, which is applicable to any other place in the tropics and to a large number of crops. Specifically we explore farmers' perceptions on drought and spatial climate variability and draw conclusions concerning basis risk with regards to the proposed methodology.

2. Study Context

Drybean is an important subsistence and cash crop of smallholder farmers throughout Central America. Surveys of smallholders in western Matagalpa department in central Nicaragua showed that they knew little about insurance of any sort [18,unpublished data]. They quickly understood how index-based crop insurance works and expressed interest in it to protect them

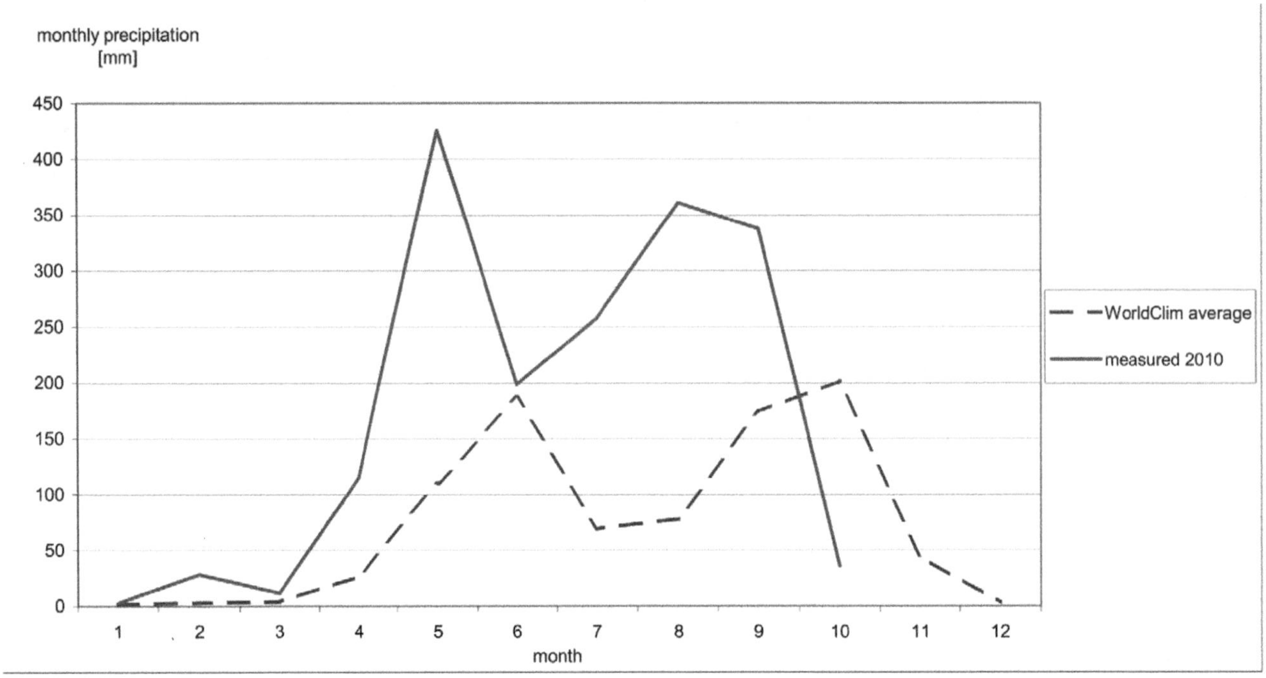

Figure 1. Precipitation January-October 2010 at weather station San Isidro (458 m). (Source: WorldClim database and [51]).

against yield loss of rainfed (not irrigated) drybean crops. Although farmers were hypothetically willing to buy insurance cover, this does not guarantee that they would buy it were it to be offered. It is noteworthy that an innovation introduced from the exterior may not be adopted, often because it does not work well at a particular locality [19].

The major problem of index-based insurance is that it exposes the insured to *basis risk* [16,17] in which the damage caused by the insured event is worse for the insured than at the point where the index is measured [20]. The insurer is covered in that the premiums are calculated on an actuarial assessment of the risk of the insured event occurring at the same point where the indemnity is assessed, the long-term meteorological station. It is the insured who bears the basis risk [16,17], which is an aspect of index-based insurance that has been neglected. If an insurance instrument has a high basis risk, it cannot fulfil the clients' expectations, and will therefore not sell [21].

The sources of basis risk in agricultural index-insurance based on precipitation are temporal risk, crop-specific risk, and spatial risk [22]. To reduce basis risk, index developers must consider (1) the temporally-varying water needs of (2) specific varieties and (3) the spatial variation of water availability due to natural heterogeneity.

The temporally-varying water needs of specific varieties ((1) and (2) above) are captured by the Decision Support System for Agrotechnology Transfer (DSSAT) simulation software [23]. DSSAT combines current knowledge of agro-ecological systems, crop agronomy, and plant physiology. It can be used to construct indices more consistently than statistical approaches [16,17]. Reducing spatial basis risk is harder, especially where there are only a few scattered meteorological stations with poor historical data, which is often the case in developing countries. It is even more difficult in heterogeneous topography. The current best solution is to extract monthly data from the 1-km-resolution climate data in the WorldClim database [24] and use them as input to the MarkSim weather generator [25,26] to produce daily weather data. It costs more to design site-specific indices, which increases premiums, but the index must be sufficiently site-specific that the buyer considers the basis risk to be negligible [27].

In this study we investigated basis risk using Díaz Nieto et al.'s methodology for index-based drought insurance for drybean [16,17,18] as a starting point. We explored farmers' perceptions and preferences on crop risk and insurance to identify what aspects of the methodology require further development. By this means, we sought to "bridge the two cultures of risk analysis" [28]: to integrate traditional numerical risk analysis with the layman's perception of risk to implement a sustainable process of risk management [29].

3. Modeling Precipitation Indices for Nicaraguan Drought Insurance

The north-central mountains of Nicaragua have an east-west precipitation gradient, with the highest annual rainfall (2000 mm) on the eastern slopes and the lowest precipitation in the valleys. A sequence of intramountain plains extends from Ciudad Dario northwest towards central Honduras forming the *franja seca* (dry fringe) with annual rainfall less than 800 mm [30]. The *franja seca* crosses the study area in the municipality of San Isidro. Rainfall is distributed bimodally with peaks in June and September/October and a relative dry spell July – August (*canícula*) [31]. There is considerable variation in precipitation in any one year (Figure 1). For example, the start of the rainy season in May and its end in November is influenced by the current status of the El Niño-Southern oscillation (ENSO). El Niño corresponds with years of extensive drought [32], while La Niña corresponds with above-average rainfall [33], such as in 2010 when a strong La Niña gave damaging high rainfall totals and in which there was no *canícula.*

Tropical rainfall mainly depends on convectional cells 100–300 km^2 in extent. These cells remain stationary for a few days and give high-intensity rainfall events that only last for a few hours.

In general, the convection cells occur randomly but they are also influenced by topography [34]. In view of this, a particular raingauge only represents a relatively small area. Indeed, two places only a few kilometers apart can have completely different rainfall patterns over periods as long as one month [34]. What, then, are the consequences for spatial basis risk of any index insurance scheme?

In the sub-humid tropics the performance of rainfed agriculture in any one year is determined primarily by the characteristics of the rainy season and temperature, which is the main driver of evapotranspiration. Temperature variations with altitude can also be important in mountainous regions. The impact of rainfall on crop production depends on daily soil water balance [34], which requires daily meteorological data, and highlights the importance of modeling soil water to obtain a valid insurance index.

The challenge in the design of an index in a heterogeneous landscape is to balance cost and benefit, that is, whether to represent the variation with more precision, and cost, to reduce the basis risk. On the other hand, even a quite elaborate index can have high basis risk if the spatial scale is not appropriate [27,35]. The DSSAT simulation model provides a cost-effective method to produce indices that take account of variable natural conditions at an appropriate scale to reduce basis risk in a transparent manner.

Díaz Nieto et al. [16,17,18] used the weather generator MarkSim and the DSSAT drybean submodel to construct precipitation indices for drought in Central America based on ten-day rainfall weights (Table 1; see also [16]). They extracted the probabilities of a given rainfall deficit from 99 years' generated daily weather data at 18-km resolution.

We used the same method, but with a spatial resolution of 1 km by extracting climate data from WorldClim [24].

Methods

1. Research Focus

Farmers are more likely to accept an index insurance product if it is designed with their participation. Although this costs more to set up, it has the advantage of exposing farmers to the new concept prior to launching the product [7,36,37]. We interviewed smallholder households, whose main crop was rainfed drybean and who are likely to be potential buyers of index insurance, with the purpose to evaluate the basis risk inherent in index insurance as described above. We focused on four areas:

(1) Exploration of farmers' notions of drought.

(2) Investigation of factors, apart from precipitation, that farmers perceive to influence yield. How do farmers perceive the importance of weather risk/drought risk compared with other risks?

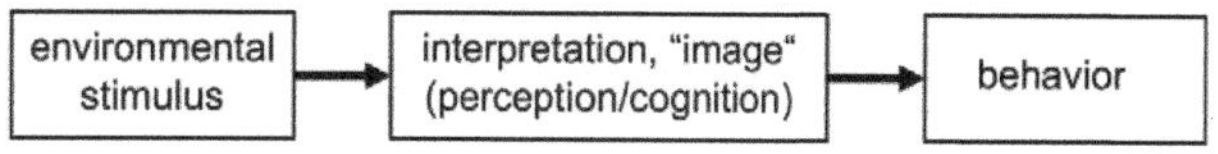

Figure 2. Behavioristic stimulus-response-model with perception component. (Source: adapted from [36]).

(3) Assessment of the plausibility of the averaged precipitation index for drybean crops developed using the Díaz Nieto et al. methodology [16,17,18]. The guiding question was how important do farmers judge the right rainfall distribution to be to obtain good drybean yields?

(4) Investigation of farmers' perceptions of spatial climate variation. What conclusions can be drawn for the assessment of spatial basis risk of an index-based insurance product?

2. Perception of Environmental Risk

In theory, perception and cognition are variables in a behavioural stimulus-response-for explaining human behaviour (Figure 2). There is a (rather conceptual) distinction between how an individual sensorially *perceives* the environment (i.e. perception) and how this information is further processed mentally, which is *cognition* [38]. As an individual's *environment* we define "the total milieu in which man lives, an environment with both physical and sociocultural attributes" [39]. Consequently, "the key intervening variables in man – environment transactions are perception and cognition – the internal mental processes by which individuals sense, perceive, interpret, and make decisions about their environment" [38].

In this study we deal with a broader understanding of *perception* "in the sense of how things are remembered or recalled by people - as with respect to 'perception' of resources or hazards" [39]. In this sense, perception denotes the mental representation one has of one's environment. It is not therefore intended to explain a certain behavior through perception, but rather to investigate a subject's mental images.

We are concerned here with farmers' perceptions of site-specific risks in rainfed drybean. "Risk perception, in general, denotes the processing of physical signals and/or information about potentially harmful events ... and the formation of a judgement about seriousness, likelihood and acceptability of the respective event" [29]. In this case we considered climate events that reduce yields.

We could not capture farmers' perceptions of basis risk directly because we judged that they could not understand the concept within the time limitation of a single interview. Thus, farmers could not evaluate the insurance model directly. Instead, we sought to employ farmers' perceptions of the environment as a

Table 1. Weightings of ten-day precipitation in the DSSAT averaged insurance index for drybean.

	Day 1-10	Day 11-20	Day 21-30	Day 31-40	Day 41-50	Day 51-60	Day 61-70	Day 71-80	Day 81-90
crop stage	**planting/ seedling**	**Seedling /flowering**	**flowering**	**Flowering /grain fill**	**grain fill**	**grain fill/ maturity**	**maturity**	**maturity**	**maturity**
index values [mm]	10	10	25	40	40	40	30	10	0
index weighting (rounded)	0.05	0.05	0.12	0.2	0.2	0.2	0.15	0.05	0

(Source: adapted from [17]).

proxy for their perception of basis risk by using qualitative methods for analysis of verbal data [40,41].

3. Research Design

The climate and topography in western Matagalpa vary considerably within short distances, making it suitable to test whether WorldClim and MarkSim represent the spatial variation of rainfall patterns and realistically model the probability of drought. We also evaluated the suitability of the averaged precipitation indices that we calculated.

The municipalities of Matagalpa, San Isidro, and San Dionisio, all within 30 km, encompass the climatic variability of the region. Within them we chose sites having contrasting climates to sample a wide range of climate risk and maximize variance [40]. Moreover, each site has an official weather station so that they could hypothetically serve as reference gauges if an index insurance scheme based on precipitation were to be established (Figure 3).

We held six meetings with drybean farmers in each municipality, with no more than two farmers from any one neighbourhood or community. The meetings were organised as "problem-centred" interviews [42], which focus on a specific problem, in this case perceptions of the effect of weather on drybean yields. Problem-centred interviews need a certain degree of structuring but they also foster narrative flow, which helps in creating confidence and lessens the interview's artificiality [42].

Data analysis and interpretation follow Mayring's "qualitative content analysis" [41] and established typologies [40], both of which interpret verbal data by building systems of thematic categories. Categories, in turn, were derived deductively from research hypotheses, which we formed in advance or alternatively created *ad hoc*, inductively if necessary [40]. In the final step we linked categories to build a system of theorizing statements (typologies).

As an example, we defined a typology of households that captured their distinctive features: (1) the extent of a household's resource scarcity, and (2) the household's attitude towards its resource scarcity. This gives four types of households, which we named *pioneers* (young families, scarce resources, optimistic attitude), *established* (families with grown-up children, sufficient resources, modest or indifferent attitude), *precarious* (drought-prone, scarce resources, ironic or pessimistic attitude), and *go-aheads* (leadership tasks in community, sufficient and diversified resources, cooperative and innovative attitude).

Results

We first present a short overview on household characteristics and farming practices for which we use data from our survey (cf. 1.). After this we elaborate the interpreted verbal data with reference to the points listed in chapter 1 of the methods section in four subsections (cf. 2.–5.).

1. Households' Farming Systems and Economic Behaviour

The farmers grow drybean in two periods during the rainy season, the *primera* (May–August) and the *postrera* (September–

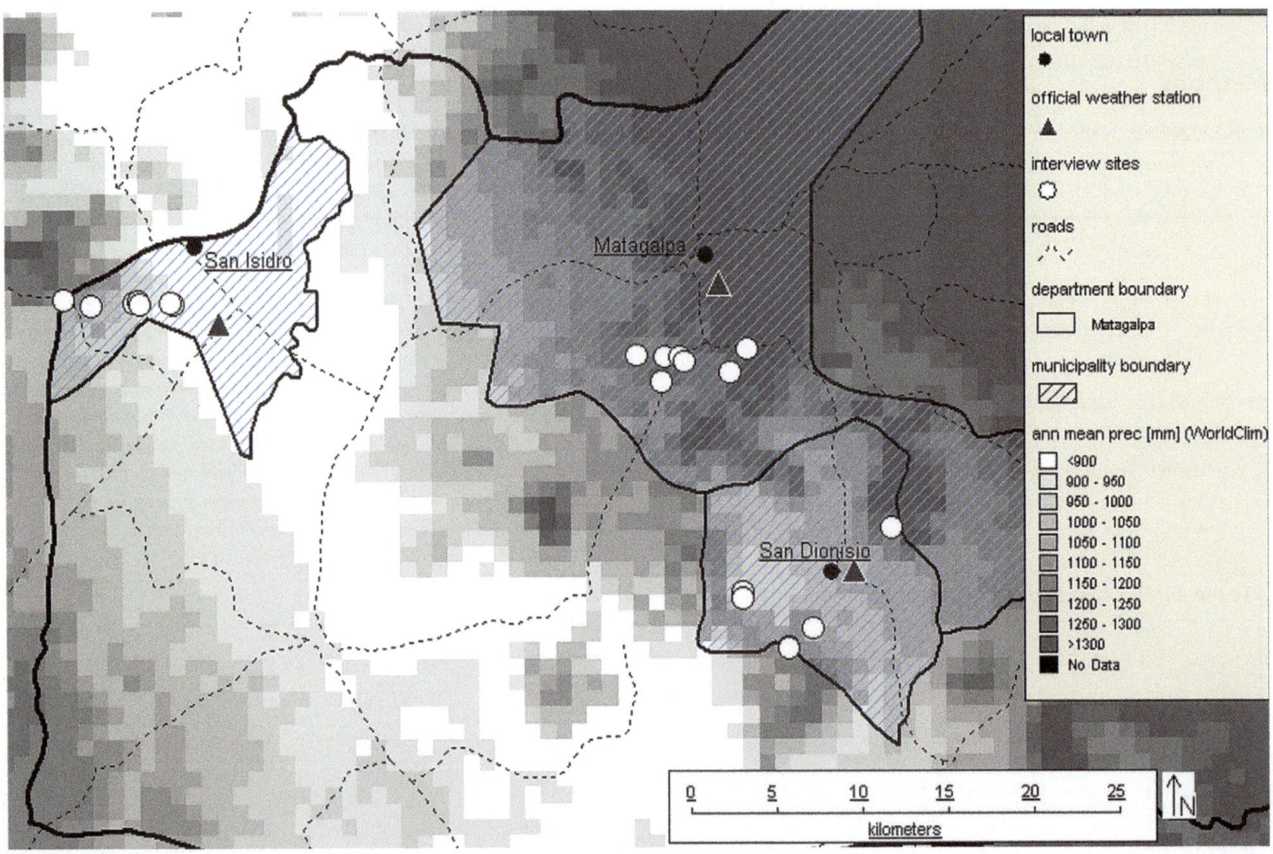

Figure 3. Mean annual precipitation in the municipalities of Matagalpa, San Isidro, and San Dionisio. (Source: WorldClim database, prepared with DIVA-GIS).

November). Normally, the *primera* crop is harvested in the *canícula* (July–August) and the *postrera* is harvested in the first weeks of the dry season (November–December). Occasionally, farmers opportunistically grow a third *apante* crop, sown in November/December and harvested in February, but yields are low, the area sown is less than 0.7 ha.

Farmers normally sow drybean 20 May–10 June (*primera*), 1 September–5 October (*postrera*), and 10 November–15 December (*apante*) [43]. Individual farmers may use narrower ranges, especially in hotter locations for the *postrera*. Farmers typically wait for three to five consecutive rainy days to have sufficient soil water to sow the crop. Harvest is 60–90 days after sowing, with crop duration shorter in warmer climates (Table 2), which are often also drier. Duration of the *postrera* crop is 10 days shorter than the primera.

Farmers' households were 3–11 persons, usually the farmer couple and their non-adult children, and occasionally other relatives. All households cultivated the basic grains, drybean and maize, and in warmer climates also sorghum, often with some livestock. In cooler climates, households typically grew coffee as a cash crop. Many households grew some citrus, while the most diversified household had additional crops such as banana, sugar cane, cassava, and rice.

Holding size varied considerably, from landless households, which rent the land they cultivated, to a maximum of 55 ha, with larger holdings having more livestock. The area cropped with drybean is typically 1.5–2 ha in each season, with a maximum of 7 ha.

All households were risk averse. Economic priorities were always: (1) Securing household members' basic food security; (2) Settling outstanding liabilities; (3) Creating a reserve for the next crop; and (4) Providing for health care, clothing, and education. All households' primary goal was to achieve food security with the basic grains and then to stabilise the household's assets. This means that any cash income from the crop covers running expenses first and leaves neither scope nor incentive for investment.

Many farmers use short-term credit, either formally from an organization or informally from a friend or relative. Credit is not invested in productive assets, but is used only to cover emergencies, often only when there was no other way to finance the new sowing. These characteristics are typical of the poverty trap [44,45] of persistent, long-term, economic stagnation. A household may accumulate some capital in the short- or mid-term, but unforeseeable (= risk as probability) and unavoidable (= risk as exposure) shocks counteract any attempt at economic expansion [45]. Economic growth is impeded by the existence of uncontrollable risks, which is how the farmers perceive them, whose potential damage exceeds a household's ability to cope.

2. Farmers' Notions of Drought

Wilhite and Glantz [46] classify drought into meteorological, hydrological, agricultural and socioeconomic drought. Here, we fit farmers' notions of drought to this scheme.

(1) Meteorological drought: For the farmers, "drought" is marked by the two dry seasons in the course of a year. The first is the long dry season December–April (*verano*), the second is the short *canícula* dry period.

(2) Hydrological drought: Farmers associated dried-up streams with a drought, concluding that there was no rain even in the uplands.

(3) Socioeconomic drought: Streams that had run dry had serious consequences for everyday life since there were shortfalls in drinking water, which constrains all economic activities.

(4) Agricultural drought: Drought just after sowing of the drybean crop gives poor germination, so that farmers delay sowing until there is sufficient soil water, because they cannot afford a failed sowing. In some localities with favorable soils, drybean crops can withstand dry periods for as much as one month after emergence. Drought at harvest is desirable to get good grain quality, while rain at harvest reduces both yield and quality. Nevertheless, most crops give some yield in high rainfall years, and coffee performs better. Farmers fear drought because it causes food insecurity and leaves no agricultural alternatives to produce income.

3. Perceptions of Yield-determining Factors

We investigated how farmers manage their drybean crops, hypothesizing that both agroecological processes and daily household routines play key roles, as well the households' perception of weather risk. There are three systems:

- Households that grow drybean as a component of food security, typically a specific variety to satisfy their household's culinary preferences of taste and cooking consistency;
- Households that only produce grain for sale, usually a specific variety that yields best or commands a market premium;
- Households that sell the surplus to their food needs. Some of these households cultivate two varieties, one for consumption with the preferred culinary characteristics, and a second for sale with better yields or higher price.

Farmers do not have information on the yield potentials of varieties suitable for marketing against which they can assess the performance of their crops. Nor do they have ready access to new varieties (see also [47]). They therefore usually refer to benchmark yields derived from their own experiences with one variety, which they have likely grown for decades. Only those households classified as "go-aheads" obtained benefit from the national extension service, which installed experimental plots of improved varieties in farmers' fields. The experiments tested the varieties' local adaptation and provided objective yield benchmarks.

Farmers know that drybean responds to fertilizers of different qualities, pesticides, soil fertility, hill slope, and variety, all of which are under their control. They know that they can choose the sort and quantity of inputs, soil conservation measures, which lots to cultivate, and which variety to grow. In reality, however, both their limited perception of agroecological processes, especially yield responses, and socioeconomic necessities limit their ability to increase yields. Moreover, people will only seek to innovate to solve a problem if they recognize that indeed there is a problem.

Farmers agreed that the distribution of rainfall in the wet season was the main factor influencing yields, indicating that they accepted having no control over climatic risk, which they did not perceive as a problem needing resolution. Reliable weather forecasts can help farmers better estimate climate risk in the short term, for example to adjust sowing and harvesting dates [48]. Surveyed farmers used traditional dates even when weather forecasts were unfavourable, indicating that farmers believe that the national meteorological agency's (INETER's) forecasts were too approximate to be reliable. Moreover, INETER's forecasts were often at odds with farmers' traditional forecasts. We could not verify that seasonal forecasts, broadcast on radio and television, influenced farmers' decisions on crop management.

Table 2. Climate and bean cultivation characteristics at 18 interview sites.

Interview site		Climate (WorldClim)			Bean cultivation characteristics			
No.	**Community (Municipality)**	**Temperature (average/ year) [°C]**	**Precipitation (May-Nov) [mm]**	**Altitude [m] (SRTM90)**	**sowing in *primera***	**sowing in *postrera***	**bean cycle [days]**	**grown varieties**
7	Santa Rosa de la Lima (San Isidro)	24.1	877	486	1.–10.5.	15.8.–10.9.	60	criollo
8	Santa Rosa de la Lima (San Isidro)	24.1	877	488	1.5.–10.6.	8.–20.9.	60	INTA rojo
13	Las Cuchillas (San Dionisio)	23.5	965	593	1.–25.6.	1.–20.9.	75	INTA Masatepe
12	El Bocón (San Isidro)	23.1	950	674	1.–31.5.	1.–15.9.	60	criollo
14	Las Cuchillas (San Dionisio)	23.1	988	641	28.5.–15.6.	28.9.–15.10.	75	INTA Masatepe
16	Ocote Arriba (San Dionisio)	23	1032	685	12.5.–15.6.	1.9.–5.10.	60	INTA Masatepe
1	Limixto (Matagalpa)	22.4	1280	918	15.5.	1.9.-begin Oct	75	criollo
17	Wibuse (San Dionisio)	22.2	1278	637	1.–31.5.	1.–20.9.	85	INTA rojo
18	Wibuse (San Dionisio)	22.2	1278	637	28.5–23.6.	15.9.–9.10.	75	INTA rojo
11	El Bocón (San Isidro)	22.1	972	680	10.5.–13.6.	1.–15.9.	75–85	criollo
15	Ocote Arriba (San Dionisio)	21.9	1142	786	1.5.- begin June	1.9.–15.10.	90	INTA rojo
3	Jucuapita (Matagalpa)	21.8	1255	828	May-June	15.9.-begin Oct	70	Estelí
							85	H
2	Jucuapa Centro (Matagalpa)	21.2	1203	889	1.5.–15.6.	1.–30.9.	85–90	H
9	Las Sidras (San Isidro)	21.1	1043	965	10.5.–30.6.	1.9.–5.10.	75	INTA Masatepe
6	El Ocotal (Matagalpa)	20.5	1229	973	10.5.–4.6.	14.9.–4.10.	80–90	INTA Palma
10	Las Sidras (San Isidro)	20.3	1179	1120	-	-	-	-
4	Nuestra Tierra (Matagalpa)	20	1390	1104	1.5.–20.6.	1.–20.9.	-	-
5	Santa Josefina (Matagalpa)	19.6	1352	1112	June	1.9.–15.10.	85	INTA Masatepe

(Source: WorldClim database, SRTM90 elevation data, own data).

4. Perceptions of an Optimal Rainfall Distribution in Drybean Cultivation

We asked farmers to sketch optimal rainfall distribution (ten-day steps) that would give good yield and classified the graphed results. We also used them to cross-check the averaged rainfall index (Table 1, Figure 4) we developed [17].

Farmers agreed that good crop yields required evenly-distributed rainfall throughout the season, with a moderate (*normal*) total, with a few days of rain alternating with a few sunny days. A *normal* wet season gives *normal* yields of drybean. *Normal* yields varied between farmers as much as twofold in the same growing season in the same year (300–600 kg/ha) whereas maximum yields could be as much as1300 kg/ha. Yields in 2010 *primera* season ranged from total crop loss to about 1000 kg/ha.

Heavy rains rather than drought caused total crop failure, while all farmers indicated that there was always enough crop to cover

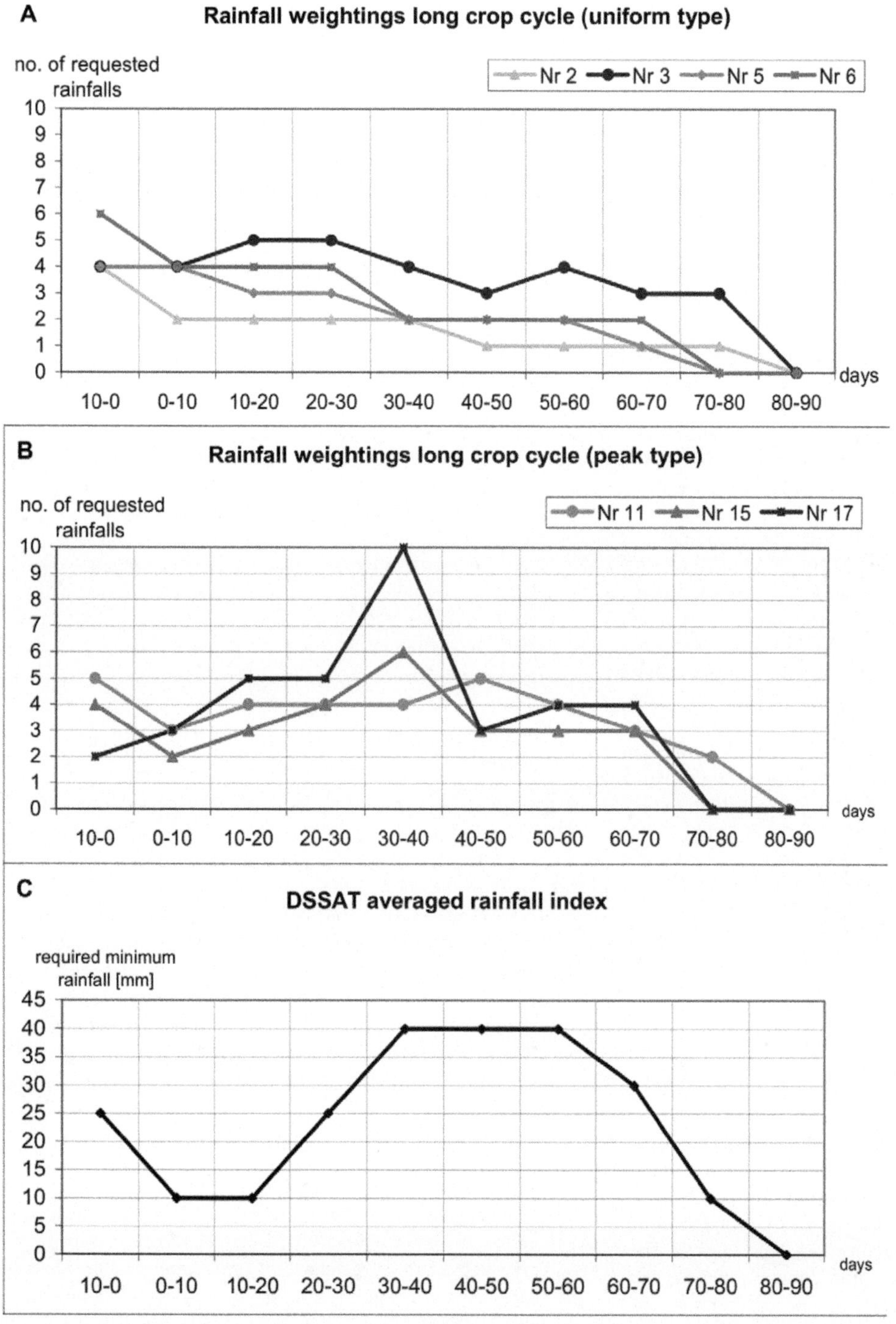

Figure 4. Optimal rainfall weightings in farmers' graphs (A and B) and DSSAT averaged insurance index (C). (Source: own data, [13]).

the household's consumption needs. Nevertheless, in drought years yields were much less than normal.

The graphical comparison showed that drybean has differing crop durations across the research area (Table 2), with long duration (85–90 days, 7 farmers), medium duration (75 days, 5 farmers) and short duration (60 days, 4 farmers). Rainfall is most important in the 10 days before sowing, while dry conditions are required at harvest.

We distinguished two weighting patterns in the farmers' graphs of crop water requirement (Figure 4). Panel B shows increasing requirement until the middle of the cycle and then declines (*peak* distribution), reflecting the physiological sensitivity of the bean crop to water deficits, which the farmers knew. Panel A shows almost constant demand (*uniform* distribution), which is consistent with crops grown on soils that have high available soil water. Farmers with this type of demand indicated that ample rainfall was important before sowing, but that a few evenly distributed rainfall events are sufficient thereafter. Yields in the drier areas were lower, requiring rainfall each 8–15 days.

5. Perceptions of Spatial Climate Variation

Climate information is available to farmers on spatial scales ranging from national, local, to their own on-farm perceptions. Farmers' concept of climate is more than a location's meteorology, however. They typically responded that the climate is *normal* or *good*, meaning that they perceive it to be normal or good for the crops they grow, which emphasizes that they see agriculture and climate from a holistic perspective. Similarly, they often linked *climate* with *soil*, referring to soil properties, such as water holding capacity, when describing a farm's climate more precisely. This indicates farmers' awareness of the linkage between rainfall and soil properties on crop performance [49,50].

We delineated farmers' spatial *frame of reference* [51] for climate. They commonly used four terms to describe climate at different locations: *helado*, *fresco*, *cálido*, and *caliente*, literally cold, fresh, warm, and hot. Farmers' use of these terms includes the concept of both temperature and humidity, in that *caliente* means both hot and dry, and *helado* means not only cold but humid as well.

To many farmers a location's climate is described in terms of its altitude relative the farmer's reference point. If a location is higher (*más arriba*) than the farmer's own farm, the climate up there will be cooler (*más fresco*). Thus farmers in Matagalpa perceive spatial climate variation as a difference in altitude, and by comparing altitudinal levels, they construct small-scale homogenous climatic zones, e.g. *zona fresca* and *zona caliente*.

Farmers perceive soils to be better in higher locations in the landscape because they have more available soil water (ASW), calling them *tierras negras* caused by their higher organic matter content and higher ASW. At the other end of the scale are *tierras blancas*, which are sandier, have less organic matter, and lower ASW.

Farmers assume that rainfall originates near a mountain and spreads out from there to lower altitudes with decreased intensity and frequency. A farmer's typical perception of spatially differing precipitation is relative (*more* or *less*) since they have little knowledge about rainfall gauging, but a perception that is linked to the perceived climate zones. They judge a weather station's precipitation data to be representative of their farm if they perceive it to be situated within the same climatic zone, e.g. *zona fresca*. When extreme events occur, such as the high rainfall totals in 2010 or a drought, farmers perceive that all locations are affected similarly. In May, 2010, however, Matagalpa's rainfall was 128 mm while San Isidro, 30 km distance, recorded 425 mm [52]. Further analysis of the meteorological records shows significant variability in monthly rainfall totals at nearby weather stations.

Farmers believe that hotter climates have more severe and more frequent droughts, which cause less loss because the hotter climates are less productive. In contrast, although less frequent, they believe that drought causes more damage within the more productive zones at higher altitudes with moderate climate.

Discussion

We discuss the results presented above in relation to the insurance model. We first discuss how the two core components, the index model and the spatial model, correspond with farmers' perceptions to highlight how farmers see basis risk as it applies to them (cf.1.–2.). We then draw on farmers' perceptions to propose measures that lessen basis risk in the research area but which are also relevant elsewhere. Finally we relate farmers' perceptions to model variables that need scrutiny and discuss some practical implications (cf. 3.).

1. Validity of the Insurance Index

Here we rely on the results of the farmers' rainfall drawing exercise and we use *validity* in the context of *plausibility* of the insurance index from the farmers' perspectives.

We assume farms classified as requiring *peak* rainfall distribution are based on some knowledge about physiological stages and related water needs of drybean. For these farmers, the averaged insurance index (Figure 4, Table 1) will probably be more plausible and hence more acceptable than for those whose farms were classified as requiring *uniform* rainfall distribution. The *uniform* distribution requirement may be interpreted in one of three ways:

(1) Farmers did not know that water demands of drybean vary according to physiological stages, and they think that equally-distributed rainfalls lead to the best yields;

(2) There is sufficient water supply at this site during the rainy season (it has low drought risk, and a cool microclimate); or

(3) The site has soils with high ASW and a microclimate with low evapotranspiration.

The averaged index (Table 1) would be plausible to farmers for cases (1) and (2), requiring some extension work to make them aware of basic crop physiology. Case (3), however, suggests that the averaged index is still too imprecise to capture spatial variation of factors that determine yield at some sites.

Farmers indicated crop durations ranging 60–90 days, depending on the microclimate of their sites because crop duration is largely under thermal control. The averaged insurance index covers a period of 90 days [16,17,18], but needs to be modified for shorter-duration crops. Further analysis of the farmers' curves of water needs curves shows higher demands on soils other than *tierras negras*, which is in line with Díaz Nieto et al.'s recommendation to tailor indices to different soils [16,17].

2. Representation of Spatially Varying Drought Risk

Farmers' perceived risk of drought coincided with their perception of spatial variability of climate. Risk of drought was highest for the drier, warmer, and lower altitude sites. A single drought causes the highest damage in the moderate zones where drought is less likely but where asset exposure is high. This perception pattern of drought risk fits well to the analytical concept of risk as the product of damage potential and probability of an adverse event.

In general, farmers' perceptions of drought risk coincided with the probabilities derived from the simulations, although farmers tended to underestimate the spatial differences compared with simulated values.

Farmers' perceptions of climate were strongly dependent on the scale at which they operate, often discriminating as little as 1–2 km, and confirming the need for spatial resolution of 1 km to keep spatial basis risk low. For example, in San Dionisio the probability of a rainfall deficit of 60 mm at Wibuse is 0.15 but 10 km away at Las Cuchillas it is 0.66, which farmers knew had a higher risk of drought.

3. Recommendations for Further Methodology Improvement

We confirm that data of farmers' perceptions and preferences support Díaz Nieto et al.'s methodology [16,17,18]. Technically, spatial basis risk can be lowered through higher resolution analysis, which from the viewpoint of the insured offers contracts tailored to their microclimate and soils.

3.1 Modeling of indices. Díaz Nieto et al.'s averaged rainfall index [16,17,18] was based on MarkSim's built-in climate surface with 10 arc-minute resolution. WorldClim's surface with 30 arc-second resolution permits the method to take account for the research area's heterogeneity. Moreover, the DSSAT crop simulations from which rainfall indices are derived need to take account of:

(1) Adjust the sowing window (15 April–15 May) to those actually used by the target farmers. Díaz Nieto identified sowing window as critically important as it has a big influence on the periodical weighting of precipitation in the index [53] (cf. Table 1).

(2) Soil characteristics are very important and the simulations should use data for the actual soils in the target area, including water characteristics, mineral composition, organic matter, and chemical data for each layer. Simulations should ideally include a range of terrain slopes and corresponding adjustments to the infiltration characteristics. Where possible the simulations should include comparison with field data to ensure satisfactory calibration. It should be tested whether offering two insurance instruments for soils of contrasting texture captures farmers' distinction between *tierras negras* and *tierras blancas* and in doing so adequately reduces basis risk compared with one instrument for an average soil.

(3) Farmers should be able to opt for indices for crop durations of 60, 75 and 85 days, depending on their sites' microclimates (cf. Table 2). It is therefore adequate to compute indices for each of the climatic zones (*fresca, cálida, caliente*) as perceived by farmers.

(4) Farmers reported that the duration of the *postrera* crop is about 10 days shorter than the *primera*. We need to know if this is a temperature effect captured by DSSAT and whether the *postrera* needs different weighting patterns, i.e. another set of indices.

(5) Farmers often mentioned the specific microclimate on their fields that influenced cultivar perfomance. DSSAT can take account of slope and aspect, which, if there was perceived to be a demand, could readily be simulated to provide the basis for an appropriate insurance product. Still, the feasibility of offering a multitude of insurance products would depend on the demand and the willingness of the insurer to meet it. Technically there is no problem.

(6) Farmers did not perceive differences in water demands between *mejoradas* and *criollas* cultivars. If there are differences, their genetic coefficients need to be determined to permit DSSAT to simulate them satisfactorily.

3.2 Calculating drought probabilities on which to base insurance premiums. MarkSim-generated deficit probabilities in 1 km spatial resolution exhibit consistent spatial patterns, which agree largely with farmers' perceptions of spatial climate variation. However, some aspects still need work:

(1) An insurer would be interested in basing calculations on validated probability values. MarkSim has been validated a number of times [18,25,54,55,56,57,58]. Local validation is desirable, but practically difficult because of the uncertain quality of the historical meteorological data.

(2) Spatial basis risk faced by a farmer is lowered by simulating site-specific probabilities at 1-km resolution and making premiums more realistic. The important task is to install sufficient rain gauges to capture the trigger event for specific farmers. This would also bring more evidence about the reliability of simulated drought probabilities. In Kenya, the Syngenta Foundation installs automatic rain gauges on cellphone towers, whose density is sufficient to provide adequate cover.

3.3 Technical aspects

(1) Farmers often grow drybean concurrently at different sites with different deficit probabilities. GPS equipment can provide the coordinates of each site to select the appropriate product. Similarly a group of neighbouring farmers doing cooperative agriculture could be provided with a group policy.

(2) Farmers synchronised sowing with adequate soil water so that sowing failures were not relevant and need not to be considered in the insurance. In contrast, failures due to heavy rain at sowing and crop damage at harvest seem important, but lay outside the scope of this study.

(3) Farmers partially knew about the ENSO phenomenon and corresponding local climate effects. MarkSim captures the variability caused by the ENSO phenomenon within its historical weather data, but does not explicitly generate weather that takes account of it. This is a clear area where refinement of MarkSim would be useful, and where adverse selection could be a problem.

(4) For maximum transparency, it is necessary that farmers understand how the index is calculated and how premiums and indemnities are determined, which requires the involvement of the agricultural extension services.

Conclusions

The study contributes to expanding rural households' risk management strategies via formal insurance. The study's approach emphasizes that perceptions and preferences of potential customers must be included in the process of developing acceptable insurance products.

Farmers' perceptions supported the expectation that spatial basis risk merits special attention in mountainous areas, which has consequences for the design of appropriate indices. Firstly, the insurance has to specify what it undertakes to cover. To make basis risk transparent to farmers, they should be allowed to choose between a reasonable number of standardized contracts that take account of microclimate and soils and offer variable indices that give different levels of coverage, and of course, premiums. In the end it is the individual farmer who will decide if basis risk, and hence the insurance product, is acceptable or not.

Acknowledgments

This research was conducted under the CGIAR Research Program on Climate Change, Agriculture and Food Security (CCAFS) with additional funding from Deutsche Gesellschaft für Internationale Zusammenarbeit, Beratungsgruppe Entwicklungsorientierte Agrarforschung (GIZ-BEAF). Many thanks also to Lesbia Rizo, Dr. Carlos Zelaya, Narisoki Castro and Armando Martínez from CIAT as well as to the *Nicaraguan Ministerio Agropecuario y Forestal (MAGFOR)* and the *Instituto Nicaragüense de Technología Agropecuaria (INTA)*, for supporting this study.

Author Contributions

Conceived and designed the experiments: AK PL SC. Performed the experiments: AK LG. Analyzed the data: AK. Wrote the paper: AK MF PL.

References

1. Miranda MJ (1991) Area-Yield Crop Insurance Reconsidered. Am J Agric Econ 73(2): 233–242. Available: http://www.jstor.org/stable/1242708. Accessed 23 September 2010.
2. Skees JR., Hazell P, Miranda M (1999) New Approaches to Crop Yield Insurance in Developing Countries. Washington, DC: IFPRI. Available: http://www.ifpri.org/sites/default/files/publications/eptdp55.pdf. Accessed 20 September 2010.
3. Hazell P, Anderson J, Balzer N, Hastrup L, Clemmensen U, et al. (2010) Potential for scale and sustainability in weather index insurance for agriculture and rural livelihoods. Rome: IFAD. Available: http://www.ifad.org/ruralfinance/pub/weather.pdf. Accessed 06 June 2011.
4. Gudger M (1991) Crop Insurance: Failure of the Public Sector and the Rise of the Private Sector Alternative. In: Holden D, Hazell P, Pritchard A editors. Risk in Agriculture. Proceedings of the 10th Agriculture Sector Symposium. Washington, DC: The World Bank. 143–159.
5. World Bank (2005) Managing Agricultural Production Risk. Innovations in Developing Countries. Washington, DC: The World Bank. Available: http://siteresources.worldbank.org/INTARD/Resources/Managing_Ag_Risk_FINAL.pdf. Accessed 20 September 2010.
6. Hardaker JB, Huirne RBM, Anderson JR, Lien G (2004) Coping with Risk in Agriculture. 2. ed. Wallingford: CABI Publ.
7. Hellmuth ME, Osgood DE, Hess U, eds (2009) Index Insurance and Climate Risk: Prospects for Development and Disaster Management. Palisades, NY: IRI (Climate and Society 2). Available: http://portal.iri.columbia.edu/portal/server.pt/gateway/PTARGS_0_5024_4201_0_0_18/Climate%20and%20Society%20Issue%20Number%202.pdf. Accessed 2011 June 6.
8. Barnett BJ, Mahul O (2007) Weather Index Insurance for Agriculture and Rural Areas in Lower-Income Countries. Am J Agric Econ 89(5): 1241–1247.
9. Anderson JR (2003) Risk in Rural Development: Challenges for Managers and Policy Makers. Agric Syst 75(2–3): 161–197.
10. Skees JR (2008) Innovations in Index Insurance for the Poor in Lower Income Countries. Agricultural and Resource Economics Review 37(1): 1–15. Available: http://ageconsearch.umn.edu/handle/44733. Accessed 2010 September 21.
11. World Bank (2007) Agriculture for Development. Washington, DC: The World Bank (World Development Report 2008).
12. Giannini A, Hansen J, Holthaus E, Ines A, Kaheil Y, et al. (2009) Designing Index-based Weather Insurance for Farmers in Central America. Palisades, NY: IRI. (IRI Technical Report, 09–01). Available: http://iri.columbia.edu/publications/id = 875. Accessed 2010 September 29.
13. El Nuevo Diario (2010) Amplían la cobertura del seguro agrícola. Available: http://www.elnuevodiario.com.ni/economia/72333. Accessed 2012 October 12.
14. El Nuevo Diario (2009) Lafise lanza seguro agrícola ante adversidades del clima. Available: http://www.elnuevodiario.com.ni/economia/48044. Accessed 2012 October 12.
15. La Prensa (2009) Lanzan nuevo seguro agrícola. Available: http://archivo.laprensa.com.ni/archivo/2009/mayo/14/noticias/economia/327226. Accessed 2012 October 12.
16. Díaz Nieto J, Cook SE, Läderach P, Fisher MJ, Jones PG (2010) Rainfall Index Insurance to Help Smallholder Farmers Manage Drought Risk. Climate and Development 2(3): 233–247.
17. Díaz Nieto J, Cook SE, Läderach P, Fisher MJ (2012) Weather indices for designing micro-insurance products for poor small-holder farmers in the tropics. PloS ONE 7(6): e38281. Available: DOI:10.1371/journal.pone.0038281. Accessed 2012 June 24.
18. Díaz Nieto J, Cook S, Lundy M, Fisher M, Sanchez D, et al. (2006) A System of Drought Insurance for Poverty Alleviation in Rural Areas. Final Report. Cali: CIAT. Available: http://gisweb.ciat.cgiar.org/sig/download/RESEARCHREPORT_Final_01JE06.pdf. Accessed 2011 June 6.
19. Fujisaka S (1994) Learning from Six Reasons Why Farmers Do Not Adopt Innovations Intended to Improve Sustainability of Upland Agriculture. Agric Syst 46(4): 409–425. Available: DOI:10.1016/0308–521X(94)90104-N. Accessed 2010 September 28.
20. Gabler Wirtschaftslexikon (2011) *"Basisrisiko"*. Available: http://wirtschaftslexikon.gabler.de/Archiv/296400/basisrisiko-v1.html. Accessed 2011 May 18.
21. Giné X, Menand L, Townsend R, Vickery J (2010) Microinsurance. A Case Study of the Indian Rainfall Index Insurance Market. Washington, DC: The World Bank. (Policy Research Working Paper 5459).
22. Miranda M, Vedenov DV (2001) Innovations in Agricultural and Natural Disaster Insurance. Am J Agric Econ 83(3): 650–655.
23. Jones JW, Hoogenboom G, Porter CH, Boote KJ, Batchelor WD, et al. (2003) The DSSAT Cropping System Model. Eur J Agron 18: 235–265.
24. Hijmans RG, Cameron SE, Parra JL, Jones PG, Jarvis A (2005) Very High Resolution Interpolated Climate Surfaces for Global Land Areas. International Journal of Climatology 25: 1965–1978. Available online: http://www.worldclim.org/worldclim_IJC.pdf. Accessed 2011 August 9.
25. Jones PG, Thornton PK (2000) MarkSim: Software to Generate Daily Weather Data for Latin America and Africa. Agron J 93(3): 445–453. Available: doi:10.2134/agronj2000.923445x. Accessed 2010 September 21.
26. Jones PG, Thornton PK, Diaz W, Wilkens PW, Jones AL (2002) MarkSim: A computer tool that generates simulated weather data for crop modeling and risk assessment.Users Manual. Cali: CIAT. Available: http://gisweb.ciat.cgiar.org/marksim/index.htm. Accessed 2011 June 6.
27. Osgood DE, McLaurin M, Carriquiry M, Mishra A, Fiondella F, et al. (2007) Designing Weather Insurance Contracts for Farmers in Malawi, Tanzania and Kenya. Final Report to the Commodity Risk Management Group, ARD, World Bank. Palisades, NY: IRI. (IRI Technical Report, 07–02). Available: http://iri.columbia.edu/%7Edeo/IRI-CRMG-Africa-Insurance-Report-6-2007/IRI-CRMG-Kenya-Tanzania-Malawi-Insurance-Report-6-2007.pdf. Accessed 2010 September 21.
28. Jasanoff S (1993) Bridging the Two Cultures of Risk Analyis. Risk Anal 13(2): 123–129.
29. Renn O (2008): Risk governance. Coping with uncertainty in a complex world. London: Earthscan (Earthscan Risk in Society Series).
30. Instituto Nicaragüense de Estudios Territoriales (INETER) (2005a) Características del Clima de Nicaragua. Managua. (Atlas Climático de Nicaragua 1971–2000). Available: http://webserver2.ineter.gob.ni/Direcciones/meteorologia/clima%20nic/caracteristicasdelclima.html. Accessed 2011 June 6.
31. Magaña V, Amador JA, Medina S (1999) The Midsummer Drought over Mexico and Central America. J Clim 12(6): 1577–1588. Available: http://journals.ametsoc.org/doi/pdf/10.1175/1520-0442%281999%29012%3C1577%3ATMDOMA%3E2.0.CO%3B2. Accessed 2010 October 25.
32. Instituto Nicaragüense de Estudios Territoriales (INETER) (2005b) Fenómeno El Niño. Managua. Available: http://webserver2.ineter.gob.ni/Direcciones/meteorologia/Desastres/index_desastres.html. Accessed 2011 April 9.
33. Instituto Nicaragüense de Estudios Territoriales (INETER) (2005c) Fenómeno La Niña. Managua. Available: http://webserver2.ineter.gob.ni/Direcciones/meteorologia/Desastres/Fenomenoelnino/lanina.html. Accessed 2011 April 9.
34. Nieuwolt S (1982) Tropical Rainfall Variability - The Agroclimatic Impact. Agriculture and Environment 7(2): 135–148. Available: http://dx.doi.org/10.1016/0304-1131(82)90003-0. Accessed 2011 June 6.
35. Baethgen W, Hansen JW, Amor IVM, Jones JW, Meinke H, et al. (2008) Contributions of Agricultural Systems Modeling to Weather Index Insurance. Workshop: Technical Issues in Index Insurance. Palisades, NY: The International Research Institute for Climate and Society (IRI).Available: http://iri.columbia.edu/publications/download = 1008. Accessed 2011 June 6.
36. Carter MR, Barrett CB, Boucher S, Chantarat S, Galarza F, et al. (2008) Insuring the Never before Insured: Explaining Index Insurance through Financial Education Games. Madison: University of Wisconsin. Available: http://www.basis.wisc.edu/basis_crsp/pubs.html. Accessed 2010 September 15.
37. Churchill C (2006) What Is Insurance for the Poor? In: Churchill C, editor. Protecting the Poor. A Microinsurance Compendium. Geneva: ILO. 13–24.
38. Golledge RG, Stimson RJ (1997) Spatial Behavior. A Geographic Perspective. New York: The Guilford Press.
39. Gold JR (1980) An Introduction to Behavioural Geography. Oxford: Oxford University Press.
40. Kelle U, Kluge S (2010) Vom Einzelfall zum Typus. Fallvergleich und Fallkontrastierung in der qualitativen Sozialforschung. 2. rev. ed. Wiesbaden: VS.
41. Mayring P (2003) Qualitative Inhaltsanalyse. Grundlagen und Techniken. 8. ed. Weinheim/Basel: Beltz.
42. Witzel A (2000) Das problemzentrierte Interview. Forum Qualitative Sozialforschung 1(1): Art. 22. Available: http://www.qualitative-research.net/index.php/fqs/article/view/1132. Accessed 2011 January 26.
43. Instituto Nicaragüense de Tecnología Agropecuaria (INTA) (2004) Cultivando Frijol con menos Riesgos. Managua. Available: http://www.inta.gob.ni/guias/frijol_mip.pdf. Accessed 2011 May 25.

44. Dercon S (2004) Growth and Shocks: Evidence from Rural Ethiopia. Journal of International Development Economics 74: 309–329.
45. Barrett CB, Barnett BJ, Carter MR, Chantarat S, Hansen JW, et al. (2007) Poverty Traps and Climate Risk: Limitations and Opportunities of Index-Based Risk *Financing*. Palisades, NY: The International Research Institute for Climate and Society (IRI). Available: http://papers.ssrn.com/sol3/papers.cfm?abstract_id = 1141933. Accessed 2010 September 15.
46. Wilhite DA, Glantz MH (1985) Understanding: the Drought Phenomenon: The Role of Definitions. Water International 10(3): 111–120. Available: http://www.informaworld.com/smpp/ftinterface~db = all~content = a908087273~fulltext = 713240930. Accessed 2010 October 29.
47. FAO (2011) Promoting the Growth and Development of Smallholder Seed Enterprises for Food Security Crops. Best Practices and Options for Decision Making. Rome: FAO. Available: http://www.fao.org/agriculture/crops/publications/detail/tr/item/49674/icode/en/. Accessed 2011 May 28.
48. Ingram KT, Roncoli MC, Kirshen PH (2002) Opportunities and Constraints for Farmers of West Africa to Use Seasonal Precipitation Forecasts with Burkina Faso as a Case Study. Agric Syst 74: 331–349.
49. DeWalt BR (1994) Using Indigenous Knowledge to Improve Agriculture and Natural Resource Management. Hum Organ 53(2): 123–131.
50. Ericksen PJ, Ardón M (2003) Similarities and Differences between Farmer and Scientist Views on Soil Quality Issues in Central Honduras. Geoderma 111(3–4): 233–248.
51. Kitchin R, Blades M (2002) The Cognition of Geographic Space. London.
52. Instituto Nicaragüense de Estudios Territoriales (INETER) (2010) Boletines Climáticos Mensuales. Managua. Available: http://webserver2.ineter.gob.ni/Direcciones/meteorologia/. Accessed: 2011 June 26.
53. Díaz Nieto J (2004a) Task Report 4: Results from the Modeling Exercise: Identifying In-Season Rainfall Weights. Cali: CIAT.
54. Jones PG, Thornton PK (1993) A Rainfall Generator for Agricultural Applications in the Tropics. Agricultural and Forest Meteorology 63: 1–19.
55. Jones PG, Thornton PK (1997) Spatial and Temporal Variability of Rainfall Related to a Third-Order Markov Model. Agricultural and Forest Meteorology 86: 127–138.
56. Mavromatis T, Hansen JW (2001): Interannual Variability Characteristics and Simulated Crop Response of Four Stochastic Weather Generators. Agricultural and Forest Meteorology 109: 283–296.
57. Hartkamp AD, White JW, Hoogenboom G. (2003) Comparison of Three Weather Generators for Crop Modeling: A Case Study for Subtropical Environments. Agric Syst 76: 539–560.
58. Díaz Nieto J (2004b) Task Report 11: Validation of MarkSim obtained probabilities. Cali: CIAT.

Long-Term Environmental Correlates of Invasion by *Lantana camara* (Verbenaceae) in a Seasonally Dry Tropical Forest

Geetha Ramaswami[1], Raman Sukumar[1,2*]

1 Centre for Ecological Sciences, Indian Institute of Science, Bangalore, India, 2 Divecha Centre for Climate Change, Indian Institute of Science, Bangalore, India

Abstract

Invasive species, local plant communities and invaded ecosystems change over space and time. Quantifying this change may lead to a better understanding of the ecology and the effective management of invasive species. We used data on density of the highly invasive shrub *Lantana camara* (lantana) for the period 1990–2008 from a 50 ha permanent plot in a seasonally dry tropical forest of Mudumalai in southern India. We used a cumulative link mixed-effects regression approach to model the transition of lantana from one qualitative density state to another as a function of biotic factors such as indicators of competition from local species (lantana itself, perennial grasses, invasive *Chromolaena odorata*, the native shrub *Helicteres isora* and basal area of native trees) and abiotic factors such as fire frequency, inter-annual variability of rainfall and relative soil moisture. The density of lantana increased substantially during the study period. Lantana density was negatively associated with the density of *H. isora*, positively associated with basal area of native trees, but not affected by the presence of grasses or other invasive species. In the absence of fire, lantana density increased with increasing rainfall. When fires occurred, transitions to higher densities occurred at low rainfall values. In drier regions, lantana changed from low to high density as rainfall increased while in wetter regions of the plot, lantana persisted in the dense category irrespective of rainfall. Lantana seems to effectively utilize resources distributed in space and time to its advantage, thus outcompeting local species and maintaining a population that is not yet self-limiting. High-risk areas and years could potentially be identified based on inferences from this study for facilitating management of lantana in tropical dry forests.

Editor: Judi Hewitt, University of Waikato (National Institute of Water and Atmospheric Research), New Zealand

Funding: The Ministry of Environment and Forests, Government of India has funded this project since 1989. The funders had no role in study design, data collection and analysis, decision to publish, or preparation of the manuscript.

Competing Interests: The authors have declared that no competing interests exist.

* E-mail: rsuku@ces.iisc.ernet.in

Introduction

The success of invasive species is generally attributed to the properties of the invading species [1,2], those of the invaded communities and environment [3,4] or propagule pressure [5,6], with recent emphasis on the synthesis of these theories [7,8]. However, the dynamics of invasive species – their population structure, impacts and interactions with environmental variables such as weather and disturbances, change with time [9]. Often invasive plants are known to persist benignly in low numbers for long periods of time before exhibiting sudden changes such as local population explosion or range expansion which could be associated with either changes in the invading species itself or in the abiotic environment or biotic interactions [10]. One way to understand these processes is to monitor populations over sufficiently long time periods. Although the need for long-term monitoring has been emphasised as a requirement for understanding invasive species dynamics and changing impacts [11], only a small proportion of studies track invasive plants over long time periods [9]. Understanding the factors that affect the populations of invasive species over time could have implications for the management and restoration of invaded ecosystems. At small spatial scales, it is possible to have detailed long-term data about the demography of invasive species (e.g. [12]) as well as their interactions with the abiotic environment (e.g. [13]).

Lantana camara L. (henceforth lantana) is a thicket-forming woody shrub of tropical American origin and is currently a widespread invasive, occurring widely across the tropical and subtropical regions [14]. Seasonally dry tropical forests (250–2000 mm annual rainfall in 4–6 months of the year), though comprising 42% of tropical ecosystems of the world, are relatively understudied compared to wetter tropical forests [15,16]. The aim of this study was to understand what spatially and temporally variable abiotic and biotic factors influenced invasion by lantana, defined as its transition from a qualitatively determined lower density category to a higher one, using a time-series data set at a fine spatial scale.

We expected the following biotic factors to influence the spread of lantana -

1. Competition – Resident species, competing for limiting resources, have been cited as a probable mechanism for resistance to invasion [17,18]. In the absence of disturbances, surplus resources that can be utilized by invading species would become unavailable. Thus, competition from resident species could possibly result in the suppression of invading species. Alternately, invasive species could persist through such 'lean

periods' at lower density and expand opportunistically under future favourable conditions [19]. The local species pool could comprise of native as well as other introduced species. We hypothesise that with increasing density of competing resident shrubs and grasses, lantana would tend to persist at low densities or go from higher to lower density categories. Once higher lantana densities are attained, it was expected that there would be density-dependent limiting of invasion. We thus expected intra-specific competition to reduce lantana invasion.

2. Basal area of trees – The presence of trees has been shown to decrease the rate of spread of shrubby invasives [13]. Invasive species like lantana have been shown to perform better in forest edges and gaps than in forest interiors, with percent cover and relative growth rate decreasing with increasing distance from the centre of a gap [20,21]. Larger gaps have also been shown to support larger sub-populations of lantana and larger individuals [20]. We expected higher basal area of trees to be associated with greater canopy cover and thus lower light availability at ground level. We thus hypothesise that with increasing basal area of trees in space and time, the rate of lantana invasion would be low.

We expected the following abiotic factors to influence the intensity of lantana invasion -

1. Fire – It has been speculated that disturbances create spurts in unused resources in a habitat that can be utilized by invading species [3]. Fire is one such disturbance and has been implicated in the spread and persistence of invasive species [22]. Invasive plants that are adapted to coppicing post-fire, rapidly gain biomass and thus become fuel for further fires, creating a positive feedback cycle resulting in their persistence. A similar framework has also been hypothesised as a mechanism for spread of lantana [23]. Although the role of fire on the spread of shrubby invasive species in general and vice-versa is relatively poorly understood [24], we nevertheless hypothesise that the occurrence of fire facilitates lantana invasion over long time scales.
2. Spatial and temporal availability of moisture – While the availability of surplus resources has been shown to support greater invasive abundance and diversity, invasive species have also been hypothesised to grow and persist better than native species under resource-poor conditions by virtue of having traits that maximise the efficiency of utilisation of available resources [25]. Topographic heterogeneity and seasonality or inter-annual variability in rainfall respectively create resource (i.e. moisture) poor conditions in space and time, respectively. Invasive species may have physiological traits that make them more tolerant to water-stress conditions as compared to native species [26]. *Lantana camara* has been shown to have drought-avoidance strategies – a deep root system and high water retention – thus avoiding damage to photosynthetic apparatus even in the driest regions of arid islands of the Galapagos [27]. This is different from the shallow rooted drought-tolerance strategies of a native congener. In spite of such physiological strategies, however, growth rates of lantana were lower in drier compared to wetter regions of the islands [27]. Soil moisture, affected by local topography, could thus influence invasion by exotic plants. Another example is the invasive herb *Alliaria petiolata* (Bieb.) Cavara & Grande, which had higher productivity and survival in high soil-moisture stream-bottom sites than in upland regions with drier soils [28]. We thus hypothesise that lantana invasion would occur more frequently in moister areas.

We used a cumulative link mixed-effects regression approach at a small but fine spatial scale using 18 years (1990–2008) of data from a permanent 50 ha tropical dry forest plot in southern India. We used indicators of competition, disturbance and available resources at a given time step to understand transition of lantana in a given grid cell from one density class to another, in the next time step. The model so developed helps us understand not only the individual influence of each factor but also interactions between factors on lantana invasion.

Materials and Methods

Study Site

The 50 ha Mudumalai Forest Dynamics Plot (MFDP) is located centrally in the tropical dry forests of Mudumalai (11°30′– 11°39′ N, 76°27′–76°43′E; 321 km^2) of southern India [29]. The MFDP received 1265±292 mm rainfall annually for the period 1990–2008. Partial as well as plot-wide ground fires have occurred in the MFDP in 1991, 1992, 1994, 1996, and 2002 (Fig. 1) during this period. The forest is characterised by an open canopy supporting the persistence of 'ground vegetation' that includes grasses such as *Themeda cymbaria* Hack. and *Cymbopogon flexuosus* (Steudel) Watson, ephemeral herbs, as well as the alien invasive species lantana and *Chromolaena odorata* (L.) King & Robinson. Although lantana infestations have been reported in the Gudalur area (15 km from the MFDP) from the early 1900s [30] and has been observed elsewhere in Mudumalai since the late 1970s (R.S. pers. obs.), its presence at high abundance in the MFDP seems fairly recent.

Data

The 1 km ×0.5 km MFDP is gridded into 20 m×20 m plots, each of which is further sub-divided into four 10 m×10 m sub-plots (our unit of measurement and henceforth referred to as 'cell') arranged in 50 rows and 100 columns. Since 1989 an observer

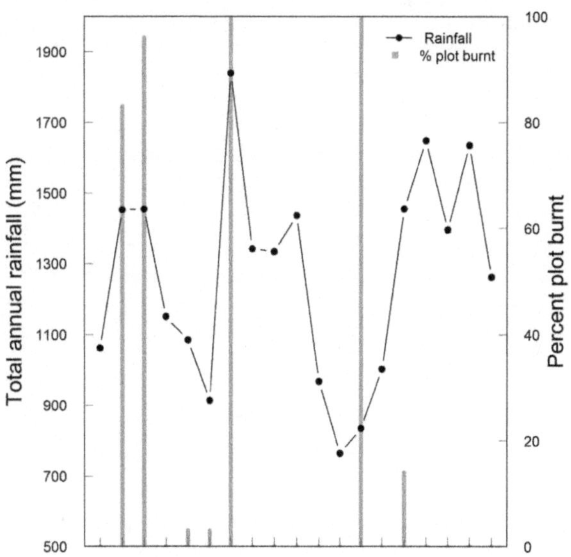

Figure 1. Interannual variability in rainfall and fire at the Mudumalai Forest Dynamics Plot. The solid line and dots represent total annual rainfall (mm) while the grey bars represent the percentage of 10 m×10 m plots burnt each year. It must be noted that fires usually occur during the dry season of each year and are not related to the total rainfall in that year.

(mostly C.M. Bharanaiah) has been visually estimating and assigning the ground vegetation groups – grasses, bamboo, ephemerals, *C. odorata* and lantana – to one of five qualitative density categories – absent, just present, common, dense, very dense (see Figure S1.1 in Appendix S1 for an example). Along with data on ground vegetation, the cells that were burnt during the dry months preceding the enumeration are also recorded. In the years 1992, 1996 and 2002, plot-wide fires had burnt all ground vegetation and data were thus not collected for these years. Excepting 1995 and 1998, the ground vegetation data are available for all years starting from 1989 to present. Permissions for field work at this site were granted by the Tamil Nadu Forest Department (Tamil Nadu state, India).

Characterizing Biotic Factors

Intraspecific and interspecific competition from ground vegetation was quantified based on the biomass of three ground vegetation groups namely – lantana, grasses and *C. odorata*. In August 2011, ten 1 m ×1 m plots corresponding to the four qualitative density classes – 'present', 'common', 'dense' and 'very dense' of lantana, grasses and *C. odorata* – were sampled outside the MFDP (n = 10 plots ×4 density categories ×3 vegetation groups = 120 sampling plots). The vegetation within each sample plot was harvested, dried and weighed. An average of the dry wt m^{-2} was used as an estimate of the biomass for each density class of each vegetation group. In this analysis, the 'absent' and the 'present' categories of lantana were combined into a single density class 'present' as these categories showed high spatial variability and could not be reliably distinguished from one another. The 'common' and 'dense' categories were also combined as the former category occurred very rarely and had biomass roughly comparable to the latter category (Table 1). The higher density categories represent established breeding plants that contribute to local intensification of invasion. We thus used three categories of lantana abundance – present, common and very dense – for this analysis. Within each vegetation group, the 'present' category was used as a reference and the ratio between this category and all other categories were used as scores for each density class (Table 1). To assess intraspecific competition, density scores for lantana based on biomass values were assigned to each cell for each year. To assess the influence of heterospecific competition on lantana spread, density scores of tall grasses, the shrub *C. odorata* and stem densities of the understorey shrub *Helicteres isora* L. were considered. The sum of density scores of tall grasses and *C. odorata* in a focal cell was used as a measure of heterospecific competition. Competition by *H. isora* was quantified as the sum of stems >1 cm dbh in a focal cell. We used the total basal area of woody stems >1 cm dbh as a measure of competition from other woody plants, as well as a proxy for shading at ground level.

Table 1. Estimated mean biomass (in g m^{-2}, ±1 standard deviation) of three vegetation categories measured in the Mudumalai Forest Dynamics Plot.

Ground vegetation	Density category	Biomass	± SD	Density score
Lantana camara	Absent	0		0
	Present	48.3	±15.6	1
	Common	503.1	±30.5	10
	Dense	858.5	±154.2	18
	Very dense	2038.5	±226.0	42
Chromolaena odorata	Absent	0		0
	Present	49.8	±11.4	1
	Common	118	±75.9	2
	Dense	728.5	±261.0	15
	Very dense	1165.9	±348.2	23
Grasses	Absent	0		0
	Present	95.5	±27.1	1
	Common	227.6	±73.8	2
	Dense	704.2	±187.5	7
	Very dense	813.2	±170.5	9

Density scores have been assigned to each category as the ratio between that category and the 'present' category.

Characterizing Abiotic Factors

In order to determine the influence of interannual variability in rainfall on lantana spread, we used yearly data from a weather station maintained by us at Kargudi, about 3 km from the MFDP. We assumed that all cells in the plot received the same amount of rain, and rainfall was therefore held constant across the plot. Since enumeration began in June of a year and extended up to December, the rainfall received between June of one year to May of the following year was considered as annual rainfall received between two consecutive enumerations. For years when plot-wide fires had occurred (1992, 1996, 2002), or when ground vegetation data were not available (1995, 1998), rainfall was averaged for the duration between two valid enumerations. To understand the effects of spatial availability of moisture on change in lantana density in the plot, we used a topography-based static measure of soil moisture – the topographic wetness index, henceforth referred to as TWI [31] (Beven and Kirkby 1979). The TWI values for each cell were derived from a digital elevation map with a resolution of 10 m×10 m from a catchment area of 20 km^2 centred around the MFDP using the software SAGA version 2.0.5 (System for Automated Geoscientific Analyses, http://www.saga-gis.org). In the MFDP, TWI values ranged from a minimum of 6.8 in the driest part of the plot to a maximum of 13.1 near streams. Since TWI is based only on the relative elevation of a cell with respect to surrounding cells, this measure remained static in time. Cell-wise information on the occurrence of fire was used to determine the influence of fire on lantana invasion. When a cell experienced consecutive fires between two time steps, the fire instances were summed. We did not have a direct measure of overall light availability at ground level for the duration of this study. However, an examination of monthly averages of light availability measured at Kargudi for the years 2005 to 2008, revealed that higher rainfall was correlated with low light availability (Spearman's rho = −0.6, Figure S1.2 in Appendix S1). We thus assumed that higher rainfall corresponded to lower global solar insolation at the plot. We also assumed that greater the area occupied by other trees in a cell, greater is the shading at ground level. Since shading at ground level was likely to be correlated with rainfall, we did not include any proxies for the same in our model.

Statistical Analyses

The density states of lantana increased in an ordinal fashion. We therefore chose a proportional odds logistic regression to assess the changes in lantana density. Since this study was conducted in one large contiguous plot which was sampled repeatedly, we used a mixed-effects framework to control for possible spatial and temporal autocorrelation in the data. We used 'clusters' of lantana as a random effect. A cluster was defined as a group of cells that had the same lantana density in the adjacent or diagonal directions; cells within a cluster were expected to be more spatially correlated than cells belonging to different clusters. In addition, cells within a cluster were expected to change more similarly in time, than cells of a different cluster. Thus, the random effect of 'clusters' was expected to account for both spatial as well as temporal autocorrelation in this data set. In order to account for unequal intervals between two consecutive enumerations, we used a 'lag' term that was equal to the number of years across which data were not available. The lantana density at a given time step, was thus modelled as a function of per-cell lantana density, heterospecific density and basal area of other woody trees, inter-census rainfall and fire occurrence in the previous time step, a lag term, TWI and a random term for spatial clusters of lantana from the previous time step. The two-way interaction between rainfall and fire and rainfall and TWI were also considered. Instances of fire for a given time step preceded the duration from which rainfall was estimated. These two abiotic factors were thus assumed to be independent of each other and an interaction between these two terms was thus considered appropriate. All variables were scaled by subtracting the mean and dividing by the standard deviation in order to bring them to comparable ranges and to make the interpretation of model coefficients easier. Since we were interested in how lantana density changed in the same location from one time step to another, we selected a random subset of ~75% of the cells to build the model and used the remaining 25% to test the model. Models were fit using the clmm2 () function of the ordinal package (Christensen 2012) in the statistical programming language R, version 2.15.1 (R Development Core Team 2012). The global model was simplified by stepwise removal of terms that had a $P \geq 0.05$, starting with the terms that had the highest P value. Significance of model terms was assessed by comparing non-simplified and simplified models using the likelihood ratio tests of the anova () function; models were considered to perform equally well at a difference of $P \geq 0.05$ and simpler models were retained. Predicted densities from the model were used for model verification using the test dataset. A given cell was expected to successfully change to a new density category with a model predicted probability obtained from the predict () function. Model accuracy was checked using a contingency table of observed frequency of occurrence of different lantana densities and those predicted by the model, and computing a gamma statistic of association [32] (Goodman and Kruskal 1954) for the same. Significant interactions were interpreted graphically by back-calculating the probability of obtaining a given density category from the estimated log-odds of getting that density category or lower (Appendix S2).

Results

The frequency of occurrence of higher density categories of lantana increased over the time period of this study, while that of the lower density classes decreased (Fig. 2). The increase in the frequency of occurrence of the 'very dense' category from 2% in 2000 to 18% in 2008 was noteworthy, as it indicated rapid intensification of invasion. The frequency of occurrence of the 'present' category has been relatively high and constant at 40 to 60% during 1990–2008. Total lantana biomass for the MFDP increased from 40 g m^{-2} in 1990 to 615 g m^{-2} in 2008 (Fig. 2).

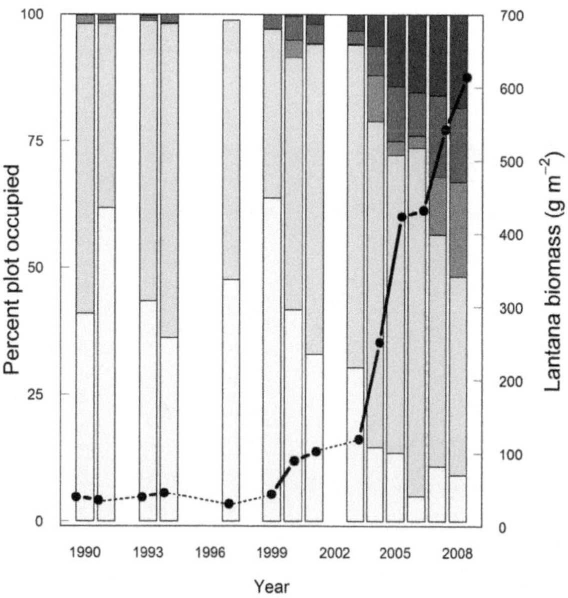

Figure 2. Frequency of occurrence (bars) and biomass (line) of lantana in the MFDP from 1990–2008. Light grey through dark grey represent the increasing density of lantana, i.e, from 'absent' through 'present', 'common' and 'dense' to 'very dense'. The five density categories of lantana occurred with varying frequencies across the time period sampled; the lower density categories occurred less frequently and the higher density categories occurred more frequently with time. Gaps and dotted lines indicate fire years or missing data. The dry weight biomass of lantana is based on a study conducted in the year 2011. Note that for the purpose of analyses, the 'absent' and 'present' categories were combined to form the low-density 'present' category while the 'common' and 'dense' categories were combined to form a moderate density 'common' category.

Model Outputs and Parameter Estimates

The random effect accounting for clustering of lantana had a large standard deviation of 2.38, indicating the considerable effect of spatial clumping in this plant. Of all the fixed effects considered in the model, only ground vegetation dropped out of the model as being insignificant. *H. isora* neighbourhood, per cell lantana density, lag, basal area of large trees and the two-way interaction between rainfall and TWI as well as between rainfall and fire were found to significantly affect transitions in lantana density (Table 2). The model had high overall predictability, with a gamma score of 0.78. However, at the level of individual density categories, the model predicted the 'present' category the best with an accuracy of 98%, while the only 35% and 40% of the denser categories were accurately predicted. The prediction was the poorest for the 'common' category, with 58% of such cells being wrongly predicted as the 'present' category.

Biotic Influences on Lantana Invasion

The term for ground vegetation dropped out of the model as being non-significant, indicating that competition from grasses and other invasive species had little or no effect on invasion by lantana. The term for lantana density in a given cell was found to be highly significant (Likelihood Ratio Test, LR statistic = 52.96, $P<0.001$), indicating that lantana density in the previous time step was an important predictor of lantana density in the next time step. The odds ratio (calculated from the coefficients in Table 2; see Appendix S2 for details) of getting a given lantana density or higher was 0.9 at lantana 'common' relative to lantana 'present'

Table 2. Output of the simplified cumulative link mixed model.

		Coef	S.E.	Wald Z	*P*
Threshold coefficients	Present to common	−2.59	0.09	−27.91	–
	Common to very dense	−0.02	0.08	−0.25	–
Fixed effects	Lantana (Common)	−0.09	0.25	−0.36	0.72
	Lantana (Very Dense)	2.32	0.35	6.52	<0.001
	Lag	−2.74	0.18	−14.84	<0.001
	Rainfall	0.38	0.12	3.19	0002
	Fire frequency	0.48	0.19	2.49	0.01
	Basal area (trees)	0.03	0.02	2.05	0.04
	Helicteres isora	−0.06	0.02	−3.44	0.001
	TWI	0.60	0.02	34.10	<0.001
	Rainfall × TWI	0.65	0.28	−2.32	0.02
	Rainfall × Fire frequency	−0.11	0.02	−6.13	<0.001
Random effect	**Variance**	**Std. Dev.**			
Spatial cluster	5.65	2.38			

All predictors, corresponding to the time step immediately before that of the current time step, were scaled by subtracting the mean and dividing by the standard deviation in order to make them comparable. The overall predictability of the model was high, with a gamma score of 0.78. TWI – topographic wetness index of a cell, '×' – interactions between factors.

category, while it was 10.2 at lantana 'very dense' relative to lantana 'present' category. Thus, cells with dense lantana in the previous time step were likely to remain so or go into a higher density state in the next time step. All else being constant, for a lag of 1 year, the odds ratio of getting a given lantana density category or higher decreased from 15.6 to 15.3 at 2 SD below and above the mean *H. isora* stem density, respectively. Thus, the probability of getting a higher lantana density class decreased with increasing *H. isora* stem density. However, since the rate of change of odds is low, this also indicates that lantana persisted at low densities ('present' category) across all densities of the competing shrub. On the other hand, for the same time lag, the odds ratio of getting a given lantana density category or higher, increased from 14.9 to 16.4 at 2 SD below and above the mean basal area of trees, indicating that lantana invasion was positively associated with basal area.

Abiotic Influences on Lantana Invasion

The interaction between rainfall and fire was graphically explored by computing the probability of lantana 'present' category remaining so, or changing into a higher density category while holding all other predictors constant. In the absence of fire, for a lag of 1 year, the probability of lantana 'present' remaining so decreased with increasing rainfall, but that of lantana going to a higher density category increased (Fig 3a). When a single fire event occurred in the interim, lantana became denser at lower rainfall values (Fig 3a). The interaction between rainfall and TWI was also a significant predictor of lantana density (Table 2). All else being constant, in drier parts of the plot (low TWI), the probability of the 'present' category remaining so decreased and that of higher density categories increased with increasing rainfall. In wetter parts of the plot (high TWI), however, the probability of finding the 'common' category was always higher than finding the 'present' category and that of the 'very dense' category increased with increasing rainfall (Fig 3b).

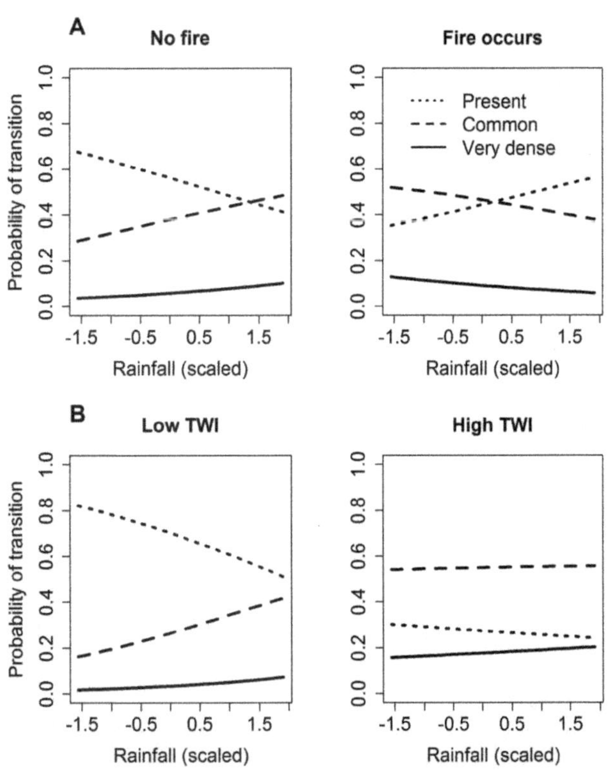

Figure 3. Model predictions for transition from lantana 'present' state to higher density categories. Over a lag of 1 year, under rainfall-fire interaction (a) and rainfall-topographic wetness index interaction (b). See text for details. Note that the x-axis (annual rainfall) is scaled by subtracting the mean and dividing by the standard deviation, thus a value of ~0 indicates the mean rainfall, values <0 indicate below-average rainfall and vales >0 indicate above-average rainfall. The limits of the x axes are the 95% quantiles of the scaled variable.

Discussion

This study examines the influences of biotic and abiotic factors on invasion by lantana in a seasonally dry tropical forest. Our observations indicate that over a span of 18 years, the transitions in lantana density have been strongly in favour of the higher density categories in the MFDP.

Biotic Influences on Lantana Invasion

The overall spread and biomass of lantana in the MFDP increased considerably during the latter half of this study (Fig. 2). Lantana in the MFDP is not self-limiting yet, as lantana 'common' cells tended to remain so or go into a higher density state in the next time step. The emergence of higher density categories, however, may be largely due to an increasing clonal population of lantana and not necessarily the emergence of new individuals. In a common garden experiment we found the germination success of lantana to be only 12% under ambient shade conditions (G. R., unpublished data). Even though low germination rates could be offset by high propagule pressure – up to 1.2 million seeds per hectare [33] – increasing lantana density in the same cell is most likely due to its coppicing ability on being top-killed (due to fire or mechanical injury) or due to vegetative propagation [34].

As expected, lower lantana density was associated with higher stem density of the competing native woody shrub *H. isora*. This is also borne out by the fact that both lantana and *H. isora* seem to have a preference for similar habitats in the MFDP (near streams). *H. isora* stem densities have been observed to increase since 2005, about 2–3 years following the expansion of lantana in the plot. Although the competitive effects of increasing stem densities of *H. isora* are likely to prevent intensification of lantana invasion, the persistence of lantana in low densities may result in further invasion in the future. Contrary to expectation, higher basal area of trees was associated with higher lantana density. At the landscape level, lantana density has been found to be negatively correlated with tree density in another dry forest of southern India [35]. We speculate that owing to the MFDP being a relatively open forest (~250 stems of >10 cm dbh per ha), it is likely that even at the high values of basal area in a cell, there is sufficient light reaching the ground to support the persistence of dense lantana.

Native grasses and exotic *C. odorata* were found to have no influence on lantana invasion although grasses themselves may be directly affected by the presence of lantana. In an independent study spanning 2 years (2008–2010) we observed that the grass cover under dense lantana was consistently low across seasons, covering an average of 34.4% sampled points outside lantana and only about 7.1% sampled points under dense lantana (Figure S3 in Appendix S3). Communities dominated by exotic species are often known to be replaced by other exotics over annual and decadal time scales [36]. We observed an increase in the 'common' and 'dense' categories of *C. odorata* occupying 40% and 31%, respectively, of all cells up to the year 2004, followed by a decrease in frequency of occurrence the following year. This decrease coincides with an increase in the frequency of occurrence of higher density classes of lantana in the MFDP.

Abiotic Influences on Lantana Invasion

Melbourne et al (2007) highlight the necessity of having both spatial and temporal variations in the environment for an invasive species to be successful. Invasive species could persist at low numbers for long times and then suddenly increase in their population by the opportunistic use of available resources [10]. It has also been hypothesised that invasive plants have high resource use efficiency, enabling high performance even in resource-poor conditions [25]. Alternatively, we find that lantana could also be opportunistically compensating for resources that are spatially unavailable but are variably abundant in time. High rainfall was found to be beneficial for lantana in the drier parts of the plot while, as expected, wetter parts of the plot always supported high densities of lantana. Increase in the previous-season's rain has been found to favour the intensification of invasion [13], a result that is supported by our study.

Changes in disturbance regimes may result in the expansion of the ranges of invasive species [37]. The MFDP has experienced considerable changes in both fire and rainfall patterns over the timeframe examined (Fig. 1). Fire has been shown to intensify invasions in certain ecosystems [22,38], although there is evidence that this is not true in all cases [39]. Contrary to what has been reported from earlier studies [40], we found that fire did not directly promote lantana invasion, at least in the short term. However, fire interacted with temporal availability of moisture. We found that in the absence of fire, transitions to higher density were more likely with increasing rainfall. When fire occurred, the transitions to higher density occurred at lower than average annual rainfall while the trend reversed at higher values of rainfall. This trend indicates that the occurrence of a disturbance event facilitated the spread of this invasive even though the availability of another resource (moisture) was reduced.

The exponential phase of lantana expansion (Fig. 3) occurred during 2002–2004 after an intense fire (in early 2002 that itself followed the onset of drought beginning 2000) followed by two drought years (2002–2003). The patterns of increase in lantana density observed in the MFDP are concurrent with the patterns observed in the dry deciduous forest of Biligiri Rangan Hills in southern India, where the mean stem density of lantana increased 10-fold in 11 years [41]. Indeed, during the decade of 2001–2010, a significant expansion of lantana also occurred over a large area in the southern Western Ghats of India, including Nagarahole, Bandipur and other parts of Mudumalai, thereby pointing to the influence of a set of common environmental factors that operated at a regional scale. The drought phase during 2000–2003 was common to all these sites and lantana may have eventually benefitted from such conditions through an interaction with fire. We thus speculate that fires coupled with droughts promote lantana invasion and that future changes in fire regimes (increasing fire frequency) and rainfall patterns (increasing number of droughts) may result in further expansion of this shrub.

Management Implications

The effects of lantana on native species have been found to be variable [41,42]. Although lantana is currently expanding its range and a large proportion of the tropics are suitable for further expansion [43], predicted ranges are likely to decrease under future climate change [44]. Worldwide, diverse control measures (physical, chemical and biological) have been largely unsuccessful in effectively curtailing the spread of lantana [43]. Traditional control methods often involve the removal of above-ground lantana biomass or burning which are temporarily effective, but result in profuse coppicing post-rain; uprooting lantana or damaging its coppicing meristem is a more effective control measure [33]. Understanding the varying environmental factors that influence the process of invasion in different ways may aid managers to determine suitable times of the year, season or phase of invasion that could be potentially managed. High-risk zones such as moister stream-sides and high risk years such as those in which droughts and fires occur simultaneously could be identified and management activities focussed accordingly.

Supporting Information

Appendix S1 Details of the study site that could not be included in the main text. Figure S1.1 gives the change in lantana density states from 1990 to 2008 in the MFDP. Figure S1.2 gives the inverse relationship between monthly rainfall and fire occurrence.

Appendix S2 Description of the method used to obtain odds ratios from a cumulative link mixed effects model.

Appendix S3 Details of a study that was conducted to understand the effects of dense lantana on the ground vegetation in the MFDP. Figure S3 gives a comparison of the proportion area occupied by grasses under and outside dense lantana over a period of two years.

Acknowledgments

We thank the Forest Department of Tamil Nadu for permissions to conduct this study. We thank a large number of field assistants and C. M. Bharanaiah for the collection of data as well as H.S. Suresh and H.S. Dattaraja for maintenance of the data over a period exceeding 20 years. Biomass data for lantana was collected by H. S. Suresh. We also thank Murali Mohan and his team at Mobiterra Solutions for providing the DEM for the MFDP and surrounding areas, and Sandeep Pulla for developing the TWI map for the MFDP and for countless discussions, suggestions and ideas with respect to the analysis of this data set, as also G. S. Bhat, Kavita Isvaran, Karpagam Chelliah, Meghana Kulkarani, Mane Meera Baban and Nandita Mondal for their feedback and support.

Author Contributions

Conceived and designed the experiments: GR RS. Analyzed the data: GR. Wrote the paper: GR RS.

References

1. Blossey B, Notzold R (1995) Evolution of increased competitive ability in invasive nonindigenous plants: A hypothesis. J Ecol 83: 887–889.
2. Callaway RM, Aschehoug ET (2000) Invasive plants versus their new and old neighbors: A mechanism for exotic invasion. Science 290: 521–523.
3. Davis MA, Grime JP, Thompson K (2000) Fluctuating resources in plant communities: a general theory of invasibility. J Ecol 88: 528–534.
4. Hierro JL, Maron JL, Callaway RM (2005) A biogeographical approach to plant invasions: the importance of studying exotics in their introduced and native range. J Ecol 93: 5–15.
5. Colautti RI, Ricciardi A, Grigorovich IA, MacIsaac HJ (2004) Is invasion success explained by the enemy release hypothesis. Ecol Lett 7: 721–733.
6. Thomsen MA, Corbin JD, D'Antonio CM (2006) The effect of soil nitrogen on competition between native and exotic perennial grasses from northern coastal California. Plant Ecol 186: 23–35.
7. Theoharides KA, Dukes JS (2007) Plant invasion across space and time: factors affecting nonindigenous species success during four stages of invasion. New Phytol 176: 256–273.
8. Catford JA, Jansson R, Nilsson C (2009) Reducing redundancy in invasion ecology by integrating hypotheses into a single theoretical framework. Diversity Distrib 15: 22–40.
9. Strayer DL, Eviner VT, Jeschke JM, Pace ML (2006) Understanding the long-term effects of species invasions. Trends Ecol Evol 21: 645–651.
10. Crooks JA (2005) Lag times and exotic species: The ecology and management of biological invasions in slow-motion. Ecoscience 12: 316–329.
11. Blossey B (1999) Before, During and After: The need for long-term monitoring in invasive plant species management. Biological Invasions 1: 301–311.
12. Buckley YM, Briese DT, Rees M (2003) Demography and management of the invasive plant species *Hypericum perforatum*. I. Using multi-level mixed-effects models for characterizing growth, survival and fecundity in a long-term data set. J Appl Ecol 40: 481–493.
13. Banasiak SE, Meiners SJ (2008) Long term dynamics of *Rosa multiflora* in a successional system. Biological Invasions 11: 215–224.
14. Richardson DM, Rejmánek M (2011) Trees and shrubs as invasive alien species – a global review. Diversity Distrib 17: 788–809.
15. Murphy PG, Lugo AE (1986) Ecology of dry tropical forests. Annual Review of Ecology and Systematics 17: 67–88.
16. Miles L, Newton AC, DeFries RS, Ravilious C, May I, et al. (2006) A global overview of the conservation status of tropical dry forests. J Biogeogr 33: 491–505.
17. Keane RM, Crawley MJ (2002) Exotic plant invasions and the enemy release hypothesis. Trends Ecol Evol 17: 164–170.
18. Mitchell CE, Agrawal AA, Bever JD, Gilbert GS, Hufbaur RA, et al. (2006) Biotic interactions and plant invasions. Ecol Lett 9: 726–740.
19. Melbourne BA, Cornell HV, Davies KF, Dugaw CJ, Elmendorf S, et al. (2007) Invasion in a heterogeneous world: resistance, coexistence or hostile takeover? Ecol Lett 10: 77–94.
20. Totland O, Nyeko P, Bjerknes A, Hegland SJ, Nielsen A (2005) Does forest gap size affects population size, plant size, reproductive success and pollinator visitation in *Lantana camara*, a tropical invasive shrub? Forest Ecol Manag 215: 329–338.
21. Raizada P, Sharma GP, Raghubanshi AS (2008) Ingress of lantana in dry tropical forest fragments: Edge and shade effects. Curr Sci 94: 180–182.
22. D'Antonio CM, Vitousek PM (1992) Biological invasions by exotic grasses, the grass/fire cycle, and global change. Annual Review of Ecology and Systematics 23: 63–87.
23. Hiremath A, Sundaram B (2005) The fire-lantana cycle hypothesis in Indian forests. Conservation and Society 3: 26–42.
24. Mandle L, Bufford JL, Schmidt IB, Daehler CC (2011) Woody exotic plant invasions and fire: reciprocal impacts and consequences for native ecosystems. Biological Invasions 13: 1815–1827.
25. Funk JL, Vitousek PM (2007) Resource-use efficiency and plant invasion in low-resource systems. Nature 446: 1079–1081.
26. Cleverly JR, Smith SD, Sala A, Devitt DA (1997) Invasive capacity of *Tamarix ramosissima* in a Mojave Desert floodplain: the role of drought. Oecologia 111: 12–18.
27. Castillo JM, Leira-Doce P, Carrión-Tacuri J, Muñoz-Guacho E, Arroyo-Solís A, et al. (2007) Contrasting strategies to cope with drought by invasive and endemic species of Lantana in Galapagos. Biodivers Conserv 16: 2123–2136.
28. Meekins JF, McCarthy BC (2001) Effect of environmental variation on the invasive success of a nonindigenous forest herb. Ecol Appl 11: 1336–1348.
29. Suresh HS, Dattaraja HS, Mondal N, Sukumar R (2011) Seasonally dry tropical forests of southern India: An analysis of floristic composition, structure and dynamics in Mudumalai Wildlife Sanctuary. In McShea MJ, Davies SJ, Bumpakhpan N, editors. The ecology and conservation of seasonally dry forests in Asia. Washington D.C.: Smithsonian Institution Scholarly Press. pp. 37–58.
30. Eagan JS (1911) The Nilgiri Guide and Directory. Mysore, India: Wesleyan Mission Press.
31. Beven KJ, Kirkby MJ (1979) A physically based, variable contributing area model of basin hydrology. Hydrological Sciences Bulletin 24: 43–69.
32. Goodman LA, Kruskal WH (1954) Measures of association for cross classification. Journal of the American Statistical Association 49: 732–764.
33. Babu S, Love A, Babu CR (2009) Ecological restoration of lantana-invaded landscapes in Corbett Tiger Reserve, India. Ecological Restoration 27: 467–477.
34. Sharma GP, Raghubanshi AS, Singh JS (2005) Lantana invasion: An overview. Weed Biol Manag 5: 157–165.
35. Kulmatiski A (2006) Exotic plants establish persistent communities. Plant Ecol 187: 261–275.
36. Moles AT, Flores-Moreno H, Bonser SP, Warton DI, Helm A, et al. (2012) Invasions: the trail behind, the path ahead, and a test of a disturbing idea. J Ecol 100: 116–127.
37. Stevens JT, Beckage B (2009) Fire feedbacks facilitate invasion of pine savannas by Brazilian pepper (*Schinus terebinthifolius*). New Phytol 184: 365–375.
38. Prasad AE (2012) Landscape-scale relationships between the exotic invasive shrub *Lantana camara* and native plants in a tropical deciduous forest in southern India. J Trop Ecol 28: 55–64.
39. Hobbs RJ, Huenneke LF (1992) Disturbance, diversity and invasion: Implications for conservation. Conserv Biol 6: 324–337.
40. Duggin JA, Gentle CB (1998) Experimental evidence on the importance of disturbance intensity for invasion of *Lantana camara* L. in dry rainforest-open forest ecotones in north-eastern NSW, Australia. Forest Ecol Manag 109: 279–292.
41. Sundaram B, Hiremath A (2012) *Lantana camara* invasion in a heterogeneous landscape: patterns of spread and correlation with changes in native vegetation. Biological Invasions 14: 1127–1141.
42. Ramaswami G, Sukumar R (2013) Performance of established native seedlings in relation to invasive *Lantana camara*, rainfall and species' habitat preferences in a seasonally dry tropical forest. Plant Ecol 214: 397–408.
43. Bhagwat SA, Breman E, Thekaekara T, Thornton TF, Willis KJ (2012) A battle lost? Report on two centuries of invasion and management of *Lantana camara* L. in Australia, India and South Africa. PLoS ONE 7 (3): e32407.
44. Taylor S, Kumar L, Reid N, Kriticos DJ (2012) Climate change and the potential distribution of an invasive shrub, *Lantana camara* L. PLoS ONE 7 (4): e35565.

Analysis of Natural Variation in Bermudagrass (*Cynodon dactylon*) Reveals Physiological Responses Underlying Drought Tolerance

Haitao Shi[1]◔, Yanping Wang[1]◔, Zhangmin Cheng[1,2], Tiantian Ye[1,2], Zhulong Chan[1]*

1 Key Laboratory of Plant Germplasm Enhancement and Specialty Agriculture, Wuhan Botanical Garden, Chinese Academy of Sciences, Wuhan, China, **2** University of Chinese Academy of Sciences, Beijing, China

Abstract

Bermudagrass (*Cynodon dactylon*) is a widely used warm-season turfgrass and one of the most drought tolerant species. Dissecting the natural variation in drought tolerance and physiological responses will bring us powerful basis and novel insight for plant breeding. In the present study, we evaluated the natural variation of drought tolerance among nine bermudagrass varieties by measuring physiological responses after drought stress treatment through withholding water. Three groups differing in drought tolerance were identified, including two tolerant, five moderately tolerant and two susceptible varieties. Under drought stress condition, drought sensitive variety (Yukon) showed relative higher water loss, more severe cell membrane damage (EL), and more accumulation of hydrogen peroxide (H_2O_2) and malondialdehyde (MDA), while drought tolerant variety (Tifgreen) exhibited significantly higher antioxidant enzymes activities. Further results indicated that drought induced cell injury in different varieties (Yukon, SR9554 and Tifgreen) exhibited liner correlation with leaf water content (LWC), H_2O_2 content, MDA content and antioxidant enzyme activities. Additionally, Tifgreen plants had significantly higher levels of osmolytes (proline level and soluble sugars) when compared with Yukon and SR9554 under drought stress condition. Taken together, our results indicated that natural variation of drought stress tolerance in bermudagrass varieties might be largely related to the induced changes of water status, osmolyte accumulation and antioxidant defense system.

Editor: Ive De Smet, University of Nottingham, United Kingdom

Funding: This research was supported by "the Hundred Talents Program", the Knowledge Innovative Key Program of Chinese Academy of Sciences (Grant No. 54Y154761O01076) to Z. Chan. The funders had no role in study design, data collection and analysis, decision to publish, or preparation of the manuscript.

Competing Interests: The authors have declared that no competing interests exist.

* E-mail: zhulongch@wbgcas.cn

◔ These authors contributed equally to this work.

Introduction

Drought is one of the most serious world-wide problems and largely affects plant growth, development and survival rate, leading to enormous crop yield loss. Plants are greatly restricted in growth field under stress condition, and have evolved many mechanisms to rapidly adapt to drought stress condition to keep growth and productivity [1,2]. Recently, more attentions have been paid to mechanisms of plant drought stress tolerance, including physiological and biochemical metabolisms, gene expression regulation, proteomic profiling and cross-talks between several hormones etc, which further helps us develop different genetic approaches to improve plant drought tolerance and prevent yield loss.

In response to drought stress, turfgrass has developed complex mechanisms such as physiological, biochemical, molecular and cellular changes to cope with limited water supply [3,4]. Comparatively, bermudagrass (*Cynodon dactylon*) is one of the most drought tolerant turfgrasses [5,6]. As a warm-season perennial grass, bermudagrass is widely used as turfgrass on sport fields, golf courses and home lawns. Drought stress is a world-wide problem, and different grass species may develop different strategies to tolerate, escape, or avoid drought stress condition [4,7,8,9,10].

To date, a lot of research groups have paid attentions to the natural variations in biotic and abiotic stress tolerances in many plant species, such as *Arabidopsis*, rice, bermudagrass, *brachypodium ditachyon*, and perennial ryegrass (*Lolium perenne*) [3,9,11,12,13,14,15]. Using these germplasms differing in drought tolerance, the major quantitative trait locis (QTL) contributing to drought tolerance and genetic networks controlling important stress processes were further dissected [4,5,9,10,15]. Previous studies have suggested that drought tolerance of bermudagrass varieties might be correlated with plant development such as leaf firing, root and shoot systems, mass production [16,17,18,19], accumulation of dehydrin [9,20], evapotranspiration [21], leaf water content (LWC), chlorophyll content, proline content, and antioxidant enzyme activities [10,22,23]. Physiological and biochemical mechanisms in response to water deficit stress have been largely studied in some bermudagrass varieties, however, most of these studies have focused on physiological level. Correlations between the natural variation and the detailed mechanisms among different species are still largely unknown.

The objectives of this study were to observe the genetic variations of several bermudagrass varieties in response to drought stress and to investigate possible mechanisms involved in drought stress tolerance variation. Nine bermudagrass varieties were collected and the natural variations of drought tolerance were evaluated. The relationships betweendrought tolerances and several physiological parameters were further comparatively dissected and discussed. These results provided some insights to understand the genetic and molecular mechanisms of bermudagrass drought tolerance.

Results

Drought tolerance evaluation of nine bermudagrass varieties

To compare the drought tolerance of nine bermudagrass varieties, healthy plants were subjected to well-watered condition and drought condition (soil water deficit) by withholding water in the soil for 28 d. After drought stress treatment, the LWC, leaf electrolyte leakage (EL) and the survival rate were examined to evaluate the drought tolerance. Under control condition, the LWC, EL and survival rate among all varieties were maintained almost at the same level (Table 1). When the drought stress was applied for 28 d, the LWC and survival rates of all varieties were largely decreased, while the ELs were increased. However, the changes of these parameters varied with varieties (Table 1). Tifgreen and Tifway had the highest LWC and survival rate but the lowest EL under drought stress condition, while Yukon and Wranger had the lowest LWC and survival rate but the highest EL. All these parameters (LWC, EL and survival rate) of other five varieties (LaPaloma, Riviera, SR9554, Laprima and Veracruz) were between the above four varieties (Table 1).

To further confirm the above results, we also detected the leaf water loss (% change in leaf fresh weight (FW)) of different bermudagrass varieties. As shown in Figure 1, with the detach time increasing from 0 h to 8 h, Tifway and Tifgreen showed the least water loss, while Yukon and Wranger exhibited the most. Other five varieties were in between the above four varieties. These results were consistent with those parameters of the stress tolerance test (Table 1).

Groups of bermudagrass varieties differing in sensitivity to drought stress

Among nine bermudagrass varieties used in this study, when subjected to drought stress for 28 d, LWC ranged from 15.93% to 54.21%, EL ranged from 49.12% to 86.74%, and survival rate ranged from 0 to 60.79%, respectively (Table 1). Additionally, the leaf water loss of these nine varieties at 8 h after detachment ranged from 44.25% to 71.25% (Figure 1). Half-life of water loss was 3–4 h for drought sensitive varieties (Yukon and Wranger), and longer than 8 h for drought tolerant ones (Tifway and Tifgreen) after detachment (Figure 1). Based on the variations of above mentioned physiological parameters (Table 1 and Figure 1), several groups differing in their sensitivities to drought stress were identified through hierarchical cluster analysis (Figure 2A–D). Interestingly, although different parameters were chosen, all nine varieties were clustered into three same groups, namely tolerant, moderately tolerant and susceptible (Figure 2A–D). Two varieties (Tifgreen and Tifway) were the most tolerant ones, with least reduction in LWC and leaf water loss *in vitro*, lowest EL but highest survival rate under drought stress condition. Five varieties (LaPaloma, Riviera, SR9554, Laprima and Veracruz) were clustered in the moderately drought tolerant group, with moderate level changes of the above mentioned physiological parameters. The other two varieties (Yukon and Wranger) were reasonably identified as the susceptible materials. These results suggested a linkage between plant physiological changes and drought tolerance in bermudagrass.

Comparison of water status and electrolyte leakage of different bermudagrass varieties under drought stress condition

To further investigate the possible mechanisms of the drought tolerance variations, three bermudagrass varieties (Tifgreen, SR9554 and Yukon) differing in sensitivities to drought stress were chosen for the following experiments. Leaf firing, also known as the change of leaf chlorosis, is a good assessment of turfgrass drought tolerance in soil [21,23]. As described previously [21,23], leaf firing was visually accessed by counting the grass leaves that turned yellow and brown. Under well-watered conditions, all of Tifgreen, SR9554 and Yukon showed very little leaf firing. After withholding water for 28 d, Yukon showed the most severe phenomena of leaf firing, while the leaf tissues of Tifgreen were

Table 1. Experiment ID, accession, drought tolerant (DT), Leaf water content (LWC), electrolyte leakage (EL) and survival rate of nine bermudagrass varieties.

			LWC (%)		EL (%)		Survival rate (%)	
ID	Accession	DT	Control	Drought	Control	Drought	Control	Drought
8	Tifgreen	T	78.37±0.34	54.27±2.25	17.26±0.93	49.12±1.37	100	58.23±2.44
9	Tifway	T	78.19±0.53	50.64±2.17	14.71±0.69	53.83±1.87	100	60.79±3.43
5	LaPaloma	M	78.94±0.48	36.61±1.38	14.43±0.93	64.17±2.84	100	15.56±6.30
7	Riviera	M	78.44±0.51	37.31±1.01	16.01±1.45	68.58±1.98	100	12.38±3.66
2	SR9554	M	79.55±0.69	30.28±1.93	14.03±1.63	64.99±2.78	100	9.60±2.47
4	Laprima	M	79.62±0.67	31.87±1.36	15.81±0.78	67.68±2.25	100	13.65±4.38
6	Veracruz	M	79.23±0.49	27.65±1.23	16.53±1.01	67.98±3.69	100	6.35±4.10
1	Wrangler	S	79.31±0.51	17.27±1.77	18.21±1.48	76.66±2.30	100	0
3	Yukon	S	78.80±0.39	15.93±1.85	16.53±1.63	86.74±2.08	100	2.02±1.02

T, tolerant; M, moderately tolerant; S, Susceptible. Data are means ± SE (n≥7).

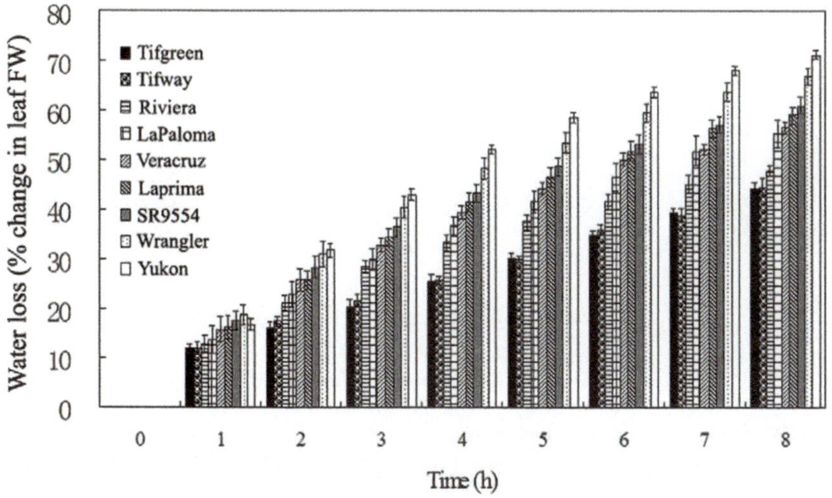

Figure 1. Analysis of leaf water loss (% change in leaf FW) in nine bermudagrass varieties. The results shown are mean ± SE (n = 8).

much greener when compared with the other two varieties (Figure 3).

Under control condition, all three bermudagrass varieties (Tifgreen, SR9554 and Yukon) maintained the LWC at about 80% (Figure 4A). After drought stress treatment, the LWC exhibited a gradual decline in all three varieties (Figure 4A). However, LWC of Yukon decreased from 66% to 42% from 7 d to 21 d after drought stress treatment, while from 69% to 47% for SR9554, and from 74% to 58% for Tifgreen, respectively, demonstrating that Tifgreen showed the slowest water loss under drought stress condition (Figure 4A).

Consistent with the change of LWC, all plants showed a gradual increase for EL after drought stress treatment. However the extent of change varied among different varieties (Figure 4B). In comparison with the others, Tifgreen exhibited significant lowest EL, while Yukon showed the highest EL after drought stress treatment from 7 d to 21 d (Figure 4B).

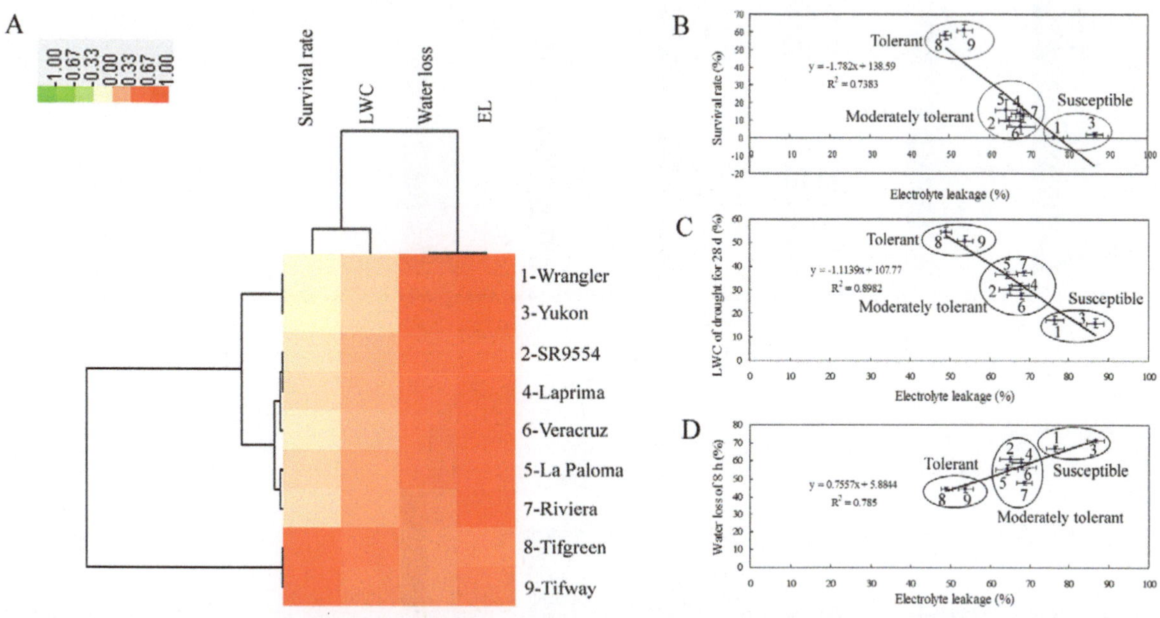

Figure 2. Groups of bermudagrass varieties differing in sensitivity to drought stress. (A) Hierarchical cluster analysis of the groups of bermudagrass varieties differing in sensitivities to drought stress based on the data of EL, survival rate, LWC after drought treatment for 28 d and water loss at 8 h after detachment. Resulting tree figure was displayed using the software package and Java Treeview. (B)–(D) Nine bermudagrass varieties were clustered to three groups differing in sensitivity to drought stress based on the data of EL and survival rate after drought treatment for 28 d (B), EL and LWC after drought treatment for 28 d (C), and EL at 28 d after drought and water loss at 8 h after detachment (D). The numbers of 1, 2, 3, 4, 5, 6, 7, 8 and 9 indicate the variety of Wranger, SR9554, Yukon, Laprima, LaPaloma, Veracruz, Riviera, Tifgreen and Tifway, respectively. The results shown are mean ± SE (n≥7).

Figure 3. Plant responses to drought stress. Three varieties representing drought tolerant, moderately tolerant and susceptible groups that were compared under drought stress condition. The plants with nearly the same crown size were subjected to control watering condition and drought condition for 28 d.

Changes of proline and sugar contents under drought stress condition

Under control condition, the internal proline level of plant was very low, but was largely induced when subjected to drought stress treatment (Figure 5A). Interestingly, Tifgreen plant had significantly higher proline content when compared with those of Yukon and SR9554 varieties under well-watered condition (Figure 5A). After drought stress treatment, proline levels were obviously up-regulated in all three varieties. However, proline level of Tifgreen was significantly higher than those of the other two varieties from 7 d to 21 d after drought treatment. SR9554 also exhibited significantly higher level of proline than Yukon after drought treatment for 21 d (Figure 5A). In addition, the contents of soluble sugars were also largely induced in all three varieties after drought treatment (Figure 5B and C). Under control condition, no significant differences of sucrose and soluble total sugar levels were observed among three varieties (Figure 5B and C). When drought stress was applied, SR9554 had significantly higher levels of both sucrose and soluble total sugars, while Tifgreen accumulated the highest content of sugars (Figure 5B and C). Together, the changes of proline content and sugars might be partially contributed to enhanced drought tolerance of Tifgreen variety.

Changes of ROS level and antioxidant enzyme activities in response to drought stress

As two major indicators for ROS level and oxidative damage, hydrogen peroxide (H_2O_2) and malondialdehyde (MDA) contents were assayed in this study [15]. As shown in Figure 6, all three varieties showed nearly the same levels of H_2O_2 and MDA under well-watered condition from 7 d to 21 d (Figure 6A and B). After drought treatment, gradual increases of H_2O_2 and MDA contents were observed in all varieties (Figure 6A and B). After 14 d and 21 d of drought stress treatments, the H_2O_2 and MDA contents in Tifgreen were significantly lower than those in Yukon (Figure 6A and B).

To address the relationship between the changes of ROS level and the antioxidant enzymes activities, five antioxidant enzymes (superoxide dismutase (SOD), catalase (CAT), peroxidase (POD), glutathione reductase (GR), glutathione peroxidase (GPX)) activities were analyzed. Under control condition, the activities of five antioxidant enzymes (SOD, CAT, POD, GR and GPX) in all three varieties (Yukon, SR9554 and Tifgreen) showed no

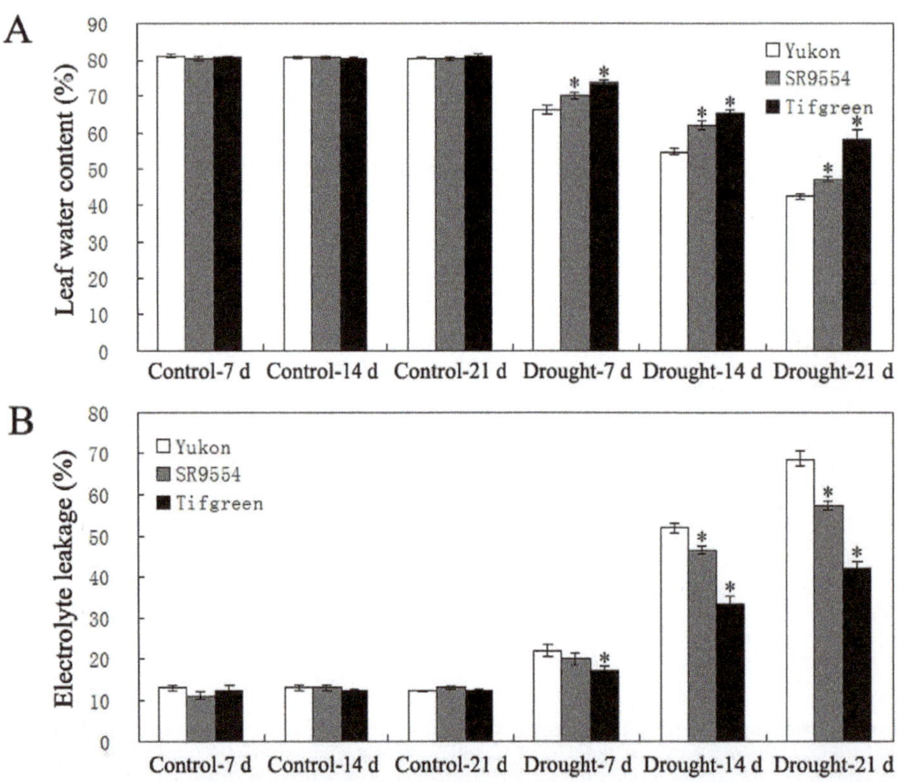

Figure 4. Water status and cell membrane damage of three bermudagrass varieties differing in drought tolerance. LWC (A) and EL (B) of Tifgreen, SR9554 and Yukon varieties during drought stress were shown here. The results shown are mean ± SE (n = 8). Asterisk symbols indicate significant differences from Yukon ($P<0.05$).

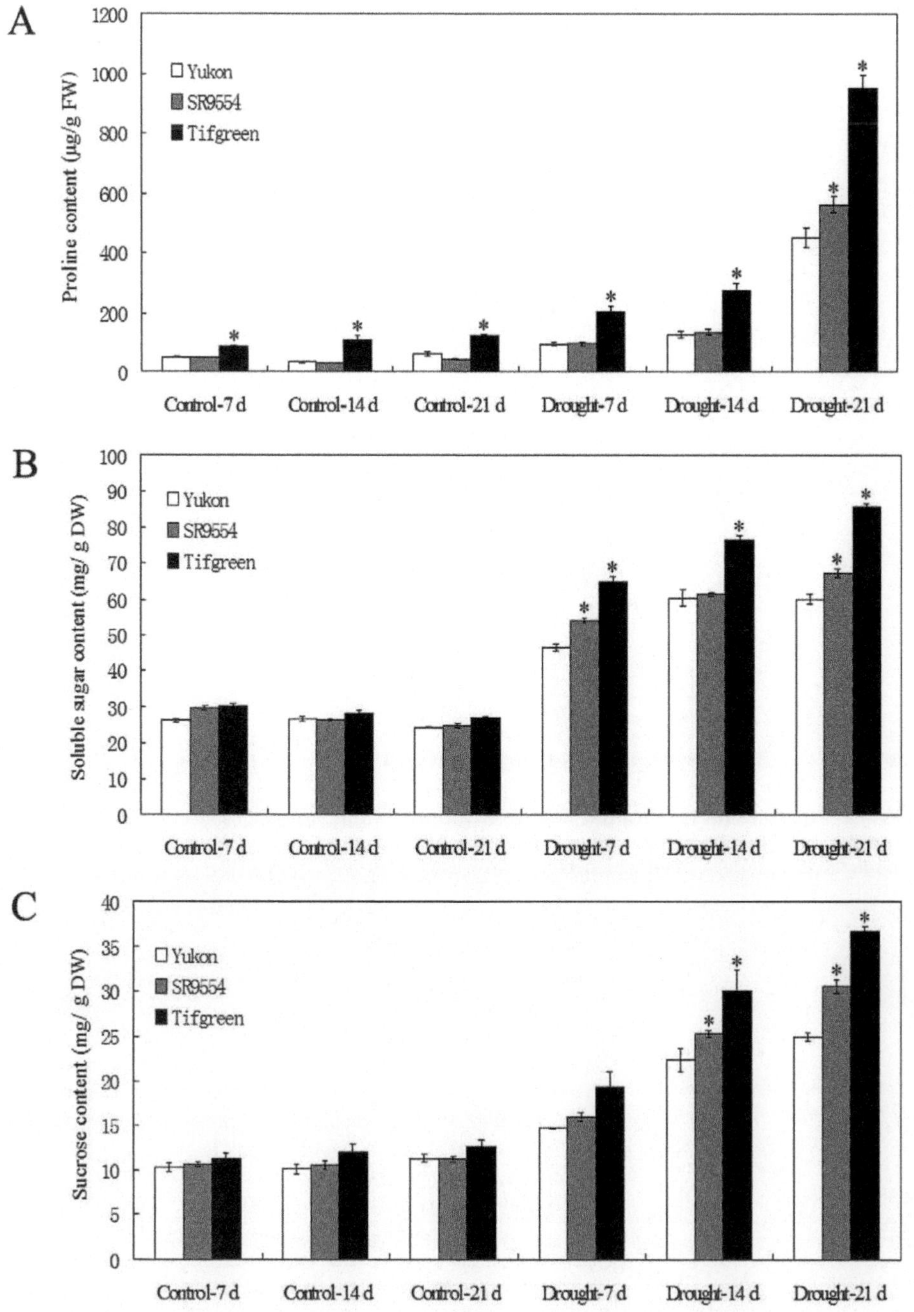

Figure 5. Accumulation of osmolytes of three bermudagrass varieties differing in drought tolerance during drought stress. Changes of proline content (A), soluble sugars (B) and sucrose content (C) of three genotypes of bermudagrass during drought stress were shown here. The results shown are mean ± SE (n = 4). Asterisk symbols indicate significant differences from Yukon ($P<0.05$).

significant differences (Figure 7A–E). After drought stress treatment, Tifgreen exhibited significantly higher activities of five antioxidant enzymes than those of Yukon and SR9554 at some timepoints (Figure 7A–E).

Correlation analyses of electrolyte leakage with other parameters under drought stress condition

To elucidate the possible mechanisms of bermudagrass in response to drought stress, we analyzed the correlations of EL with other physiological parameters of three tested varieties (Yukon, SR9554 and Tifgreen) under drought stress condition. The LWC showed negative linear correlation with EL under drought stress condition ($R^2 = 0.95$) (Figure 8A), suggesting that LWC was associated with drought induced cell injury in bermudagrass. However, there were no significant linear correlations between EL and the accumulations of osmolytes, including proline, soluble sugars and sucrose under drought stress condition ($R^2 = 0.42$, 0.55, 0.66, respectively) (Figure 8B–D). Additionally, the MDA content, H_2O_2 content, CAT, POD and GPX activity of Yukon, SR9554 and Tifgreen were positively correlated with EL under drought stress condition ($R^2 = 0.96$, 0.86, 0.76, 0.73 and 0.66, respectively) (Figure 8E–F, H–I, K). So the leaf water status, ROS level and

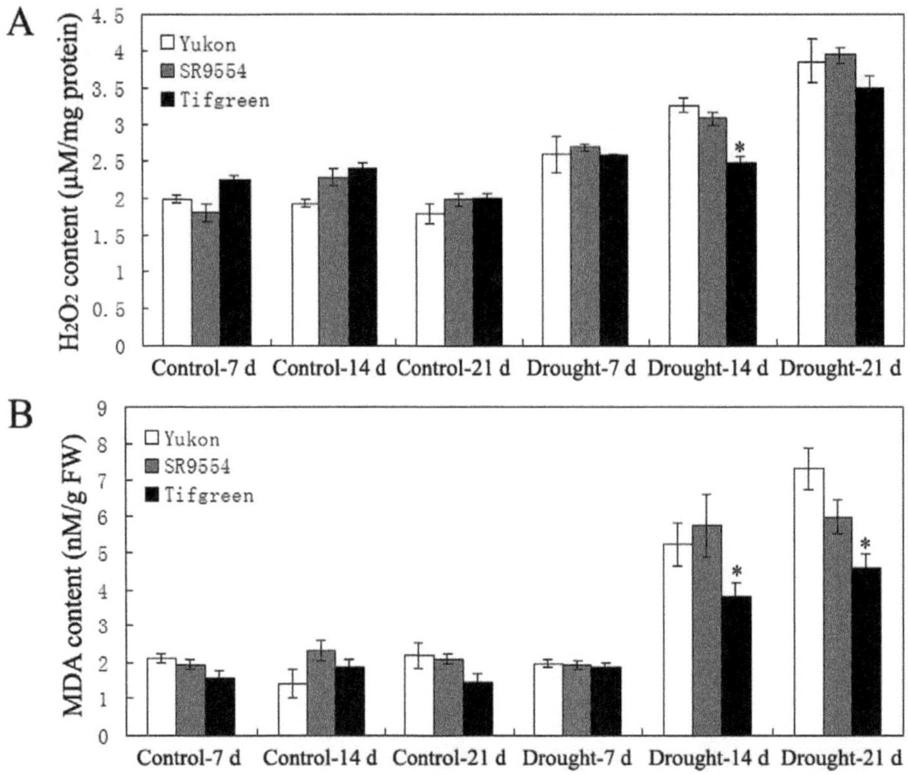

Figure 6. ROS level of three bermudagrass varieties differing in drought tolerance during drought stress. Changes of H_2O_2 content (A) and MDA level (B) of three varieties (Tifgreen, SR9554 and Yukon) differing in drought tolerance during drought stress were shown here. The results shown are mean ± SE (n = 4). Asterisk symbols indicate significant differences between Yukon and other genotypes ($P<0.05$).

activities of antioxidant enzymes might be associated with drought tolerance of bermudagrass.

Discussion

Drought is one of the major abiotic stresses, which largely limits plant growth and yield, and induces various changes at cellular, physiological and metabolic levels. Selection of natural stress tolerant genotypes among different species is the lowest cost, the most efficient and rapidest method for plant breeding [3,11,12,14]. To make full use of such strategy and develop new options, a comparative study of plant drought stress tolerances on the basis of physiological, biochemical and molecular changes is required [24]. Bermudagrass is a warm-season creeping grass and has natural variation in the drought response [5,6]. Under drought stress condition, the upper parts of bermudagrass die off, but the grass will keep growing from its rhizomes [5,6]. Additionally, bermudagrass grows and reproduces rapidly, and all these fundamental specifications make bermudagrass a potential target for genetic engineering of stress tolerant plants. In this study, we identified a wide range of variations of bermudagrass drastically differing in drought tolerance, and thoroughly detected some important parameters which were related with drought tolerance. Since turfgrasses consume large amounts of water around the world, these observations will give us some clues to breed drought-tolerant turfgrasses, as well as provide good opportunities to dissect drought stress response in other plant species. More importantly, because of its easy and quick reproduction, only litter survived bermudagrass (such as drought tolerant variety Tifgreen) during dry season can become a field of grasses during rainfall season.

Drought stress has direct impact on the disturbance of cell membrance [6,9,10]. EL, as one indicator of cell membrane stability, has been widely used to evaluate the extent of cell injury when subjected to various environment stresses [6,9,10,15,22,23]. The LWC *in vivo* and leaf water loss *in vitro* implicate the water status of plant. The slower water loss during drought stress reflects the maintenance of plant water content [6,9,10,15,22,23]. In this study, nine natural varieties of bermudagrass showed large variations in drought tolerance by evaluating cell membrane damage, water status and survival rate (Fig. 1; Table 1). Three groups of accessions differing in the drought tolerance were generally identified, with two tolerant varieties (Tifgreen and Tifway), five moderately tolerant varieties (LaPaloma, Riviera, SR9554, Laprima and Veracruz) and two susceptible varieties (Yukon and Wranger) (Fig. 2). Our results were consistent with Hu et al. [9] who found that Tifway was a drought tolerant bermudagrass material compared with the others. Besides, we also found that Tifgreen might have comparable drought tolerance with Tifway, and identified several new susceptible varieties (Yukon and Wranger) under drought condition. The difference of drought stress tolerance between Tifgreen and Yukon might be much more significant than previous identified bermudagrass varieties [9,22,23]. Therefore, the Yukon variety is also a very important material to reveal the mechanism of drought tolerance. The drastic natural variation can be further used to investigate the molecular, genetic, proteomic, metabolic basis of the drought response of bermudagrass. Further studies to identify drought-responsive candidate genes and examine transcriptomic and proteomic changes in response to drought stress using varieties of Tifgreen and Yukon are in progress. Compar-

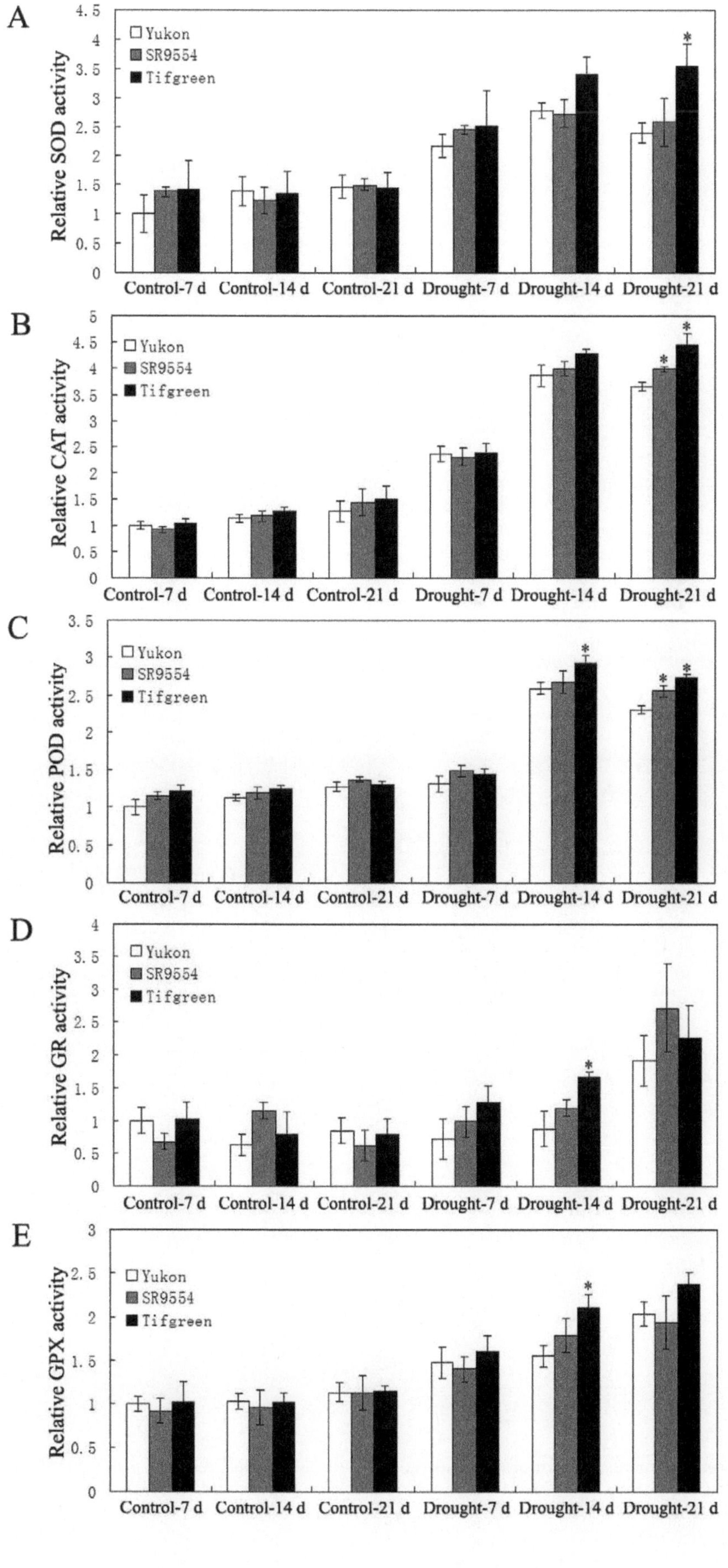
A
Yukon
SR9554
Tifgreen
Relative SOD activity
Control-7 d
Control-14 d
Control-21 d
Drought-7 d
Drought-14 d
Drought-21 d
B
Yukon
SR9554
Tifgreen
Relative CAT activity
Control-7 d
Control-14 d
Control-21 d
Drought-7 d
Drought-14 d
Drought-21 d
C
Yukon
SR9554
Tifgreen
Relative POD activity
Control-7 d
Control-14 d
Control-21 d
Drought-7 d
Drought-14 d
Drought-21 d
D
Yukon
SR9554
Tifgreen
Relative GR activity
Control-7 d
Control-14 d
Control-21 d
Drought-7 d
Drought-14 d
Drought-21 d
E
Yukon
SR9554
Tifgreen
Relative GPX activity
Control-7 d
Control-14 d
Control-21 d
Drought-7 d
Drought-14 d
Drought-21 d

Figure 7. Antioxidant enzyme activities of three bermudagrass varieties differing in drought tolerance during drought stress. (A)–(E) Comparisons of SOD (A), CAT (B), POD (C), GR (D) and GPX (E) activities three represent drought tolerant, moderately tolerant and susceptible accessions under control condition and drought stress were shown here. The relative activities were quantified as fold change in comparison with Yukon under control conditions for 7 d. The results shown are mean ± SE (n = 4). Asterisk symbols indicate significant differences between Yukon and other genotypes (*P*<0.05).

isons of transcriptomic and proteomic profilings would provide new clues to dissect additional mechanisms responsible for the drought tolerance in bermudagrass.

Complex mechanisms such as physiological, biochemical, molecular and cellular changes might contribute to drought response of bermudagrass [6,9,22,23]. Recently, Hu et al. [9] indicated that differential accumulation of dehydrins was positively connected with drought tolerance in bermudagrass. Zhao et al. [6] reported that the water-deficit tolerance of bermudagrass might be largely associated with the maintenance of photosynthesis protein metabolism and antioxidant defense. However, relationship among osmotic adaptation, ROS metabolism and antioxidant defense system has not yet been comparatively investigated in bermudagrass under drought condition. Thus, we thoroughly dissected the physiological and cellular changes among three bermudagrass varieties which showed drastically differences in drought tolerance (Yukon, SR9554 and Tifgreen). Proline is the most common compatible metabolite in the organelle and cytoplasm to regulate cell membrane stability and balance osmotic pressure of cytoplasm and environment, so higher proline content provides a significant advantage for plants to advent different stresses [6,23]. The soluble total sugars are also important osmolytes and sucrose is the major soluble sugar in plant [6,23]. In this study, the results indicated that proline and soluble sugar accumulations were significant higher in drought tolerant bermudagrass (Tifgreen) than those of the other varieties under water-deficit condition, leading the cells to improve the drought tolerance by osmotic adjustment (Fig. 5). Increased proline content might increase the osmotic pressure inside plant cells and cause more water uptake to keep a significantly increase in LWC. This result was consistent with previous researches which showed that proline accumulation and soluble sugars were positive related with drought tolerance in many plant species [23,24,25]. Moreover, water loss is a key for plant survival during drought stress condition. The results of rapid leaf water loss (Fig. 1) were consistent with changes of long term leaf water content (Fig. 4), i.e. drought tolerant bermudagrass variety exhibited lower leaf water loss, and therefore higher leaf water content. Higher amount of

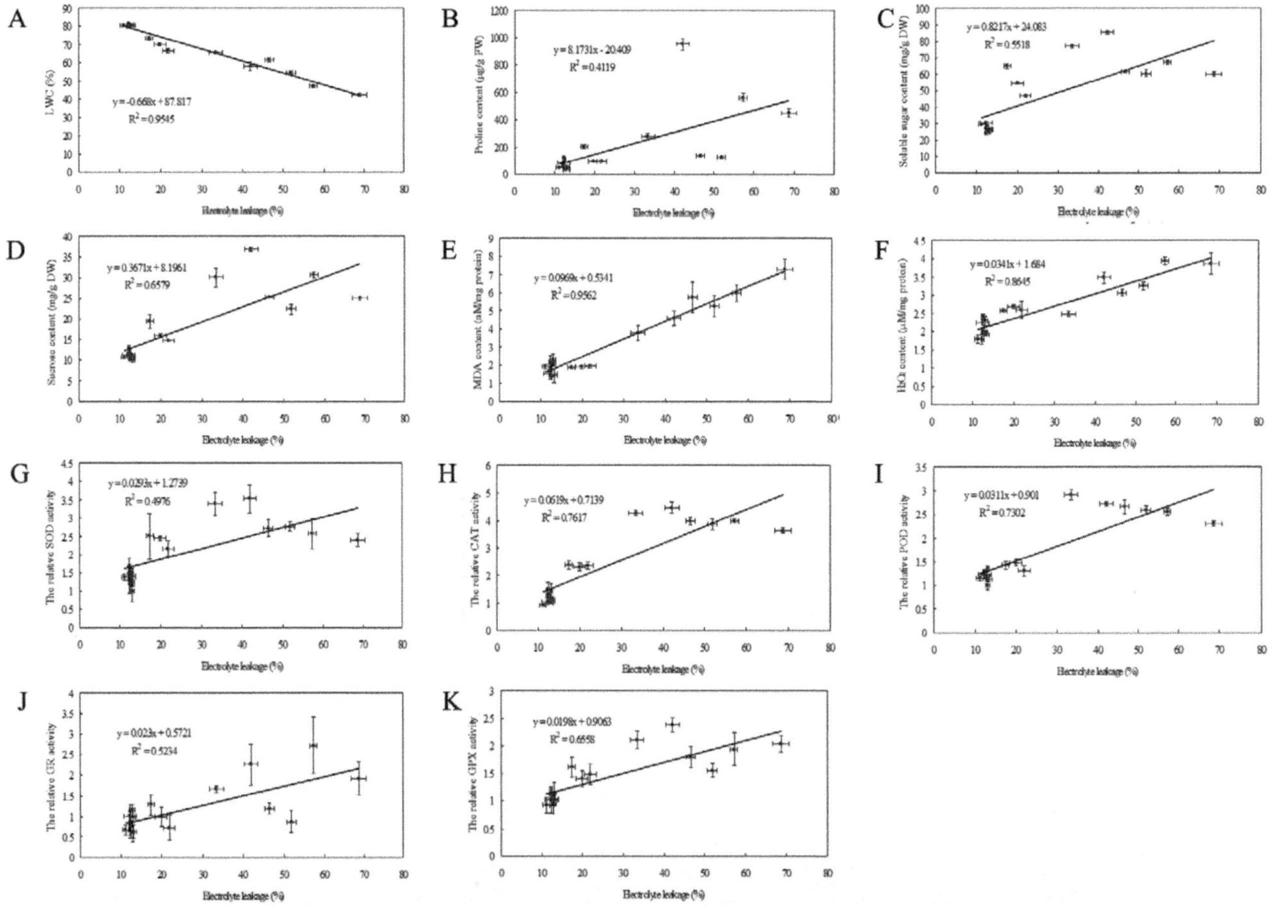

Figure 8. Correlations analysis of EL with other parameters during drought stress. (A)–(I) The linear correlation of EL with LWC (A), MDA content (B), Proline content (C), soluble sugars (D), sucrose content (E), H2O2 content (F), SOD activity (G), CAT activity (H), POD activity (I), GR activity (J) and GPX activity (K) of three represent cultivars (Yukon, SR9554 and Tifgreen) during drought stress. The results shown are mean ± SE (n≥4).

osmolytes accumulation in tolerant varieties might regulate plant stomatal movement to reduce water loss.

Drought stress can cause oxidative stress via rapid and excessive production of ROS, and H_2O_2 is the most important one [26,27,28]. The over-production of H_2O_2 can lead to oxidative damages by oxidizing proteins, damaging nucleic acids and causing lipid peroxidation (MDA), etc [26,27,28]. To scavenge the over-production of ROS and protect plant cells from ROS damage, plant developed complex antioxidant defense systems, including antioxidant enzymes like ascorbate peroxidase (APX), SOD, CAT, POD, GR, GPX, dehydroascorbate reductase (DHAR), monodehydroascorbate reductase (MDHAR), etc [6,26,27,28]. SOD functions as the first stage of the antioxidant defense system by catalysing O_2^{-}. into H_2O_2 and O_2, while APX, CAT, POD, GR and GPX are also essential for break down H_2O_2 through different pathway [26,27,28]. Using 2-DE and MS, Zhao et al. [6] found that the abundance levels of antioxidant defense proteins (APX, SOD, DHAR and MDHAR) showed a greater extent of increase in response to water-deficit stress in Tifway (tolerant variety) than in C299 (sensitive variety), indicating that antioxidant defense systems might play a critical role in the drought stress response of bermudagrass. Considering the various types of oxidative thiol modifications that may affect the antioxidant enzyme activities especially under oxidative stress conditions [29,30], we directly measured antioxidant enzymes activities and ROS level in this study. In the drought tolerant Tifgreen plant, significant lower H_2O_2 and MDA levels and higher SOD, CAT, POD, GR and GPX activities were observed than those of drought sensitive Yukon variety after drought stress treatment, indicating Tifgreen showed more capacity to remove ROS levels than Yukon. Therefore, the lower ROS level (including H_2O_2 and MDA) and higher antioxidant enzymes activities during water-deficit period might partially contribute to the enhanced drought tolerance of Tifgreen. These investigations have extended the expression pattern of antioxidant enzymes to enzyme activity level as well as ROS level which are more directly related to drought tolerance in bermudagrass.

Additionally, we analyzed the relationship of all the above parameters. Among these parameters, the LWC and MDA content showed the highest linear correlation with EL (Figure 8). The linkage between LWC and EL indicated that the regulation of plant water status might be very important for maintaining of plant cell membrane stability. Based on correlation analysis, we found that oxidative stress triggered by drought was closely related to cell membrane stability. Therefore, the cell injury during drought stress might be largely attributed to ROS production and antioxidant defense system in bermudagrass. Osmolyte like trehalose plays a direct role in eliminating H_2O_2 and superoxide anions ($O_2\bullet^{-}$) through protecting antioxidant enzymes (such as SOD) in wheat under heat stress [31]. We speculate that higher accumulation of osmolytes including proline and total sugars in tolerant variety (Tifgreen) (Fig. 5) might regulate ROS level through protecting of antioxidant enzymes, and the effective osmotic adjustment might alleviate drought induced cell injury. Following gradual loss of LWC under drought stress condition, more osmolyte accumulation in tolerant Tifgreen variety alleviated oxidative stress through activation of antioxidant defense system, resulting in lower lipid peroxidation, less cell injury, and higher plant survival rate.

Besides bermudagrass, there are also complex mechanisms which contribute to natural variation of drought tolerance in other species including Arabidopsis [14,32] and *Brachypodium distachyon* [3]. Arabidopsis showed interesting phenotypic variations in response to mild water deficit [14]. The comparison of drought tolerant variety (C24) and drought sensitive varieties (Col-0, Ws-0 and Ws-2) suggested that in Arabidopsis, increasing the leaf number with a concomitant reduction of individual leaf area and increasing in leaf temperature with reduction in stomatal conductance, are associated with the enhanced drought tolerance [32]. In *B. distachyon*, the drought tolerant varieties showed little leaf wilting and fewer reduction in chlorophyll fluorescence (Fv/Fm) and LWC, while the most susceptible varieties had opposite effect in these parameters [3]. Although drought stress significantly increased the content of total soluble sugars, no significant differences were observed among different varieties of *B. distachyon* [3]. Based on these researches and our observations, different plant species might activate similar physiological and cellular responses under drought stress condition, like maintaining of LWC. However, other parameters such as total soluble sugars might contribute to different extend of variation of drought tolerance in some plant species, like bermudagrass and *B. distachyon*. As shown in this study, natural variation of drought stress tolerance in bermudagrass varieties might be largely related to the induced changes of some important parameters, especially LWC, ROS level and antioxidant defense system.

In summary, three groups of bermudagrass germplasm resources differing in drought tolerance were characterized in this study. Three varieties of Yukon, SR9554 and Tifgreen showing relatively different drought tolerance were chosen for further study to reveal the relationships between drought tolerance and changes of physiological parameters. The Tifgreen was more effective than Yukon and SR9554 in alleviating drought stress induced EL level, H_2O_2 content and lipid peroxidation by enhancing the accumulation of osmolytes and activities of antioxidant enzymes, and by maintaining relatively higher water status. These results suggested that changes of water status, osmolyte accumulation and antioxidant defense system during drought stress may contribute to the natural variation of bermudagrass drought tolerance. The identified bermudagrass tolerant and susceptible varieties, as well as characterized physiological parameters, can be further used in plant breeding programs to enhance drought tolerance in bermudagrass and other turfgrass species.

Materials and Methods

Plant materials and growth conditions

The following nine bermudagrass varieties were used in this study: Wrangler, SR9554, Yukon, LaPaloma, LaPrima (Blend of LaPaloma and SR9554), Veracruz, Riviera, Tifgreen and Tifway. Among these varieties, the seeds of Wrangler, LaPaloma, Veracruz and Riviera were kindly provided by American Johnston Seed Company (http://www.jeinc.com/seed), and those of SR9554, Yukon and LaPrima were obtained from American Seed Research of Oregon Company (http://www.sroseed.com/). Tifgreen and Tifway materials were from Wuhan Green Garden Turfgrass Company (http://www.whcp66.com/).

The seeds were stratified at 4°C for 4 days in darkness and then cultured in the sands in the growth room controlled at an irradiance of about 150 μmol quanta m^{-2} s^{-1}, 25±2°C, 65%–75% relative humidity, and 16 h light and 8 h dark cycles for 60 d. During the 60 d growth, the plants were given irrigated nutrient solution two times every week. The cutting shoot-tips from the above nine bermudagrass varieties were placed in the growth room for about 1 month, and healthy cuttings with nearly the same crown size were chosen and replanted for stress treatment.

Drought treatment and experimental design

To assess potential drought tolerance of nine varieties of bermudagrass, the healthy cuttings in pots with nearly the same crown size were subjected to watering condition (control) and drought condition (soil water deficit) by withholding water in the soil for 28 d. More than four pots of plants from each variety were used as replicates in each independent experiment, and all these pots of plants were conducted in a randomized complete block design with the same growth condition. To minimize the environment effects, the pots with soil and plants were rotated daily. The survival rate of stressed bermudagrass was recorded at 7 d after re-watering. The leaf samples were harvest at 7 d, 14 d, 21 d and 28 d after treatments for physiological parameter analyses.

Assay of electrolyte leakage

For the EL assay, about 0.1 g leaves were incubated in 10 ml of deionized water, and shaken at room temperature for 6 h. The initial conductivity (C_i) was measured using a conductivity meter (Leici-DDS-307A, Shanghai, China). The samples were then boiled for 20 min to completely induce all electrolytes. After cooling to room temperature, the conductivity of the killed tissues (C_{max}) was determined. Relative EL (%) = $(C_i/C_{max}) \times 100$.

Assessment of water loss and leaf water content

To access leaf water status, leaf water loss *in vitro* and LWC *in vivo* were conducted in this study. Leaf water loss was expressed as % change in leaf FW *in vitro*, and LWC was the measurement of leaf water potential *in vivo*.

To compare the water loss between nine varieties of bermudagrass, the detached leaves grown under normal conditions were placed on the weighing paper inside the same growth room. FW of the detached leaves were quantified every 1 hour intervals for up to 8 h. The water loss was then calculated from the decrease in the rate of FW at designated time intervals [25].

For LWC analysis, the leaf samples were harvest at different timepoints (7 d, 14 d, 21 d and 28 d) under control and drought treatment conditions. The FW was determined immediately after harvest and the dry weight (DW) was determined after 16 h incubation at 80°C. LWC was measured according to previously described method using the following formula: LWC (%) = (FW-DW)/FW × 100 [3].

Measurement of proline content

Proline content was estimated using described method with L-proline as standard [25]. Briefly, 0.5 g leaf samples were extracted in 3% (w/v) sulfosalicylic acid before 2 ml of ninhydrin reagent and 2 ml of glacial acetic acid were added. Well mixed solutions were boiled at 100°C for 40 min. After cooling to room temperature, the proline level of samples was calculated at 520 nm absorbance by making the specification curve with known concentration of L-proline.

Measurement of sucrose and soluble total sugars

The sucrose and soluble total sugars were determined using the anthrone method as previous described with some modifications [23]. Briefly, 100 mg dried leaf samples were extracted in 5 ml of 80% (v/v) ethanol at 80°C for 40 min and centrifuged at 12000 rpm for 10 min. The pellets were further extracted twice with another 5 ml of 80% (v/v) ethanol. The combined supernatants were dipigmented by litter activated charcoal at 80°C for 30 min. The filtrated solution was brought to 10 ml for further analysis.

For sucrose content assay, 0.2 ml mixture (0.1 ml of the above extracts mixed with 0.1 ml of 2 M NaOH) was boiled at 100°C for 10 min. After cooling down, 1.4 ml of 30% (w/v) HCl and 0.4 ml of 0.1% (w/v) hydroxyphenol (0.1 g hydroxyphenol was dissolved in 100 ml of 80% (v/v) ethanol) were added and mixed thoroughly, and the mixture was heated at 80°C for 10 min. After cooling to room temperature, the sucrose content of sample was calculated at 480 nm of absorbance by making the specification curve with known concentration of surcose.

For the determination of soluble total sugars, the mixture of 0.1 ml of the extracts and 3 ml of 0.15% (w/v) anthrone reagent (0.3 g anthrone was dissolved in 200 ml of 7.74 M H_2SO_4) was heated at 90°C for 20 min. Then soluble total sugar level was examined at 620 nm of absorbance by making the specification curve with known concentration of glucose.

Determination of lipid peroxidation

The lipid peroxidation was determined by measuring the amount of MDA, with the thiobarbituric acid (TBA) as reported previously [25]. Totally 0.5 g plant leaves were ground in 2.5 ml of regent (0.25% (w/v) TBA in 10% (w/v) trichloroacetic acid), and then boiled at 100°C for 20 min. The MDA content was determined by subtracting the non-specific absorption at 600 nm from the absorbance of the sample supernatant at 532 nm.

Protein extraction and quantification

For plant protein extraction, about 1 g fresh leaves were ground with liquid nitrogen and then homogenized in extraction buffer (50 mM sodium phosphate buffer, pH 7.8). After centrifuging at 12000 rpm for 30 min at 4°C, the supernatant was used for protein content, H_2O_2 content and antioxidant enzyme activity determination [10]. The protein content was quantified using Bradford method with bovine serum albumin (BSA) as standard.

Determination of H_2O_2 level

The H_2O_2 content was determined using method described by Hu et al. [15]. Briefly, 1 ml of the above supernatant was mixed with 1 ml of 0.1% titanium sulphate in 20% H_2SO_4 (v/v) thoroughly for 10 min. After centrifuged at 12000 rpm for 10 min at room temperature, the absorbance of the supernatant was measured at 410 nm using known concentration of H_2O_2 as control.

Assay of antioxidant enzyme activities

The SOD (EC 1.15.1.1), CAT (EC 1.11.1.6), GR (EC 1.6.4.2), GPX (EC 1.11.1.9) activities in the samples were determined using Total SOD Assay Kit with WST-1 (S0102, Beyotime, China), CAT Assay Kit (S0051, Beyotime, China), GR Assay Kit (S0055, Beyotime, China) and Total GPX Assay Kit (S0058, Beyotime, China), respectively, according to the manufacturer's instructions. The POD (EC 1.11.1.7) activity was measured with Plant POD Assay Kit (A084-3, Nanjing Jiancheng Bioengineering Institute, China) as the instruction described.

For the determination of SOD activity, 2-(4-iodophenyl)-3-(4-nitrophenyl)-5-(2, 4-disulfophenyl)-2H-tetrazolium (WST-1) method was used [33]. WST-1 can couple with xanthine oxidase (XO) to generate O_2^- and formazan dye, however, this reaction can be inhibited by SOD by catalysing O_2^- into H_2O_2 and O_2. Therefore, the SOD activity can be calculated by measuring the absorbance of formazan dye at 450 nm.

The CAT activity was assayed using CAT Assay Kit (S0051, Beyotime, China) as previous described [34]. Briefly, the protein supernatants were treated with excess H_2O_2 for decomposition by

CAT for 5 min, and the remaining H_2O_2 coupled with a substrate was treated with POD to generate a red product, N-4-antipyryl-3-chloro-5-sulfonate-p-benzoquinonemonoimine, which can be examined at 520 nm. CAT activity can be determined by measuring the decomposition of H_2O_2.

For the GR activity assay, GR Assay Kit (S0055, Beyotime, China) was used. The reaction mixture included 20 μl of sample, 100 μl of GSSG solution, 70 μl of GR assay solution and 10 μl of 2 mM NADPH, and the blank was set without sample. Glutathione disulfide (GSSG) can be catalyzed to reduced glutathione (GSH) by GR in the present of NADPH. Then the GR activity can be determined by measuring the reduction of NADPH from the absorbance at 340 nm.

The GPX activity was determined using Total GPX Assay Kit (S0058, Beyotime, China) as described by Wang et al. [35]. Briefly, the reaction mixture contained 10 μl of sample supernatant, 176 μl GPX assay buffer, 10 μl GPX assay working solution (4.8 mM NADPH, 40.4 mM GSH, and GR solution supplied by the kit), 4 μl of 15 mM cumene hydroperoxide (Cum-OOH), and two controls were set without sample and without Cum-OOH, respectively. The GPX activity was calculated by measuring the reduction of NADPH to $NADP^+$ at 340 nm of absorbance.

The POD activity was assayed with Plant POD Assay Kit (A084-3, Nanjing Jiancheng Bioengineering Institute, China) as the instruction described based on the guaiacol oxidation [36]. The POD activity was determined by examining the absorbance of reaction buffer at 420 nm.

The relative activities of the above antioxidant enzymes were quantified as fold change in relative to Yukon under control condition for 7 d.

Cluster analysis

To identify the groups of bermudagrass varieties differing in sensitivities to drought stress, the data of EL, survival rate, LWC after drought treatment for 28 d, and water loss at 8 h after detachment from all nine bermudagrass varieties were chosen for cluster analysis. Hierarchical cluster analyses were performed using the CLUSTER program (http://bonsai.ims.u-tokyo.ac.jp/~mdehoon/software/cluster/) [37] by the uncentred matrix and complete linkage method. Resulting tree figures were displayed using the software package, Java Treeview (http://jtreeview.sourceforge.net/) as described by Chan et al. [38].

Statistical analysis

All experiments in this study were repeated at least three times, and the results shown are mean ± SE of these independent experiments. For each independent experiment, at least 30 plants were used for each variety and treatment combination. Asterisk symbols above the columns in the figures indicate significant differences from Yukon at $P<0.05$ (Student's *t*-test).

Acknowledgments

We thank American Johnston Seed Company, American Seed Research of Oregon Company and Wuhan Green Garden Turfgrass Company for kindly providing the *Cynodon dactylon* seeds or plants. We also thank the editor and the reviewers for their valuable comments and suggestions that help improve the manuscript. Thanks are also given to Ms. Ruijie Liu for help in the English language.

Author Contributions

Conceived and designed the experiments: HS Z. Chan. Performed the experiments: HS YW Z. Cheng TY. Analyzed the data: HS Z. Chan. Contributed reagents/materials/analysis tools: HS YW Z. Cheng TY Z. Chan. Wrote the paper: HS Z. Chan.

References

1. Yamaguchi-Shinozaki K, Shinozaki K (2006) Transcriptional regulatory networks in cellular responses and tolerance to dehydration and cold stresses. Annu Rev Plant Biol 57: 781–803.
2. Zhu JK. (2002) Salt and drought stress signal transduction in plants. Annu Rev Plant Biol 53: 247–273.
3. Luo N, Liu J, Yu X, Jiang Y (2011) Natural variation of drought response in Brachypodium distachyon. Physiol Plantarum 141: 19–29.
4. Xu C, Huang B (2010) Differential proteomic responses to water stress induced by PEG in two creeping bentgrass varieties differing in stress tolerance. Journal of Plant Physiol 167: 1477–1485.
5. Kim C, Lemke C, Paterson AH (2009) Functional dissection of drought-responsive gene expression patterns in Cynodon dactylon L. Plant Mol Biol 70:1–16.
6. Zhao Y, Du H, Wang Z, Huang B (2011) Identification of proteins associated with water-deficit tolerance in C_4 perennial grass species, *Cynodon dactylon*×*Cynodon transvaalensis* and *Cynodon dactylon*. Physiol Plantarum 141: 40–55.
7. Carmo-Silva AE, Bernardes da Silva A, Keys AJ, Parry MAJ, Arrabaça MC (2008) The activities of PEP carboxylase and the C_4 acid decarboxylases are little changed by drought stress in three C_4 grasses of different subtypes. Photosynth Res 97: 223–233.
8. Carmo-Silva AE, Keys AJ, Beale MH, Ward JL, Baker JM, et al. (2009) Drought stress increases the production of 5-hydroxynorvaline in two C_4 grasses. Phytochemistry 70: 664–671.
9. Hu L, Wang Z, Du H, Huang B (2010) Differential accumulation of dehydrins in response to water stress for hybrid and common bermudagrass genotypes differing in drought tolerance. J Plant Physiol 167: 103–109.
10. Hu L, Wang Z, Huang B (2010) Diffusion limitations and metabolic factors associated with inhibition and recovery of photosynthesis from drought stress in a C_3 perennial grass species. Physiol Plantarum 139: 93–106.
11. Alonso-Blanco C, Asrts MGM, Bentsink L, Keurentjes JJB, Reymond M, et al. (2009) What has natural variation taught us about plant development physiology and adaptation? Plant Cell 21: 1877–1896.
12. Benfey PN, Mitchell-Olds T (2008) From genotype to phenotype: systems biology meets natural variation. Science 320: 495–497.
13. Bian S, Jiang Y (2009) Reactive oxygen species, antioxidant enzyme activities and gene expression patterns in leaves and roots of Kentucky bluegrass in response to drought stress and recovery. Sci Hortic 120: 264–270.
14. Bouchabke O, Chang FQ, Simon M, Voisin R, Pelletier G, et al. (2008) Natural variation in *Arabidopsis thaliana* as a tool for highlighting differential drought responses. PLoS One 3: e1705.
15. Hu L, Li H, Pang H, Fu J (2012) Responses of antioxidant gene, protein and enzymes to salinity stress in two genotypes of perennial ryegrass (*Lolium perenne*) differing in salt tolerance. J Plant Physiol 169: 146–156.
16. Carrow RN (1996) Drought tolerance aspects of turfgrass in the southeast: root-shoot responses. Crop Sci 36: 687–694.
17. Huang B, Duncan RR, Carrow RN (1997) Drought-resistance mechanisms of seven warm-season turfgrasses under surface soil drying: II. Root aspects. Crop Sci 7: 1863–1869.
18. Huang B, Duncan RR, Carrow RN (1997) Drought-resistance mechanisms of seven warm-season turfgrasses under surface soil drying: I. Shoot response. Crop Sci 37: 1858–1863.
19. Qian Y, Fry JD, Upham WS (1997) Rooting and drought avoidance of warm-season turfgrasses and Tall Fescue in Kansas. Crop Sci 47: 2162–2169.
20. Campbell SA, Close T (1997) Dehydrins: genes, proteins, and associations with phenotypic traits. New Phytol 137: 61–74.
21. Carrow RN (1995) Drought tolerance aspects of turfgrass in the southeast: evapotraspiration and crop coefficients. Crop Sci 35: 1685–1690.
22. Lu S, Peng X, Guo Z, Zhang G, Wang Z, et al. (2007) In vitro selection of salinity tolerant variants from triploid bermudagrass (*Cynodon transvaalensis*×*C. dactylon*) and their physiological responses to salt and drought stress. Plant Cell Rep 26: 1413–1420.
23. Lu S, Chen C, Wang Z, Guo Z, Li H (2009) Physiological responses of somaclonal variants of triploid bermudagrass (*Cynodon transvaalensis*×*Cynodon dactylon*) to drought stress. Plant Cell Rep 28: 517–526.
24. Shukla N, Awasthi RP, Rawat L, Kumar J (2012) Biochemical and physiological responses of rice (*Oryza sativa* L.) as influenced by *Trichoderma harzianum* under drought stress. Plant Physiol Biochem 54: 78–88.
25. Shi HT, Li RJ, Cai W, Liu W, Wang CL, et al. (2012) Increasing nitric oxide content in Arabidopsis thaliana by expressing rat neuronal nitric oxide synthase resulted in enhanced stress tolerance. Plant Cell Physiol 53: 344–357.

26. Apel K, Hirt H (2004) Reactive oxygen species: metabolism, oxidative stress, and signal transduction. Annu Rev Plant Biol 55: 373–399.
27. Mittler R (2002) Oxidative stress, antioxidants and stress tolerance. Trends Plant Sci 7: 405–410.
28. Mittler R, Vanderauwera S, Gollery M, Van Breusegem F (2004) Reactive oxygen gene network of plants. Trends Plant Sci 9: 490–498.
29. Thamsen M, Jakob U (2011) The redoxome proteomic analysis of cellular redox networks. Curr Opin Chem Biol 15:113–119.
30. Lindermayr C, Saalbach G, Durner J (2005) Proteomic identification of S-Nitrosylated proteins in Arabidopsis. Plant Physiol 137:921–930.
31. Luo Y, Li WM, Wang W (2008) Trehalose: Protector of antioxidant enzymes or reactive oxygen species scavenger under heat stress? Environ Exp Bot 63: 378–384.
32. Bechtold U, Lawson T, Mejia-Carranza J, Meyer RC, Brown IR, et al. (2010) Constitutive salicylic acid defences do not compromise seed yield, drought tolerance and water productivity in the Arabidopsis accession C24. Plant Cell Environ 33: 1959–1973.
33. Alexander VP, Christine CW (2000) A microtiter plate assay for superoxide dismutase using a water-soluble tetrazolium salt (WST-1). Clin Chim Acta 293: 157–166.
34. He Z, Sun X, Mei G, Yu S, Li N (2008) Nonclassical secretion of human catalase on the surface of CHO cells is more efficient than classical secretion. Cell Biology Intern 32: 367–373.
35. Wang XL, Yang CP, Xu K, Qin OJ (2010) Selenoprotein W depletion *in vitro* might indicate that its main function is not as an antioxidantive enzyme. Biochemistry (Moscow) 75: 201–207.
36. Maehly AC, Chance B (1954) The assay of catalases and peroxidases. Methods Biochem Anal 1: 357–424.
37. de Hoon MJL, Imoto S, Nolan J, Miyano S (2004) Open source clustering software. Bioinformatics 20: 1453–1454.
38. Chan Z, Bigelow PJ, Grumet R, Loescher W (2012) Comparison of salt stress resistance genes in transgenic Arabidopsis thaliana indicates that extent of transcriptomic change may not predict secondary phenotypic or fitness effects. Plant Biotechnol J 10: 284–300.

21

Increased Drought Impacts on Temperate Rainforests from Southern South America: Results of a Process-Based, Dynamic Forest Model

Alvaro G. Gutiérrez[1,2*¤], Juan J. Armesto[3,4], M. Francisca Díaz[5], Andreas Huth[1]

1 Department of Ecological Modeling, Helmholtz Centre for Environmental Research (UFZ), Leipzig, Germany, **2** Forest Ecology Group, Institute of Terrestrial Ecosystems, Department of Environmental Sciences, Swiss Federal Institute of Technology (ETH Zürich), Zürich, Switzerland, **3** Instituto de Ecología y Biodiversidad (IEB), Santiago, Chile, **4** Departamento de Ecología, Facultad de Ciencias Biológicas, Universidad Catolica de Chile, Santiago, Chile, **5** Departamento de Ciencias Biológicas, Facultad de Ciencias Biológicas, Universidad Andrés Bello, Santiago, Chile

Abstract

Increased droughts due to regional shifts in temperature and rainfall regimes are likely to affect forests in temperate regions in the coming decades. To assess their consequences for forest dynamics, we need predictive tools that couple hydrologic processes, soil moisture dynamics and plant productivity. Here, we developed and tested a dynamic forest model that predicts the hydrologic balance of North Patagonian rainforests on Chiloé Island, in temperate South America (42°S). The model incorporates the dynamic linkages between changing rainfall regimes, soil moisture and individual tree growth. Declining rainfall, as predicted for the study area, should mean up to 50% less summer rain by year 2100. We analysed forest responses to increased drought using the model proposed focusing on changes in evapotranspiration, soil moisture and forest structure (above-ground biomass and basal area). We compared the responses of a young stand (YS, ca. 60 years-old) and an old-growth forest (OG, >500 years-old) in the same area. Based on detailed field measurements of water fluxes, the model provides a reliable account of the hydrologic balance of these evergreen, broad-leaved rainforests. We found higher evapotranspiration in OG than YS under current climate. Increasing drought predicted for this century can reduce evapotranspiration by 15% in the OG compared to current values. Drier climate will alter forest structure, leading to decreases in above ground biomass by 27% of the current value in OG. The model presented here can be used to assess the potential impacts of climate change on forest hydrology and other threats of global change on future forests such as fragmentation, introduction of exotic tree species, and changes in fire regimes. Our study expands the applicability of forest dynamics models in remote and hitherto overlooked regions of the world, such as southern temperate rainforests.

Editor: Bruno Hérault, Cirad, France

Funding: This study was supported by grants from Millennium Scientific Initiative (P05-002) and CONICYT PFB-23 to the Institute of Ecology and Biodiversity. AGG was funded by a DAAD doctoral fellowship, by a Marie Curie Intra European Fellowship within the 7th European Community Framework Programme (PIEF-GA-2010-274798) and CONICYT-PAI grant (82130046). Additional funding was provided by the ERC advanced grant 233066. The funders had no role in study design, data collection and analysis, decision to publish, or preparation of the manuscript.

Competing Interests: The authors have declared that no competing interests exist.

* Email: bosqueciencia@gmail.com

¤ Current address: Instituto de Conservación Biodiversidad y Territorio, Facultad de Ciencias Forestales y Recursos Naturales, Universidad Austral de Chile, Valdivia, Chile

Introduction

Climate and forests are dynamically linked through the spatial and temporal variability of soil moisture [1], with climate system effects on ecological processes which are still poorly understood. Forest dynamics models, particularly those based on interactions among individual trees (i.e. gap models [2]), provide a simple, and general framework to assess the impacts of climate on forest dynamics. These models simulate the fate of single trees on the basis of species' life-history traits and limited resource availability (e.g. soil moisture), thereby facilitating the analysis of climate-forest interactions [3].

Forest gap models use a variety of approaches to model forest hydrology. While some gap models use a simple bucket water balance model [3,4], others include physiology-based representations of plant and soil controls on water uptake and evapotranspiration [5,6]. Regardless of the level of detail used to model forest hydrology, it seems necessary that forest gap models address water availability (i.e. soil moisture) as an integrating factor, with effects on canopy transpiration [7]. Changes in rainfall regimes, summarized by changes in the duration and frequency of periods of water stress during the year, should influence soil moisture dynamics limiting plant productivity [8]. Introducing dynamic linkages of ecological processes with soil moisture variation in gap models will contribute to predict drought-induced changes in forest dynamics. Such model improvements are increasingly relevant to understanding how forests can adapt to climate change [6,9].

Forest gap models have successfully simulated the dynamics of a variety of forest types including temperate rainforests of the southern hemisphere [10,11]. In southern South America (SSA, 37–43°S), the progressive loss, fragmentation and subsequent degradation of temperate rainforests due to unsustainable logging and fire is threatening the integrity of ecosystem functions [12,13] and modifying their hydrological balance [14,15]. Annual precipitation has decreased in the same region by about 40% in the last century (time period 1901–2005, [16]) and summer rainfall is expected to decrease up to 50% by the year 2100 [17,18]. SSA forests share similar structural characteristics with temperate rainforests of the Pacific Northwest of North America, Tasmania, and New Zealand [19]. In addition, SSA forests represent the largest area of temperate forest remaining in the southern hemisphere [20]. Floristic richness is the highest among evergreen temperate rainforests worldwide and the high concentration of endemism has given this region a unique global conservation value [20,21]. The global relevance of SSA forests and climate trends predicted for the coming decades make it urgent to expand model applications into this region, as a tool to predict temperate rainforest responses to impending declines in rainfall.

This study introduces a forest gap model specifically designed for assessing the responses of temperate rainforests in southern South America to increased drought. The model provides accurate estimates of forest water fluxes and incorporates dynamical linkages among rainfall regimes, soil moisture, and individual tree growth. We assessed model performance by comparing the results with detailed field measurements of water cycling in a stand located on northern Chiloé Island, Chile (41°50′S). We also conducted a sensitivity analysis of the response of current forests to drought, i.e. when rainfall is decreased. Model predictions of forest hydrology (evapotranspiration and soil moisture) and structure (above-ground biomass and basal area) under increased drought predicted for 2100 in the study area were compared for a young-secondary (YS) and an old-growth (OG) forest stand to analyze differences in responses to expected changes in rainfall.

Materials and Methods

Study area

The study was conducted on northern Chiloé Island, Chile (41°50′ S, Fig. 1) at the private protected area *Estación Biológica Senda Darwin* (EBSD), with permission granted by the owner. Fragments of secondary and primary forests occur over rolling hills of low altitude (50–100 m) dispersed in a matrix of bogs, shrublands and grazing pastures. The present landscape has been shaped by a history of widespread use of fire to clear land for pastures since the late 1800s, followed by selective logging of remaining forest patches [22]. Soils are generally thin (<0.5 m), originated from moraine fields and outwash plains from the last glaciation, often with poor drainage [23]. Soils have high organic matter content, soil texture loam to silty loam, and a 2–4 mm thick iron silicate layer or hardpan (found at ca. 52 cm depth), where roots cannot penetrate [24]. The prevailing climate is wet-temperate with strong oceanic influence [25]. Rainfall occurs throughout the year, with an annual average of 2158 mm (25% occurring in summer). Mean annual temperature is 9.1°C. Maximum and minimum monthly temperatures are 13.9°C (January) and 4.2°C (July) [26].

Floristically, forests of the study area belong to the North Patagonian temperate rainforest type [27]. The canopy is dominated by evergreen trees, mainly *Podocarpus nubigena* (Podocarpaceae), *Drimys winteri* (Winteraceae) and *Nothofagus nitida* (Nothofagaceae), with the common presence of *Tepualia stipularis* (Myrtaceae) and several Myrtaceae tree species in the understory. Ferns (e.g. *Hymenophyllum* spp., *Hymenoglossum cruentum, Polypodium feullei)* and angiosperms (e.g. Gesneriaceae and Bromeliaceae) growing epiphytically are frequent. Detailed

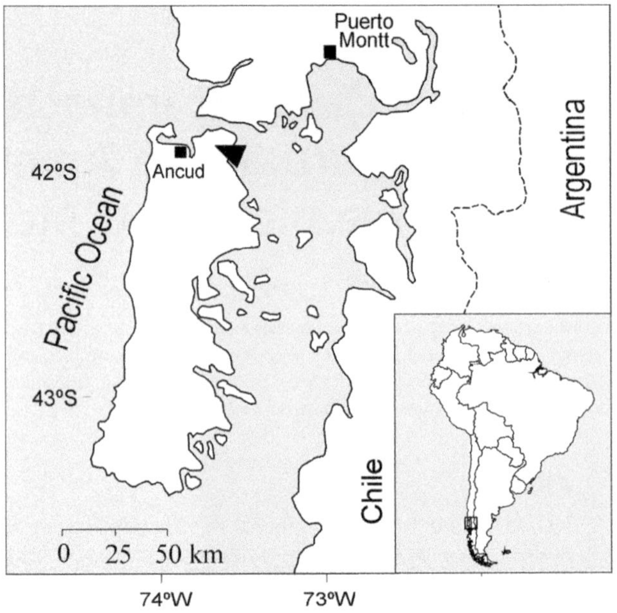

Figure 1. Location of study site (triangle) on northern Chiloé Island, Chile.

descriptions of structure and dynamics of this forest type have been previously published [26,28]. The study did not involve endangered or protected tree species.

The forest model

Here, we introduce an individual-oriented dynamic forest model (FORMIND-CL v.1.0) that includes calculations of hydrologic balance. The model is based on FORMIND, a forest model comprehensively tested to simulate the dynamics of temperate rainforests in SSA [11,13]. FORMIND is a generalized forest growth model that simulates the spatial and temporal dynamics of uneven-aged, mixed species forest stands [29–31]. The model simulates forest dynamics (in annual time steps, *t*) as a mosaic of interacting forest patches of 20×20 m, which is the approximate crown size of a large mature tree in the forest. Within these patches, stand dynamics is driven by competition for light and space following the gap model approach [2]. For the explicit modeling of the competition for light, each patch is vertically divided into height layers of 0.5 m, where leaf area is summed up and the light environment under the canopy is calculated via a light extinction law. The carbon balance of each individual tree is modeled explicitly, including the main physiological processes (photosynthesis and respiration [13]). Allometric functions and geometrical relations are used to calculate above-ground biomass, tree height, crown diameter and stem volume from the stem diameter at 1.3 m height of the tree (*dbh*). Tree mortality can occur either through self-thinning in densely populated stands, tree senescence, gap formation by large falling trees, slow tree grow, or external disturbances (e.g. windthrow). Gap formation links neighboring forest areas. Tree regeneration rates are formulated as maximum rates of recruitment of small trees at *dbh* threshold of 1 cm, with seed loss through predation and seedling mortality being incorporated implicitly [13]. Maximum recruitment rates are reduced by shading. Nutrient availability is considered to be homogeneous at the stand scale. A description of the core model and its equations is given elsewhere [11,13]. We focus below on the extensions added to incorporate forest hydrology.

Table 1. Structure and composition of the young-secondary North Patagonian forest stand (YS) in Senda Darwin Biological Station, compared to values reported for other secondary forest stands, elsewhere in Chiloé Island, Chile.

	Study site		Literature*			
Species	**Density (trees ha^{-1})**	**Basal area (m^2 ha^{-1})**	**Density (trees ha^{-1})**		**Basal area (m^2 ha^{-1})**	
Drimys winteri	1550	32.3	1800	4532	24.7	61.4
Eucryphia cordifolia	175	0.9	112	480	1.4	14.7
Myrtaceae†	1014	3.5	252	827	2.3	4.4
Nothofagus nitida	713	17.7	4	260	0.1	4
Podocarpaceae¥	38	0.1	80		0.5	
Weinmannia trichosperma	88	0.3				
Others	287.5	1.9				
Total	3575	54.8	2431	7950	46	86.9

*Source: [26,77–80] only forest stands <100 years old. Only listed tree species present in YS.
†*Tepualia stipularis, Myrceugenia spp.* and *Amomyrtus spp.;*
¥*Podocarpus nubigena, Saxegothaea conspicua.*

The hydrologic submodel

Soil moisture dynamics is described at a daily timescale, treating soil as a reservoir with an effective storage capacity that is intermittently filled by rainfall events. Soil water losses occur via transpiration, interception by the forest canopy, and drainage below the root zone. We neglected lateral water flow, thus the model applies mainly to flat terrains. This is a reasonable assumption in forests of the study area because during the rainy season soils tend to be saturated and accumulated water cannot infiltrate the soil.

Soil moisture s (dimensionless, $0 \leq s \leq 1$), vertically averaged over the soil depth z (mm), was considered as central state variable [8]. Thus, the water balance equation for a given point in the forest can be expressed as [1]:

$$n \cdot z \cdot \frac{ds}{dd} = Pnet_d - Tr_d - Q(s,d) \quad (1)$$

where d is the Julian day of the year, n is the porosity (volume of voids/total volume of soil, i.e. dimensionless, vertically averaged); $Pnet_d$ is the net precipitation falling to the soil surface (mm day^{-1}); Tr_d is the transpiration rate (mm day^{-1}); and $Q(s,d)$ is the soil drainage (mm day^{-1}). Both n and z are assumed to be time-invariant parameters [1]. The volumetric water content (θ, m^3 water/m^3 soil, i.e. dimensionless) can be calculated as follows [1]:

$$\theta = s \cdot n \quad (2)$$

The normalized version of equation (1) is used through the text where all terms are divided by $n{\bullet}z$. Both the local vertical and horizontal variability of soil moisture are considered negligible at the daily timescale, assuming an equal propagation of the wetting front and equal soil moisture redistribution over the rooting zone [8,32].

Net precipitation. Daily net precipitation falling to the soil surface ($Pnet_d$) is described by,

$$Pnet_d = P_d - Ec_d \quad (3)$$

where, Ec_d is the canopy interception (mm day^{-1}), defined here as the total daily rainfall (P_d, mm day^{-1}) that is retained by the canopy and is evaporated so that it does not reach the ground. Following [33], we assumed that Ec_d asymptotically approaches the canopy retention capacity and can be modeled at daily intervals as:

$$Ec_d = S_t \cdot (1 - e^{-\alpha_h \cdot P_d}) \quad (4)$$

where S_t is the canopy water retention capacity of the stand at year t and α_h is a parameter describing the slope of the saturation curve. The parameter α_h represents, in a simplified terms, the complex process of water partitioning into throughfall and stem flow [34]. S_t depends on leaf area index of the forest patch at simulated year t (LAI_t) and is calculated by the expression [34,35]:

$$S_t = \frac{LAI_t}{LAI_{\max}} \cdot f_h \cdot \log(1 + LAI_t) \quad (5)$$

where, LAI_{max} is the maximum leaf area index of the forest and f_h is a shape parameter. We avoided unrealistic canopy interception values in the model by setting $Ec_d = P_d$ when $Ec_d > P_d$.

Table 2. Parameter descriptions and parameterization methods used for running simulations in FORMIND-CL v.1.0.

	Description	Value	Units	Method	Reference
Weather generator					
$1/\lambda$	Mean interval time between rainfall events[¥]	†	days	a	This study
η	Mean depth of rainfall events[¥]	†	mm day^{-1}	a	This study
T_μ	Mean daily temperature[¥]	†	°C	a	This study
T_σ	Standard deviation daily temperature[¥]	†	°C	a	This study
Rg_μ	Daily global radiation above canopy[¥¶]	†	μmol (photons) $m^{-2}\,s^{-1}$	a	This study
Rg_σ	Standard deviation of Rg_μ [¥¶]	†	μmol (photons) $m^{-2}\,s^{-1}$	a	This study
Hydrologic submodel					
n	Vertically averaged porosity of the soil	0.757	-	b	[53]
z	Soil depth	520	mm	b	[14]
k_{soil}	Saturated hydraulic conductivity	4	mm day^{-1}	b	[36,54]
α_h	Slope of the canopy saturation curve	0.7	-	b, c	[34]
f_h	Parameter of the relationship LAI and canopy storage capacity	3	mm day^{-1}	c	[35]
LAI_{max}	Maximum LAI of the studied forest	5.5	$m^2\,m^{-2}$	b	[81]
WUE	Water-use efficiency	9	g CO_2 $kg^{-1}H_2O$	e	[14]
γ	Psychometer constant	65	Pa K^{-1}	b	[43,44]
L	Latent heat of vaporization of water	2.56×10^6	J kg^{-1}	e	[43,44]
Δ	Rate of change of saturated vapor pressure with temperature	c	Pa K^{-1}	d	[43,44]
θ_{wp}	Wilting point of the soil	0.125	-	b	[36]
θ_{fc}	Field capacity of the soil	0.3	-	b	[36]

Method refers to a: calculated from daily meteorological data from *Senda Darwin Biological Station,* period 1998–2009, b: from literature, c: calibrated with field data, d: calculated, e: calibrated using literature.
†values indicated in Table 3.
¥per season.
¶calculated for dry ($Pd<1$ mm) and wet days ($Pd\geq1$ mm).

Soil moisture modeling. Drainage out of the root zone ($Q(s,d)$) was modeled according to [1]. When the soil is saturated ($s = 1$), soil water is permitted to percolate at a rate equivalent to the saturated hydraulic conductivity of the soil (k_{soil}, mm day^{-1}, [36]). Runoff occurs when the soil is saturated and no more water can be held in place. The excess of water is assumed to leave the system, which is reasonable to assume given the large rainfall intensity in the study area. When $s < 1$, soil deep percolation rate is calculated using the empirical relationship of Neilson [37],

$$Q(s,d) = k_{soil} \cdot s^2 \tag{6}$$

Transpiration. Water-use efficiency describes the proportion of water used for the assimilation of a unit of carbon in the photosynthesis [38,39]. This concept can be used to estimate transpiration of trees (Tr, mmol H_2O $m^{-2}\,s^{-1}$) from:

$$Tr = \frac{PB}{WUE} \tag{7}$$

where, PB is the gross biomass production of the tree (μmol carbon dioxide $m^{-2}\,s^{-1}$), and WUE is a parameter denoting water-use efficiency at stand level. PB is obtained from the rate of single-leaf photosynthesis following [40], which is integrated over the total LAI of the tree to account for self-shading [13]. The resulting photosynthetic rate is then multiplied by the crown area of the tree to obtain PB (see also equations in Appendix S1 and [13]). Daily transpiration (Tr_d) of trees is obtained from equation 8 and dividing Tr by the length of the active photosynthetic period per year.

The daily potential evapotranspiration (PET_d, mm day^{-1}) describes a physical limit for the amount of water that can be held and transported away from the canopy under given climatic conditions. Evaporation is neglected in the model; therefore, it is assumed that maximum water losses by vegetation are limited by the difference between PET_d and the canopy interception of the day (Ec_d), as follows:

$$PET_d - Ec_d \geq Tr_d \tag{8}$$

PET_d is calculated using a modified Penman-Monteith expression in case of aerodynamic conductance [41,42] and determined by the variation of the daily net radiation flux (Rn_d, J m^{-2} day^{-1}):

$$PET_d = \left(\frac{\Delta}{\Delta+\gamma}\right) Rn_d / L \tag{9}$$

where γ is the psychometric constant (ca. 65 Pa K^{-1}, slightly depends on temperature), L is the latent heat of vaporization of

Table 3. Parameter values used to run the weather generator under different climatic scenarios.

	Current climate			Future scenarios		
Rainfall	Seasonal sum (mm, average ± sd)	1/λ (days)	η (mm day^{-1})	Seasonal sum (mm, average)	1/λ (days)	η (mm day^{-1})
DJF	284.6±132.7	0.9	8.3	238–131	0.99–1.36	4.2–7.5
MAM	543.6±138.6	0.55	12.8	813.4	0.55	12.8
JJA	813.4±197.2	0.29	16	543.6	0.29	16
SON	424.5±144.7	0.54	9.5	382–212	0.59–0.8	4.7–8.5
Annual sum (mm, average ± sd)	2094.8±353.8					
Temperature	T_μ (°C)	T_σ (°C)			T_μ (°C)	
DJF	12.5	2.6			16.5	
MAM	10.1	3.6			13.1	
JJA	8.4	4.4			10.4	
SON	9.8	3.3			11.8	
Radiation	Rg_μ (Rg_σ)	Rg_μ (Rg_σ)				
	[P_d<1 mm]	[P_d≥1 mm]				
DJF	1413.9 (317.0)	986.9 (378.2)				
MAM	701.2 (294.1)	398.8 (261.6)				
JJA	408.9 (159.1)	229.7 (139.4)				
SON	1065.5 (348.7)	640.8 (320.1)				

Current climate indicate parameters used to run the model under current climate based on instrumental records (weather station at *Senda Darwin Biological Station*, period 1998–2009). Radiation describes parameters daily global radiation Rg_μ and Rg_σ (the latter in brackets, μmol(photons) m^{-2} s^{-1}). Temperature is mean daily air temperature. *Future scenarios* are the range of climatic parameters that were varied to run the model under increased drought (36 scenarios in total, see *Methods* for details). *DJF:* December to February (austral summer, growing season); *MAM*: March to May (austral autumn); *JJA*: June to August (austral winter), *SON*: September to November (spring, growing season). sd: standard deviation.

water (ca. 2.56×10^6 J kg^{-1} slightly depends on temperature). The rate of change of saturated vapor pressure with temperature (Δ, Pa K^{-1}) is calculated as [43,44]:

$$\Delta=\frac{2.503x10^6 e^{17.269\frac{T_d}{237.3+T_d}}}{(237.3+T_d)^2} \tag{10}$$

Rn_d was calculated from latitude, day of the year, sunshine hours and daily air temperature (T_d, °C) following [42,45].

Soil moisture impact on tree biomass production. The dependence of water uptake for tree biomass production on soil moisture is described by a function representing a reduction factor due to water scarcity ($\omega(s)$, $0\leq \omega(s) \leq 1$ [46]). This factor accounts indirectly for the impact of water demand on potential photosynthetic production (i.e. possible to achieve under competition for light). $\omega(s)$ is implemented as a daily reduction factor due to water scarcity by,

$$\omega(s)=0 \text{ if } s<\theta_{wp}$$

$$\omega(s)=\frac{s-\theta_{wp}}{\theta_{msw}-\theta_{wp}} \text{ if } \theta_{wp}<s\leq\theta_{msw}$$

$$\omega(s)=1 \text{ if } s>\theta_{msw} \tag{11}$$

where, θ_{wp} is the wilting point, and θ_{msw} represents a threshold when enough soil moisture is available for potential tree biomass production. We calculated θ_{msw} from:

$$\theta_{msw}=\theta_{wp}+\frac{1}{3}(\theta_{fc}-\theta_{wp}) \tag{12}$$

where, θ_{fc} is the soil field capacity. θ_{wp}, θ_{msw}, and θ_{msw} are expressed as normalized soil moisture. In the model, the wilting point (θ_{wp}) determines the minimum soil moisture content necessary for tree biomass production. Thus, we assumed a linear reduction of biomass production when soil water content was between θ_{msw} and θ_{wp}. The water required for biomass production of trees is completely removed from the soil compartment when soil moisture reaches θ_{msw} (i.e., $\omega(s)=1$), after the calculation of maximum possible transpiration of trees. Both biomass production and water supply are reduced until the water needed for biomass production corresponds with θ_{wp}. The calculated rate of biomass production influences tree respiration rate through maintenance and growth respiration, which are calculated subsequently in the model (see [13] for equations and Fig. S1 for a diagram). All calculations are performed for every tree and pooled together to calculate the stand-level values.

Weather generator. Rainfall time series, representing the frequency and depth of rainfall events, were constructed as series of random numbers generated by probability distributions. The interval between rainfall events, τ (day) can be expressed as an exponential distribution given by [47].

$$f_T(\tau)=\lambda\cdot e^{-\lambda\tau}, \text{ for } \tau\geq0 \tag{13}$$

Table 4. Model estimates of water balance components for a young secondary forest (YS, 60 years-old) and an old-growth North Patagonian forest (OG, >500 years-old) located in northern Chiloé Island, Chile, under current climate.

Variable	This study		Literature	
	mm year^{-1} ± sd	%	Value	Source*
Young secondary stand (YS)				
Canopy Interception	381.5±30.3	20.0	20–40%	1
Deep percolation	980.7±156.8	49.9	47%	1
ET	573.6±35.1	30.3	45.2%	1
Net precipitation	1617.2±434.6	80.0	60–80%	1
Runoff	722.9±397.7	33.3		
Soil moisture	-	62.5		
Transpiration	192.1±18.8	10.3	22%	1
Old-growth stand (OG)				
Canopy Interception	378.3±33.6	19.9	17.8%	2, 3
Deep percolation	907.4±142.4	46.5	66.5%	2
ET	648.0±44.1	34.3	19.9–33.3%	2
Net precipitation	1591.8±368.6	80.1		
Runoff	665.2±321.9	31.8	30–55%	4
Soil moisture	-	55.3		
Transpiration	269.7±26.1	14.4		
Potential evapotranspiration	769±4		576–724 mm year^{-1}	5

Model results are the average of 100 simulations per stand (see *Methods* for details), with annual sum of rainfall averaging 1970–2000 mm year^{-1}. Literature refers to values reported by independent studies in Chilean temperate rainforests and comparable forests elsewhere. sd: standard deviation; %: percentage of total annual rainfall; ET: Evapotranspiration (sum of canopy interception and transpiration). sd: standard deviation.
*(1) Data for other broad-leaved evergreen forests, ca. 200 years old. Annual rainfall 2500 mm year^{-1} [82,83]. (2): Mixed deciduous-broad-leaved old-growth forest. Annual rainfall 2400 mm year^{-1}[12]. (3) Mixed broad-leaved and conifer forest, ca. 200 years old in New Zealand. Annual rainfall 3400 mm year^{-1}[84]. (4) Annual rainfall 1700–4500 mm year^{-1}, data from evergreen, broad-leaved forests with 90% cover [85]. (5) Annual rainfall of 2427–3991 mm year^{-1} [86], weather stations of Castro and Punta Corona.

where $1/\lambda$ is the mean time interval between rainfall events (days). Total daily rainfall (P_d) depends on the amount of rain of each event (h, mm day^{-1}), which is also assumed to be an independent random variable, expressed by an exponential probability density function [47]:

$$P_d(h) = \frac{1}{\eta} e^{-\frac{1}{\eta}h}, \text{ for } h \geq 0 \tag{14}$$

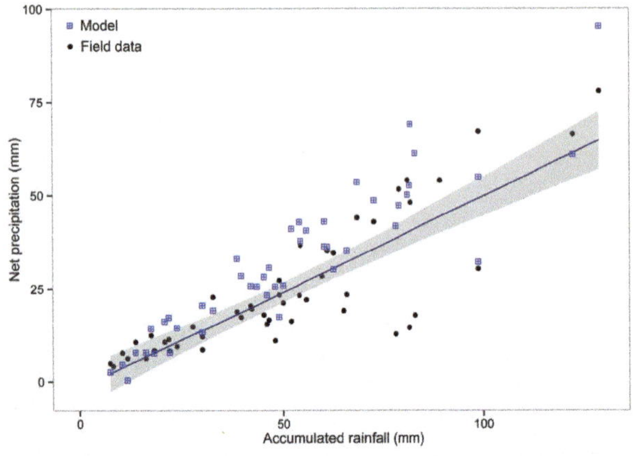

Figure 2. Comparison between measured and modeled net precipitation for 50 rain events recorded in a young secondary stand in northern Chiloé Island, Chile, for the time period 2007–2010. The line represents a linear regression between field measured net precipitation and accumulated rainfall during each event. The gray area represents 0.95 confidence intervals. Model results are for a forest patch are of 400 m^2, with a LAI = 5.

where η is the mean depth of rainfall events (mm day^{-1}). Both $1/\lambda$ and η parameters are calculated for each season of the year. We obtained daily global radiation (Rg_d) from EBSD instrumental records (period from May 1998 to February 2009). Rg_d varied among seasons in relation to daily rainfall (T-test, $p<0.001$). Therefore, in the model, Rg_d was distributed as a Gaussian variable whose mean and standard deviation depended on P_d (mean, Rg_μ, and standard deviation, Rg_σ). Values of Rg_μ and Rg_σ were obtained from instrumental records and varied depending on a threshold value of 1 mm of P_d and the season of the year. Daily temperature (°C, T_d) was simulated by a Gaussian random variable with parameters (mean, T_μ, and standard deviation, T_σ) that varied according to season of the year.

Field data

Stand structure. We characterized stand structure in terms of tree species dominance (basal area, m^2 ha^{-1}) and size (dbh) distribution in a young secondary and an old-growth North Patagonian forest stand (hereafter YS and OG respectively) found in a flat forested area at EBSD (Fig. 1). The YS stand was initiated from a stand-replacing fire about 60 years ago and currently presented no evidence of logging. In 2007, we set up two 20×20 m

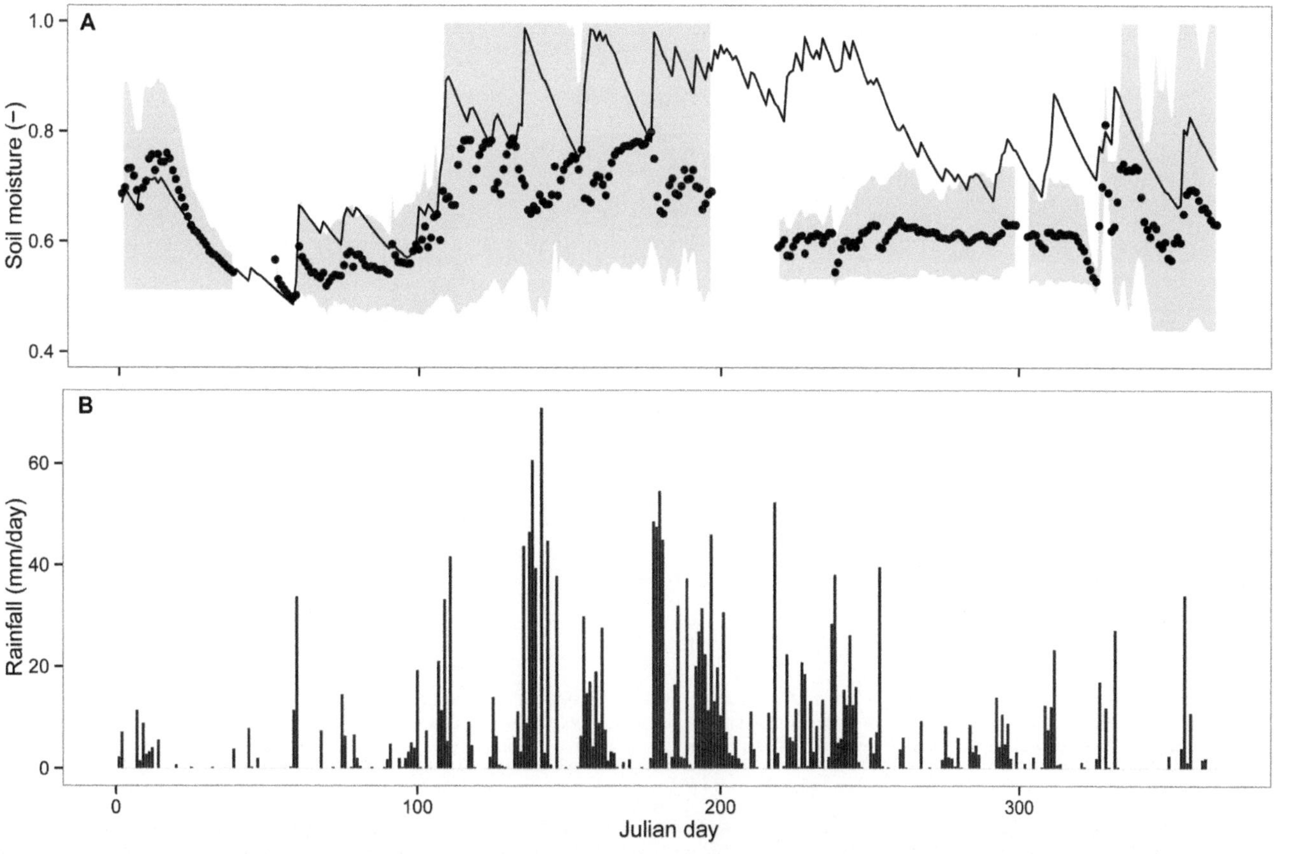

Figure 3. Variation in soil moisture during 2008 in a young secondary, North Patagonian forest stand (YS) in northern Chiloé Island, Chile. a) Comparison between observed and simulated soil moisture (normalized data, dimensionless). Observed data were obtained from five sensors randomly placed inside a 400-m^2 plot in YS (daily means denoted by filled dots, gray area showing the range of data). Simulated soil moisture (line) is for one-hectare forest with a successional age of 60 years (LAI = 4.5). Soil parameters are the same as described in Table 2 (see also *Methods*). b) Rain events recorded instrumentally during 2008.

plots to measure the hydrologic balance and stand structure. All trees rooted within each plot with stems >1.3 m height and > 5 cm *dbh* were permanently marked with numbered aluminum tags, identified to species, and their *dbh* measured to the nearest cm. Structure and composition of YS is comparable to that described for young-secondary stands elsewhere on northern Chiloé Island (Table 1). OG was an unmanaged forest stand >590 years old, without evidence of recent human disturbance, and representative of old-growth North Patagonian forest on Chiloé Island and elsewhere in the region [26]. Sampling methods, stand history, species composition and structure of OG is described by [28]. OG had total basal area of 72 m^2 ha^{-1} and density of 2610 trees ha^{-1}, with a mixed dominance of Podocarpaceae species, *D. winteri, N. nitida* and Myrtaceous tree species in the understory [28].

Hydrologic measurements. We estimated net precipitation using throughfall measurements taken in two YS forest plots (i.e. rainfall falling through canopy gaps plus canopy drip), adding stemflow (water running down the stems). We conducted these measurements of volume accumulated during rainfall events occurred between June 2007 and December 2010. During this time period, we also analyzed hourly records of rainfall from the meteorological station at EBSD to obtain daily incident rainfall above the canopy. Rainfall events considered in the analysis occurred with a separation of at least two hours without rain to allow for full drip from the forest canopy. Stemflow collectors consisted of a 2 mm thick smooth polycarbonate sheet molded around the stem to form a funnel. A hose led from the lowest point of the funnel to a 25 l polythene container, where the stemflow volume was collected after each rain event. Stemflow collectors were placed in 10 randomly selected trees of the two main canopy species, *D. winteri* and *N. nitida* (*dbh* >10 cm), in each plot. We eliminated two trees from our stemflow estimates that died during the study period. We converted the volume of collected water to millimeters of rain assuming that the surface of the collectors equals the projected tree crown area. Crown area was approximated by the area of an ellipse. Throughfall collectors, 0.12×2 m long (0.7 m^2 total area per plot) gutters were held, with a slight inclination, 0.5 m above the ground at three different locations within each plot. Collectors were connected with a funnel to a 25 l polythene container. Soil matric potential was measured every 30 minutes with four sensors per plot (*WatchDog Data Loggers* 450 and 800) placed approximately in every quarter of each plot, beneath the canopy and at ca. 15 cm soil depth. Continuous soil moisture measurements were obtained for the period January 2007 to March 2009.

Model parameterization

We used a previous model parameterization for North Patagonian forests including all main canopy tree species (11 tree species) occurring in the studied forests. The calibration, validation and robustness of this parameter set to reproduce forest stand

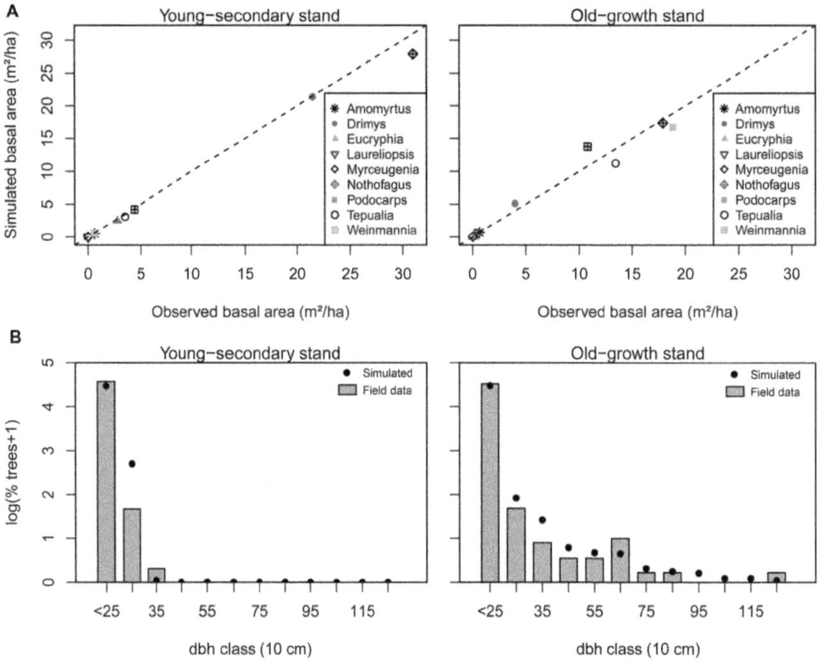

Figure 4. Comparison of forest structure between (a) observed (field data) and simulated basal area of tree species (Spearman's r^2>0.9, p<0.01 in both cases) and (b) *dbh* distributions for the young secondary (YS) and old-growth (OG) North Patagonian forest stands studied in Chiloé Island, Chile. Simulated OG structure was obtained initializing the model with inventory data.

structure is discussed in detail by [11]. Here, we describe calibration of parameters related to the inclusion of hydrologic balance into the model. New parameters needed to run `FORMIND-CL v.1.0` and their values are shown in Table 2.

The parameter f_h describing the relationship between leaf area index and canopy water storage capacity was calibrated following [34] and assuming that storage capacity reaches 4.9 mm day^{-1} at a leaf area index of 5.0 as measured by [14]. LAI_{max} was set to 5.5 following the maximum value observed in other Chilean temperate rainforests [48]. The slope of the saturation curve of the canopy rain retention capacity (α_h) was set according to common values for broad-leaved temperate trees [34]. To the best of our knowledge no estimation exists for water-use efficiency at stand scale in forests of the study area. Therefore we calibrated *WUE* using transpiration estimates of Díaz *et al.* [14] in Chiloé Island and the potential canopy photosynthetic rate estimated by the model for the study area under current climate ($Tr = 296$ mm year^{-1} and $PB = 32.9$ tC ha^{-1}. The selected *WUE* was then confirmed by comparison with reported values from other temperate rainforests [49–52]. Soil characteristics (porosity and depth) followed field descriptions from Chiloé Island [53]. We set water-retention and percolation properties of the soil (parameters θ_{wp}, θ_{fc} and k_{soil}) to average values [36,54] using texture classes (loam to silty loam) described for soils in the study area [53]. Daily records of rainfall from the EBSD weather station (60 m a.s.l, period from May 1998 to February 2009) were directly used to calculate rainfall parameters for the current climate simulations (Table 3). EBSD is the nearest and most representative weather station for the climate at the study site. We calculated the mean time interval between rainfall events from the duration (days) of rain events occurring in each season. The mean depth of rainfall events was calculated by dividing seasonal rainfall sum by the amount of wet days (P_d >0). These calculations were done only for seasons with >85 daily records. We avoided the potential overestimation of annual rainfall by normalizing predicted seasonal rainfall sum by prescribed seasonal rainfall averages (Table 3). The weather generator reproduced well the seasonal fluctuations in rainfall, temperature and radiation during the year with no significant departures from observed climatic records (Fig. S2, S3). Calibrated climatic parameters were assumed representative for growing conditions of North Patagonian forests on Chiloé Island and neighboring regions in the mainland.

Analyses

Model verification. We compared field measurements of net precipitation with model predictions at daily temporal scale. For this analysis we selected 50 rain events, representing field measurement intervals <20 days long, for which accumulated rainfall during the event (i.e. sum of daily rainfall from first to last day of the event) was correctly measured by EBSD weather station, and for which <50% of containers were filled completely during the rain event. Hydrologic parameters used for model estimation of net precipitation are indicated in Table 2. LAI measurements were unavailable for YS, thus we set LAI to comparable, averaged values reported by Diaz *et al.* [14] for the same forest type. We qualitatively compared daily variation in soil moisture produced by the model with field observations. For this analysis soil matric potential obtained for an entire year was transformed to soil moisture contents using a water retention curve [55]. We set parameters of the water retention curve following [56] for loam to silty loam texture classes and particle density of the soil type under study (ca. 2.0 g/cm^3, [53]). Model simulations were run using rainfall data for the same period of field measurements conducted over a whole year, i.e. 2008. Note that for both analyses spatial scale of model results (>1 ha) differed from the scale of field measurements (400 m^2 plots). We also compared model results with field measurements of hydrologic

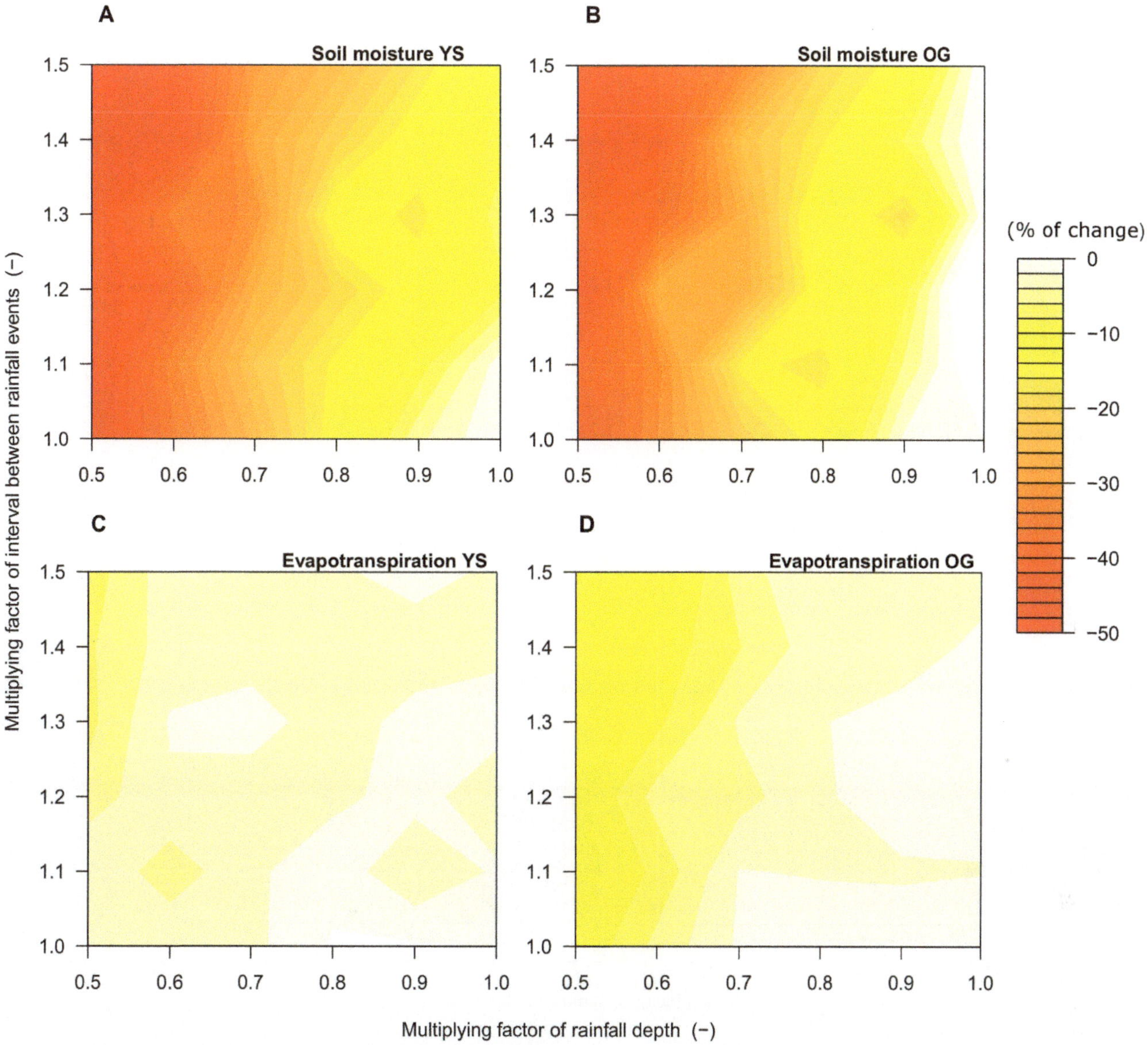

Figure 5. Sensitivity of soil moisture (a, b) and evapotranspiration (c, d) predicted by the model under increased drought in a young-secondary (YS) and an old-growth (OG) North Patagonian forest. Results represent the difference between the average under current climate (indicated in Table 4) and the average under future scenarios. A value of 0% indicates no change. The y-axis represents the variation of mean interval time between rainfall events (parameter $1/\lambda$) and the x-axis represents the variation in the mean depth of rainfall events (η) under increased drought. To represent increased drought scenarios, the parameter $1/\lambda$ was multiplied by a factor ranging from 1 (current climate) to 1.5 whereas the parameter η was multiplied by a factor ranging from 1 (current climate) to 0.5. The axes of the figures correspond to these multiplying factors of rainfall parameters. Results are the averages of 30 simulations per scenario for YS and OG.

balance at yearly temporal scales in temperate rainforests elsewhere in Chile (i.e. independent studies).

We tested model performance to reproduce forest structure under current climate. For this analysis, we compared tree basal areas and stand *dbh* distributions predicted by the model with measured structure of both YS and OG North Patagonian stands. Model comparison for the YS was performed after 60 years of succession, with succession initiated from a treeless state. To compare OG forest structure, we initiated simulations with stand inventory data and run the model for 1000 years to allow the simulated stand to reach dynamic equilibrium. We compared data at the end of the simulations with the known OG structure [28]. For each stands, we ran 100 simulations of 1 ha (i.e. 25×0.04 ha forest patches, 2500 patches in total) using current climate parameters and model parameters listed in Table 2 and 3. Demographic and species parameters were taken from [11] (site Tepual). To assess the consequences of the hydrologic submodel for the simulated forest composition, we compared species-specific basal areas reported by [11] with data from the simulations using the model version described in this study. Simulations were conducted following methods outlined by [11] and results were compared at the corresponding successional ages of YS and OG.

Simulations under increased drought. We tested the sensitivity of model predictions for forest structure and hydrological balance to changes in rainfall regimes. Climatic scenarios were selected to represent the potential range of expected climate

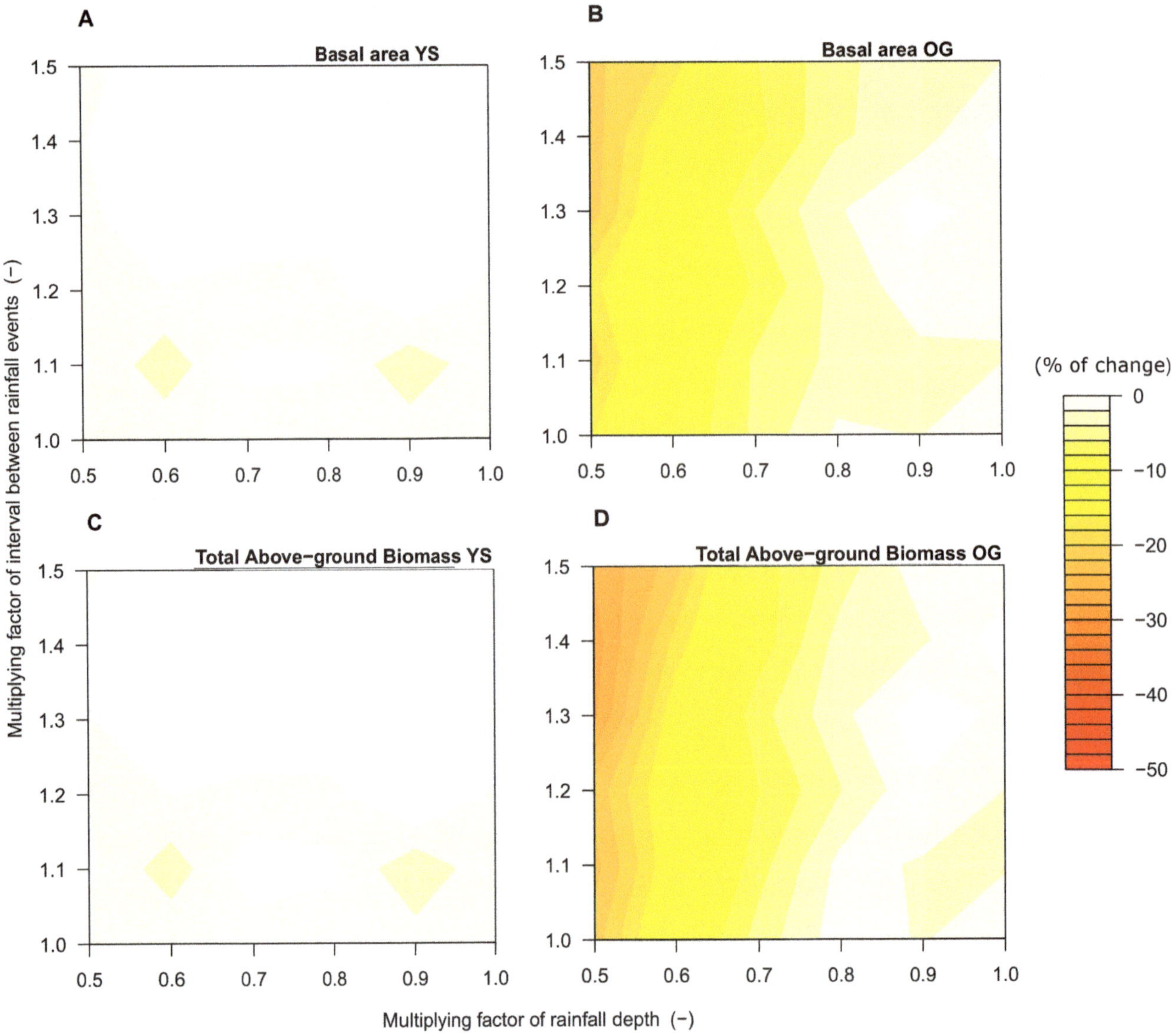

Figure 6. Sensitivity of total basal area (a, b) and above-ground biomass (c, d) predicted by the model under increased drought in a young-secondary (YS) and an old-growth (OG) North Patagonian forest. Results are the percentage of change between the average under current climate (indicated in Table 4) and the average under future scenarios. Results are the average of 30 simulations per scenario for YS and OG. The axes of the figures are as in Figure 5.

change (here increased drought) predicted for this century in the study area. To this end, we used a regional climate model downscaled for Chilean landscapes (PRECIS-DGF model, [17]). The business-as-usual scenario provided by PRECIS-DGF suggests a 50% decrease in rainfall during the growing season by year 2100. We used the daily scale output of the PRECIS-DGF model for year 2100 to calculate seasonal climatic parameters for year 2100. We linearly interpolated this seasonal rate of change between 1998–2009 and 2100. We used this scenario to set the limit of change in rainfall by year 2100 and developed climatic scenarios covering the change from current climate to the business-as-usual scenario. Climate scenarios were developed by gradually changing current climate parameters $1/\lambda$ and η until they reached the estimated value for year 2100, i.e. 50% of the current value. First, we reduced η (mean depth of rainfall events) multiplying current values by 0.9 to 0.5 in steps of 0.1. Then, we increased the parameter $1/\lambda$ (interval between events) multiplying current values by 1.1 to 1.5 in steps of 0.1. These two parameters were first varied separately (keeping the second parameter constant) and then both together. Parameter variations produced a total of 36 climate scenarios, i.e. six levels of each of the two rainfall parameters, including the current climate scenario (Table 3). In addition, we assessed the impact of warming trends on potential evapotranspiration, and its influence on other hydrologic components and forest structure. To this end, we ran an additional scenario including temperature changes expected for year 2100, with parameters $1/\lambda$ and η kept at their current values (Table 3). We considered four output variables computed at yearly temporal scale: total basal area, above-ground biomass, evapotranspiration (computed as the sum of canopy transpiration and interception) and soil moisture. The latter is dynamically linked to forest processes such as annual gross biomass production of each tree (*PB*, cf. Eq. 8, see also Fig. S1). We described changes in forest structure at different successional stages based on simulations for

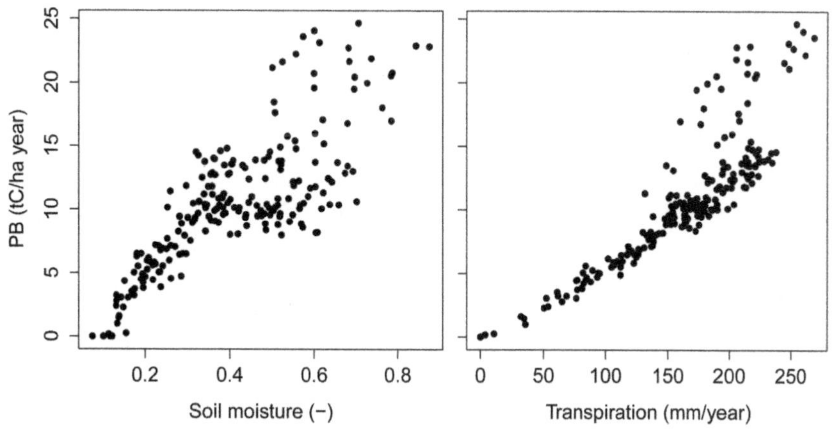

Figure 7. Shifts in annual gross biomass production (*PB*) of the YS North Patagonian forest stand in response to soil moisture and canopy transpiration changes in northern Chiloé Island, Chile. Results are simulations for one hectare of forest with a successional age of 60 years.

both YS and OG forest stands. Simulations were initialized and ran under the same conditions described in section *Model verification*. We ran 30 simulations per scenario and for each forest stand. To assess the impacts of increased drought on forests we compared the differences between means of the studied variable under current climate and means of the same variable under climate change scenarios. Finally, we checked soil moisture influence on *PB* and *Tr* by running simulations under different soil moisture conditions. We ran 216 simulations, at a scale of one ha, by monotonically varying current rainfall depth parameter ($1/\lambda$) from 0 to 100% of the current value (Table 2). All other parameters were kept constant. We assessed results at the corresponding successional ages of YS and OG, with succession initiated from a treeless state. All statistical analyses were done in R [57].

Results

Model results under current climate

Model predictions for the major components of forest hydrologic balance at a yearly temporal scale were similar to reported values for broad-leaved temperate rainforests in Chile and elsewhere (Table 4, independent studies). At a daily scale, the hydrologic model captured a large portion of net precipitation variability measured in the field ($N = 50$, $r^2 = 0.8$, $P<0.001$, Fig. 2). However, for rain events accumulating >100 mm, the model predicted higher net precipitation than recorded in the field (Fig. 2). Modeled daily variation in soil moisture resembled the daily pattern of soil matric potential measured in the field during the year (Fig. 3, $r^2 = -0.65$, $p<0.001$). Soil moisture increased during austral fall and winter (Fig. 3a, 59< julian day <242) due to a higher frequency and depth of rainfall events and reduced transpiration of trees. In contrast, during the growing season, soil moisture gradually decreased according to the model and field data (Fig. 3). Forest structure simulated by the model with the inclusion of the hydrologic submodel was qualitatively similar to field data (Fig. 4a), but with some departures from observed data for small dbh-classes in YS (<35 cm, Fig. 4b). Total basal area of YS simulated by the model was similar to basal area observed in the field but total basal area of OG was underestimated (Fig. S4). However, model predictions for forest structure including the hydrologic submodel varied in the same manner across species as in the original model version and resembled field data for both stands ($r^2 > 0.97$, $P<0.001$, Fig. S4).

Net precipitation and runoff predictions under current climate did not differ between YS and OG stands ($P>0.12$, two-sample Wilcoxon test, Table 4). Water loss through deep percolation was lower in OG than in YS ($P<0.001$, two-sample Wilcoxon test, Table 4). Evapotranspiration was higher in OG than YS, mainly due to higher canopy transpiration rates in OG (Table 4, $P<0.001$, two-sample Wilcoxon test). Modeled soil moisture was lower in OG than in YS (0.55 and 0.63 respectively, $P<0.001$, two-sample Wilcoxon test, Table 4).

Simulations under increased drought

The model predicted changes in hydrological components and forest structure when simulations were run under expected trends of increasing drought (Fig. 5, 6). In both forest stands, YS and OG, we found similar responses of soil moisture to changes in rainfall parameters, with declines of soil moisture up to 50% (Fig. 5a, b). Reducing η to values <80% of the current value (Table 2, e.g. multiplying factor of 0.8) consistently reduced soil moisture by 20% (Fig. 5a, b). The influence of $1/\lambda$ (mean interval time between rainfall events) was negligible when η was kept constant at its current value (Fig. 5a, b). The model consistently predicted less evapotranspiration (hereafter ET) relative to current values (Fig. 5c, d). In YS, decreases in ET under drier climate were < 50 mm year^{-1} (<8% reduction relative to current values, Fig. 5c). In OG forest, ET was reduced up to 94.4 mm year^{-1} from the current value (15% reduction relative to current values Fig. 5d). Such decreases in ET occurred when current value of η was multiplied by 0.6 and $1/\lambda$ was multiplied by 1.2 (Fig. 5d). An increase in PET was predicted by the model when simulations included warming trends ($P<0.001$, two-sample Wilcoxon test, Fig. S5). Only transpiration and ET in YS changed in response to warming trends and increased drought ($P<0.05$, two-sample Wilcoxon test, Fig. S5). Moreover, in both forest stands, changes in PET due to warming trends did not transfer to changes in forest structure ($P>0.05$, two-sample Wilcoxon test, see also Fig. S5).

We did not find distinct differences in basal area and aboveground biomass (AGB) in YS attributable to changes in rainfall parameters (Fig. 6). In contrast, in OG forest increased drought produced decreases in basal area by 21% of the current value (Fig. 6b, current basal area of 63.8 m^2 ha^{-1}) and decreases in AGB by 27% of the current value (Fig. 6d, current AGB of

309.5 tC ha^{-1}). Main changes in basal area and AGB of OG stand were predicted when current η and $1/\lambda$ parameters were multiplied by 0.6 and 1.3, respectively (Fig. 6b, d).

Discussion

Model performance

We developed and evaluated the performance of an individual- and process-based dynamic forest model that incorporates detailed calculations of water cycling for temperate rainforests of southern South America. The model allows for the investigation of dynamic linkages between rainfall trends and forest processes at stand scale. Parameters selected for this study were taken from the literature or calibrated using our own field data (Table 2), thus they can be considered as empirically based. The model incorporates the main known hydrological controls on forest processes in temperate rainforests of southern South America (Table 2, 4).

The accuracy of model predictions regarding forest composition and structure were comparable to previous results (Fig. S4), which were performed using a previous model version without hydrologic balance calculations [11]. Model prediction of total basal area of forests decreased with the inclusion of the hydrologic module, mainly due to fewer large trees (>100 cm dbh) predicted by the model in OG stands (Fig. 4b). However, model predictions of forest structure were considered realistic at small spatial scales (< 0.2 ha) used for field sampling, compared to model results (25 ha, see also discussion in [11]). Some departures between observed and predicted daily values of net precipitation could also be attributed to canopy heterogeneity operating at different spatial scales of model results and empirical data. These discrepancies may appear when comparing forest hydrologic balance estimated for 1 ha against net precipitation measured within considerably smaller areas (<1 m^2). Further, model predictions of daily net precipitation were based on average estimated LAI values for YS because of the lack of field measurements. LAI is a relevant variable to understanding biogeochemical cycles [58], and it should therefore be incorporated in future hydrologic analyses. Additionally, sampling errors during field measurements cannot be disregarded. Despite these limitations, we considered model predictions of forest hydrology acceptable at daily temporal scale for YS (Fig. 2, 3).

At yearly temporal scale, the model reproduced the main hydrologic components reported for similar evergreen, broad-leaved temperate rainforests in southern South America and elsewhere (Table 4, independent studies). Differences in deep percolation predicted by the present model compared to values reported in Table 4 might be accounted for specific physical characteristics of glacially originated soils on northern Chiloé Island. Deep percolation in the model is mainly depending on k_{soil}, which was calibrated specifically for soils from northern Chiloé Island. Applying this model to forests developing on other soil types in SSA (e.g. volcanic originated soils) will require a site-specific calibration of soil parameters (Table 2). Moreover, we compared model results against a different forest type (Table 4), with shared dominance by evergreen and deciduous tree species, and differing in LAI and annual net precipitation. Despite these broad differences, Table 4 suggests that selected hydrologic parameters (e.g. α_{soil}, f_h, and LAI_{max}, see Table 1) yielded reasonable values for canopy water retention capacity of broad-leaved temperate rainforests in SSA.

Forest responses to increased drought

According to predictions from our model, North Patagonian forests are likely to be altered by increased drought predicted for this century by climate change models. In our modeling study, the simulation of 50% reduction in summer rainfall predicted for the study area (business-as-usual scenario, [17]) can induce changes in both hydrological balance (up to ca. 100 mm $year^{-1}$ decrease in ET, Fig. 5) and forest structure (up to 83 tC ha^{-1} decrease in current AGB, Fig. 6), even without considering the potential ecological effects of concomitant global warming. A direct interpretation of changes in rainfall regimes is possible because the model accounts explicitly for changes in frequency and depth of rainfall events [8,32,47]. Decreasing in the depth of rainfall alone can induce some structural changes in the studied forest type (Fig. 6), but simultaneous changes in the frequency and depth of rainfall produced the strongest changes in hydrology and structure of stands (Fig. 6). These results highlight the impact on forest structure and growth of the duration and frequency of water limitation periods.

Reductions of basal area due to increased drought (Fig. 6) conform to empirical findings that drought increases the likelihood of mortality of large trees [59–61]. Our model formulation implies that trees assimilating greater biomass will have increasing demands of water for growth (Eq. 7, see also [62]). For example, annual gross biomass production increased with greater soil moisture availability during the year, a mechanism triggered by increased canopy transpiration (Fig. 7). Consequently, big trees that occur primarily in old-growth forests (Fig. 4b) experience increased stress-induced mortality due to greater hydrologic limitations during dry years. Trees that die under increased drought produce a decrease in both stand basal area and above-ground biomass (AGB, Fig. 6). Moreover, the model predicted that the OG forest has a higher *PB* than the young secondary stand (31.6±2.3 vs. 18.2±1.2 tC ha^{-1} $year^{-1}$, both obtained under current climate), and consequently a higher water demand for biomass production. Under increased drought, water demand for biomass production in OG forest is not fully covered by soil moisture supply, which causes the predicted decline of AGB and basal area (Fig. 6). We propose that the contrasting ET and structural patterns between YS and OG predicted for the coming decades under increasing drought are mainly due to significant limitation of available soil moisture for biomass production in OG, with lower impact on YS.

Research needs and model application

Here, we focused on developing an accurate model for assessing the influence of hydrologic processes involved in forest dynamics. Using the model, we quantitatively demonstrated the relative importance of soil moisture on forest structure (Fig. 6). We excluded other hydrologic processes to keep model formulation simple and results tractable with empirical information available. Processes such as water table dynamics, root dynamics, increased run off in slopes, and soil moisture dynamics across multiple soil layers can be incorporated as more empirical data becomes available for model calibration and validation. However, we strongly suggest that future model applications prioritize processes known to have an influence on the system under study [63]. In our study area, variations in the height of the water table may interfere with ecological processes such as tree establishment and mortality [14,24]. Our model provides a convenient starting point to incorporate water table dynamics into the analysis of climate change impacts, and to explore its effects on long-term forest dynamics. Our results also highlight the need of further fieldwork and experimental research on less known, mechanistic soil parameters (e.g. k_{soil}).

The rise of atmospheric carbon dioxide concentration and ensuing climate change are influencing water-use efficiency of forests [64,65]. WUE under future climate is likely to differ among forests [66], local scales [67], and species [68]. Forest water-use efficiency is a sensitive parameter in our model (e.g. Fig. S6) and illustrates the need for detailed studies of the expected variations in this parameter under drought, and their connection to gas exchange capacity of trees in different topographic settings and for a large set of species in SSA. It was beyond the scope of the present study to discuss model behavior along drought-to-moist gradients operating at regional scales because such analyses require an accurate and quantitative assessment of species-specific water-use efficiency [4]. In SSA, WUE variations have been experimentally tested in few study sites and for only three tree species included in our study [49,69]. Additional experimental research should address the complex interaction between photosynthetic carbon assimilation and water loss via transpiration with declining water supply [70]. Future research can address this question by studying changes in photosynthetic parameters along climatic gradients. Incorporating WUE as a species-specific parameter or a state dependent variable in our model is straightforward based on further empirical information.

Here, we deliberately excluded the influence of expected regional changes in temperature on forest processes (e.g. tree growth) to rather emphasize direct impacts of drought on model results. However, the model suggests that changes in PET due to expected warming trends in the study region are negligible compared to the strong impacts of increased drought (Fig. S5). Our simulated scenarios have been done using the common assumption of vegetation dynamic models that climate-forest interactions under inter-annual variation of climatic conditions can be used as proxy also for impacts of long-term climatic variations. In the later case it would be possible that some tree species would show adaptations effects. However, tree species adaptation to climate variability is still poorly understood in SSA. Moreover, warming can modify individual tree growth by affecting photosynthesis and both plant and soil respiration [38] [71], nutrient dynamics [72] and tree establishment [73]. An undergoing study analyzes the combined effect of temperature and rainfall changes on tree demography in the study area [74].

To the best of our knowledge, this is the first application of a forest gap model in temperate rainforests of SSA that integrates dynamic calculations of forest hydrology. The present work uses the best information available to ensure that climate patterns were directly comparable to hydrologic field measurements used for model calibration. However, our results on climate change impacts should be interpreted with caution because our baseline climate is constrained to a short period (1998–2009) within a long-term trend of rainfall (time period 1901–2005). Long-term monitoring of forest hydrology and dynamics can corroborate our results. To date, long-term monitoring (>10 years) of forest hydrology is lacking in forests of SSA. As more empirical data becomes available, the model can be revised and updated. The model developed here allows for the analysis of multiple environmental factors driving forest dynamics. For example, our model can help us understand the long-term responses of regional forest to drought events induced by El Niño Southern Oscillation that amplify background tree mortality rates in *Nothofagus* forest of SSA [75]. Moreover, increased regional drought is likely to interact with other drivers of global change such as changes in fire regimes, massive introduction of exotic forestry species, and forest fragmentation. To date, the implications of these drivers in the context of a changing climate remain poorly understood in SSA. These interacting growing threats demand from ecologists to understand and integrate multiple dimensions of global change on forest functioning. The model presented here is a particularly suitable tool for analyzing broad global change questions in forests of SSA because it also includes logging and fragmentation submodels [13,76]. Model-based experiments can also contribute to develop sound management strategies that anticipate forest responses to increasing drought and other drivers of climate change.

Conclusions

We developed and applied a forest dynamic model to analyze the impact of climate-driven increased drought on ecological and hydrological processes. The developed model was accurate for depicting forest hydrology at stand scales (i.e. <100 ha) and allowed the analysis of the dynamical linkages among rainfall regimes, soil moisture variation, and individual tree growth. Using the model we demonstrated that evergreen, broad-leaved temperate rainforests in southern South America are expected to be highly sensitive to future climate change, particularly increases in drought during parts of the year. Increased summer drought predicted for this century will likely impair biomass carbon accumulation, and amplify background tree mortality rates in this region. The developed model expands the range of applicability of gap models to assess climate change impacts in remote and understudied regions of the world, such as temperate forests of the southern hemisphere. It also represents an advance in the development of simple, general models to account for complex and dynamical processes operating at multiple spatial scales in forests.

Supporting Information

Figure S1 A diagram of the hydrologic submodel of FORMIND-CL v.**1.0.** Interaction between processes and variables in the hydrologic submodel, and their respective time scales of calculations. Arrows indicate whether the results of a model calculation influence the calculations of another submodel. Blue boxes represent analyzed variables of this study. All calculations are done in yearly time steps in the model, excepting the ones indicated in the dashed box. Variable notations follow the text. AGB: Above-ground biomass, LAI: leaf area index.

Figure S2 Weather generator results. Density functions of daily rainfall and daily mean temperature predicted by the weather generator compared to observed weather records from EBSD weather station.

Figure S3 Weather generator results. Comparison between simulated and observed climatic patterns during the year. Simulations were run for 100 years using parameters in Table 3. Daily data were averaged by seasons (mean daily temperature and daily radiation). Rainfall is the amount of rainfall during each season. Observed weather data are from EBSD weather station and seasons according to table 3.

Figure S4 Forest composition predictions. Model results for forest composition using different model versions compared to field data. Simulations run under the same conditions detailed in Methods section.

Figure S5 Drought induced simulations with warming included. Changes in hydrologic components and forest structure when warming and increased drought was considered. PET: Potential evapotranspiration (mm year^{-1}), T: transpiration (mm $year^{-1}$), Ec: canopy interception (mm $year^{-1}$), ET: evapotranspiration (mm $year^{-1}$), BAT: total basal area (m^2 ha^{-1}), BT: Total biomass (tC ha^{-1}). Result of a two-sample Wilcoxon test is shown on the upper right of each panel. Pink lines, drought induced simulations with warming included, blue lines drought induced simulations without warming, circles represent the values of simulation results. Note different scales for the axes.

Figure S6 Sensitivity of evaporatranspiration. Changes in evapotranspiration (ET) of the old-growth stand under current climate when using different water-use efficiency values (WUE). Simulations run under the same conditions detailed in Methods section.

Appendix S1 Calculation of canopy photosynthetic rate in FORMIND-CL **v.1.0.**

Acknowledgments

We appreciate the useful comments offered by Hans Pretzsch, and anonymous reviewers in previous versions of the manuscript. This is a contribution to the Research Program of the Chilean LTSER network.

Author Contributions

Conceived and designed the experiments: AGG AH JJA MFD. Performed the experiments: AGG JJA MFD. Analyzed the data: AGG AH MFD. Contributed reagents/materials/analysis tools: AGG JJA AH. Wrote the paper: AGG.

References

1. Rodriguez-Iturbe I, Porporato A, Ridolfi L, Isham V, Cox DR (1999) Probabilistic modelling of water balance at a point: the role of climate, soil and vegetation. Proceedings of the Royal Society a-Mathematical Physical and Engineering Sciences 455: 3789–3805.
2. Botkin DB, Wallis JR, Janak JF (1972) Some Ecological Consequences of a Computer Model of Forest Growth. Journal of Ecology 60: 849–872.
3. Bugmann H (2001) A review of forest gap models. Climatic Change 51: 259–305.
4. Bugmann H, Cramer W (1998) Improving the behaviour of forest gap models along drought gradients. Forest Ecology and Management 103: 247–263.
5. Bugmann HKM, Wullschleger SD, Price DT, Ogle K, Clark DF, et al. (2001) Comparing the performance of forest gap models in North America. Climatic Change 51: 349–388.
6. Reynolds JF, Bugmann H, Pitelka LF (2001) How much physiology is needed in forest gap models for simulating long-term vegetation response to global change? Challenges, limitations, and potentials. Climatic Change 51: 541–557.
7. Asbjornsen H, Goldsmith GR, Alvarado-Barrientos MS, Rebel K, Van Osch FP, et al. (2011) Ecohydrological advances and applications in plant-water relations research: a review. Journal of Plant Ecology-Uk 4: 3–22.
8. Porporato A, Daly E, Rodriguez-Iturbe I (2004) Soil water balance and ecosystem response to climate change. American Naturalist 164: 625–632.
9. Vose JV, Sun G, Ford CR, Bredemeier M, Otsuki K, et al. (2011) Forest ecohydrological research in the 21st century: what are the critical needs? Ecohydrology 4: 146–158.
10. Hall GMJ, Hollinger DY (2000) Simulating New Zealand forest dynamics with a generalized temperate forest gap model. Ecological Applications 10: 115–130.
11. Gutierrez AG, Huth A (2012) Successional stages of primary temperate rainforests of Chiloe Island, Chile. Perspectives in Plant Ecology Evolution and Systematics 14: 243–256.
12. Echeverría C, Newton AC, Lara A, Benayas JMR, Coomes DA (2007) Impacts of forest fragmentation on species composition and forest structure in the temperate landscape of southern Chile. Global Ecology and Biogeography 16: 426–439.
13. Rüger N, Gutiérrez AG, Kissling WD, Armesto JJ, Huth A (2007) Ecological impacts of different harvesting scenarios for temperate evergreen rain forest in southern Chile - A simulation experiment. Forest Ecology and Management 252: 52–66.
14. Díaz MF, Bigelow S, Armesto JJ (2007) Alteration of the hydrologic cycle due to forest clearing and its consequences for rainforest succession. Forest Ecology and Management 244: 32–40.
15. Little C, Lara A, McPhee J, Urrutia R (2009) Revealing the impact of forest exotic plantations on water yield in large scale watersheds in South-Central Chile. Journal of Hydrology 374: 162–170.
16. Trenberth KE, Jones PD, Ambenje P, Bojariu R, Easterling D, et al. (2007) Observations: Surface and Atmospheric Climate Change. In: Solomon S, Qin D, Manning M, Chen Z, Marquis M et al., editors. Climate Change 2007: The Physical Science Basis Contribution of Working Group I to the Fourth Assessment Report of the Intergovernmental Panel on Climate Change. Cambridge, United Kingdom and New York, NY, USA.: Cambridge University Press. 336
17. DGF CONAMA (2006) Estudio de la variabilidad climática en Chile para el siglo XXI. Santiago: Departamento de Geofisica. Universidad de Chile. 63 p.
18. Christensen JH, Hewitson B, Busuioc A, Chen A, Gao X, et al. (2007) Regional Climate Projection. In: Solomon S, Qin D, Manning M, Chen Z, Marquis M et al., editors. Climate Change 2007: The Physical Science Basis Contribution of Working Group I to the Fourth Assessment Report of the Intergovernmental Panel on Climate Change. Cambridge, United Kingdom and New York, NY, USA: Cambridge University Press. 94
19. Alaback PB (1991) Comparative Ecology of Temperate Rain-Forests of the America Along Analogous Climatic Gradients. Revista Chilena de Historia Natural 64: 399–412.
20. Armesto JJ, Smith-Ramírez C, Carmona MR, Celis-Diez JL, Díaz I, et al. (2009) Old-growth temperate rain forests of South America: Conservation, plant-animal interactions, and baseline biogeochemical processes. In: Wirth C, Gleixner G, Heimann M, editors. Old-growth forests: Function, fate and value. Berlin, Heidelberg: Springer New York. 367–390
21. Armesto JJ, Rozzi R, Smith-Ramírez C, Arroyo MTK (1998) Conservation targets in South American temperate forests. Science 282: 1271–1272.
22. Willson MF, Armesto JJ (1996) The natural history of Chiloé: on Darwin's trail. Revista Chilena de Historia Natural 69: 149–161.
23. Holdgate MW (1961) Vegetation and soils in the South Chilean Islands. Journal of Ecology 49: 559–580.
24. Díaz MF, Armesto JJ (2007) Physical and biotic constraints on tree regeneration in secondary shrublands of Chiloe Island, Chile. Revista Chilena De Historia Natural 80: 13–26.
25. di Castri F, Hajek E (1976) Bioclimatología de Chile. Santiago: Universidad Católica de Chile.
26. Gutiérrez AG, Armesto JJ, Aravena JC, Carmona M, Carrasco NV, et al. (2009) Structural and environmental characterization of old-growth temperate rain-forests of northern Chiloe Island, Chile: Regional and global relevance. Forest Ecology and Management 258: 376–388.
27. Veblen TT, Schlegel FM, Oltremari JV (1983) Temperate broad-leaved evergreen forest of South America. In: Ovington JD, editor. Temperate Broad-Leaved Evergreen Forest. Amsterdam: Elsevier Science Publishers. 5–31
28. Gutiérrez AG, Armesto JJ, Aravena JC (2004) Disturbance and regeneration dynamics of an old-growth North Patagonian rain forest in Chiloe Island, Chile. Journal of Ecology 92: 598–608.
29. Köhler P (2000) Modelling anthropogenic impacts on the growth of tropical rain forests-using an individual-oriented forest growth model for the analyses of logging and fragmentation in three case studies [PhD thesis http://hdl.handle.net/10013/epic.15101].Osnabrück, Germany: Center for Environmental Systems Research and Department of Physics. University of Kassel. Der Andere Verlag.
30. Köhler P, Chave J, Riera B, Huth A (2003) Simulating the long-term response of tropical wet forests to fragmentation. Ecosystems 6: 114–128.
31. Köhler P, Huth A (1998) The effects of tree species grouping in tropical rainforest modelling: Simulations with the individual-based model FORMIND. Ecological Modelling 109: 301–321.
32. Kumagai T, Katul GG, Saitoh TM, Sato Y, Manfroi OJ, et al. (2004) Water cycling in a Bornean tropical rain forest under current and projected precipitation scenarios. Water Resources Research 40: W01104.
33. Rutter AJ, Morton AJ (1977) Predictive Model of Rainfall Interception in Forests.3. Sensitivity of Model to Stand Parameters and Meteorological Variables. Journal of Applied Ecology 14: 567–588.
34. Wattenbach M, Hattermann F, Weng R, Wechsung F, Krysanova V, et al. (2005) A simplified approach to implement forest eco-hydrological properties in regional hydrological modelling. Ecological Modelling 187: 40–59.
35. Rey JM (1999) Modelling potential evapotranspiration of potential vegetation. Ecological Modelling 123: 141–159.
36. Maidment DR (1993) Handbook of Hydrology; McGraw-Hill, editor: McGraw-Hill.
37. Neilson RP (1995) A Model for Predicting Continental-Scale Vegetation Distribution and Water-Balance. Ecological Applications 5: 362–385.
38. Lambers H, Chapin FS, Pons TL (1998) Plant physiological ecology. New York: Springer. 540 p.
39. Bazzaz FA (1979) Physiological ecology of plant succession. Annual Review of Ecology and Systematics 10: 351–371.

40. Thornley HMJ, Johnson IR (1990) Plant and Crop Modelling – A mathematical approach to plant and crop physiology. Oxford, UK.: Clarendon Press.
41. Gerten D, Schaphoff S, Haberlandt U, Lucht W, Sitch S (2004) Terrestrial vegetation and water balance - hydrological evaluation of a dynamic global vegetation model. Journal of Hydrology 286: 249–270.
42. Venevsky S, Maksyutov S (2007) SEVER: A modification of the LPJ global dynamic vegetation model for daily time step and parallel computation. Environmental Modelling & Software 22: 104–109.
43. Prentice IC, Sykes MT, Cramer W (1993) A Simulation-Model for the Transient Effects of Climate Change on Forest Landscapes. Ecological Modelling 65: 51–70.
44. Haxeltine A, Prentice IC (1996) BIOME3: An equilibrium terrestrial biosphere model based on ecophysiological constraints, resource availability, and competition among plant functional types. Global Biogeochemical Cycles 10: 693–709.
45. Prentice IC, Cramer W, Harrison SP, Leemans R, Monserud RA, et al. (1992) A Global Biome Model Based on Plant Physiology and Dominance, Soil Properties and Climate. Journal of Biogeography 19: 117–134.
46. Dingman SL (2002) Physical hydrology. New Jersey: Prentice Hall. 646 p.
47. Laio F, Porporato A, Ridolfi L, Rodriguez-Iturbe I (2001) Plants in water-controlled ecosystems: active role in hydrologic processes and response to water stress - II. Probabilistic soil moisture dynamics. Advances in Water Resources 24: 707–723.
48. Lusk CH (2001) When is a gap not a gap? Light levels and leaf area index in bamboo-filled gaps in a Chilean rain forest. Gayana Botanica 58: 25–30.
49. Piper FI, Corcuera LJ, Alberdi M, Lusk C (2007) Differential photosynthetic and survival responses to soil drought in two evergreen Nothofagus species. Annals of Forest Science 64: 447–452.
50. Zuñiga R, Alberdi M, Reyes-Diaz M, Olivares E, Hess S, et al. (2006) Seasonal changes in the photosynthetic performance of two evergreen Nothofagus species in south central Chile. Revista Chilena de Historia Natural 79: 489–504.
51. Cunningham SC (2005) Photosynthetic responses to vapour pressure deficit in temperate and tropical evergreen rainforest trees of Australia. Oecologia 142: 521–528.
52. Brodribb T, Hill RS (1998) The photosynthetic drought physiology of a diverse group of southern hemisphere conifer species is correlated with minimum seasonal rainfall. Functional Ecology 12: 465–471.
53. Janssen I, Kruemmelbein J, Horn R, Ellies AS (2004) Physical and hydraulic properties of the ñadi soils in south Chile - Comparison between untilled and tilled soil. Revista de la Ciencia del Suelo y Nutrición Vegetal 4: 14–28.
54. Sitch S, Smith B, Prentice IC, Arneth A, Bondeau A, et al. (2003) Evaluation of ecosystem dynamics, plant geography and terrestrial carbon cycling in the LPJ dynamic global vegetation model. Global Change Biology 9: 161–185.
55. van Genuchten MT (1980) A closed-form equation for predicting the hydraulic conductivity of unsaturated soils. Soil Science Society of America Journal 44: 892–898.
56. Seki K (2007) SWRC fit-a nonlinear fitting program with a water retention curve for soils having unimodal and bimodal pore structure. Hydrology and Earth System Sciences Discussions 4: 407–437.
57. R-Development-Core-Team (2012) R: A language and environment for statistical computing. Vienna, Austria.: R Foundation for Statistical Computing.
58. Breda NJJ (2003) Ground-based measurements of leaf area index: a review of methods, instruments and current controversies. Journal of Experimental Botany 54: 2403–2417.
59. Condit R, Hubbell SP, Foster RB (1995) Mortality-Rates of 205 Neotropical Tree and Shrub Species and the Impact of a Severe Drought. Ecological Monographs 65: 419–439.
60. Allen CD, Macalady AK, Chenchouni H, Bachelet D, McDowell N, et al. (2010) A global overview of drought and heat-induced tree mortality reveals emerging climate change risks for forests. Forest Ecology and Management 259: 660–684.
61. Phillips OL, van der Heijden G, Lewis SL, Lopez-Gonzalez G, Aragao LEOC, et al. (2010) Drought-mortality relationships for tropical forests. New Phytologist 187: 631–646.
62. Larcher W (2001) Ökophysiologie der Pflanzen. Stuttgart: 6th edition, Verlag Eugen Ullmer.
63. Grimm V, Railsback SF (2005) Individual-based modeling and ecology. New Jersey: Princeton University Press.
64. Penuelas J, Canadell JG, Ogaya R (2011) Increased water-use efficiency during the 20th century did not translate into enhanced tree growth. Global Ecology and Biogeography 20: 597–608.
65. Keenan TF, Hollinger DY, Bohrer G, Dragoni D, Munger JW, et al. (2013) Increase in forest water-use efficiency as atmospheric carbon dioxide concentrations rise. Nature 499: 324–+.
66. Silva LCR, Anand M (2013) Probing for the influence of atmospheric CO_2 and climate change on forest ecosystems across biomes. Global Ecology and Biogeography 22: 83–92.
67. Penuelas J, Hunt JM, Ogaya R, Jump AS (2008) Twentieth century changes of tree-ring delta C-13 at the southern range-edge of Fagus sylvatica: increasing water-use efficiency does not avoid the growth decline induced by warming at low altitudes. Global Change Biology 14: 1076–1088.
68. Levesque M, Saurer M, Siegwolf R, Eilmann B, Brang P, et al. (2013) Drought response of five conifer species under contrasting water availability suggests high vulnerability of Norway spruce and European larch. Global Change Biology 19: 3184–3199.
69. Figueroa JA, Cabrera HM, Queirolo C, Hinojosa LF (2010) Variability of water relations and photosynthesis in *Eucryphia cordifolia* Cav. (Cunoniaceae) over the range of its latitudinal and altitudinal distribution in Chile. Tree Physiology 30: 574–585.
70. Chaves MM, Pereira JS, Maroco J, Rodrigues ML, Ricardo CPP, et al. (2002) How plants cope with water stress in the field. Photosynthesis and growth. Annals of Botany 89: 907–916.
71. Ryan MG, Law BE (2005) Interpreting, measuring, and modeling soil respiration. Biogeochemistry 73: 3–27.
72. Aerts R, Chapin FS (2000) The mineral nutrition of wild plants revisited: A re-evaluation of processes and patterns. Advances in Ecological Research, Vol 30 30: 1–67.
73. Hobbie SE, Chapin FS (1998) An experimental test of limits to tree establishment in Arctic tundra. Journal of Ecology 86: 449–461.
74. Gutiérrez AG (2010) Long-term dynamics and the response of temperate rainforests of Chiloé Island (Chile) to climate change [Phd Thesis]. Freising, Germany: Technische Universität München. 170 p.
75. Suarez ML, Ghermandi L, Kitzberger T (2004) Factors predisposing episodic drought-induced tree mortality in Nothofagus-site, climatic sensitivity and growth trends. Journal of Ecology 92: 954–966.
76. Groeneveld J, Alves LF, Bernacci LC, Catharino ELM, Knogge C, et al. (2009) The impact of fragmentation and density regulation on forest succession in the Atlantic rain forest. Ecological Modelling 220: 2450–2459.
77. Donoso PJ, Soto DP, Bertin RA (2007) Size-density relationships in *Drimys winteri* secondary forests of the Chiloe Island, Chile: Effects of physiography and species composition. Forest Ecology and Management 239: 120–127.
78. Navarro C, Donoso C, Sandoval V (1999) Los renovales de Canelo. In: Donoso C, Lara A, editors. Silvicultura de los bosques nativos de Chile. Santiago, Chile: Editorial Universitaria.
79. Naulin P (2002) Estimacion de la biomasa en un renoval de canelo (Drimys winteri J.R. et Foster) en la Comuna de Ancud. Santiago: Universidad de Chile. 45 p.
80. Aravena JC, Carmona MR, Perez CA, Armesto JJ (2002) Changes in tree species richness, stand structure and soil properties in a successional chronosequence in northern Chiloe Island, Chile. Revista Chilena De Historia Natural 75: 339–360.
81. Lusk CH (2001) Leaf life spans of some conifers of the temperate forests of South America. Revista Chilena De Historia Natural 74: 711–718.
82. Huber A, Iroume A (2001) Variability of annual rainfall partitioning for different sites and forest covers in Chile. Journal of Hydrology 248: 78–92.
83. Oyarzun CE, Godoy R, Staelens J, Donoso PJ, Verhoest NEC (2011) Seasonal and annual throughfall and stemflow in Andean temperate rainforests. Hydrological Processes 25: 623–633.
84. Barbour MM, Hunt JE, Walcroft AS, Rogers GND, McSeveny TM, et al. (2005) Components of ecosystem evaporation in a temperate coniferous rainforest, with canopy transpiration scaled using sapwood density. New Phytologist 165: 549–558.
85. Lara A, Little C, Urrutia R, McPhee J, Alvarez-Garreton C, et al. (2009) Assessment of ecosystem services as an opportunity for the conservation and management of native forests in Chile. Forest Ecology and Management 258: 415–424.
86. CIREN (1994) Delimitación y descripción de microregiones para la transferencia tecnológica de INDAP. Santiago: CORFO. 99 p.

Genetic Dissection of Drought and Heat Tolerance in Chickpea through Genome-Wide and Candidate Gene-Based Association Mapping Approaches

Mahendar Thudi[1], Hari D. Upadhyaya[1], Abhishek Rathore[1], Pooran Mal Gaur[1], Lakshmanan Krishnamurthy[1], Manish Roorkiwal[1,2], Spurthi N. Nayak[1], Sushil Kumar Chaturvedi[3], Partha Sarathi Basu[3], N. V. P. R. Gangarao[4], Asnake Fikre[5], Paul Kimurto[6], Prakash C. Sharma[2], M. S. Sheshashayee[7], Satoshi Tobita[8], Junichi Kashiwagi[9], Osamu Ito[10], Andrzej Killian[11], Rajeev Kumar Varshney[1]*

1 International Crops Research Institute for the Semi-Arid Tropics (ICRISAT), Hyderabad, Andhra Pradesh, India, **2** Guru Gobind Singh Indraprastha University, Delhi, India, **3** Indian Institute of Pulses Research (IIPR), Kanpur, Uttar Pradesh, India, **4** International Crops Research Institute for the Semi-Arid Tropics (ICRISAT), Nairobi, Kenya, **5** Ethiopian Institute of Agricultural Research (EIAR), Debre Zeit, Ethiopia, **6** Egerton University (EU), Egerton, Kenya, **7** University of Agricultural Sciences- Bangalore, Bangalore, Karnataka, India, **8** Japan International Research Center for Agricultural Sciences (JIRCAS), Tsukuba, Japan, **9** Hokkaido University, Sapporo, Japan, **10** United Nations University, Yokohama, Japan, **11** DArT Pty. Ltd., Yarralumla, Australia

Abstract

To understand the genetic basis of tolerance to drought and heat stresses in chickpea, a comprehensive association mapping approach has been undertaken. Phenotypic data were generated on the reference set (300 accessions, including 211 mini-core collection accessions) for drought tolerance related root traits, heat tolerance, yield and yield component traits from 1–7 seasons and 1–3 locations in India (Patancheru, Kanpur, Bangalore) and three locations in Africa (Nairobi, Egerton in Kenya and Debre Zeit in Ethiopia). Diversity Array Technology (DArT) markers equally distributed across chickpea genome were used to determine population structure and three sub-populations were identified using admixture model in STRUCTURE. The pairwise linkage disequilibrium (LD) estimated using the squared-allele frequency correlations (r^2; when $r^2<0.20$) was found to decay rapidly with the genetic distance of 5 cM. For establishing marker-trait associations (MTAs), both genome-wide and candidate gene-sequencing based association mapping approaches were conducted using 1,872 markers (1,072 DArTs, 651 single nucleotide polymorphisms [SNPs], 113 gene-based SNPs and 36 simple sequence repeats [SSRs]) and phenotyping data mentioned above employing mixed linear model (MLM) analysis with optimum compression with P3D method and kinship matrix. As a result, 312 significant MTAs were identified and a maximum number of MTAs (70) was identified for 100-seed weight. A total of 18 SNPs from 5 genes (*ERECTA*, 11 SNPs; *ASR*, 4 SNPs; *DREB*, 1 SNP; *CAP2* promoter, 1 SNP and *AMDH*, 1SNP) were significantly associated with different traits. This study provides significant MTAs for drought and heat tolerance in chickpea that can be used, after validation, in molecular breeding for developing superior varieties with enhanced drought and heat tolerance.

Editor: Pilar Hernandez, Institute for Sustainable Agriculture (IAS-CSIC), Spain

Funding: This study was undertaken as a part of Tropical Legumes I (TLI- Objective 4: Chickpea) funded by Bill & Melinda Gates Foundation (BMGF) through CGIAR Generation Challenge Programme (http://www.generationcp.org/). This work has been undertaken as part of the CGIAR Research Program on Grain Legumes. ICRISAT is a member of CGIAR Consortium. DArT Pty. Ltd. provided support in the form of a salary for author AK, but did not have any additional role in the study design, data collection and analysis, decision to publish, or preparation of the manuscript. The specific roles of the authors are articulated in the "author contributions" section. The funders had no role in study design, data collection and analysis, decision to publish, or preparation of the manuscript.

Competing Interests: AK is an employee of DArT Pty. Ltd. There are no patents, products in development or marketed products to declare.

* E-mail: r.k.varshney@cgiar.org

Introduction

Chickpea (*Cicer arietinum* L) is the second most important grain legume cultivated mostly on residual soil moisture in the arid and semi-arid regions of the world. It is a diploid member of family Leguminosae with basic chromosome number eight and genome size 738 Mb [1]. Globally it is cultivated on over 13.2 Mha with an annual production of 11.6 million metric tons [2]. Chickpea is a rich source of proteins, essential amino acids and vitamins such as riboflavin, niacin, thiamin, folate and the vitamin A precursor β-carotene [3]. Based on seed size and color chickpea is grouped into two market classes namely *desi* and *kabuli*. Globally about 80% of total production is contributed by *desi* cultivars. Further, among chickpea growing countries India alone contributes to 70% of the world's total production [2]. Other major chickpea producing countries include Pakistan, Turkey, Australia, Myanmar, Ethiopia, Iran, Mexico, Canada and the USA. Although the chickpea production potential is high, it is not fully realized owing to several abiotic and biotic stresses. Among abiotic stresses that affect the chickpea production, drought and heat are considered as major constraints. Annually, 40–50% reduction in yield has been reported worldwide as a result of terminal drought in chickpea [4]. Further, the damage due to drought is compounded by heat stress in the warmer Mediterranean regions and regions like South Asia where temperature increases towards flowering [5] and it is difficult to differentiate between the damage caused by the individual stresses. Nevertheless as a result of drought stress, the growing season may be shortened affecting yield components, i.e., total biomass, pod number, seed number, seed weight and quality and yield plant^{-1} [6]. Flowering and seed set are the most critical growth stages affected by drought in chickpea. Drastic reductions in chickpea seed yields were observed when plants were exposed to high (35°C) temperatures at flowering and pod development stages [7]. Heat stress also adversely affects pollen viability, fertilization and seed development leading to a reduced harvest index. The identification of genomic regions associated with the drought and heat tolerance would enable breeders to develop improved cultivars with increased drought and heat tolerance using molecular breeding.

Large scale genomic resources are essential for understanding the genetics of complex abiotic stresses like drought, heat tolerance etc. In the case of chickpea, during last five years >3,000 microsatellites or simple sequence repeats (SSRs) [8–10], Diversity Array Technology (DArT) arrays [10] and single nucleotide polymorphism (SNP) [11] markers were developed. Further these marker resources were used for linkage map construction [9,10] as well as trait mapping. Majority of trait mapping studies were focused on biotic stresses [12]. Recently sequencing of *desi* and *kabuli* chickpea genomes has been completed [1,13] and a genome-wide physical map (http://probes.pw.usda.gov:8080/chickpea/) was developed [14]. Despite the availability of large scale genomic resources, most of the studies to understand the genetics of complex traits were limited to quantitative trait loci (QTL) mapping studies. Moreover, the family based QTL studies were mostly limited to biotic stresses like *Fusarium* wilt, *Ascochyta* blight and *Botrytis* gray mold [12]. Very few studies were conducted to understand the genetics of drought tolerance [15] and salt tolerance [16] in chickpea.

Despite the continuous efforts to enhance the productivity of chickpea, climate changes during past two decades had tremendous influence on the production and productivity [17]. Global warming, coupled with increased temperatures in arid and semi-arid regions has necessitated development of crop varieties that can sustain and yield high in harsh climatic conditions by virtue of being resilient to warmer temperatures. Further, the quantitative inheritance of drought and heat; their interaction with environment have been posing challenges to our understanding of the genetic basis of these traits. Although conventional breeding has substantial impact in marginal chickpea growing environments [18], future genetic gains will require a more systematic use of physiological and genetic approaches, facilitated by the rapid increase in genome knowledge and understanding. Thus the knowledge generated through advances in genomics during past two decades, have enormous potential in enhancing the tolerance to these stresses.

During last two decades, molecular markers have provided greater insights into complex traits in several crop species and the research endeavors in crop improvement shifted from quantitative to molecular genetics with emphasis on QTL identification and adoption of marker-assisted selection (MAS). However, only modest results have been witnessed due to several factors including absence of tight linkage between makers and QTLs, non-availability of mapping populations, and substantial time needed to develop such populations. Further, the QTL mapping approaches cannot make use of the huge variation present across the germplasm available in the genebanks. In addition, the resolution achieved through linkage mapping based on bi-parental mapping population is low compared to population linkage disequilibrium (LD) based association mapping. The genome-wide association study (GWAS) makes it possible to simultaneously screen a very large number of accessions for genetic variation underlying diverse complex traits. In fact, the association studies in other crop species especially in cereals [19–21] have revealed that the linkage based QTL analyses can be complemented by LD based association studies. Association mapping studies in legumes are limited to soybean [22], *Medicago* [23] and common bean [24]. Most of these association studies either have deployed GWAS or candidate gene-sequencing approach. In some recent studies, however, the combined approach of GWAS and candidate-gene sequencing has been shown as more powerful approach than the individual approach [25]. However, to date there is no report on association studies in the case of chickpea. Recently a diverse set of 300 accessions, called as 'reference set' [26] that included 'mini core collection' [27] has been used to analyze sequence diversity for 10 drought responsive genes [28].

In the present study, both genome-wide and candidate gene-sequencing based association mapping approaches were employed to understand genetics of the two most important complex abiotic stresses "drought" and "heat" and establish significant marker-trait associations (MTAs). For this the chickpea reference set/mini-core collection was genotyped for 1,813 marker loci (SSRs, DArTs and SNPs) and phenotyped at three locations in Africa (Nairobi and Egerton in Kenya; and Debre Zeit in Ethiopia) and three locations in India (Patancheru, Kanpur and Bangalore). Besides significant MTAs, the present study also provides an in-depth understanding of the genetic diversity, population structure in the reference set and linkage disequilibrium (LD) in genome of chickpea.

Results and Discussion

The chickpea reference set, comprising of 300 diverse accessions of *Cicer* spp. including 293 accessions from *C. arietinum*, 4 accessions from *C. reticulatum* and 3 accessions from *C. echinospermum* (Table S1), from Asia (198), Africa (21), Europe (3), Mediterranean (56), Americas (10), CIS (6), and 6 accessions with unknown geographical origin, was chosen for association studies. This set was evaluated for 34 traits (root, morphological, phenological, transpiration efficiency related traits, yield and yield related traits) under drought and heat stress environments with 2–3 replications and 1–7 years (Table S2). Large phenotypic variation was observed for all traits phenotyped on the reference set or mini-core collection (Table S3).

Genome-wide marker profiling

We report the first comprehensive characterization of the chickpea reference set using 16,046 markers (35 SSRs, 15,360 DArT features and 651 SNPs) in the present study. A total of 917 alleles were detected by 35 SSR markers with an average of 26.2 alleles/marker locus and the polymorphism information content (PIC) values ranged from 0.48–0.96 (Table S4). Further, major allele frequency and the overall mean heterozygosity was 0.21 and 0.04 respectively. The DArT markers developed by Thudi et al. [10] were used for genotyping the reference set, as a result 1,156 DArT loci were found polymorphic with PIC values in the range of 0.01–0.38 (Table S5). Among 651 SNPs genotyped using KASPar assays, 381 SNPs were polymorphic and the PIC values ranged from 0.007–0.375 (Table S6).

Allele mining in candidate drought responsive genes

A total of 10 candidate drought responsive genes (Table S7) were amplified on the reference set/mini-core collection. Of these, 5 genes (Abscisic acid stress and ripening, *ASR*; *CAP2* gene; *ERECTA*; sucrose synthase, *SUSY*; Sucrose phosphate synthase, *SPS*) were amplified on the reference set. The number of genotypes for which good quality sequences were obtained varied from 79 (*ERECTA* fragment obtained from 7f–5r primer pairs) to 236 genotypes (*SPS* gene) out of 300 genotypes attempted [28]. The longest sequence obtained was for *SUSY* gene which was about 1,600 bp and the shortest sequence was obtained for *SPS* gene with 312 bp. Varying number of SNPs were identified in each gene using *DIVEST* tool. SNPs were inspected manually for possible sequencing errors and SNPs having clear peaks were considered as true SNPs. For instance, highest number of SNPs (34) was obtained for *ASR* gene while no SNPs were found in the case of *CAP2* gene. In total, 33 SNPs were identified in case of *ERECTA* gene, of which 13 SNPs (9 transitions and 4 transversions) were in *ERECTA* (7f–5r) gene fragment and 20 SNPs (10 transitions and 10 transversions) in case of *ERECTA* (8f–8r) gene fragment. Only 1 indel was observed in case of *ERECTA* (7f–5r) gene fragment. One indel and 3 SNPs were observed in case of *SPS* gene sequence (Table S7). In addition to above mentioned 5 drought responsive genes, 5 abiotic stress responsive genes (AKIN-SNF1 related protein kinase, *AKIN*; Aminoaldehyde dehydrogenase gene, *AMADH*; Dehydrin, *DHN*; Dehydration responsive element binding protein, *DREB* and Myb transcription factor, *Myb*) were used for allele mining across mini-core collection (Table S7). Out of 211 genotypes screened, number of genotypes yielding high quality sequence varied from 191 (*DREB*) to 209 (*AMADH*). Highest number of SNPs (14) was obtained for *DREB* gene (8 transitions and 6 transversions). Apart from SNPs, 23 indels were also detected. *AKIN* gene was found to be most conserved with just 2 SNPs (transitions) and 2 indels. A total of 13 SNPs were identified (6 transitions and 7 transversions) in case of *AMADH* with 3 indels among 209 high quality sequences, while in case of *DHN* gene 7 SNPs (5 transitions and 2 transversions) were identified among 198 sequence analyzed. For *Myb* gene only 6 SNPs (1 transition and 5 transversions) were reported with 2 indels across *Myb* sequences. Average PIC value of SNPs ranged from 0 (*CAP2* gene) to 0.43 (*CAP2* promoter) (Table S7).

Population structure and genetic relationships

In order to assess the population structure and number of sub-populations, 85 evenly distributed DArT loci on chickpea genome [10] were used on the reference set employing STRUCTURE 2.3.4 [29]. Based on the maximum likelihood and delta K (ΔK) values, three sub-populations (Group I, Group II and Group III) were determined in the reference set (Fig 1). However, different K values are possible (Fig S1); nevertheless, these do not qualitatively affect the results. As the K value is increasing the allelic admixture among the sub-populations is more clearly demarcated. Using a membership probability threshold of 0.60, 109 accessions were assigned to sub-population 1; 154 accessions to sub-population 2; 26 accessions to sub-population 3 and 11 accessions were retained in the admixed group (Table S1).

To understand the genetic diversity in the reference set neighbor joining (NJ) trees were constructed using allelic data from 35 SSRs, 1,156 DArT loci and 651 SNP markers independently as well as all markers combined together, which revealed three major clusters (Fig. S1). However, the grouping of the accessions in each cluster for different marker systems was not similar. This can be attributed to the nature of markers used and the genomic regions sampled by these marker systems. Nevertheless, the SNP markers could more clearly demarcate the accessions into smaller sub-groups compared to SSRs and DArT loci (Fig S1). This indicates SNP markers have more potential application for plant genome analysis compared to other markers. However, none of the three marker systems could group the association panel into distinct groups based on either the market class or the biological status of the germplasm set analyzed. This indicates that genetic variation in the reference set is very high and it is an ideal germplasm panel for association studies. The wild genotypes were grouped in third cluster and these are distributed away from the cultivated genotypes in the cluster indicating that the wild species genotypes are more distinct (Fig S1).

Linkage disequilibrium (LD) decay

A consensus genetic map was constructed using the two inter-specific genetic maps developed by [10] and [11] based on mapping population ICC 4958× PI 489777. A total of 1,358 markers (706 DArT loci, 622 SNPs and 30 SSRs) mapped on to the consensus map were used for estimating the LD decay. Estimation of LD decay is essential to determine the number of markers required for association mapping of complex traits like drought and heat tolerance in chickpea. In general, the genetic distance at which r^2 decays to 0.1–0.2 is considered to be the extent of LD in a species [30]. In the present study, the pairwise LD estimated using the squared-allele frequency correlations (r^2; when $r^2<0.20$) was found to decay rapidly with the genetic distance of 5 cM (Fig 2), when $r^2<0.1$ LD decay was found to decay at a genetic distance of 20 cM. The results suggest that, LD decay over short distances will facilitate fine mapping of QTL, while LD decay over longer distances will facilitate initial association of trait data with haplotypes in chromosome regions. Further, researchers can use the LD map as a reference to find target QTL and genes for positional cloning [31].

Genome-wide and candidate gene-sequencing based association

In order to reduce the number of false positive associations, both population structure and relative kinship information were employed. Mixed linear model (MLM) with optimum compression and P3D in TASSEL 2.01 software was employed for establishing MTAs. Further, to eliminate the false positives Bonferroni correction was used and as a result, a total of 312 highly significant MTAs were identified for 25 agronomically important traits (Table 1). Phenotypic variance explained (PVE) for these MTAs ranged from low (4.14%) to very high (96.55%) and thus MTAs detected with high PVE for desired traits can be improved through molecular breeding.

Drought tolerance root traits. Drought is the major limiting factor to crop production and especially chickpea experiences various kinds of drought stresses depending on the timing and intensity of the water stress relative to the reproductive stage of the crop. Terminal drought alone has been leading to more than 50% yield losses in chickpea. In the context of receding soil moisture, the breeding strategies should focus to enhance the maximum utilization of available soil moisture efficiently; hence breeding efforts should focus on improving root traits that enhance the efficient extraction of soil moisture. A total of 8 root traits were phenotyped for three seasons at Patancheru for gaining insights into the root system (Table S2). Association analysis identified 15 markers significantly associated with 5 root traits (Root dry weight, RDW; root length density, RLD; root surface area, RSA; root volume, RV and rooting depth, RDp) with PVE ranging from 8.25–22.41%. Among them, 7 markers showed significant association with single trait (RDp) and 2 markers (NCPGR7,

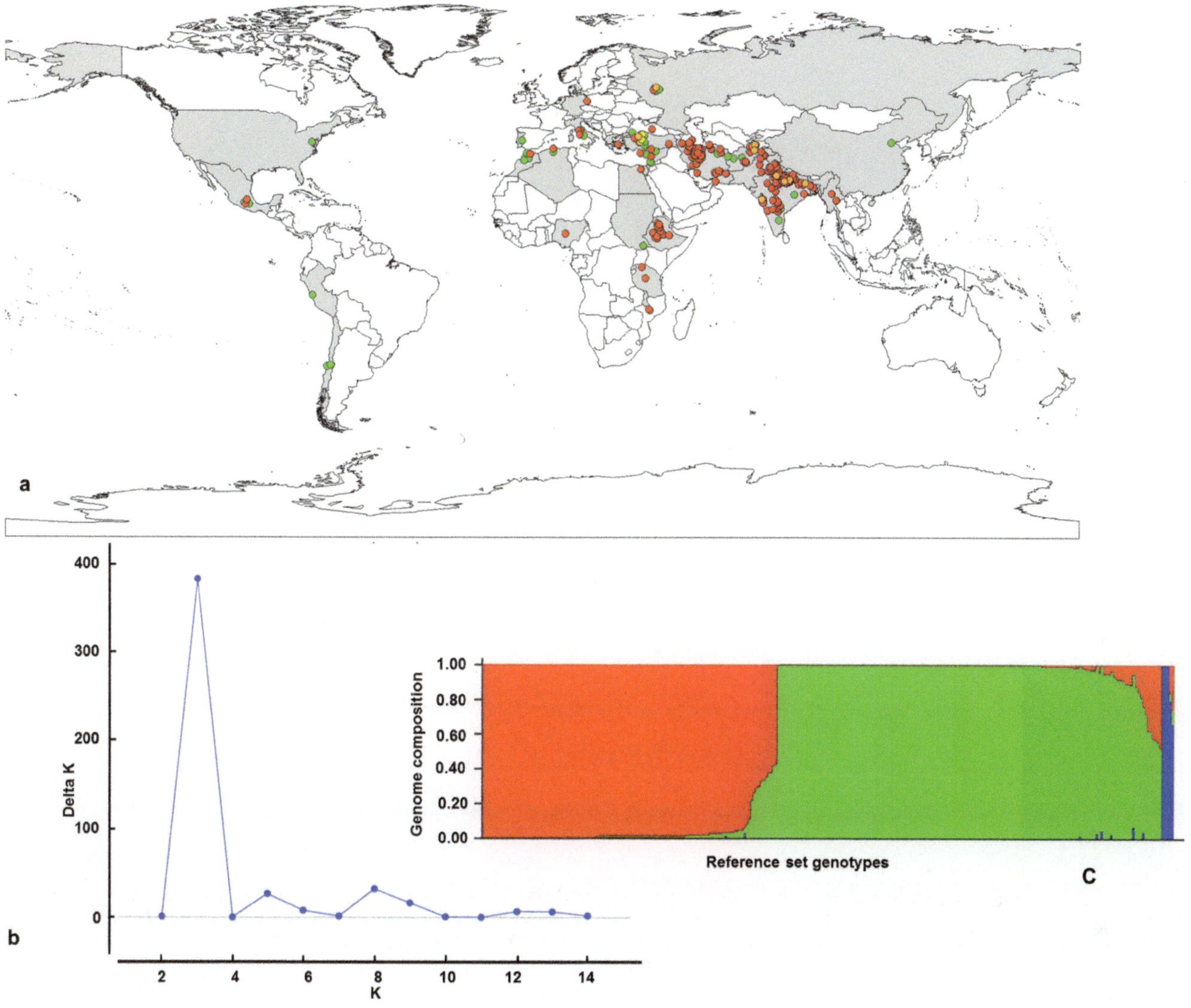

Figure 1. Geographic origin and population structure of chickpea reference set. a) the distribution of chickpea reference set, *desi* in red, *kabuli* in green, pea-shaped in orange and wild in yellow color dots b) ΔK is function of k from the structure run, the plateau at k = 3 indicates number of sub-populations in the reference set; c) Clustering of chickpea set genotypes into three groups (Group I, Group II and Group III).

DR_237) showed associations with more than one trait (Table 1). Hence, these two markers are believed to be associated with co-localized/pleiotropic QTLs. The co-localization of specific genes/QTLs could be a better way to understand the molecular basis of drought tolerance or of traits related to drought response. The presence of several co-localized/pleiotropic QTLs verified the complex quantitative nature of drought tolerance in chickpea and allowed the identification of some important genomic regions for traits related to drought tolerance. The markers associated with more than one trait may be efficiently utilized in improvement of more than one trait simultaneously through marker assisted selection (MAS). Till date there are no reports of association studies in the case of chickpea, however the association studies in other crop species especially in cereals such as maize [32], barley [33], sorghum [34] and wheat [35] have revealed that the linkage based QTL analyses can be complemented by LD based association studies.

Morphological traits. Among morphological traits, significant marker trait associations were identified only for plant height (PHT), shoot dry weight (SDW), apical secondary branches (ASB) and basal primary branches (BPB). A total of 47 significant MTAs were identified in the present study for the morphological traits that explained 7.77–38.01% phenotypic variation. Notably, 35 (74.49%) MTAs identified were for PHT (Table 1). Among 35 markers (2 SNPs, 2 SSRs, 29 DArT loci and 2 gene-based SNPs) that were associated with PHT, TA28 locus explained maximum phenotypic variation (38.01%).

Phenolological traits. Three phenological traits days to 50% flowering (DF; refers to the day when more than 50% of the plants in a plot initiated flowering), days to maturity (DM) and flowering days (FD; refers to the number of days from the start of flowering to cessation of flowering) were phenotyped for 7–11 seasons in 1–5 locations (Kanpur, Patancheru, Debre Zeit, Egerton and Nairobi). The variation in crop duration is known to affect the seed yield under both drought and heat stresses [5]. Among three traits, 5 significant MTAs were identified for DM and 2 significant MTAs for FD. CaSTMS2 marker associated with FD explained highest phenotypic variation (96.55%; Table 1). Among 4 markers (TA14,

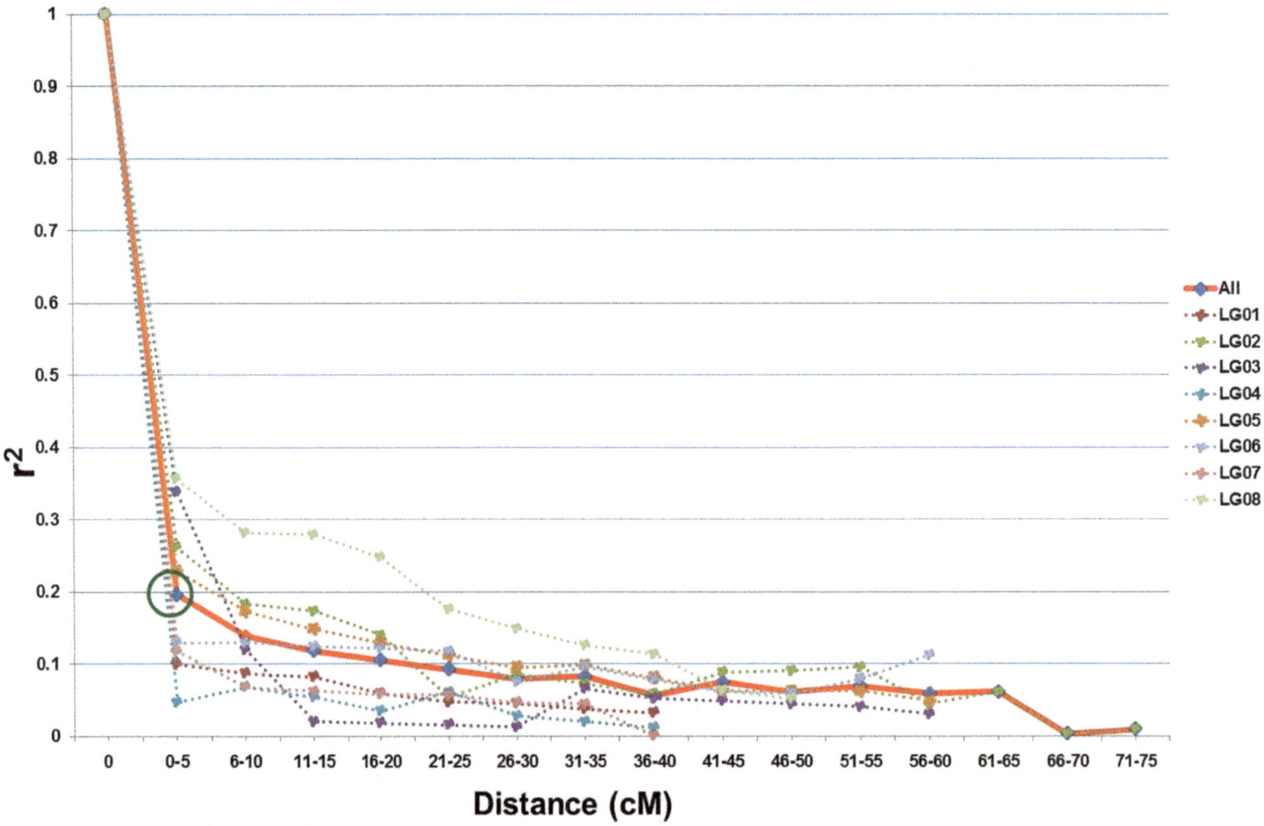

Figure 2. Linage disequilibrium (LD) decay across all linkage groups. The overall LD decay across the genome is at 5 cM.

CKaM1056, cpPb-489599, ASR391(C/T) and ASR447(C/T)), associated with DM, TA14 explained maximum phenotypic variation (79.31%). Further, among phenological traits high heritability was observed in case of DF followed by DM and FD (Table S3), indicating the possibility of selecting phenotypes required for various target environments. Early maturity is one of the crop adoption strategies to escape drought and heat stresses [36]. Markers that are associated with DM and have significant negative effect on the trait will serve in selection of genotypes with early maturity, thus enhance drought tolerance. Conventionally, late sowing is recommended to overcome heat stress [36]; however, the unpredictable climatic condition may lead to yield losses significantly in case of late planting. Hence, identification of markers associated with DF and DM can be deployed for developing lines that escape drought.

Yield and yield related traits. A total of 13 yield and yield related traits (100-seed weight, 100SDW; Biomass, BM; Harvest index, HI; Yield, YLD; Podm^{-2}, POD; Podsplant^{-1}, PPP; Seedm^{-2}, SPM, Seedpod^{-1}, SPP; shoot day^{-1}, PDS and Total dry matter weight, TDM) were studied in 1–7 seasons, 1–2 environments at two location in India (Patancheru and Kanpur) and three locations in Africa (Debre Zeit, Nairobi and Egerton). A total of 221 significant MTAs were identified in the present study for the yield and yield component traits that explained 4.23–55.42% phenotypic variation. Notably, 70 (22.43%) MTAs identified were for 100SDW (Table 1). Among 26 markers (9 SNPs, 6 SSRs, 6 DArT loci and 5 gene-based SNPs) that were associated with 100SDW, TA71 locus explained maximum phenotypic variation (88.34%). Further, among 5 gene-based SNP markers [AM_192, ASR_192_290, Ca_Cap2promo, CAP2-promo98(C/G) and DR_237], CAP2promo98(C/G) explained maximum phenotypic variation (36.95%). The heritability of 100SDW was more than 0.9 across all environments and locations, except for Kanpur (0.671) under heat stress environment (Table S3). Thirty four significant MTAs were found in case of SPM (8.06–55.42%, PVE), 32 in case of YLD (11.43–29.03%, PVE), 16 in case of HI (4.23–15.53%, PVE), 11 in case of BM (16.34–18.99% PVE), 13 in case of SPP (7.72–17.75% PVE), 10 for POD (9.18–22.05% PVE), 9 for TDM (8.84–12.88% PVE), 6 for PPP (8.27–50.44% PVE), and 1 for PDS (10.09%, PVE).

Transpiration efficiency related traits. SPAD chlorophyll meter reading (SCMR) and $\delta^{13}C$ are the transpiration efficiency related traits phenotyped in the present study. However, 22 significant MTAs were identified for $\delta^{13}C$ explaining phenotypic variation ranging from 7.81–34.77%. The $\delta^{13}C$ is considered as an indirect measure of transpiration efficiency and MTA identified in the present study can be used for improving transpiration efficiency. The heritability values ranges from 0.65 to 0.71 for $\delta^{13}C$, indicating that this can be an ideal surrogate to breed for transpiration efficiency in chickpea.

Interestingly, the MTAs reported in the present study for $\delta^{13}C$ (Fig 3a) and 100SDW (Fig 3b) were falling in the "*QTL-hotspot*" region reported on CaLG04 of intra-specific genetic map ([37]; Fig 3c), that reemphasizes the significance of QTLs detected that can be deployed in molecular breeding for trait improvement.

Candidate gene associations

A number of genes involved in plant drought responses and tolerance have been identified in model crops [38]. The candidate genes for drought tolerance defined by sequence variants and the

Table 1. Significant marker-trait associations (MTAs) identified for different traits

Trait	Number of marker trait associations	P- value range	Phenotypic variation (%)
Root traits			
Root dry weight (RDW, g plant^{-1})	2	1.16×10^{-6}–3.67×10^{-6}	8.25–10.344
Root length density (RLD, cm cm^{-3})	2	2.89×10^{-10}–9.19×10^{-8}	9.96–11.17
Root surface area (RSA, cm^2 plant^{-1})	2	5.73×10^{-7}–7.41×10^{-7}	11.95–21.71
Root volume (RV, cm^3 plant^{-1})	2	2.80×10^{-6}–2.87×10^{-6}	10.60–19.74
Rooting depth (RDp, cm)	7	1.56×10^{-11}–1.15×10^{-8}	13.12–22.41
Morphological traits			
Plant height (PHT, cm)	35	2.48×10^{-14}–4.58×10^{-6}	12.76–38.01
Shoot dry weight (SDW, g)	3	5.45×10^{-11}–2.09×10^{-6}	7.77–19.61
Apical secondary branches	5	3.46×10^{-8}–1.97×10^{-7}	12.36–17.38
Basal primary branches	4	1.44×10^{-10}–1.34×10^{-6}	11.53–19.44
Phenological traits			
Days to maturity (DM)	5	9.84×10^{-17}–3.81×10^{-6}	4.14–79.31
Flowering days (FD)	2	1.71×10^{-34}–1.61×10^{-6}	11.49–96.55
Yield related traits			
100-seed weight (100SDW, g)	70	9.53×10^{-14}–9.77×10^{-6}	8.73–36.95
Biomass (BM)	11	8.43×10^{-10}–1.22×10^{-8}	16.34–18.99
Harvest index (HI, %)	16	5.33×10^{-17}–8.24×10^{-7}	4.23–15.53
Yield (YLD)	32	6.09×10^{-16}–5.09×10^{-6}	11.43–29.03
Pod m^{-2} (POD)	10	3.27×10^{-8}–1.86×10^{-6}	9.18–22.05
Pods plant^{-1} (PPP)	6	8.12×10^{-19}–3.26×10^{-6}	8.27–50.44
Seed m^{-2} (SPM)	34	4.02×10^{-12}–4.73×10^{-6}	8.06–55.42
Seed pod^{-1} (SPP)	13	2.82×10^{-12}–4.85×10^{-6}	7.72–17.75
Per day shoot (PDS)	1	1.22×10^{-6}	10.09
Production (PROD)	3	8.12×10^{-19}–3.26×10^{-6}	18.00–50.44
Heat tolerance index (HTI)	9	1.77×10^{-12}–3.42×10^{-6}	11.17–30.59
Rate of partitioning coefficient	7	5.19×10^{-8}–4.32×10^{-6}	5.03–14.99
Total dry matter weight (TDM, g/m^2)	9	2.01×10^{-8}–4.56×10^{-6}	8.84–12.88
Transpiration efficiency			
Delta Carbon ratio ($\delta^{13}C$)	22	2.59×10^{-16}–4.68×10^{-6}	7.81–34.77
Total MTAs	312		

phenotypic data on drought and heat tolerance related traits obtained on the reference set was employed to find the association between trait and genes. A total of 113 candidate gene-based SNPs identified in 10 candidate genes were used for association analysis as result, 18 SNPs from 5 genes (*ERECTA*, 11 SNPs; *ASR*, 4 SNPs; *AMADH*, 1 SNP; *CAP2* promoter, 1 SNP and *DREB*, 1 SNP), were significantly associated with different traits. Interestingly all 11 SNPs in ERECTA gene were significantly associated with 100SDW. Furthermore, the sole SNP found in case of CAP2 promoter was detected be associated with 100SDW. AMADH192 gene-based SNP marker is significantly associated with 100SDW, explaining 25.71% phenotypic variation. The role of the gene *AMADH*, in response to stress caused by mechanical damage was evaluated by Petřivalský et al. [39]. This gene is also expected to play a role in physiological processes related to polyamine degradation, converting 4-aminobutanal to GABA [38]. Protective role of *AMADH* in response to abiotic stress is also confirmed during the present study. The *AMADH* gene is found to be associated with four stress related root traits, deciphering the importance of this gene in abiotic stress response.

The gene-based SNP markers with significant MTAs were used to identify the coordinates on the chickpea genome and the amino acid changes due the SNPs were identified (Table 2). Among 11 SNPs in ERECTA7f fragment, 2 SNPs (ERECTA7f_33 and ERECTA7f_682) were found in the CDS regions, however, none of the SNPs had any effect on the change in amino acid. While in case of *ASR* gene all three SNPs that had significant associations were in the CDS region. Two SNPs ASR_209 and ASR_261 were associated with δ13C altered the amino acid they code for. In the case of ASR_209, asparagine is changed to glutamic acid while in the case of ASR_261 valine is changed to lysine. Interestingly, asparagine synthase has been reported to be negatively correlated with most performance parameters under drought stress in case of rice [40] and the presence of glutamic acid has been reported to increase the drought tolerance. Further, enhanced lysine content has been demonstrated to possess enhanced drought tolerance in maize (http://www.echocommunity.org/resource/resmgr/a_to_z/azch3gra.htm). In addition lysine increases the chlorophyll content in leaves and thus enhances drought tolerance. Among the candidate genes used in the present study, genes like *ERECTA* and

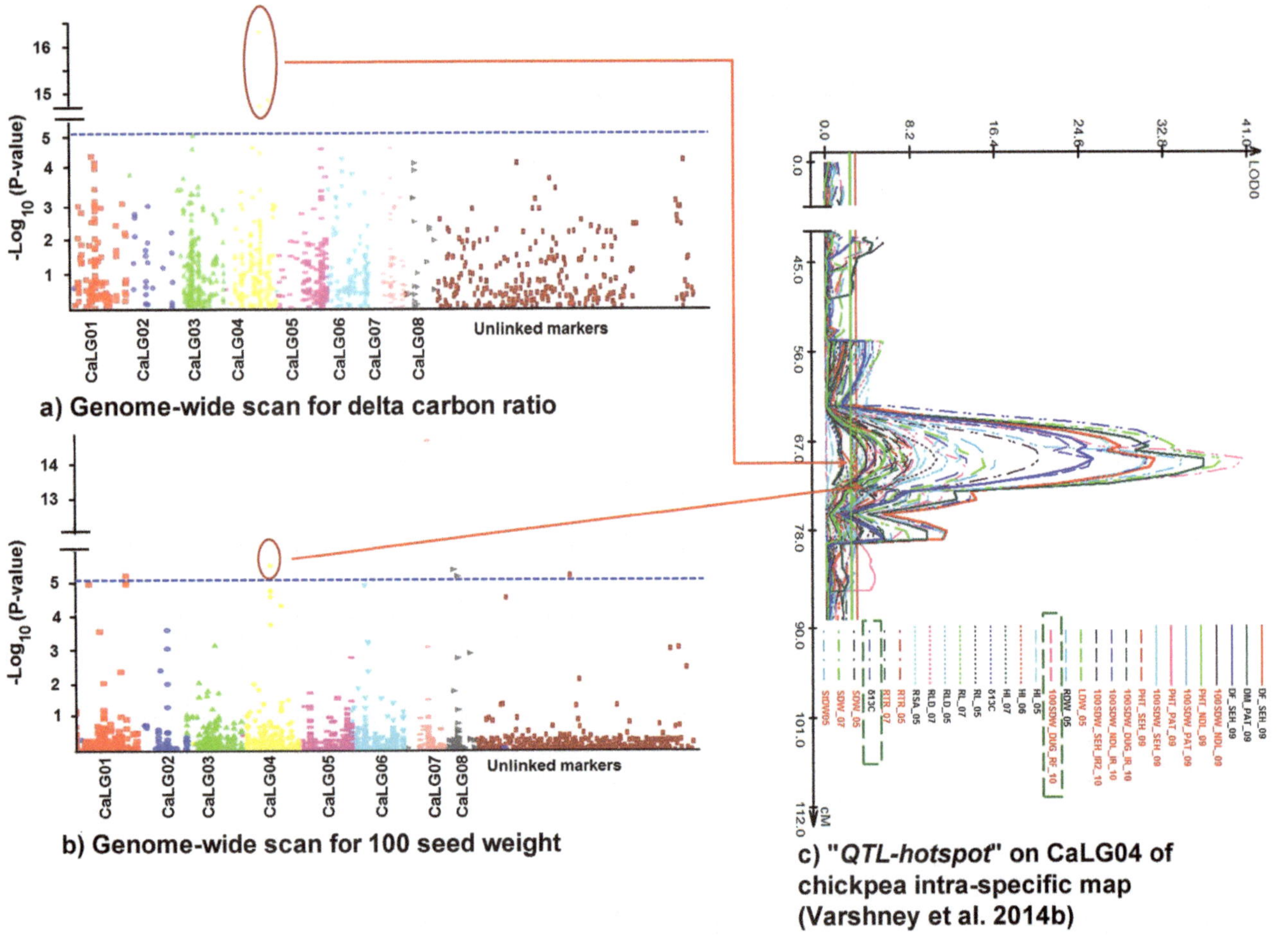

Figure 3. Significant marker trait associations (MTAs) for $\delta^{13}C$ and 100 seed weight mapped on to "*QTL-hotspot*" on CaLG04 of intra-specific map of chickpea. (a) Genome wide association scan for δ13C; the Y-axis represent $-\log_{10}(P)$ values of the P-value of the MTAs, while linkage groups are indicated on X-axis. (b) Genome wide association scan for 100SDW. (c) "*QTL-hotspot*" on CaLG04 of chickpea intra-specific genetic map harboring QTLs for drought tolerance related traits. Significant MTAs for 100SDW and δ13C falling in the QTL region are indicated using the arrows in red, the traits are indicated using dotted rectangles in green.

DREB were reported in the "*QTL-hotspot*" region on CaLG04 of chickpea [14]. In addition *ERECTA* gene was also reported to enhance transpiration efficiency [41] and water use efficiency and transpiration efficiency in *Arabidopsis* [42] and *DREB* has been reported to play important role in abiotic stress response in *Medicago* [43]. Hence, the availability of the physical map [14], consensus genetic map [15] and the MTAs established in the present study can be effectively deployed for cloning of these important genes for enhancing the drought tolerance in chickpea.

Marker-trait associations for molecular breeding

The ultimate aim of breeding is to obtain higher yields under stress conditions. In the present study, a total of 159 MTAs were identified with PVE >25% for 4 important traits. A total of 38 significant allele effects for these 8 traits were identified associated with 9 markers showing significant impact on these traits while 9 markers were found to be associated with multiple traits (Table 3). All these associated markers and identified genotypes with favorable alleles can be deployed after validation for improving above mentioned traits through molecular breeding.

Conclusion

To understand the genetic basis of tolerance to drought and heat stresses in chickpea, a comprehensive association mapping approach has been undertaken. As a result, 312 MTAs were identified and maximum number of MTAs (70) was identified for 100-seed weight. In the present study, the pairwise LD estimated using the squared-allele frequency correlations (r^2; when $r^2<0.20$) was found to decay rapidly with the genetic distance of 5 cM (Fig 2), when $r^2<0.1$ LD decay was found to decay at a genetic distance of 20 cM. Among 113 gene-based SNPs, 6 SNPs in *ASR*, 3 SNPs each in *DHN* and *DREB* were also found to have significant associations with traits like 100-seed weight, $\delta^{13}C$, plant height, root dry weight, pods per plant and yield under stressed conditions. This study provides significant MTAs for drought and heat tolerance in chickpea that can be used, after validation, in molecular breeding for developing superior varieties with enhanced drought and heat tolerance.

Table 2. Candidate gene- based marker trait associations

Trait	Marker associated	SNP position	Major allele	Minor allele	Chromos-ome	Position on Genome	Feature	Amino acid	Altered amino acid
100SDW	ERECTA7f_33	33	A	G	Ca4	44785692	CDS	Leucine	Leucine
	ERECTA7f_187	187	C	A	Ca4	44785846	Intron	-	-
	ERECTA7f_304	304	C	T	Ca4	44785949	intron	-	-
	ERECTA7f_317	317	T	C	Ca4	44785962	intron	-	-
	ERECTA7f_424	424	G	C	Ca4	44786069	intron	-	-
	ERECTA7f_558	558	G	T	Ca4	44786203	intron	-	-
	ERECTA7f_587	587	A	C	Ca4	44786232	intron	-	-
	ERECTA7f_601	601	C	T	Ca4	44786246	intron	-	-
	ERECTA7f_682	682	T	C	Ca4	44786327	CDS	Phenyl alanine	Phenyl alanine
	ERECTA7f_741	741	A	G	Ca4	44786386	intron	-	-
	ERECTA7f_883	883	A	G	Ca4	44786528	intron	-	-
RV, RSA, PODM and 100SDW	DREB_237	237	C	G	Scaffold134	275138	Non-genic	-	-
100 SDW	AMDH_192	192	C	T	Ca7	9152243	Intron	-	-
FD	CAP2prom_166	185	C	T	Scaffold134	275092	Non-genic	-	-
$\delta^{13}C$	ASR_209	209	G	A	Ca4	11451303	CDS	Asperagine	Glutamic acid
$\delta^{13}C$	ASR_261	261	G	A	Ca4	11451251	CDS	Valine	Lysine
DM	ASR_315	315	C	G	Ca4	11451197	CDS	Serine	Aspartic acid

Table 3. Significant marker trait associations with >25% phenotypic variations and their effects on the trait

Trait	Season	Marker		Allele	PVE	Effect	Mean phenotypic value		
		Name	Type				Genotypes with +ve locus	Genotypes with -ve locus	Population mean
100SDW	EIAR_RF_2008	cpPb-677136	DArT	0	35.72	3.36	43.12	19.64	19.62
100SDW	EIAR_RF_2009	Ca_Cap2promo	GB-SNP*	C:C	35.41	−9.21	25.17	16.47	19.07
100SDW	EIAR_RF_2009	CKaM0804	SNP	T:T	35.41	−8.97	25.80	16.17	19.03
100SDW	EIAR_RF_2009	DREB_237	GB-SNP	C:C	35.41	−10.90	16.51	25.36	18.33
100SDW	PAT_RF_2003	CKaM0804	SNP	C:C	31.45	7.45	21.90	14.61	16.76
100SDW	PAT_IR_2001	Ca_Cap2promo	GB-SNP	G:G	31.27	6.91	20.44	13.95	15.50
100SDW	PAT_IR_2001	CKaM0804	SNP	C:C	31.27	6.92	20.84	13.29	15.47
100SDW	PAT_IR_2008	DR_237	GB-SNP	C:C	30.47	−6.19	13.15	18.35	14.14
100SDW	PAT_RF_2001	CKaM0804	SNP	C:C	29.46	7.81	23.70	15.38	17.81
100SDW	EIAR_RF_2008	DR_237	GB-SNP	C:C	28.96	−8.18	17.28	22.90	18.47
100SDW	PAT_RF_2001	TA200	SSR	286:286	27.80	−2.80	16.38	18.21	17.91
100SDW	100SDW_RF_2008	Ca_Cap2promo	GB-SNP	C:C	27.67	−5.85	20.85	14.48	16.13
100SDW	100SDW -RF	CAP2prom98(C/G)	GB-SNP	C:C	27.30	−6.90	19.66	15.91	16.92
100SDW	100SDW_RF_2009	DR_237	GB-SNP	C:C	26.18	−5.21	15.12	22.79	16.34
PROD	PROD_MINI_ENV5	CKaM1144	SNP	A:A	50.45	1.79	5.60	3.52	5.62
SPM	SeedNom-2_IR_2008	ERECTA7f_741(A/G)	GB-SNP	A:A	55.43	975.59	2519.50	2421.90	2498.73
YLD	YKGH_MINI_ENV5	CKaM1144	SNP	A:A	28.10	148.10	643.65	430.03	641.54
YLD	Yldkghaday_RF_2009	CKaM0923	SNP	C:C	28.08	5.94	33.52	28.63	33.28
YLD	YKGH_MINI_ENV5	cpPb-491124	DArT	1	27.53	−138.96	653.49	646.54	646.46

*GB-SNP = Gene based SNP.

Materials and Methods

Plant material

A chickpea reference set, comprising of 300 accessions (including 267 landraces, 13 advanced lines and cultivars, 7 wild *Cicer* accessions, and 13 accessions with unknown biological status) defined based on molecular characterization of global composite collection [26], captured 1,315 (78%) of the 1,683 composite collection alleles was employed in the present study (Table S1). The mini-core collection of chickpea, is a subset of reference set, comprises of 211 accessions [27].

DNA isolation and quantification

The DNA was isolated from the tender leaf tissues of 15 day old seedlings as per Cuc and colleagues [44]. The quality and quantity of DNA was checked on 0.8% agarose gel using λ-DNA standard. The DNA was normalized to 5 ng/ µl for further use.

Genotyping of reference set

In addition to the genotyping data for 35 SSR markers generated previously [26], 15,360 DArT loci and 651 SNP markers using KASPar assays were genotyped on the reference set.

DArT genotyping

The reference set was genotyped with the 15,360-clone DArT arrays developed by Thudi et al. [10]. In brief, genomic representations for genotyping were prepared by the complexity reduction method described by Yang and colleagues [45]. Briefly, ca. 100 ng of DNA of reference set were digested with restriction enzymes *Pst*I and *Hae*III (New England Biolabs, USA) and the *Pst*I adapter was simultaneously ligated. One µl of restriction/ligation reaction served as a template in a 50 µl amplification reaction with *Pst*I+0 primer. Adaptor and primer sequences and cycling conditions are given in the earlier study [46]. Arrays were hybridized with fluorescently labeled targets from all genotypes used for the array development [45,46]. After overnight hybridization at 62°C, the slides were washed and scanned with a Tecan LS300 confocal laser scanner (Grödig, Salzburg, Austria). Individual samples were processed identically to samples for marker discovery and with similar marker quality thresholds in DArTsoft analysis [10].

SNP genotyping

Based on high PIC values and distribution of SNPs on the chickpea genome, 651 KASPar assays for the targeted SNPs were selected from [11] for genotyping at LGC Genomics, UK. Details on principle and procedure of the assays are available at http://www.kbioscience.co.uk/reagents/KASP_manual.pdf and http://www.kbioscience.co.uk/download/KASP.swf. SNPViewer was employed for SNP calling.

Candidate gene selection, sequencing and SNP identification

The candidate drought responsive genes were genotyped either on reference set or on the mini-core collection (Table S7). The candidate genes like abscisic acid stress and ripening gene (*ASR*), drought responsive element binding protein (*DREB*) gene, *ERECTA*, sucrose synthase (*SuSy*), sucrose phosphate synthase (*SPS*), dehydrin (*DHN*), aminoaldehyde dehydrogenase (*AMADH*), *AKIN* (SNF1 related protein kinase), MYB transcription factor, responsible for drought tolerance were identified based on prior information of involvement of the genes in drought tolerance mechanism in other crop species. In order to identify the above mentioned candidate genes in chickpea, the sequences were downloaded either from chickpea or from related legume species like *Medicago*, which is phylogenetically related to chickpea.

The candidate genes were amplified on the reference/mini-core collection. The PCR amplicons were purified using Exonuclease I and 1 U of shrimp alkaline phosphatase (SAP) per 5 µl of PCR product. The Exo/SAP added PCR products were products were incubated for 45 min at 37°C followed by denaturing at 80°C for 15 min in the thermal cycler for deactivating unused Exonuclease enzyme. The Exo/SAP treated amplicons were mixed with 1 µl of BigDye Terminator V3.1 (Applied Biosystems, California, USA), 2 µl of 5× sequencing dilution buffer and 3.2 µM of primer (forward and reverse separately) and the volume was made to 10 µl. The sequencing PCR profile included an initial denaturation of 96°C for 10 sec, 50°C for 5 sec, and 60°C for 4 min. Before sequencing, the PCR products were treated with 2.5 µl of 125 mM EDTA and 25 µl of absolute ethanol and incubated for 15 min at room temperature to precipitate the DNA. The plate containing the PCR product was centrifuged at 4000 rpm for 30 min at 4°C. The Ethanol/EDTA mix was poured off by inverting the plate, without losing the pellet. To each well, 60 µl of 70% ethanol was added and again spun at 4000 rpm for 20 min at 4°C. The ethanol was poured off as earlier. The plate was air-dried and 10 µl of HiDi formamide (Applied Biosystems, California, USA) was added and the products were denatured (94°C for 5 min, then immediately cooled to 4°C for 5 min) and sequenced using an ABI3700 automated sequencer (Applied Biosystems, California, USA). The large-scale sequencing of candidate genes across 300 genotypes of reference set was carried out at MACROGEN, Korea using BigDye terminator cycle sequencing chemistry.

The gene sequences were subjected to BLAST against chickpea reference genome assembly [1] and the genome coordinates of these genes were determined. Using the genome coordinates for each genes, the features and the amino acid changes due to the SNPs were determined using SNPeff tool (http://snpeff.sourceforge.net/).

Phenotyping of reference set/mini-core collection. In the present study, the reference set/mini-core collection was phenotyped for 34 traits in 5 locations (Patancheru, Kanpur, Nairobi, Debre Zeit and Egerton) in three countries (India, Ethiopia and Kenya), in 1–5 environments (cylinder culture, CC; rainfed, RF; irrigated, IR; normal and heat stress environments) in 2–3 replications. The detailed information on the traits phenotyped, locations and seasons are provided in Table S2.

Root traits

The reference set was phenotyped for drought tolerance related root traits (root length, RL, cm plant^{-1}; root length density, RLD, cm-cm^{-3}; root dry weight, RDW, g plant^{-1}; rooting depth, RDp, cm; root surface area, RSA, cm^{2} plant^{-1}; root volume, RV, cm^{3}; ratio between RDW, and total dry weight, RTR, %; Average diameter, AVD, mm; projected area, PRA, cm^{2} plant^{-1}) for three seasons (2007–08, 2008–09, 2010–11) in cylinder culture in three replications using semi-automated high-throughput precise phenotyping facility at ICRISAT, Patancheru as described earlier [47].

Morphological traits

The reference set/mini core collection was phenotyped for morphological traits like shoot dry weight (SDW, g plant^{-1}), for three seasons (2007–08, 2008–09, 2010–11) in cylinder culture in three replications at Patancheru. In addition, morphological traits like plant height (PHT, cm), plant width (PWD, cm), apical primary branches (APB), apical secondary branches (ASB), basal primary branches (BPB), basal secondary branches (SBS), Tertiary branches (TB) in 1–5 locations and 1–13 seasons (Table S2) in rainfed and or irrigated environments in 2–3 replications.

Phenological traits

Phenological traits like days to 50% flowering (DF), days to maturity (DM), and flowering days (FD) were phenotyped on reference/mini-core collection in 1–5 locations and 1–7 seasons (Table S2).

Yield and yield component traits

The reference set/mini-core collection was phenotyped for yield and yield related traits like seeds pod^{-1} (SPD), pods plant^{-1} (PPP), 100-seed weight (100SDW, g or kg/ha), yield (YLD, kg/ha), yield plant^{-1} (YPP, g), biomass (BM, kg), harvest index (HI, %), seeds m^{-2} (SPM), total dry matter weight (TDM, g).

In addition transpiration efficiency related traits like delta carbon ratio (δ^{13}C), and SPAD chlorophyll meter reading (during 2004–05 and 2005–06) specific leaf area were also phenotyped (Table S2).

Heat tolerance phenotyping

A set of 280 accessions from reference set were phenotyped during the post-rainy 2009–10 in two sowing dates (normal and late sowing) on a Vertisol at Patancheru and in an Inceptisol (sandy loam) at Kanpur. The alpha lattice design was adopted and evaluated in three replications at both the locations.

Data Analysis

Statistical analysis

For each trait, data was analyzed using analysis of variance (ANOVA) by using SAS General Linear Model (GLM) procedure [48] considering all effects as fixed. Along with ANOVA results, least square means (LSM; genotype as fixed effect), standard error of differences (SED), least significant difference (LSD) and descriptive statics like coefficient of variation (CV) and grand mean (GM) were calculated. Considering genotype as random, best linear unbiased predictors (BLUPs) were estimated by using SAS MIXED procedure [48]. Correlation coefficients among different traits were calculated by Karl Pearson's method using SAS CORR procedure [48]. Considered genotype as random in the statistical model and variance components were estimated for each effect, which are used for calculating heritability.

Marker attributes and diversity analysis

The marker attributes like major allele frequency, gene diversity and PIC value for all markers was computed using PowerMarker ver. 3.25 [49]. Neighbor joining trees were constructed using SSR, DArT and SNP markers data independently and combining all the data using DARwin ver. 5.0.158 [50] (http://darwin.cirad.fr/darwin). Neighbor joining trees were viewed employing Dendroscope ver. 3.2.2 [51] (http://ab.inf.uni-tuebingen.de/data/software/dendroscope3).

Population structure

A set of 85 DArT loci uniformly distributed across the chickpea genome [10] were used to understand the genetic structure and number of sub-populations in the reference set employing STRUCTURE version 2.3.1 [29] (http://pritch.bsd.uchicago.edu/structure.html) was employed. For this, the number of sub-populations (K) was presumed as 1 to 15, and each was repeated two times. For each run, burn-in and iterations were set to 1,00,000 and 2,00,000 respectively, and admixture and correlated allele frequencies was used. The run with maximum likelihood was used to assign individual genotypes into sub-population.

Association analysis

The SNP, DArT and SSR markers data set from the reference set was used to generate a matrix of similarity between each pair of genotypes in the study (the K matrix) using the program TASSEL. The Q and K matrices were used to correct for the effects of population substructure in the association panel which can cause false positive associations. Using the Q and K matrices as a covariate, markers were tested for association with each phenotype using the program TASSEL (http://www.maizegenetics.net). A Mixed Linear Model (MLM) analysis with optimum compression with P3D method [52] was used in regression, to allow for multiple testing effects. A mixed model approach implemented in EMMA 22 was used to correct the confounding of population structure.

Supporting Information

Figure S1 Population structure and genetic relationships among the chickpea reference set. a) Structure of sub-populations at different K values ranging from 2–15 b) Comparison of genetic relationships revealed by SSR, DArT and SNP markers clearly indicated three major clusters.

Table S1 Information on the association panel used in the present study.

Table S2 Summary of phenotyping data generated on reference set/mini-core collection.

Table S3 Summary statistics of root traits, morphological, phenological, transpiration efficiency related traits, yield and yield component traits evaluated on reference set and/mini-core collection.

Table S4 Major allele frequency, gene diversity and PIC content of 35 SSR loci used in the study.

Table S5 Major allele frequency, gene diversity and PIC content of 1,156 DArT loci used in the study.

Table S6 Number of alleles and PIC values of 381 SNP markers used in the study.

Table S7 Candidate genes used for association mapping.

Author Contributions

Conceived and designed the experiments: RKV. Performed the experiments: MT HDU AR PMG LK MR SNN SKC PSB NVPRG AF PK PCS MSS ST JK OI AK RKV. Analyzed the data: MT AR HDU RKV. Contributed reagents/materials/analysis tools: HDU PMG LK SKC NVPRG PK ST OI RKV. Wrote the paper: MT RKV.

References

1. Varshney RK, Song C, Saxena RK, Azam S, Sheng Y, et al. (2013) Draft genome sequence of chickpea (*Cicer arietinum*) provides a resource for trait improvement. Nat Biotechnol 31: 240–246.
2. FAOSTAT (2013) http://faostat.fao.org/site/567/DextopDEfault.aspx?PageID = 567#ancor
3. Jukanti AK, Gaur PM, Gowda CLL, Chibbar RN (2012) Chickpea: Nutritional properties and its benefits. Brit J Nutr 108:S11–S26
4. Ahmad F, Gaur PM, Croser J (2005) Chickpea (*Cicer arietinum* L.) In: Singh RJ, Jauhar PP eds. Genetic resources, chromosome engineering, and crop improvement - grain legumes. CRC Press, pp 187–217.
5. Krishnamurthy L, Kashiwagi J, Upadhyaya HD, Gowda CLL, Gaur PM, et al. (2013) Partitioning coefficient - A trait that contributes to drought tolerance in chickpea. Field Crops Res 145: 354–365.
6. Toker C, Lluch C, Tejera NA, Serraj R, Siddique KHM, et al. (2007) Abiotic stresses In: Yadav SS, Redden R, Chen W, Sharma B eds. Chickpea breeding and management. CAB International, pp 474–496.
7. Wang J, Gan YT, Clarke F, McDonald CL (2006) Response of chickpea yield to high temperature stress during reproductive development. Crop Sci 46: 2171–2178.
8. Varshney RK, Hiremath P, Lekha P, Kashiwagi J, Balaji J, et al. (2009) A comprehensive resource of drought- and salinity responsive ESTs for gene discovery and marker development in chickpea (*Cicer arietinum* L.). BMC Genomics 10: 523.
9. Nayak SN, Zhu H, Varghese N, Datta S, Choi HK, et al. (2010) Integration of novel SSR and gene-based SNP marker loci in the chickpea genetic map and establishment of new anchor points with *Medicago truncatula* genome. Theor Appl Genet 120: 1415–1441.
10. Thudi M, Bohra A, Nayak SN, Varghese N, Shah TM et al. (2011) Novel SSR markers from BAC-end sequences, DArT arrays and a comprehensive genetic map with 1,291 marker loci for chickpea (*Cicer arietinum* L.). PLoS ONE 6:e27275.
11. Hiremath PJ, Kumar A, Penmetsa RV, Farmer A, Schlueter JA, et al. (2012) Large-scale development of cost-effective SNP marker assays for diversity assessment and genetic mapping in chickpea and comparative mapping in legumes. Plant Biotechnol J 10: 716–732.
12. Varshney RK, Kudapa H, Roorkiwal M, Thudi M, Pandey MK, et al. (2012) Advances in genetics and molecular breeding of three legume crops of semi-arid tropics using next generation sequencing and high-throughput genotyping technologies. J Biosci 37: 811–820.
13. Jain M, Misra G, Patel RK, Priya P, Jhanwar S, et al. (2013) A draft genome sequence of the pulse crop chickpea (*Cicer arietinum* L.). Plant J 74: 714–719.
14. Varshney RK, Mir RR, Bhatia S, Thudi M, et al. (2014a) Integrated physical map with the genetic maps and reference genome sequence for chickpea (*Cicer arietinum* L.) improvement. Funct Integr Genomics DOI 10.1007/s10142-014-0363-6.
15. Varshney RK, Thudi M, Nayak SN, Gaur PM, Kashiwagi J, et al. (2014) Genetic dissection of drought tolerance in chickpea (*Cicer arietinum* L.) Theor Appl Genet 127: 445–461.
16. Vadez V, Krishnamurthy L, Thudi M, Anuradha C, Colmer TD, et al. (2012) Assessment of ICCV 2× JG 62 chickpea progenies shows sensitivity of reproduction to salt stress and reveals QTLs for seed yield and seed number. Mol Breed 30: 9–21.
17. Devasirvatham V, Tan DKY, Gaur PM, Raju TN, Trethowan RM (2012) High temperature tolerance in chickpea and its implications for plant improvement. Crop Pasture Sci 63: 419–428.
18. Yadav SS, Sandhu JS (2007) Area, production and distribution In: Yadav SS, Redden RJ Chen W eds. Chickpea breeding and management CABI International, pp 167–178.
19. Zhao K, Tung CW, Eizenga GC, Wright MH, Ali ML, et al. (2011) Genome-wide association mapping reveals a rich genetic architecture of complex traits in *Oryza sativa*. Nat Commun 2: 467.
20. Pasam RK, Sharma R, Malosetti M, van Eeuwijk FA, Haseneyer G, et al. (2012) Genome-wide association studies for agronomical traits in a worldwide spring barley collection. BMC Plant Biol 12: 16.
21. Morris GP, Ramu P, Deshpande SP, Hash CT, Shah T, et al. (2013) Population genomic and genome-wide association studies of agroclimatic traits in sorghum. Proc Natl Acad Sci 110: 453–458.
22. Li Y-H, Smulders MJM, Chang RZ, Qiu L-J (2011) Genetic diversity and association mapping in a collection of selected Chinese soybean accessions based on SSR marker analysis. Conserv Genet 12: 1145–1157.
23. Sakiroglu M, Sherman-Broyles S, Story A, Moore KJ, Doyle JJ, et al. (2012) Patterns of linkage disequilibrium and association mapping in diploid alfalfa (*M. sativa* L.). Theor Appl Genet 125: 577–590.
24. Shi C, Navabi A, Yu K (2011) Association mapping of common bacterial blight resistance QTL in Ontario bean breeding populations. BMC Plant Biol 11: 52.
25. Wilkening S, Chen B, Bermejo JL, Canzian F (2009) Is there still a need for candidate gene approaches in the era of genome-wide association studies? Genomics 93: 415–419.
26. Upadhyaya HD, Dwivedi SL, Baum M, Varshney RK, Udupa SM, et al. (2008) Genetic structure, diversity, and allelic richness in composite collection and reference set in chickpea (*Cicer arietinum* L.). BMC Plant Biol 8: 106.
27. Upadhyaya HD, Ortiz R (2001) A mini core subset for capturing diversity and promoting utilization of chickpea genetic resources in crop improvement. Theor Appl Genet 102: 1292–1298.
28. Roorkiwal M, Nayak SN, Thudi M, Upadhyaya HD, Brunel D, et al. (2014) Allele diversity for abiotic stress responsive candidate genes in chickpea reference set using gene based SNP markers. Front Plant Sci (Accepted)
29. Pritchard JK, Stephens M, Donnelly P (2000) Inference of population structure using multilocus genotype data. Genetics 155: 945–959.
30. Zhu C, Gore M, Buckler ES, Yu J (2008) Status and prospects of association mapping in plants. The Plant Genome 1: 15–20.
31. Somers DJ (2005) Molecular breeding and assembly of complex genotypes in wheat. Frontiers of Wheat Bioscience. The 100 memorial issue of wheat information service 235–246.
32. Lu Y, Yan J, Guimaraes CT, Taba S, Hao Z, et al. (2009) Molecular characterization of global maize breeding germplasm based on genome-wide single nucleotide polymorphisms. Theor Appl Genet 120: 93–115.
33. Cockram J, Jon W, Fiona JL, Vincent JL, Elena C et al. (2008) Association mapping of partitioning loci in barley. BMC Genet 9: 16.
34. Shehzad T, Hiroyoshi I, Kazutoshi O (2009) Genome-wide association mapping of quantitative traits in sorghum [*Sorghum bicolor* (L.) Moench] by using multiple models. Breeding Sci 59: 217–227.
35. Neumann K, Kobiljski B, Denčić S Varshney RK, Börner A (2011) Genome-wide association mapping: a case study in bread wheat (*Triticum aestivum* L.). Mol Breed 27: 37–58.
36. Gaur PM, Kumar J, Gowda CLL, Pande S, Siddique KHM, et al. (2008) Breeding chickpea for early phenology: Perspectives progress and prospects. In: Kharkwal MC ed. Food legumes for nutritional security and sustainable agriculture. Indian Society of Genetics and Plant Breeding, pp 39–48
37. Varshney RK, Thudi M, Nayak SN, Gaur PM, et al. (2014b) Genetic dissection of drought tolerance in chickpea (*Cicer arietinum* L.). Theor Appl Genet 127: 445–462.
38. Campo S, Peris-Peris C, Montesinos L, Peñas G, Messeguer J, San Segundo B (2012). Expression of the maize ZmGF14-6 gene in rice confers tolerance to drought stress while enhancing susceptibility to pathogen infection. J Exp Bot 63: 983–999.
39. Petřivalský M, Brauner F, Luhová L, Gagneul D, Sebela M (2007) Aminoaldehyde dehydrogenase activity during wound healing of mechanically injured pea seedlings. J. Plant Physiol 164: 1410–1418.
40. Degenkolbe T, Do PT, Kopka J, Zuther E, Hincha DK, et al. (2013) Identification of drought tolerance markers in a diverse population of rice cultivars by expression and metabolite profiling. PLoS ONE 8:e63637.
41. Masle J, Gilmore SR, Farquhar GD (2005) The ERECTA gene regulates plant transpiration efficiency in *Arabidopsis*. Nature 436: 866–870.
42. Xing HT, Guo P, Xia XL, Yin WL (2011) PdERECTA, a leucine-rich repeat receptor-like kinase of poplar, confers enhanced water use efficiency in *Arabidopsis*. Planta 234: 229–241.
43. Li D, Zhang Y, Hu X, Shen X, Ma L, Su Z, Wang T, Dong J (2011) Transcriptional profiling of *Medicago truncatula* under salt stress identified a novel CBF transcription factor MtCBF4 that plays an important role in abiotic stress responses. BMC Plant Biol 11: 109.
44. Cuc LM, Mace E, Crouch J, Quang VD, Long TD, et al. (2008) Isolation and characterization of novel microsatellite markers and their application for diversity assessment in cultivated groundnut (*Arachis hypogaea*). BMC Plant Biol 8: 55.
45. Yang SY, Saxena RK, Kulwal PL, Ash GJ, Dubey A, et al. (2011) The first genetic map of pigeonpea based on diversity arrays technology (DArT) markers. J Genet 90: 103–109.
46. Yang S, Pang W, Ash G, Harper J, Carling J, et al. (2006) Low level of genetic diversity in cultivated pigeonpea compared to its wild relatives is revealed by diversity arrays technology (DArT). Theor Appl Genet 113: 585–595.
47. Kashiwagi J, Krishnamurthy L, Crouch JH, Serraj R (2006) Variability of root length density and its contributions to seed yield in chickpea (*Cicer arietinum* L.) under terminal drought stress. Field Crops Res 95: 171–181.
48. SAS Institute Inc. (2008) SAS/STAT 9.2 User's Guide. Cary, NC: SAS Institute Inc.
49. Liu K, Muse SV (2005) PowerMarker: integrate analysis environment for genetic marker data. Bioinformatics 21: 2128–2129.
50. Perrier X, Flori A, Bonnot F (2003) Data analysis methods. In: Hamon P, Seguin M, Perrier X, Glaszmann JC eds. Genetic diversity of cultivated tropical plants. Enfield, Montpellier, France, pp 3–76.
51. Huson DH, Richter DC, Rausch C, Dezulian T, Franz M, et al. (2007) Dendroscope: An interactive viewer for large phylogenetic trees. BMC Bioinformatics 8: 460.
52. Zhang Z, Ersoz E, Lai C-Q, Todhunter RJ, Tiwari HK, et al. (2010) Mixed linear model approach adapted for genome-wide association studies. Nat Genet 42: 355–360.

23

Physiological Stress in Koala Populations near the Arid Edge of Their Distribution

Nicole Ashley Davies[1]*, Galina Gramotnev[2], Clive McAlpine[2], Leonie Seabrook[2], Greg Baxter[2], Daniel Lunney[3,4], Jonathan R. Rhodes[2], Adrian Bradley[1]

1 School of Biomedical Sciences, The University of Queensland, St Lucia, Queensland, Australia, **2** Landscape Ecology and Conservation Group, School of Geography, Planning and Environmental Management, The University of Queensland, St Lucia, Queensland, Australia, **3** Office of Environment and Heritage New South Wales, Hurstville, New South Wales, Australia, **4** School of Veterinary and Life Sciences, Murdoch University, Perth, Western Australia, Australia

Abstract

Recent research has shown that the ecology of stress has hitherto been neglected, but it is in fact an important influence on the distribution and numbers of wild vertebrates. Environmental changes have the potential to cause physiological stress that can affect population dynamics. Detailed information on the influence of environmental variables on glucocorticoid levels (a measure of stress) at the trailing edge of a species' distribution can highlight stressors that potentially threaten species and thereby help explain how environmental challenges, such as climate change, will affect the survival of these populations. Rainfall determines leaf moisture and/or nutritional content, which in turn impacts on cortisol concentrations. We show that higher faecal cortisol metabolite (FCM) levels in koala populations at the trailing arid edge of their range in southwestern Queensland are associated with lower rainfall levels (especially rainfall from the previous two months), indicating an increase in physiological stress when moisture levels are low. These results show that koalas at the semi-arid, inland edge of their geographic range, will fail to cope with increasing aridity from climate change. The results demonstrate the importance of integrating physiological assessments into ecological studies to identify stressors that have the potential to compromise the long-term survival of threatened species. This finding points to the need for research to link these stressors to demographic decline to ensure a more comprehensive understanding of species' responses to climate change.

Editor: Adam J. Munn, The University of Wollongong, Australia

Funding: Funding was provided by The Australian Koala Foundation, Australian Research Council and South West NRM. The funders had no role in study design, data collection and analysis, decision to publish, or preparation of the manuscript.

Competing Interests: The authors have declared that no competing interests exist.

* E-mail: n.davies1@uq.edu.au

Introduction

Stressors are pervasive and include environmental and ecological disturbances that have profound effects on the ecology and evolution of organisms [1–3]. The sources of stress include biotic factors (predation, competition, social dynamics), extremes in physical factors (temperature, salinity) and climatic factors (drought, storms) [3,4]. Shifts in the environment can be labile (i.e. eventually subside, such as cold, storms, heat and drought) or permanent, such as global climate change, urbanization, habitat degradation, pollution, with permanent changes requiring more than just temporary assimilation and they can result in changes in range, adaptations or local extinctions [5]. Stressors can impact on both physical and biotic components of an organism's environment and, depending on their pervasiveness, magnitude and frequency, can profoundly influence the fitness of individuals via costs to health, reproduction and survival [6]. Ultimately, stressors can have profound effects on population viability, species' distribution and extinction risk [7–9]. Unfavourable conditions trigger physiological responses that result in hypothalamic-pituitary-adrenal (HPA) axis activation and glucocorticoid secretion (cortisol in marsupials) by the adrenal cortex [10–13]. These glucocorticoid hormones affect physiological and behavioural traits that regulate the responses of vertebrates to stressors, and potentially offer a metric for evaluating species risk to global change [6]. Thus, physiological investigations of wild animals are vital to identify specific conservation concerns and identify threat status. Depending on the type of sample collected, they can also offer a potentially non-invasive way to monitor the efficacy of management strategies [14].

In response to a stressor, the HPA axis is activated, resulting in the increased secretion of glucocorticoids (primarily cortisol and corticosterone) from the adrenal gland (stress response) [15–19]. Glucocorticoid receptors are distributed widely in the body, thus glucocorticoid hormones can initiate a wide range of responses affecting behaviour, reproduction, growth, the immune system, and metabolism and energy allocation [20–23]. The stress response seeks to reduce the negative effects of a stressor and may cause temporary suppression or interruption of an individuals normal life history stage (such as delays in reproduction, development, migration) to redirect resources into an 'emergency' state to facilitate immediate survival [4,24–27]. When an emergency life-history stage is triggered, it allows an individual to temporarily move away from the source of disturbance (adjustment) or endure it while adopting energy-saving physiological strategies (adaptation) that allow it to cope [24]. As a consequence, glucocorticoid secretion resulting from acute stress (e.g. predator attack) improves the fitness of an organism by mobilizing energy during the stressful situation [22,28–31]. However, a stress response involving excessive glucocorticoid secretion, often termed allostatic overload or chronic stress, arises when an individual's HPA axis is challenged by excessive stressor exposure or pervasiveness resulting in fitness loss [32,33]. Evidence from the biomedical literature shows that sustained, high levels of glucocorticoids can cause deleterious effects on the physiological health of an animal and lead to greater susceptibility to disease and reduced fecundity and survivorship [12,13,20,34,35]. Accordingly, variation in the release of glucocorticoids, as a measure of the stress response, is increasingly being used in ecological and conservation studies to infer the health of animals [23] and to enable us to understand how both natural (e.g. predators, conspecifics, weather) and imposed environmental challenges (e.g. climate change, relocation and habitat disturbance) impact on individuals and populations [36–40].

While the deleterious effects of chronic stress are theoretically well conceptualized from biomedical evidence, there are few empirical studies showing fitness loss or pathology and ensuing demographic consequences of allostatic overload in wild populations [6,41]. Exceptions include a few studies on wild populations showing predation pressure causing indirect and negative glucocorticoid-mediated effects on individual fitness, resulting in decreased prey population growth [42–44]. Boonstra [41] contends that although chronic stress does occur in nature, the impacts of chronic stressors as laid out in the biomedical literature (i.e. pathology) do not occur in nature. Boonstra [41] argues that both demographic (the indirect effects on reproduction) and hormonal evidence indicate that although some wild populations may be chronically stressed by predators, there is no evidence of pathology or an HPA axis with an inability to respond. Boonstra [41] suggests that chronic stress is not the inevitable result of exposure to a persistent, severe stressor, but is part of the normal experience of many animals in nature and, although there are fitness costs, it is an evolved adaptive response to particular ecological and habitat pressures – i.e. trade-off reproduction for survival but not to such an extent that survival itself is at risk. Different views of the impacts of chronic stressors, as laid out in the biomedical literature relating to pathology and the impacts within natural populations, must be taken into consideration when assessing glucocorticoid concentrations from wild populations.

Non-invasive methods for measuring glucocorticoid metabolites in faeces and hair have become a widely accepted tool for gaining important information about an animal's endocrine status [12,45–48]. Faecal glucocorticoid metabolite analysis to assess physiological stress has been validated using different methods and employed in studies of several marsupial species including the honey possum (*Tarsipes rostratus*) [49], tammar wallaby (*Macropus eugenii*) [50], Gilbert's potoroo (*Potorous gilbertii*) [51], the southern hairy-nosed wombat (*Lasiorhinus latifrons*) [52], the numbat (*Myrmecobius fasciatus*) [53] and the koala (*Phascolarctos cinereus*) [54,55].

Unlike measurements of plasma and serum hormone concentrations that measure the actual hormone, non-invasive methods measure various metabolic end products of the hormone after it has been cleared from the circulation and extensively modified by bacteria in the gut [47,56–59]. A specific hormone is typically broken down into various metabolites during this process [59], and it is these glucocorticoid metabolites that are measured in faeces (faecal glucocorticoid metabolites) and reflect the free (unbound) glucocorticoid fraction of total glucocorticoids [60,61]. However, there are a number of caveats relating to sex, diet, metabolic rate and individual differences in hormone metabolite formation that are the source of complications that need to be considered for a meaningful interpretation of hormone metabolite data (for a review see [62,63]). Despite these caveats, it has been demonstrated that faecal glucocorticoid metabolite levels can reflect both baseline glucocorticoid levels and an animal's stress response [61], making this non-invasive technique a valuable alternative to invasive blood sampling in field studies [63]. Faecal glucocorticoid measures also provide an integrated measure of fluctuating blood concentrations in the preceding few days, thus providing a stable measure of an individual's average hormone secretion [64–66]. Furthermore, faecal glucocorticoid measures can be obtained easily without handling or observing the animal, and thereby not elevating stress levels [12,65,67,68].

As a consequence of a changing climate, distributional changes are occurring at the boundaries of geographic ranges (i.e. at the range limit) of many species, including boundary expansions at the leading edge of a range and contraction at the trailing edge [69]. Populations, in particular of folivores, at the trailing edge of their range, are likely to be most vulnerable to climate change through physiological stress associated with the decline in the nutrient richness of their food source [70,71]. Trailing edge populations can be critical to the long-term survival of species because they may contain individuals that can adapt to changing climatic conditions [72–74].

The koala (*Phascolarctos cinereus*) is an arboreal, marsupial folivore, which feeds almost exclusively on a limited variety of *Eucalyptus*, *Corymbia* and *Angophora* species. It is widely distributed, with its range extending across 30 bioregions from tropical Queensland to temperate Victoria and South Australia. The koala was listed in May 2012 under the Commonwealth's *Environmental Protection and Biodiversity Conservation Act 1999* as threatened in Queensland, the Australian Capital Territory and New South Wales. Previous research in southwestern Queensland, near the arid edge of their range, has determined that its distribution is directly affected by climatic factors (high temperature and low rainfall), which supports the view that these western koalas are near their physiological limits [75,76]. Therefore, western koalas can be expected to cope poorly with climate change.

Significant effects of season or weather conditions (e.g. temperature, humidity, food and water availability) on glucocorticoid concentrations have been shown for several mammalian species, with most studies reporting higher levels during harsher conditions in winter, extreme heat, or during the dry season [66,77–82]. Furthermore, a number of studies have attributed population declines to the effects of chronic stress in free-living mammals [36,83,84]. Additionally, a study on a long-lived seabird, the black-legged kittiwake, found that initial increases in allostatic load can evoke increased foraging behaviour and consumption, triggered by increasing glucocorticoid levels, which resulted in the accumulation of energy stores [85]. However, the metabolic effects of increased glucocorticoid levels experienced under high allostatic loads lead to a reduction of energy stores (indicating high physiological stress), with observations of decreased fitness (increased mortality, decreased reproduction) with high glucocorticoid levels in this species [85]. Throughout the koala's distribution, droughts and heatwaves have been responsible for population crashes and local extinctions [71,86–88]. In southwestern Queensland, there was an 80% decline in koala numbers between 1995 and 2009, with drought being the main contributor to the decline [88]. Climate change predictions for this region indicate that the intensity and frequency of droughts and heatwaves will increase, and moisture availability will decrease [89–91]. Therefore measuring glucocorticoid concentrations of western koala populations will help us understand how climatic variables impact these populations, and allow us to anticipate

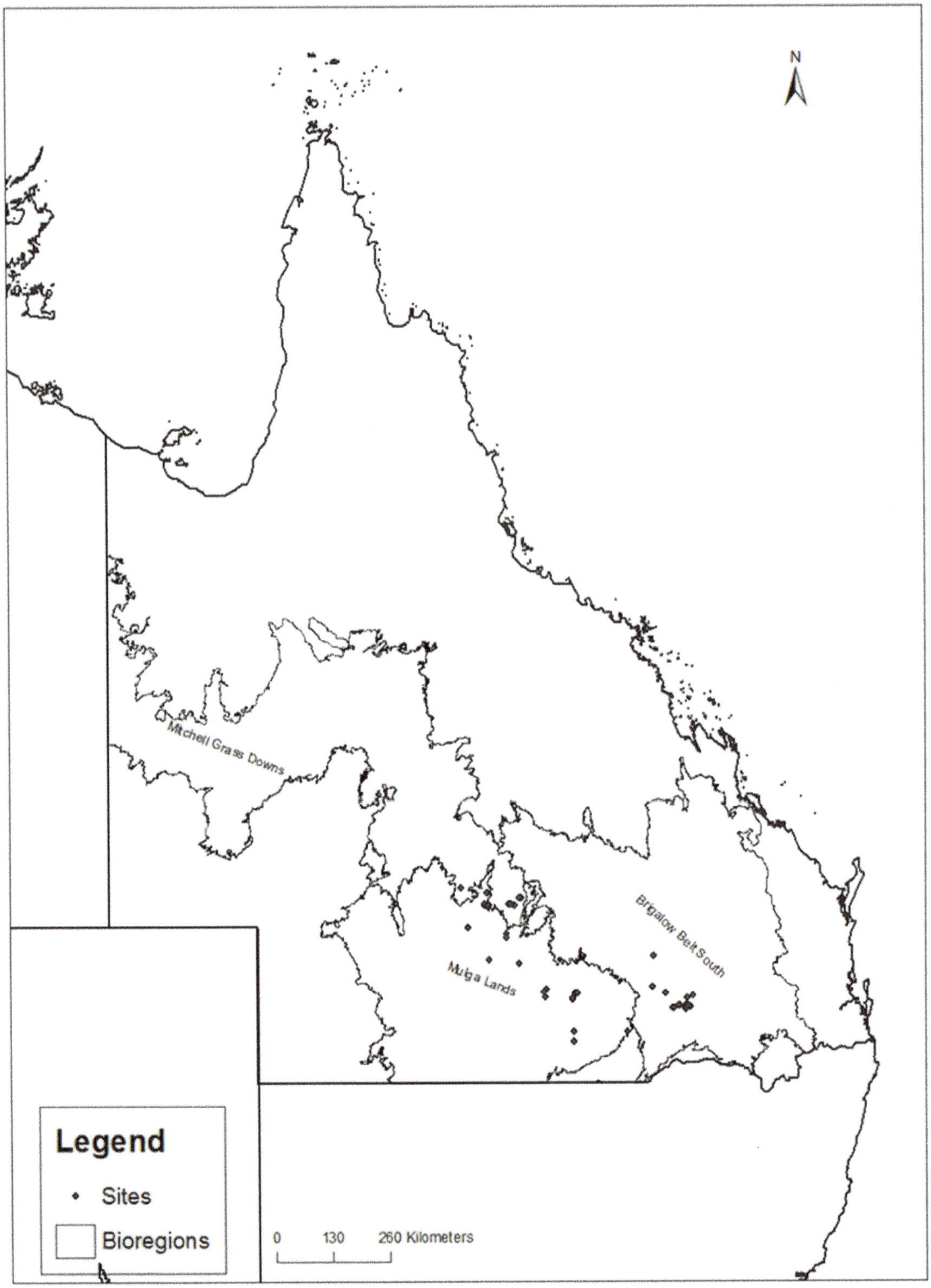

Figure 1. Study area within the Mulga Lands, Mitchell Grass Downs and Brigalow Belt South bioregions, Queensland, Australia. 'Sites' denote locations where fresh faecal pellets were collected.

where problems will arise and thereby highlight specific conservation concerns [14].

The aim of this study was to investigate the effect of climate variables on the physiological stress of koala populations of southwestern Queensland to predict the potential response of these populations to future climate change. Koalas in this region are at the western limits of their geographic range and form a trailing-edge population [88]. We used non-invasive techniques to measure glucocorticoid levels. Mixed effects modeling was used to relate the current physiological conditions of koalas to current climate and environmental conditions. These relationships were then used to estimate potential physiological conditions based on predicted future climates.

Materials and Methods

The approved protocol for this project was approved by the Animal Ethics Committee at The University of Queensland (GPA/603/08/ARC) and the Environmental Protection Agency (now the Department of Environment and Resource Management) (Scientific Purposes Permit WISP05343008).

Study Area

This study was conducted in semi-arid, southwestern Queensland, comprising portions of the Mulga Lands bioregion, the Mitchell Grass Downs bioregion and the Brigalow Belt South bioregion (Figure 1). Annual average rainfall ranges from 750 mm in the east declining to 250 mm in the west. Rain falls mainly in summer and is highly variable [92]. The hottest month is January with mean monthly maximum and minimum temperatures of approximately 35°C and 19°C respectively. The coolest month is July with mean maximum and minimum temperatures of approximately 19°C and 4°C respectively [92].

The Mulga Lands bioregion is dominated by flat to undulating plains and low ranges supporting *Acacia aneura* (mulga) shrubland and low woodlands [93]. The woodlands that dominate waterways and associated floodplains are predominantly comprised of *Acacia* spp., *Eucalyptus populnea* (poplar box), *E. camaldulensis* (river red gum), *E. coolabah* (coolabah) and *E. orchophloia* (yapunyah) [93].

The western portion of the Brigalow Belt South bioregion is predominantly comprised of *Acacia harpophylla* (brigalow), *Casuarina cristata* (belah) and *E. populnea* open-forest woodland [93]. Riparian vegetation is dominated by *E. camaldulensis*, *E. coolabah* and *E. largiflorens* (black box) [94].

The Mitchell Grass Downs bioregion is dominated by treeless plains of *Astrebla spp.* (Mitchell grass) with some occasional ridges, rivers and gorges. Patches of low open woodland of *E. coolabah* occur across the region in association with low-lying plains and drainage lines [94].

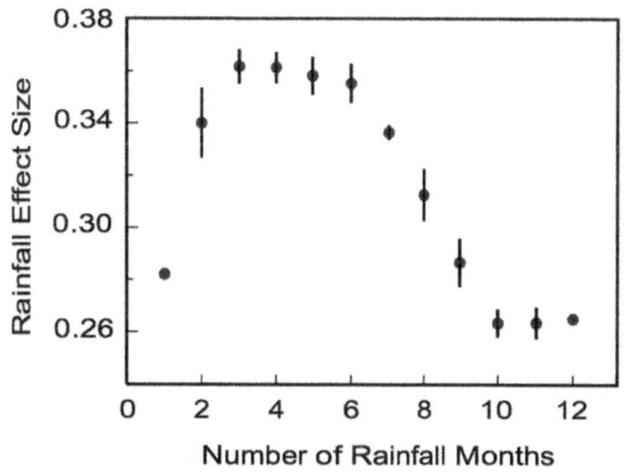

Figure 2. The moving average dependence of the effect size (standardized coefficients in Table 2) for the rainfall variable versus number of months with rain prior to data collection. The moving interval over which averaging is undertaken contains three consecutive values of the coefficients from Table 2. The first and the last points are added to this dependence directly from Table 2 (because these edge points are not included into the moving average dependence with the three-point moving interval). Error bars calculated as the errors of the mean within the moving interval are shown for each of the moving average points. Note, this figure shows the magnitude of the effect size, and the actual effect of rainfall on cortisol levels is negative.

Faecal Pellet Collection

Searches were conducted to collect fresh koala faecal pellets for cortisol analysis in 29 sites across the three bioregions (Mulga Lands 13 sites; Brigalow Belt South 11 sites; Mitchell Grass Downs 5 sites) (Figure 1) between August 2009 and June 2011. Morning (07:00–10:00 hours), afternoon (16:00–18:00 hours) and night (spotlighting, 19:00–04:00 hours) searches were conducted in a systematic manner along transects (1 km by 80 m) within riparian

Table 1. Explanatory variables used to examine variation in faecal cortisol concentrations.

Variable	Units	Full description
Sex	Categorical	Sex of each koala from which faecal samples were collected.
Total Nitrogen	Kjeldahl	Soil total nitrogen content at each site - derived from geology mapping of Queensland (The Queensland combined soils dataset).
Phosphorous	Kjeldahl	Soil phosphorous content at each site - derived from geology mapping of Queensland (The Queensland combined soils dataset)
Annual rainfall	Millilitres	Average annual rainfall (1990–2011) for each site.
Rainfall 1–12 months	Millilitres	Total rainfall with monthly resolution at each site for 1–12 months prior to each faecal sample collected.
Temperature	Degrees Celsius	Average minimum and maximum temperatures at each site the day prior to sample collection.
Bioregion	Categorical	Bioregion (Mulga Lands, Mitchell Grass Downs, Brigalow Belt South)
Catchment	Categorical	Catchment (Mungallala, Warrego, Condamine, Moonie)
Basin	Categorical	Basin (Warrego, Moonie, Balonne-Condamine)
Drought or post-flood	Categorical	Categorized based on whether each sample was collected during the drought (2009) or post-flood event (2010 onwards).
Season	Categorical	Categorized based on whether each sample was collected during the koala breeding season (September to March) or non-breeding season (April–August).

Table 2. Statistical effects of rainfall for one to 12 months prior to sample collection on faecal cortisol concentration (Coefficients from standardized models (effect sizes)).

Independent Variable	Coefficient	Standard error	P
Rainfall 1 month prior	−0.28	0.11	0.02
Rainfall 2 months prior	−0.391	0.095	<0.001
Rainfall 3 months prior	−0.347	0.096	<0.001
Rainfall 4 months prior	−0.348	0.097	<0.001
Rainfall 5 months prior	−0.39	0.11	<0.001
Rainfall 6 months prior	−0.34	0.10	0.001
Rainfall 7 months prior	−0.34	0.11	0.002
Rainfall 8 months prior	−0.34	0.11	0.001
Rainfall 9 months prior	−0.27	0.11	0.013
Rainfall 10 months prior	−0.26	0.10	0.013
Rainfall 11 months prior	−0.26	0.10	0.01
Rainfall 12 months prior	−0.27	0.10	0.01

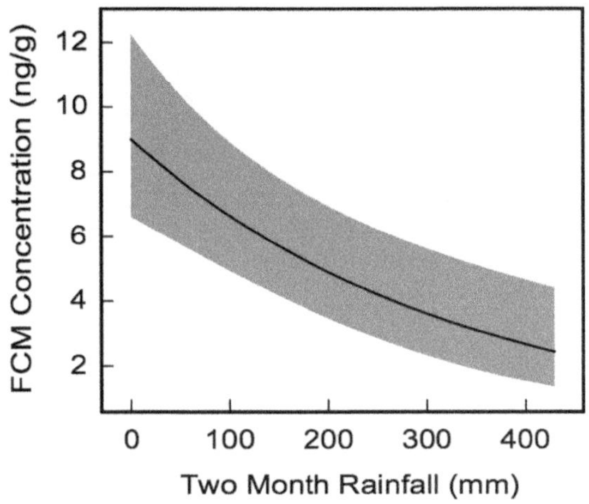

Figure 3. Faecal cortisol metabolite (FCM) concentration versus rainfall during the two months before the collection. The grey band shows the 95% prediction interval. The total rainfall for the previous 4 months (months 3 to 6) is assumed to equal its average value for that period of 310 mm.

habitats. Previous research had identified riparian habitats as the primary resource used by koalas in southwestern Queensland [88,95]. If a koala was sighted along a 1 km transect, then an extra 1 km was surveyed to locate additional koalas. Once a koala was located, sex was recorded and faecal pellets were collected by searching the entire area beneath the canopy of the tree in which the koala was located. Fresh, intact pellets were placed in clean containers using forceps and stored in the field in a liquid nitrogen dry shipper (MCE SC 20/12 V Vapour Shipper, In Vitro Technologies Pty. Ltd) for transport to the laboratory. The use of faecal cortisol analysis in the assessment of the activity of the HPA axis in freshly collected koala faecal pellets has previously been validated using the methods described below [54]. That research showed an excretion time lag with faecal cortisol metabolite concentrations corresponding to the release of glucocorticoids from the adrenal cortex at least 36 hours prior to deposition [54]. Other studies have also successfully assayed cortisol metabolites in koala faeces (with an ACTH challenge), with time lags of 24 and 48 hours in females and males respectively [55].

Extraction and Analysis

Faecal pellets from individual koalas were oven dried at 70°C for 10 hrs. The dried samples were crushed between two 10 mm PVC plates, passed through a 1 mm sieve to remove leaf particles, and collected into 5 ml flat bottom polypropylene screw capped vials. For each sample, 200 mg (±1 mg accuracy) of dry faecal powder was weighed into a 16×100 mm glass test tube. Borate buffer 2 ml (pH 6.5 0.1 M) was added to the dry powder, vortex mixed and then 50 μl of β glucuronidase (β-D-Glucuronoside glucuronosohydrolase, EC 3.2.1.31, Sigma Chemical Co. USA) containing approximately 4,000 units was added to each test tube to cleave glucuronide and sulfatase polar groups. Test tubes were then incubated for 4 hrs at 37°C on an orbital mixer. Redistilled diethyl ether 3 ml was added to each test tube and vortexed for 2 minutes, then the test tubes were allowed to stand for 2 minutes. The lower aqueous phase was frozen by briefly dipping the test tubes into liquid nitrogen and the supernatant ether containing the partitioned steroid was decanted into 12×75 mm glass test tubes and evaporated to dryness at 40°C in a hot block evaporator in a fume hood. The residue containing extracted steroid was re-dissolved in 200 μl of diluted zero cortisol calibration solution (Demeditec Saliva Free Cortisol Kit - diluted 1:10) and placed on an orbital mixer at 37°C for 60 minutes, followed by short, high-speed vortex (20 seconds). 100 μl of test samples, standards and controls were then pipetted into 96 well plates of Saliva Free Cortisol Kits (Demeditec) and run as per kit directions. The efficiency of the extraction process was progressively tracked by addition of 30,000 dpm ^{3}H-cortisol (1,2,6,7 ^{3}H cortisol 160 Ci/mmol Perkin-Elmer Life Sciences) and the final cortisol assay concentration was corrected for this efficiency. This assay allowed us to measure cortisol after first cleaving polar bonds in the major known metabolites (faecal cortisol metabolite: FCM), because it has been reported that there are no native, unmetabolized glucocorticoids present in faeces [60].

The specificity of the assay was: cortisol 100%, corticosterone 0.38%, cortisone 1.85%, 11-deoxycortisol 0.88%, prednisolone 9.89%, with all other steroids of similar structure <0.20%. Inter-assay variation was 5.4% CV, while intra-assay variation was 5.5% CV. Analytical sensitivity of the assay is 0.014 ng/ml or 4.2 pg per loaded sample. Serial dilutions of glucuronidase-treated koala faecal extracts run against Demeditec assay kit calibrator standards gave a satisfactory degree of parallelism for the assay. Assay data were analysed employing a four parameter logistic fit using MyAssays Analysis Software Solutions (www.myassays.com).

Explanatory Variables

Table 3. Statistical effects of rainfall 2 months prior and rainfall 3, 4, 5 and 6 months prior (combined) on faecal cortisol concentration (standardized).

Independent Variable	Coefficient	Z	P
Rainfall 2 months prior	−0.2647747	−4.11	0.000
Combined rainfall for months 3, 4, 5 & 6 prior	−0.1715594	−2.51	0.012

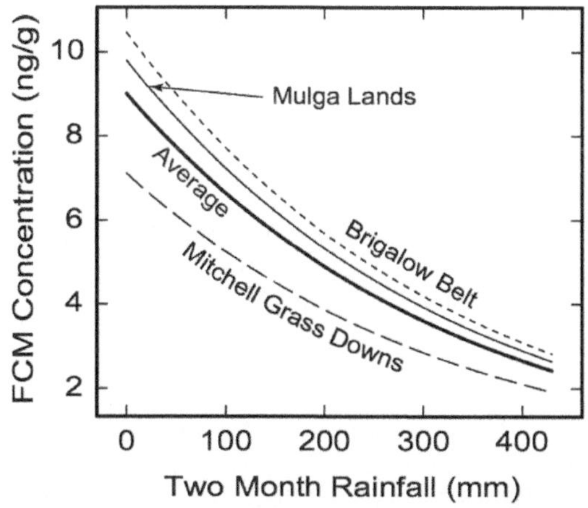

Figure 4. Regional variation in the relationship between faecal cortisol metabolite (FCM) concentration and rainfall during the two months before sample collection. The thick solid curve shows the cortisol levels averaged over all three bioregions combined (the same as the curve in Figure 3). The other three curves represent the average FCM levels for each of the bioregions: Brigalow Belt (small dashed curve), Mitchell Grass Downs (dashed curve), and Mulga Lands (thin solid curve). The total rainfall for the previous 4 months (months 3 to 6) is assumed to equal its average value of 310 mm.

The explanatory variables included sex, climate data, soil characteristics and location (Table 1). Climate data from the Bureau of Meteorology (1990–2011) were examined for stations closest to each site (within the same bioclimatic region and <30 km apart). Minimum and maximum average temperatures were considered for each sample corresponding to the day prior to sample collection to take into account the excretion time lag [54] (Table 1). Monthly rainfall was calculated because of time lags between rainfall and the response by vegetation [96–98]. Soil properties (e.g. total nitrogen and phosphorus) at each site were derived from geology mapping of Queensland (The Queensland Combined Soils dataset) at a scale of 1:250,000 to 1:200,000,000 (Table 1). Each site was also categorized by bioregion, catchment, basin, drought or post-flood event and season (Table 1).

Statistical Analysis

Statistical analyses were conducted using the Stata statistical software package [99] to determine the dependencies between the following variables: FCM concentration (ng/g) (dependent variable) and rainfall, temperature, soil nitrogen, soil phosphorus, sex, time of the year (month), and season (breeding/non-breeding) (independent variables – Table 1). As was demonstrated from the Shapiro-Wilk test for normality of data, the FCM (dependent) variable was not distributed normally ($p<0.001$). Therefore, the FCM variable was transformed logarithmically to achieve a normal distribution of the dependent variable. The explanatory variables were all standardized to have a zero mean and a standard deviation of one to allow comparison between the model parameter estimates.

Data were collected at 29 sites across the three bioregions. Therefore, we assumed a hierarchically structured population with random sampling of the regions, sites and subjects within the sites. Multilevel (three-level) mixed-effects linear regressions with random intercepts were used to assess the effect of each explanatory variable on koala FCM levels. The three levels within the considered model were associated with: (1) individual koala; (2) the site of data collection; (3) the region where the site was located. The random effects model determined the dependency of FCM levels on a particular rainfall variable (e.g., two month rainfall) for each site and/or region. Thus the random effects within the model are random intercepts and/or slopes for the respective regression lines of FCM levels versus a rainfall variable for each region and/or site. Though the model does not directly give the values of the random effects (random intercepts and/or slopes), their best linear unbiased predictions, also known as empirical Bayes predictions, can be calculated for both region and site levels [100].

Results

All explanatory variables (Table 1), other than the covariates related to rainfall, were omitted from the considered model, as they had no significant impact on the levels of FCM ($p>0.1$). FCM concentration was assessed against rainfall for 1 to 12 months prior to sample collection (Table 2). The moving average dependence of the magnitudes of the standardized coefficients (Table 2) on the number of rainfall months is illustrated by Figure 2. Changes in FCM levels have a typical time lag of about 2–6 months after rain. This follows from the maximum of the curve in Figure 2, between ~2 and 6 months. Further, the points corresponding to 4, 5 and 6 months in Figure 2 show the effect size for the 2 months rainfall. This means that the effect size for months 4, 5 and 6 is enhanced by the effect size for the 2 months rain. Thus, the strongest impact on FCM levels comes from the rainfall between ~2–3 months prior to the collection time.

Therefore, the model for the impact of rainfall on levels of FCM included two independent variables: rainfall for two months prior to the time of collection of the faecal pellets and rainfall for four months (3 to 6 months) prior to the time of collection. The variable of rainfall for months 3 to 6 prior to time of collection was included because it takes into account the decaying (but still extant) impact of rainfall during earlier months before the two-month period to the time of collection and its statistical significance (with the standardized coefficient being –0.25 and $p = 0.012$) identifies the long period over which the effect is evident (Table 3).

We did not find significant, non-linear quadratic terms in the dependence of FCM levels on two-month rainfall. Further, since comparing the fit of the random intercept model to that of the multiple regression model, using the likelihood ratio test yielded $p = 0.01$, we can thus reject the null hypothesis that the intercept is the same across all the regions and sites, and the random intercept model is valid and essential. At the same time, the likelihood ratio tests comparing the random slope model with the model that contains only random intercepts did not show a statistically significant difference for both the region or site levels (the p-values are close to 1 for all these tests). As a consequence, the consideration of random slopes was not necessary, so we limited our model only to the random intercepts.

The intra-class correlation, interpreted as a ratio of the variance of FCM levels due to different sites (or bioregions) to the total variance of FCM, determines the contribution of sites (or bioregions) to the total variance of FCM. That the intra-class correlation is equal to ~0.26 for the site level, and ~0.11 for the bioregion level means that the contribution of different sites to the variability of cortisol was significantly larger than that of different bioregions, indicating particularly strong clustering of the data with respect to the collection sites.

The dependence of the average FCM level on rainfall for the two-month period prior to the collection, adjusted for rainfall during the following months 3 to 6, is significant and demonstrates a monotonic decrease of FCM levels with increasing two-month rainfall (Figure 3). For a two-month rainfall of 400 mm, with the total rainfall for the previous 4 months being equal to its average value of 310 mm, the average FCM level is ~2.66 ng/g. If the two-month rainfall decreased from 400 mm to 50 mm, the FCM levels increased ~3 times to ~7.73 ng/g (Figure 3).

The typical differences between the cortisol dependences for different bioregions are characterized by significantly lower levels of FCM in koalas at any level of rainfall (Figure 4). The model of the influence of rainfall on FCM concentration in koalas can account for ~10% of the total FCM variability. The relatively small fraction of the total FCM variability, explained by the rainfall variables, does not mean that this effect is insignificant. As demonstrated, the results are statistically significant with low p-values. Further, the contribution of the rainfall variables is comparable with the variability of FCM by region (also ~10%) and by site (~14%). This means that the other ~66% of FCM variability is due to other unexplained factors.

To further assess the fit of the models, 12 quantile-quantile (Q-Q) plots of the standardized residuals were considered for rainfall during the periods from 1 month to 12 months prior to sample collection. These plots further demonstrated the validity of the considered models (Figure S1).

Discussion

Chronic stress can reduce resistance to disease and affect population performance (including impacting negatively on survival and reproduction) [35,101–105], thus understanding the physiological responses of populations exposed to environmental changes is critical to conservation. We show that higher FCM levels in koala populations near the arid edge of their range in western Queensland were associated with lower rainfall levels, consistent with physiological stress when moisture levels are low. These results are of particular concern to the continued survival of western koala populations, especially when a number of studies have attributed population declines of free-living mammals to the effects of chronic stress [36,83,84]. Droughts and heatwaves have been responsible for population crashes and local extinctions in a number of koala populations [71,86–88]. Predictions indicate that the intensity and frequency of droughts and heatwaves will increase and moisture availability will decrease [89–91]. Our findings show that western koalas, at the trailing edge of their geographic range, will cope poorly with changes in rainfall patterns resulting from climate change.

Arid and semi-arid landscapes are temporally and spatially variable, heterogeneous ecosystems where ecological processes are primarily driven by climate and nutrients [106]. Previous research within these arid and semi-arid landscapes has determined that the distribution, density, habitat preferences and home range sizes of koalas are affected either by water availability (including leaf moisture) and/or rainfall [75,76,86,107,108]. The findings from the current research have also demonstrated the impact of rainfall on koala populations by showing that low rainfall negatively impacts on FCM levels. In particular, rainfall from the previous two months had the strongest influence on FCM levels. Time lags can occur between rainfall events and the subsequent response by vegetation, with previous studies showing that total rainfall over the preceding two months has a strong effect on vegetation growth [96–98]. This indicates that the influence of rainfall on FCM levels in koalas is likely to be caused by the improved leaf moisture and nutrition in the form of new growth resulting from the rainfall. The delay of the new growth after a rainfall causes the consequential time lag of about two months between the rainfall and the maximum reduction in FCM levels in koalas. Foliar moisture supplies most of the koala's water requirements. It has been proposed that in dry environments or during drought water, rather than leaf nutrients, influences tree selection by koalas [107,109]. Therefore, rainfall two months prior is an essential determinant of leaf moisture content, and hence food quality, which in turn impacts on FCM concentrations of southwestern Queensland koalas. Results from this study are consistent with studies on African elephants, muriqui monkeys and spider monkeys [77,79,110]. These studies reported FCM concentrations significantly higher in the dry versus the wet season as food and/or water availability declined [77,79,110]. For example, FCM concentrations of free-ranging African elephants were significantly higher in the dry season as food, water availability and body condition declined and were inversely correlated with rainfall across seasons [79]. The authors also noted that elephant diets change from high-quality grass in the wet season to less nutritious and more sparsely distributed bark and leaves in the dry season [111,112]. Spider monkeys in conserved forests also show seasonal variation in faecal cortisol concentrations, with higher concentrations measured in the dry season compared to the wet season, when there is a reduction in the number of fruiting trees and hence food availability [110]. The same study also found that mean cortisol concentrations of spider monkeys living in fragmented habitat remained high throughout the year, indicating that individuals within these habitats may suffer from long-term stress [110]. In addition, serum cortisol was significantly higher in camels in drought-affected areas than camels in other areas [113].

When comparing cortisol concentrations in relation to prior rainfall, FCM concentrations of koalas from the Mitchell Grass Downs were below the combined bioregion average, whereas FCM concentrations of koalas from the Mulga Lands and Brigalow Belt South were above the average. The dominant soil type of sites within the Mitchell Grass Downs (Vertosol) has high water holding capacity and very high soil fertility compared to the dominant soils of the Mulga Lands (Kandosol) and Brigalow Belth South (Sodosol, Chromosol and Vertosol) bioregions [114–116]. The difference in soil water holding capacity and fertility among the regions could explain why the Mitchell Grass Downs koalas had lower FCM concentrations, i.e. trees within this region are likely to have higher leaf moisture and forage quality as a result of the higher water holding capacity and fertility of the soil. A previous study within southwestern Queensland found that there was a significant relationship between leaf moisture and soil types, although once soil types were divided into groups based on water availability, the authors suggested the relationship could be ascribed to water availability in sites [117]. Furthermore, in this study, soil properties had no significant impact on the levels of FCM. Therefore, although soil water-holding capacity and/or fertility may explain the lower FCM concentrations of the Mitchell Grass Downs compared to the other two bioregions, rainfall two months prior is most likely the principal determinant of leaf moisture content and quality, regardless of soil type.

Significant effects of temperature on glucocorticoid concentrations have been shown for several mammalian species, with most studies reporting higher levels during harsher conditions in winter or during extreme heat [118,119]. For example, a significant negative relationship was found between minimum ambient temperature and FCM levels of red deer [118]. In Marwari sheep, both hot ambient temperatures (maximum of 43–46°C) and drought conditions resulted in a significant increase in serum cortisol [119]. However, we found no impact of minimum and maximum ambient temperatures on FCM levels of koalas. Behavioural adjustment and physiological adaptations of koalas living near the arid edge of their range help mitigate the physiological stress of extreme temperatures. In central Queensland, the temperature in trees used by koalas for shelter in summer can be around 2°C cooler than the ambient temperature [71], and in Gunnedah, in north-western New South Wales, tree choice was found to be mediated by temperature [120]. Furthermore, the pelt of the koala has the highest recorded insulation (including a high resistance to wind) for a marsupial [121]. The dorsal surface is more densely furred and less reflective of solar radiation than the ventral surface [121]. Degabriele [121] suggested that different postural adjustments, particularly in relation to wind direction, allow for insulation capabilities, and that the field water influx and low metabolic rate of koalas are likely to be adaptations to their arboreal, low-energy folivorous lifestyle that extends over a wide latitudinal range. However, throughout the koala's distribution, droughts and heatwaves have been responsible for population crashes and local extinctions [71,86–88]. Although the metabolic and thermoregulatory adaptations of koalas help mitigate the effect of temperature on physiological stress, the studies to date identify that there is a threshold that, once crossed, cause population crashes. In this study, due to the relatively small number of samples collected at the lower and higher temperature extremes, this influence of temperature was not detected.

Variation in the physiological stress response, that influences the concentration of fecal GCMs, has been reported to reflect reproductive status or phase (i.e. oestrous, gestation, lactation) for a number of mammalian species [81,122–124]. Glucocorticoids are also known to rise significantly near term in mammalian species because they trigger the cascade resulting in parturition [12,125]. While these large changes in plasma cortisol occur in most eutherian mammals during gestation, the different endocrine control of parturition following the well known short gestation period in marsupials is not associated with any large changes in maternal plasma cortisol [126]. Consistently we did not detect differences between sexes or reproductive condition in this study. It is possible that during drought years, reproduction might have been suspended or reduced in southwestern Queensland koala populations, while as rainfall increases a normal pattern of reproduction might have been found. Indeed anecdotal evidence suggests such a scenario with only one mother and joey found during koala searches during the drought compared with numerous observations of mothers with joey's post-flood. However, we found the opposite with higher FCM during low rainfall periods (consistent with moisture stress) compared with periods of increased rainfall when females might be in reproductive condition. Narayan *et al* [55] reported that FCM levels of captive koalas significantly differed by sex, reproductive condition (lactating versus non-lactating koalas) and handling groups; however, *post hoc* comparisons revealed that although FCM levels were significantly different between the sexes and handling groups, there was no significant differences between the reproductive conditions. Narayan *et al* [55] also reported no significant difference in FCM levels between wild and captive koalas, with *post hoc* comparisons further supporting the suggestion that within each environment (captive or wild) there was a significant effect of sex but not reproductive condition on the levels of FCM. Sex differences in either plasma free cortisol or fecal cortisol were not apparent in a study on captive koalas by Davies *et al* [54], however, these animals were not lactating and not known to be pregnant at the time of the study. Because of the large difference between eutherian and metatherian reproductive endocrine control, the large changes in maternal plasma cortisol that occur in most eutherians during pregnancy do not occur during the short gestation in marsupials and fetal cortisol is more important for initiating parturition and enabling development and maturation of organs (in particular the lungs) in preparation for parturition [126]. Since this study measures a downstream excretory product of hepatic clearance of plasma free cortisol, not total cortisol, sex differences are likely to be small. This adds further support to the suggestion that rainfall, and the subsequent leaf moisture and/or leaf nutritional content, is the principal determinant of FCM levels of southwestern koala populations. However, future studies should address possible differences between sexes and reproductive status and to do this thoroughly, steroid partitioning in the blood plasma must be assessed enabling calculation of free cortisol.

Sustained high levels of glucocorticoids from chronic stress (e.g. from long-term climate change) can be deleterious to the physiological health of an animal and lead to greater susceptibility to disease, as well as reduced fecundity and survivorship [12,13,20,34,35]. While the deleterious effects of chronic stress are well conceptualized from biomedical evidence, there is some doubt whether or not these deleterious effects, especially relating to pathology, occur in nature [41,127]. Boonstra [41] argues that both demographic (the indirect effects on reproduction) and hormonal evidence indicate that although some wild populations may be chronically stressed by predators, there is no evidence of pathology or an HPA axis with an inability to respond. Boonstra [41] suggests that chronic stress is not the inevitable result of a persistent, severe stressor, but is part of the normal experience of many animals in nature and, although there are fitness costs, it is an evolved adaptive response to particular ecological and habitat pressures. While this may be the case for predation stressors, there is no evidence in wild populations whether this adaptive response can be applied to other severe chronic stressors of a different nature, such as climate change or habitat loss and fragmentation. For species that have populations threatened by disease, chronic stress acting via well known effects on the immune system, can be expected to increase susceptibility to disease and threaten the long-term viability of populations. Koalas are subject to a range of diseases, which can lead either to death or infertility [128,129]. Infectious diseases in koalas are recognised to threaten the long-term viability of populations in Queensland and New South Wales [129]. In northwestern New South Wales, the prevalence of the disease chlamydiosis was found to increase dramatically following intense stress from extreme hot weather (intense heatwaves) during a drought [87]. Following a review of studies of chlamydial disease and population decline, Lunney *et al.* [87] also linked chlamydiosis to koala populations under stress from habitat loss and fragmentation. Furthermore, Gordon [86] reported that on dry stretches of creek in southwestern Queensland koalas showed poorer health (poor condition, anemia, high tick loads) and suffered much higher mortality, i.e. population crashes, during drought compared to sites where trees were not affected by the drought (mainly on large permanent waterholes). We found low rainfall resulted in high cortisol levels, demonstrating that droughts have the potential to impact on the physiological stress of individual koalas as a result of reduced food and water availability. It is increasingly likely exposure to these environmental stressors will increase their susceptibility to disease. However, although there is anecdotal evidence reporting that stress and disease prevalence in various koala populations throughout its distribution are related, the link between chronic stress, disease and ensuing demographic impact needs detailed examination before any causal links can relate chronic stress to pathology. Our results provide new insights into the response of koalas to stressors, such as drought and heatwaves, and highlight the conservation concerns for the continued survival of western koala populations in light of predicted climate change.

Implications for Conservation

Drought and heatwaves have the potential to impact on physiological stress in individual koalas as these extreme climatic conditions can affect the quality of nutrients and moisture available in the diet of entire populations [130,131]. Observed FCM concentration patterns in relation to rainfall levels strongly suggest that physiological stress in western koalas is being adversely affected by limited access to food and water. These western Queensland koala populations are at the semi-arid edge of their distribution, where drought and heatwaves have resulted in population declines [86,88], which confirms that koalas in this semi-arid region are near their physiological tolerance limits. Climate change predictions for this region indicate that the intensity and frequency of droughts and heatwaves will increase, evapotranspiration will increase, and that foliar moisture will decrease [89–91]. This highlights conservation and management concerns for the continued survival of koala populations of western Queensland.

Investigating FCM levels at the trailing edge of a widely distributed species boundary, and how these levels vary with climatic factors, allows us to non-invasively identify changes likely to cause stress. It also allows us to identify the habitats that are least susceptible to climate change as priority sites for protection, as well as to facilitate management decisions to prevent further contractions in a species' distribution, and monitor the efficacy of management strategies. In southwestern Queensland, riparian habitat is the primary resource used by koalas [88,95,108], based on evidence of populations declining and contracting to riparian habitats during drought and heatwaves [86,88]. It follows that since the moisture content of leaves in trees within riparian habitats would be higher, it makes them higher quality habitat for koalas. Considering that low rainfall, and hence low leaf moisture, resulted in higher FCM levels, it places even more importance on maintaining and improving the quality of riparian habitats for continued survival of these western koala populations, particularly during extreme weather. Historical land management practices have diminished riparian habitats from the silting of previously-permanent water-holes [86]. For the long-term conservation of western koala populations, fencing of riparian habitats to keep out feral animals and domestic stock should be considered. This will help to reduce the silting of waterholes and allow the recruitment of food trees. However, reduced grazing could also increase weed species, which could negatively impact on koalas and other native wildlife. This points to the need for close monitoring and control of weed infestations. Previous studies have also found water availability and rainfall to be the main factors defining habitat preference of koalas in arid and semi-arid regions [75,107]. Maintaining access to freestanding water, such as farm dams, and planting favoured tree species, particularly those that tend to have higher leaf moisture content, such as *E. camadulensis* and *E. coolabah* [117]), close to dams would also increase the availability of high quality trees, as these trees are likely to have higher leaf moisture levels. Therefore, to ensure the continued survival of western Queensland koala populations in extreme weather, the most important actions for koala conservation within these semi-arid landscapes are the maintenance of the quality and quantity of riparian habitats, expanding the availability and accessibility of freestanding water, such as around farm dams, and providing specific trees as food resources.

Conclusion

Ecological studies are increasingly using the release of glucocorticoids as a measure of the stress response in animals. Using non-invasive methods to measure the glucocorticoid metabolites in faeces, we found that koalas had higher FCM levels during low rainfall periods. An analysis of mechanisms for the impact of rainfall on FCM levels in koalas is beyond the scope of this paper, but one mechanism could be through the new growth and higher foliar moisture content of eucalypt leaves during periods of rainfall. This would provide koalas with an abundance of food and water causing reductions of their cortisol levels. These results provide new insights into the effect of droughts and heat waves on physiological stress levels and highlight conservation concerns for the continued survival of western Queensland koala populations in the face of predicted climate change. Our results demonstrate the importance of integrating a physiological dimension into ecological studies to ensure a more comprehensive understanding of species' responses to habitat loss, particularly riparian habitat, and the increasing threat of relentless climate change.

Supporting Information

Figure S1 Q-Q plots of the standardized residuals for the models for (a) rainfall 1 month prior to sample collection, (b) rainfall 2 months prior to sample collection, (c) rainfall 3 months prior to sample collection, (d) rainfall 4 months prior to sample collection, (e) rainfall 5 months prior to sample collection, (f) rainfall 6 months prior to sample collection, (g) rainfall 7 months prior to sample collection, (h) rainfall 8 months prior to sample collection, (i) rainfall 9 months prior to sample collection, (j) rainfall 10 months prior to sample collection, (k) rainfall 11 months prior to sample collection, and (l) rainfall 12 months prior to sample collection.

Acknowledgments

We are extremely grateful to all of the volunteers that have assisted with the koala surveys, with special thanks to Grace Davies, Stuart Davies, Will Goulding, Kailas Wild, Ladislas Parraud and Huiying Wu. We would like to thank all of the landholders who allowed koala surveys to be conducted on their land, their hospitality and knowledge was very much appreciated.

Author Contributions

Conceived and designed the experiments: ND CM LS GB DL JR AB. Performed the experiments: ND AB. Analyzed the data: ND GG. Contributed reagents/materials/analysis tools: AB. Wrote the paper: ND GG CM LS GB DL JR AB.

References

1. Badyaev AV (2005) Stress-induced variation in evolution: from behavioral plasticity to genetic assimilation. Proceedings of the Royal Society of London Series B: Biological Sciences 272: 877–886.
2. Badyaev AV (2005) Role of stress in evolution: from individual adaptability to evolutionary adaptation. In: Hallgrimsson B, Hall BK, editors. Variation: A Central Concept in Biology. Amsterdam: Elsevier Academic Press. 277–302.
3. Hoffmann AA, Hercus MJ (2000) Environmental stress as an evolutionary force. BioScience 50: 217–226.
4. Wingfield JC, Maney DL, Breuner CW, Jacobs JD, Lynn S, et al. (1998) Ecological bases of hormone-behavior interactions: The "emergency life history stage". American Zoology 38: 191–206.
5. Wingfield JC (2013) Ecological processes and the ecology of stress: the impacts of abiotic environmental factors. Functional Ecology 27: 37–44.
6. Jessop TS, Woodford R, Symonds MRE (2013) Macrostress: do large-scale ecological patterns exist in the glucocorticoid stress response of vertebrates? Functional Ecology 27: 120–130.

7. Bijlsma R, Loeschcke V (2005) Environmental stress, adaptation and evolution: an overview. Journal of Evolutionary Biology 18: 744–749.
8. Hoffmann AA, Parsons PA (1991) Evolutionary Genetics and Environmental Stress. Oxford, UK: Oxford University Press.
9. Hoffmann AA, Parsons PA (1997) Extreme Environmental Change and Evolution. Cambridge, UK: Cambridge University Press.
10. Seyle H (1935) A syndrome produced by diverse nocuous agents. Nature 138: 32–33.
11. Axelrod J, Reisine TD (1984) Stress hormones: their interaction and regulation. Science 224: 452–459.
12. Mostl E, Palme R (2002) Hormones as indicators of stress. Domestic Animal Endocrinology 23: 67–74.
13. Stewart PM (2003) The adrenal cortex. In: Larsen PR, Kronenberg HM, Melmed S, Polonsky KS, editors. Williams Textbook of Endocrinology. Philadelphia: Saunders. 491–551.
14. Wikelski M, Cooke SJ (2006) Conservation physiology. Trends in Ecology and Evolution 21: 38–46.
15. Turner CD, Bagnara JT (1976) General Endocrinology. 6th edition. Philadelphia: W. B. Saunders.
16. Moberg GP (1987) Problems in defining stress and distress in animals. Journal of American Veterinary Medical Association 191: 1207–1211.
17. McEwen BS, Schmeck HM (1994) The hostage brain. New York: Rockefeller University Press.
18. Matteri RL, Carrol JA, Dyer CJ (2000) Neuroendocrine responses to stress. In: Moberg GP, Mench JA, editors. The Biology of Animal Stress: CAB International. 43–76.
19. Keay JM, Singh J, Gaunt MC, Kaur T (2006) Fecal glucocorticoids and their metabolites as indicators of stress in various mammalian species: a literature review. Journal of Zoo and Wildlife Medicine 37: 234–244.
20. Sapolsky RM, Romero LM, Munck AU (2000) How do glucocorticoids influence stress responses? Integrating permissive, suppressive, stimulatory, and preparative actions. Endocrine Reviews 21: 55–89.
21. Dallman MF, Bhatnagar S (2001) Chronic stress and energy balance: role of the hypo-thalamo-pituitary axis. In: McEwen BS, Goodman HM, editors. Handbook of Physiology: Section 7: The Endocrine System, Coping with the Environment: Neural and Endocrine Mechanisms. New York: Oxford University Press. 179–210.
22. Sapolsky RM (2002) Endocrinology of the stress response. In: Becker JB, Breedlove SM, Crews D, McCarthy MM, editors. Behavioural Ecology. Cambridge: The MIT Press. 409–450.
23. Romero LM (2004) Physiological stress in ecology: lessons from biomedical research. Trends in Ecology and Evolution 19: 249–255.
24. Wingfield JC (2003) Control of behavioural strategies for capricious environments. Animal Behaviour 66: 807–816.
25. Wingfield JC (2005) The concept of allostasis: coping with a capricious environment. Journal of Mammalogy 86: 248–254.
26. Wingfield JC, Kitaysky AS (2002) Endocrine responses to unpredictable environmental events: stress or anti-stress hormones? Integrative and Comparative Biology 42: 600–609.
27. Romero LM, Butler LK (2007) Endocrinology of stress. International Journal of Comparative Psychology 20: 89–95.
28. Raynaert R, De Paepe M, Peeters G (1976) Influence of stress, age and sex on serum growth hormone and free fatty acids in cattle. Hormone Metabolism Research 8: 109–114.
29. Lee AK, Cockburn A (1985) Evolutionary Ecology of Marsupials. Cambridge: Cambridge University Press.
30. Kenagy GJ, Place NJ (2000) Seasonal changes in plasma glucocorticosteroids of free-living yellow pine chipmunks: effects of reproduction and capture and handling. Endocrinology 117: 189–199.
31. Mateo JM (2006) Developmental and geographic variation in stress hormones in wild Belding's ground squirrels (*Spermophilus beldingi*). Hormones and Behavior 50: 718–725.
32. McEwen BS, Wingfield JC (2003) The concept of allostasis in biology and biomedicine. Hormones and Behavior 43: 2–15.
33. McEwen BS, Wingfield JC (2010) What is in a name? Integrating homeostasis, allostasis and stress. Hormones and Behavior 57: 105–111.
34. Lane J (2006) Can non-invasive glucocorticoid measures be used as reliable indicators of stress in animals? Animal Welfare 15: 331–342.
35. Munck A, Guyre PM, Holbrook NJ (1984) Physiological functions of glucocorticoids in stress and their relation to pharmacological actions. Endocrinology Review 5: 25–44.
36. Boonstra R, Singleton GR (1993) Population declines in the snowshoe hare and the role of stress. General and Comparative Endocrinology 91: 126–143.
37. Creel S, Fox JE, Hardy A, Sands J, Garrott B, et al. (2002) Snowmobile activity and glucocorticoid stress responses in wolves and elk. Conservation Biology 16: 809–814.
38. Sheriff MJ, Bosson CO, Krebs CJ, Boonstra R (2009) A non-invasive technique for measuring fecal cortisol metabolites in snowshoe hares (*Lepus americanus*). Journal of Comparative Physiology B 179: 305–313.
39. Thiel D, Jenni-Eiermann S, Braunisch V, Palme R, Jenni L (2008) Ski tourism affects habitat use and evokes a physiological stress response in capercaillie *Tetrao urogallus*: a new methodological approach. Journal of Applied Ecology 45: 845–853.
40. Wasser SK, Bevis K, King G, Hanson E (1997) Noninvasive physiological measures of disturbance in the northern spotted owl. Conservation Biology 11: 1019–1022.
41. Boonstra R (2013) Reality as the leading cause of stress: rethinking the impact of chronic stress in nature. Functional Ecology 27: 11–23.
42. Boonstra R, Hik D, Singleton GR, Tinnikov A (1998) The impact of predator-induced stress on the snowshoe hare cycle. Ecological Monographs 68: 371–394.
43. Creel S, Winnie JA, Christianson D (2009) Glucocorticoid stress hormones and the effect of predation risk on elk reproduction. Proceedings of the National Academy of Sciences USA 106: 12388–12393.
44. Zanette LY, White AF, Allen MC, Clinchy M (2011) Perceived predation risk reduces the number of offspring songbirds produce per year. Science 334: 1398–1401.
45. Davenport MD, Tiefenbacher S, Lutz CK, Novak MA, Meyer JS (2006) Analysis of endogenous cortisol concentrations in the hair of rhesus macaques. General and Comparative Endocrinology 147: 255–261.
46. Koren L, Mokady O, Karaskov T, Klein J, Koren G, et al. (2002) A novel method using hair for determining hormonal levels in wildlife. Animal Behaviour 63: 403–406.
47. Palme R, Rettenbacher S, Touma C, El-Bahr S, Mostl E (2005) Stress hormones in mammals and birds: comparative aspects regarding metabolism, excretion, and noninvasive measurement in faecal samples. Annals of the New York Academy of Sciences 1040: 162–171.
48. Young KM, Walker SL, Lantheir C, Waddell WT, Monfort SL, et al. (2004) Noninvasive monitoring of adrenocortical activity in carnivores by fecal glucocorticoid analyses. General and Comparative Endocrinology 137: 148–165.
49. Oates JE, Bradshaw FJ, Bradshaw SD, Stead-Richardson EJ, Philippe DL (2007) Reproduction and embryonic diapause in a marsupial: insights from captive honey possums, *Tarsipes rostratus* (Tarsipedidae). General and Comparative Endocrinology 150: 445–461.
50. McKenzie S, Deane EM (2005) Faecal corticosteroid levels as an indicator of well-being in the tammar wallaby, *Macropus eugenii*. Comparative Biochemistry and Physiology, Part A 140: 81–87.
51. Stead-Richardson EJ, Bradshaw D, Friend T, Fletcher T (2010) Monitoring reproduction in the critically endangered marsupial, Gilberts potoroo (*Potorous gilbertii*): preliminary analysis of faecal oestradiol-17B, cortisol and progestagens. General and Comparative Endocrinology 165: 155–162.
52. Hogan LA, Johnston SD, Lisle AT, Keeley T, Wong P, et al. (2011) Behavioural and physiological responses of captive wombats (*Lasiorhinus latifrons*) to regular handling by humans. Applied Animal Behavioral Science 134: 217–228.
53. Hogan LA, Lisle AT, Johnston SD, Robertson H (2012) Non-invasive assessment of stress in captive numbats, *Myrmecobius fasciatus* (Mammalia: Marsupialia), using faecal cortisol measurement. General and Comparative Endocrinology 179: 376–383.
54. Davies N, Gillett A, McAlpine C, Seabrook L, Baxter G, et al. (2013) The effect of ACTH upon faecal glucocorticoid excretion in the koala. Journal of Endocrinology 219: 1–12.
55. Narayan EJ, Webster K, Nicolson V, Mucci A, Hero J-M (2013) Non-invasive evaluation of physiological stress in an iconic Australian marsupial: The Koala (*Phascolarctos cinereus*). General and Comparative Endocrinology 187: 39–47.
56. Taylor W (1971) The excretion of steroid hormone metabolites in bile and feces. Vitamins and Hormones 29: 201–285.
57. Goymann W (2005) Non-invasive monitoring of hormones in bird droppings: biological validations, sampling, extraction, sex differences, and the influence of diet on hormone metabolite levels. Annals of the New York Academy of Sciences 1046: 35–53.
58. Klasing KC (2005) Potential impact of nutritional strategy on non-invasive measurements of hormones in birds. Annals of the New York Academy of Sciences 1046: 5–16.
59. Palme R (2005) Measuring faecal steroids: guidelines for practical application. Annals of the New York Academy of Sciences 1046: 75–80.
60. Touma C, Palme R (2005) Measuring faecal glucocorticoid metabolites in mammals and birds: the importance of validation. Annals of the New York Academy of Sciences 1046: 54–74.
61. Sheriff MJ, Krebs CJ, Boonstra R (2010) Assessing stress in animal populations: Do fecal and plasma glucocorticoids tell the same story? General and Comparative Endocrinology 166: 614–619.
62. Goymann W (2012) On the use of non-invasive hormone research in uncontrolled, natural environments: the problem with sex, diet, metabolic rate and the individual. Methods in Ecology and Evolution 3: 757–765.
63. Sheriff MJ, Dantzer B, Delehanty B, Palme R, Boonstra R (2011) Measuring stress in wildlife: techniques for quantifying glucocorticoids. Oecologia 166: 869–887.
64. Goymann W, Mostl E, Van't Hof T, East ML, Hofer H (1999) Noninvasive fecal monitioring of glucocorticoids in spotted hyenas, *Crocuta crocuta*. General and Comparative Endocrinology 114: 340–348.
65. Harper JM, Austad SN (2000) Fecal glucocorticoids: A nonivasive method of measuring adrenal activity in wild and captive rodents. Physiological and Biochemical Zoology 73: 12–22.
66. Harper JM, Austad SN (2001) Effect of capture and season on fecal glucocorticoid levels in Deer Mice (*Peromyscus maniculatus*) and Red-Backed

Voles (*Clethrionomys gapperi*). General and Comparative Endocrinology 123: 337–344.
67. Millspaugh JJ, Washburn BE (2003) Within-sample variation of fecal glucocorticoid measurements. General and Comparative Endocrinology 132: 21–26.
68. Monfort SL, Mashburn KL, Brewer BA, Creel S (1998) Evaluating adrenal activity in African wild dogs (*Lycaon ictus*) by fecal corticosteroid analysis. Journal of Zoo and Wildlife Medicine 29: 129–133.
69. Parmesan C (2006) Ecological and evolutionary responses to recent climate change. Annual Review of Ecology Evolution and Systematics 37: 637–669.
70. DeGabriele JL, Moore BD, Marsh KJ, Foley WJ (2010) The effect of plant secondary metabolites on the interplay between the internal and external environments of marsupial folivores. Chemoecology 20: 97–108.
71. Ellis W, Melzer A, Clifton ID, Carrick F (2010) Climate change and the koala *Phascolarctos cinereus*: water and energy. Australian Zoologist 35: 369–377.
72. Hampe A, Petit RJ (2005) Conserving biodiversity under climate change: the rear edge matters. Ecology Letters 8: 461–467.
73. Thomas CD, Franco AMA, Hill JK (2006) Range retractions and extinction in the face of climate warming. Trends in Ecology and Evolution 21: 415–416.
74. Wilson RJ, Thomas CD, Fox R, Roy DB, Kunin WE (2004) Spatial patterns in species distributions reveal biodiversity change. Nature 432: 393–396.
75. Sullivan BJ, Baxter GS, Lisle AT (2003) Low-density koala (*Phascolarctos cinereus*) populations in the mulgalands of south-west Queensland. III. Broad-scale patterns of habitat use. Wildlife Research 30: 583–591.
76. Sullivan BJ, Norris WM, Baxter GS (2003) Low-density koala (*Phascolarctos cinereus*) populations in the mulgalands of south-west Queensland. II. Distribution and diet. Wildlife Research 30: 331–338.
77. Strier KB, Ziegler TE, Wittwer DJ (1999) Seasonal and social correlates of fecal testosterone and cortisol levels in wild male muriquis (*Brachyteles arachnoides*). Hormone Behaviour 35: 125–134.
78. Beehner JC, McCann C (2008) Seasonal and altitudinal effects on glucocorticoid metabolites in a wild primate (*Theropithecus gelada*) Physiology and Behavior 95: 508–514.
79. Foley CAH, Papageorge S, Wasser S (2001) Noninvasive stress and reproductive measures of social and ecological pressures in free-ranging African Elephants. Conservation Biology 15: 1134–1142.
80. Megahed G, Anwar M, Wasfy S, Hammadeh M (2008) Influence of heat stress on the cortisol and oxidant-antioxidants balance during oestrous phase in buffalo-cows (*Bubalus bubalis*): thermo-protective role of antioxidant treatment. Reproduction in Domestic Animals 43: 672–677.
81. Weingrill T, Gray D, Barrett L, Henzi S (2004) Fecal cortisol levels in free ranging female chacma baboons: relationship to dominance, reproductive state and environmental factors. Hormones and Behavior 45: 259–269.
82. Ziegler T, Scheffler G, Snowdon C (1995) The relationship of cortisol levels to social environment and reproductive functioning in female cotton-top tamarins, *Saguinus oedipus*. Hormones and Behavior 29: 407–424.
83. Charbonnel N, Chaval Y, Bertheir K, Deter J, Morand S, et al. (2008) Stress and demographic decline: a potential effect mediated by impairment of reproduction and immune function in cyclic vole populations. Physiological and Biochemical Zoology 81 63–73.
84. Pride ER (2004) High faecal glucocorticoid levels predict mortality in ring-tailed lemurs (*Lemur catta*). Biology Letters 1 60–63.
85. Schultner J, Kitaysky AS, Welcker J, Hatch S (2013) Fat or lean: adjustment of endogenous energy stores to predictable and unpredictable changes in allostatic load. Functional Ecology 27: 45–55.
86. Gordon G (1988) A koala (*Phascolarctos cinereus* Goldfuss) population crash during drought and heatwave conditions in south-western Queensland. Australian Journal of Ecology 13: 451–461.
87. Lunney D, Crowther MS, Wallis I, Foley WJ, Lemon J, et al. (2012) Koala populations and climate change: a case study on the Liverpool Plains, north-west NSW. In: Lunney D, Hutchings P, editors. Wildlife and Climate Change: Towards Robust Conservation Strategies for Australian Fauna. Mosman, NSW, Australia: Royal Zoological Society of NSW. 150–168.
88. Seabrook L, McAlpine C, Baxter G, Rhodes J, Bradley A, et al. (2011) Drought-driven change in wildlife distribution and numbers: a case study of koalas in south west Queensland. Wildlife Research 38: 509–524.
89. CSIRO (2007) Climate change in Australia - Technical Report 2007. Canberra: Commonwealth Scientific and Industrial Research Organisation.
90. Cai W, Crimp S, Jones R, McInnes KL, Durack PJ, et al. (2005) Climate change in Queensland under enhanced greenhouse conditions: Report, 2004–2005.: CSIRO Marine and Atmospheric Research report for the Queensland Government.
91. IPCC (2007) Climate Change 2007: The Physical Science Basis. Contribution of Working Group I to the Fourth Assessment Report of the Intergovernmental Panel on Climate Change - Summary for Policymakers.
92. Bureau of Meteorology (2013) Maps of average conditions. Canberra: Australian Government Bureau of Meteorology.
93. Sattler P, Williams R (1999) The conservation status of Queensland's bioregional ecosystems. Brisbane: Environmental Protection Agency, Queensland Government.
94. Thackway R, Cresswell ID (1995) An Interim Biogeographic Regionalisation for Australia: a framework for establishing the national system of reserves, Version 4.0. Canberra: Australian Nature Conservation Agency.
95. Smith AG, McAlpine C, Rhodes JR, Seabrook L, Baxter G, et al. (2012) At what spatial scales does resource selection vary? A case study of koalas in a semi-arid region.. Austral Ecology doi: 10.1111/j.1442-9993.2012.02396.x.
96. Malo AR, Nicholson SE (1990) A study of rainfall dynamics in the African Sahel using normalized difference vegetation index. Journal of Arid Environment 19: 1–24.
97. Davenport ML, Nicholson SE (1993) On the relation between rainfall and the normalized vegetation index for diverse vegetation types in East Africa. International Journal of Remote Sensing 14: 2369–2389.
98. Karabulut M (2003) An examination of relationships between vegetation and rainfall using maximum value composite AVHRR-NDVI data. Turkish Journal of Botany 27: 93–101.
99. StataCorp (2011) Stata Statistical Software: Release 12. College Station, TX: StataCorp LP.
100. Henderson CR (1975) Best linear unbiased estimation and prediction under a selection model. Biometrics 31: 423–447.
101. Moberg GP (1985) Animal Stress; Moberg GP, editor. Bethesda, Maryland: American Physiological Society.
102. Liptrap RM (1993) Stress and reproduction in domestic animals. Annual New York Academic Science 697: 275–284.
103. Dobson H, Smith RF (1995) Stress and reproduction in farm animals. Journal of Reproductive Fertility Supplement 49: 451–461.
104. Millspaugh JJ, Washburn BE (2004) Use of fecal glucocorticoid metabolite measures in conservation biology research: considerations for application and interpretation. General and Comparative Endocrinology 138: 189–199.
105. Busch SD, Hayward LS (2009) Stress in a conservation context: A discussion of glucocorticoid actions and how levels change with conservation-relevant variables. Biological Conservation 142: 2844–2853.
106. Stafford Smith M, McAllister RRJ (2008) Managing arid zone natural resources in Australia for spatial and temporal variability - an approach from first principles. Rangeland Journal 30: 15–27.
107. Munks SA, Corkrey R, Foley WJ (1996) Characteristics of Arboreal Marsupial habitat in the semi-arid woodlands of northern Queensland. Wildlife Research 23: 185–195.
108. Davies N, Gramotnev G, Seabrook L, Bradley A, Baxter G, et al. (2013) Movement patterns of an arboreal marsupial at the edge of its range: a case study of the koala. Movement Ecology 1.
109. Melzer A, Lamb D (1996) Habitat utilisation by a central Queensland koala colony. In: Gordon G, editor. Koalas: research for management. Brisbane, Queensland: World Koala Research. 17–22.
110. Rangel-Negrin A, Alfaro JL, Valdez RA, Romano MC, Serio-Silva JC (2009) Stress in Yucatab spider monkeys: effects of environmental condition on fecal cortisol levels in wild and captive populations. Animal Conservation 12: 496–502.
111. Western D, Lindsay WK (1984) Seasonal herd dynamics of a savanna elephant population. African Journal of Ecology 22: 229–244.
112. Spinage CA (1994) Elephants. London: T & A D Poyser.
113. Kataria N, Kataria AK (2004) Use of blood analytes in assessment of stress due to drought in camel. Journal of Camel Practice and Research 11: 129–133.
114. Australian Soil Resource Information System (2011) Atlas of Australian Soils. Australia: CSIRO.
115. Department of Environment and Resource Management (2012) Soils in Queensland. Queensland Government.
116. Gary JM, Murphy BW (2002) Predicting Soil Distribution, Joint Department of Land and Water Conservation (DLWC) and Australian Society for Soil Science Technical Poster. Sydney: DLWC.
117. Wu Y, McAlpine CA, Seabrook LM (2012) The dietary preferences of koalas, *Phascolarctos cinereus*, in southwest Queensland under wet winter conditions preceded by a severe drought. Australian Zoologist 36: 93–102.
118. Huber S, Palme R, Arnold W (2003) Effects of season, sex, and sample collection on concentrations of fecal cortisol metabolites in red deer (*Cervus elaphus*). General and Comparative Endocrinology 130: 48–54.
119. Kataria N, Kataria AK (2006) Endocrine and metabolic responses of Marwari sheep in arid tract. Slovenian Veterinary Research 43: 135–142.
120. Crowther MS, Lunney D, Lemon J, Stalenberg E, Wheeler R, et al. (In press) Climate-mediated habitat selection in an arboreal folivore. Ecography.
121. Degabriele R, Dawson TJ (1979) Metabolism and heat balance in an arboreal marsupial, the koala. Journal of Comparative Physiology B 134: 293–301.
122. Cavigelli SA (1999) Behavioural patterns associated with faecal cortisol levels in free-ranging female ring-tailed femurs, *Lemur catta*. Animal Behavior 57: 935–944.
123. Cavigelli SA, Monfort SL, Whitney TK, Mechref YS, Novotny M, et al. (2005) Frequent serial fecal corticoid measures from rats reflect circadian and ovarian corticosterone rhythms. Journal of Endocrinology 184: 153–163.
124. Goymann W, East ML, Wachter B, Honer OP, Mostl E, et al. (2001) Social, state-dependent and environmental modulation of faecal corticosteroid levels in free-ranging female spotted hyenas. Proceedings of the Royal Society of London, B Biological Science 268: 2453–2459.
125. Möstl E, Maggs JL, Schrötter G, Besenfelder U, Palme R (2002) Measurement of cortisol metabolites in faeces of ruminants. Veterinary Research Communications 26: 127–139.
126. Young IR, Renfree MB, Mesiano S, Shaw G, Jenkin G, et al. (2011) The Comparative Physiology of Parturition in Mammals: Hormones and Parturi-

tion in Mammals. In: Norris DO, Lopez KH, editors. Hormones and Reproduction of Vertebrates, Volume 5 - Mammals. USA: Academic Press.
127. Boonstra R (2013) The ecology of stress: a marriage of disciplines. Functional Ecology 27: 7–10.
128. Gordon G, McGreevy D, Lawrie B (1990) Koala populations in Queensland: major limiting factors. In: Lee AK, Handasyde KA, Sanson GD, editors. Biology of the Koala. Chipping Norton, NSW Australia: Surrey Beatty. 85–95.
129. Hanger J, Loader J (2009) Infectious disease in koalas: implications for conservation. Paper presented at the Lismore Koala Conservation Conference. Lismore.
130. Cork SJ, Braithwaite LW (1996) Resource availability, eucalypt chemical defences, and habitat quality for leaf eating marsupials. In: Gordon G, editor. Koalas: Research for Management Brisbane Koala Symposium 22–23 September 1990. Brisbane: World Koala Research. 9–16.
131. Moore BD, Foley WJ (2000) A review of feeding and diet selection in koalas (*Phascolarctos cinereus*). Australian Journal of Zoology 48: 317–333.

Geographical Pattern and Environmental Correlates of Regional-Scale General Flowering in Peninsular Malaysia

Shinya Numata[1]*, Masatoshi Yasuda[2], Ryo O. Suzuki[3], Tetsuro Hosaka[1], Nur Supardi Md. Noor[4], Christine D. Fletcher[4], Mazlan Hashim[5]

1 Department of Tourism Science, Tokyo Metropolitan University, Hachiouji, Tokyo, Japan, **2** Kyushu Research Center, Forestry and Forest Products Research Institute, Kumamoto, Kumamoto, Japan, **3** Sugadaira Montane Research Center, University of Tsukuba, Ueda, Nagano, Japan, **4** Forestry Division, Forest Research Institute Malaysia, Kepong, Selangor, Malaysia, **5** Institute of Geospatial Science and Technology, Universiti Teknologi Malaysia, UTM, Skudai, Johor Bahru, Malaysia

Abstract

In South-East Asian dipterocarp forests, many trees synchronize their reproduction at the community level, but irregularly, in a phenomenon known as general flowering (GF). Several proximate cues have been proposed as triggers for the synchronization of Southeast Asian GF, but the debate continues, as many studies have not considered geographical variation in climate and flora. We hypothesized that the spatial pattern of GF forests is explained by previously proposed climatic cues if there are common cues for GF among regions. During the study, GF episodes occurred every year, but the spatial occurrence varied considerably from just a few forests to the whole of Peninsular Malaysia. In 2001, 2002 and 2005, minor and major GF occurred widely throughout Peninsular Malaysia (GF2001, GF2002, and GF2005), and the geographical patterns of GF varied between the episodes. In the three regional-scale GF episodes, most major events occurred in regions where prolonged drought (PD) had been recorded prior, and significant associations between GF scores and PD were found in GF2001 and GF2002. However, the frequency of PD was higher than that of GF throughout the peninsula. In contrast, low temperature (LT) was observed during the study period only before GF2002 and GF2005, but there was no clear spatial relationship between GF and LT in the regional-scale episodes. There was also no evidence that last GF condition influenced the magnitude of GF. Thus, our results suggest that PD would be essential to trigger regional-scale GF in the peninsula, but also that PD does not fully explain the spatial and temporal patterns of GF. The coarse relationships between GF and the proposed climatic cues may be due to the geographical variation in proximate cues for GF, and the climatic and floristic geographical variations should be considered to understand the proximate factors of GF.

Editor: Kamal Bawa, University of Massachusetts, United States of America

Funding: This study was financially supported by a grant from the Environment Research and Technology Development Fund of Ministry of the Environment of Japan (RFd-1101). The funders had no role in study design, data collection and analysis, decision to publish, or preparation of the manuscript.

Competing Interests: The authors have declared that no competing interests exist.

* E-mail: nmt@tmu.ac.jp

Introduction

General flowering (GF) is a supra-annual community-level synchronization of reproduction of many plant taxa in South-East Asian rain forests. In this region, diverse tree species, including dipterocarps, synchronize their reproduction over 5–7 months at irregular intervals of 1–10 years [1–6]. For example, 70% of emergent trees and 40% of canopy trees flowered only during GF periods in a Bornean rain forest [3]. GF results in a massive number of fruits and extensive seedling establishment. Conversely, most dipterocarps and many of the other species flower little or not at all in the years between GF episodes. GF can be seen across different forest types (excluding montane and peat swamp forests) at the landscape level [7], and therefore plays a central role in the regeneration of dipterocarp forests in South-East Asia [1,8].

Several hypotheses have been proposed to explain the mechanism of synchronization at the community level [2,4,5]. Some specific climatic conditions, such as low temperature (LT) and prolonged drought (PD), have been proposed as triggers [9]. LT, defined as drops in daily minimum temperature below 20°C during dry spells [1], has been considered over the past two decades as a plausible trigger [3,4,10]. In contrast, the importance of severe droughts has been emphasized by many studies since the early 20th century [8,11–13]. In Borneo, irregular prolonged droughts (defined as 30-day running total rainfall of <40 mm) were observed before all GF episodes, although some GFs occurred without a preceding LT [8,13]. However, the cues and mechanisms of GF remain controversial, because the effects of PD may be obscured by differences in site drainage and plant coverage [1].

Furthermore, intervals of reproduction of trees may influence the pattern of GF. The resource-matching hypothesis explains that the intervals between reproductions are controlled by the consumption of internal resources for flowering and fruiting [14–16]. In South-East Asia, the accumulation of phosphorus might be important in determining the occurrence and frequency of reproduction of dipterocarp trees [17]. Isagi *et al.* (1997) suggested that intermittent synchronous production of large seed crops could be tied to the resource balance of plants, even without any inter-annual environmental fluctuations [15]. Satake & Iwasa (2002) also hypothesized that synchronized reproduction can occur over a wide range without environmental fluctuations [16]. Therefore, magnitude of a GF may influence that of the following GF if GF consecutively occurs in a forest.

GF is a community-level phenomenon, but its spatial scale and distribution would be highly variable. Flowering years and seasons can differ among forests and species within a region [7,18–20]. GF can synchronize not only within a region, but also among regions [4,10], while local GF episodes without inter-regional synchronization are also often observed [1,21]. Therefore, it has been assumed that global climatic phenomena such as El Niño are associated with severe droughts, which stimulate inter-regional GF [22–23]. If Asian dipterocarp species share common climatic cues for synchronized reproduction within and among forests, the spatial and temporal patterns of the cues would be consistent with those of GF. Yasuda *et al.* (1999) found a significant spatial relationship between the magnitude of GF and preceding low night-time temperature during dry spells in the Malay Peninsula in 1996 [4]. However, information on the relationship between spatiotemporal patterns of GF and climatic conditions at the large scale is surprisingly limited due to the lack of dataset of phenology at large-scale.

Understanding of environmental correlates to large-scale GF phenomenon is important for predicting fruit supply for large-scale forest restoration and negative impacts of global climate change on tropical rainforest ecosystem [8,10,24]. We monitored the density of fruiting trees in forests to evaluate magnitude of GF phenomenon throughout Peninsular Malaysia from 2001 to 2005. In this study, we hypothesized that: (1) geographical pattern and scale of GF would vary among GF episodes, (2) geographical pattern of GF would be explained by spatial occurrences of last GF, PD and/or LT. To evaluate the spatial and temporal variations of GF, geographical patterns of GF were compared among different episodes. As proximate factors affecting spatial and temporal pattern of GF, we examined the effects of last GF magnitude and two plausible meteorological factors (PD and LT) on the magnitude of GF. Finally, we also examined the characteristics of the critical climatic conditions to discern geographical variations in cues and to predict the effects of global climate change.

Materials and Methods

Geographical Distribution of General Flowering in Peninsular Malaysia

Using the methods of Yasuda *et al.* (1999) and Numata *et al.* (2003), we monitored forests from November 2001 to June 2005 in 10 states (except Perlis state) in Peninsular Malaysia [4,10]. To determine the geographical distribution pattern of GF, we surveyed the density of fruiting dipterocarp trees in forest reserves, protected forests and old secondary forests, from lowlands to hills accessible by vehicles. We established 52–84 observation points for the survey. The latitude, longitude and elevation of each point was recorded on a GPS receiver. The location of the observation point did not necessarily correspond with that of a forest, because observations were made from up to 1 km away with binoculars. No specific permissions were required for our observations because most of the observations were made at forest parks opened to the public or the rest were made from a distance (on the fringe of a forest).

The observation time is important in quantifying the magnitude of GF [4]. Numata et al. (2003) suggested two potential GF seasons (spring and autumn types) characterized by the annual patterns of rainfall and low temperature in Peninsular Malaysia when GF is likely to occur [10]. They assumed that many flowerings occur in both the second (spring type) and fourth (autumn type) quarters in Peninsular Malaysia. Dipterocarp species have a short-flowering period (2–3 week) and do not flower simultaneously with one another in a local species assemblage, but disperse their mature fruits synchronously [25]. Therefore, if spring or autumn type GF occurs, we anticipate that conspicuous red or light yellowish immature fruits can be observed in forests. Consequently, we made observations in June–July (the spring type) and November–December (the autumn type) of 2001–2005. The use of two or more observers minimized observation errors. The GF status of each forest was evaluated by eye and scored on the basis of the density of fruiting dipterocarps as "none", "fruiting" (a few dipterocarps bearing fruits), "minor" (up to 50%) and "major" (50% or more of large dipterocarps bearing fruits).

Meteorological Data

The daily minimum temperatures (°C) and daily rainfall (mm) of 14 principal meteorological stations in Peninsular Malaysia from 1981 to 2008 were provided by the Malaysian Meteorological Department. Previous studies have demonstrated that beginning of flowering in GF occurs approximately 2 months after abnormal meteorological condition [1–4]. Numata *et al.* (2003) also implied that the dry and cool seasons in Peninsular Malaysia may correspond to trigger period of GF [10]. Based on the rainfall seasonality in Peninsular Malaysia, we focused on relatively dry seasons (3 months) before the beginning of GF episode as a putative trigger period: January–March in 2002–2005 for the spring type GF and June–August in 2001–2004 for the autumn type GF. During the putative trigger periods, the occurrences of abnormal weather conditions, PD (prolonged drought: 30-day moving total rainfall <40 mm [3]) and LT (low temperature: daily minimum temperature <20°C [1]) were determined from the meteorological data. We did not calculate the 30-day moving total of rainfall when rainfall did not occur in the preceding 29 days. To quantify intensity of climatic cues, we calculated numbers of PD days and LT days for each putative trigger period. We also determined monthly occurrences of LT and PD during the putative periods of floral induction from 2001 to 2005. To determine the ENSO (El Niño Southern Oscillation) phase, we used the Niño 3 index and ENSO information from 2001 to 2005 provided by the Japan Meteorological Agency.

Data Analysis

All statistical analyses were conducted in R v. 2.9.2 statistical software (R Foundation for Statistical Computing, 2010). To examine geographical pattern of GF and factors affecting GF magnitude and distribution, we used a generalized linear mixed model (GLMM) with binomial errors for each GF episode using glmmML package in R. As data for the models, we used numerical transformed GF scores (none: 0; sporadic: 1; minor: 3; major: 5) for each episode. To specify GF status in a proportional response variable, we bind two vectors into a single object: ("GF scores" and "5 - GF scores"). We used forest id as a random effect in the models. The fixed effects of geography (latitude, longitude and elevation) on GF status were examined to determine the geographical pattern of GF. To examine effects of proximate factors on GF status, GLMM was used to test the fixed effects of last GF status and preceding occurrences of PD and LT. For the analyses, forests within 50 km of meteorological stations were chosen. The level of significance was assessed by the Wald test. To analyze correlations between climatic characteristics and frequencies of LT and PD among 14 meteorological stations, we used Spearman's rank correlation test. To compare frequencies of LT and PD, Fisher's exact test and Ryan's multiple comparison method (MCM) were used.

Results

Spatial and Temporal Patterns of General Flowering in Peninsular Malaysia

Different spatial scales of GF were observed during the study period (Figure 1). In 2001, 2002 and 2005, minor and major GF occurred widely throughout Peninsular Malaysia (Figure 1): 44% of forests in November 2001 (GF2001), 59% in June 2002 (GF2002) and 46% in June 2005 (GF2005). Local-scale GF also occurred in November 2003 (7%), June 2004 (9%) and November–December 2004 (5%). We discerned three spatial scales of GF episodes: regional-scale GFs across six or more states (November 2001, June 2002 and June 2005), local-scale GFs with sparse GF in a few states (November 2003, June 2004 and November 2004) and sporadic fruiting episodes (November 2002 and June 2003).

In GF2001, major and minor GF occurred in central and south-eastern Peninsular Malaysia, and the result of GLMM analysis suggested significant latitudinal effect on GF status (Table 1). In GF2005, major and minor GF occurred throughout the region except along the north-eastern coast, and significant latitudinal and elevational effects suggested a distribution at higher elevations. In GF2002, in contrast, major and minor GF occurred throughout Peninsular Malaysia except along the central west coast, but no significant geographical effects were found.

Temporal Pattern of Prolonged Drought and Low Temperature

Although temperatures on the Malay Peninsula are generally uniform throughout the year, there were differences in meteorological variables among the meteorological stations (Table S1 in File S1). Annual rainfall in 1981–2008 ranged from 1810 mm (Sitiawan) to 2969 mm (Kuantan). Rainfall seasonality (CV of

Table 1. Effects of geography (longitude, latitude and elevation) on GF status in regional-scale GF episodes examined by generalized linear mixed models with binomial error structure.

	GF2001	GF2002	GF2005
(Intercept)	1.35e+02	−1.82e+01	2.80e+01
Elevation	−6.42e-04	9.43e-04	4.61e-03*
Latitude	−1.40e+00*	4.54e-01	1.02e-00***
Longitude	−1.28e+00	1.61e-01	−4.94e-01

*p<0.05,
***p<0.001,
The forest id was used as a random effect in the models.

monthly rainfall) also varied from 0.49 (Ipoh) to 1.20 (Kota Bahru). Mean daily minimum temperature in 2001–2005 ranged from 22.9°C (Senai and Kuala Krai) to 25.1°C (Langkawi). Mean 30-day moving total rainfall in 2001–2005. 30-day moving total rainfall ranged from 145.2 mm (Temerloh) to 264.3 mm (Kuantan).

Although frequencies of PD and LT varied greatly among the meteorological stations, PD was much more frequent than LT during the study. PD occurred every year during the study period while LT was not observed during the study period except before GF2002 and GF2005 (Figure 2). In 2001–2005, 2137 days with PD and 43 days with LT were observed at the 14 meteorological stations. 85.8% of PD occurred in the putative trigger period (July–August and January–March). All of LT occurred in the putative trigger period in 2001–2005. Among the 14 stations, frequencies of LT and PD during 2001–2005 were significantly positively correlated with those during 1981–2008 (LT: $r_s = 0.72$, $P<0.01$; PD: $r_s = 0.93$, $P<0.0001$; Table 2), suggesting consistency

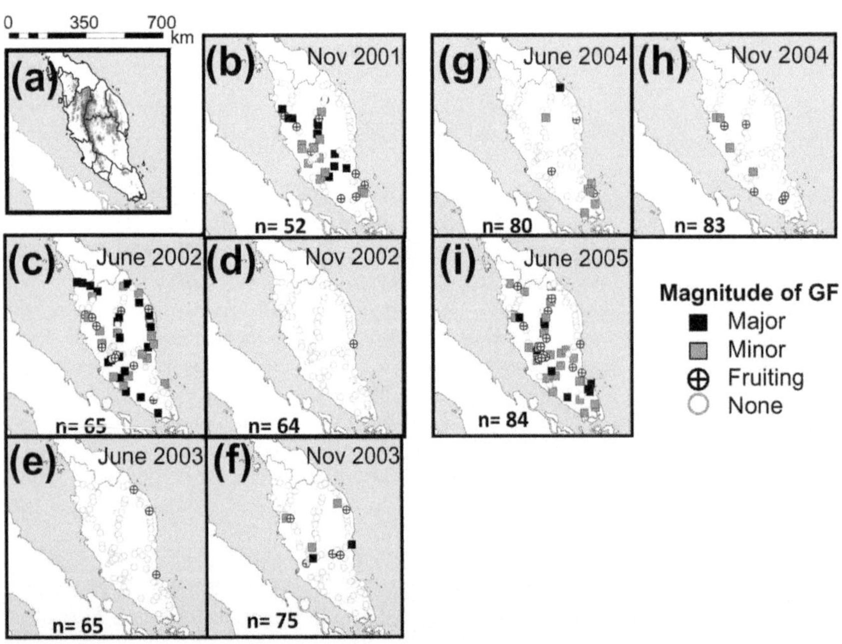

Figure 1. Geographical distribution of GF from November 2001 to June 2005. Each site was scored for the density of fruiting dipterocarp trees (major GF, minor GF, sporadic fruiting or none). (a) Topography of Peninsular Malaysia. (b, d, f, h) GF scores in November–December 2001–2005 (the autumn type). (c, e, g, i) GF scores in June-July 2002–2005 (the spring type).

of our results with long-term regional trends. The long-term averages (±SD) of the frequencies of LT and PD at the 14 stations were 0.0035±0.0053 day^{-1} and 0.0781±0.0508 day^{-1}, respectively. This means that PD was >22 times as frequent as LT. The long-term frequency of LT varied greatly among the stations, from 0.0001 day^{-1} (Ipoh and Kluang) to 0.0205 day^{-1} (Kuala Krai) (Table 2). The long-term frequency of PD also varied greatly, from 0.0181 day^{-1} (Ipoh) to 0.1784 day^{-1} (Langkawi).

There were significant differences in the frequencies of PD and LT between the 1980s, 1990s and 2000s, and the frequencies tended to decrease with time. Across the 14 stations, the frequencies of LT were 0.0053 day^{-1} in the 1980s, 0.0037 day^{-1} in the 1990s and 0.0014 day^{-1} in the 2000s, and they differed significantly between decades ($P<0.0001$; test of proportions with Ryan's MCM; Table S2 in File S1). Frequencies of PD at the 14 stations were 0.0861, 0.0803 and 0.0640 day^{-1}, respectively, and they differed significantly between decades ($P<0.0001$; Table S3 in File S1).

During the survey period, an El Niño episode occurred from summer 2002 to winter 2002–03, and La Niña from winter 2005 to spring 2006, and all three regional-scale GF episodes began during the absence of either (Figure 2). There were also significant differences in the frequencies of PD and LT among ENSO conditions (El Niño, La Niña and neutral): The frequency of LT during El Niño (0.0018 day^{-1}) was significantly lower than those during La Niña (0.0039 day^{-1}) or neutral conditions (0.0036 day^{-1}) ($P<0.0001$; test of proportions with Ryan's MCM; Table S4 in File S1). The frequency of PD during El Niño (0.1084 day^{-1}) was significantly higher than that during La Niña (0.0480 day^{-1}) and neutral conditions (0.0835 day^{-1}) ($P<0.0001$; Table S5 in File S1).

Rainfall seasonality (coefficient of variation of monthly rainfall) at a station was positively correlated with the frequency of PD across all 14 stations ($r_s = 0.54$, $P<0.05$), but not with that of LT ($r_s = 0.12$, $P>0.05$; Figure S1 in File S1). There were no significant correlations between annual rainfall (1981–2008) and frequencies of LT or PD across the 14 stations (vs LT: $r_s = -0.12$, $P>0.05$; vs PD: $r_s = -0.26$, $P>0.05$; Figure S1 in File S1). Seven stations showed significant positive correlations between the monthly lowest 30-day total rainfall and monthly minimum temperature (Table 2).

Spatial Relationship between General Flowering, Prolonged Drought and Low Temperature

We compared the spatial patterns of GF during the three regional-scale episodes and meteorological stations with PD or LT (Figure 3). In June, July and August 2001, before GF2001, PD was recorded at six stations (Ipoh, Kota Bahru, Kuala Krai, Muadzam Shaw, Sitiawan and Temerloh) in northern and central Peninsular Malaysia, but no LT was recorded. These results suggest that GF tended to occur in areas with PD except in the north in GF2001. In January, February and March 2002, before GF2002, PD was recorded throughout Peninsular Malaysia except at two stations in the central west (Ipoh and Sitiawan) and two in the south (Mersing and Senail), but LT was recorded at only four stations (Kuala Terengganu, Kuala Krai, Temerloh and Muadzam Shaw), in the centre. These results imply that GF tended to occur in areas with PD. In January, February and March 2005, before GF2005, LT

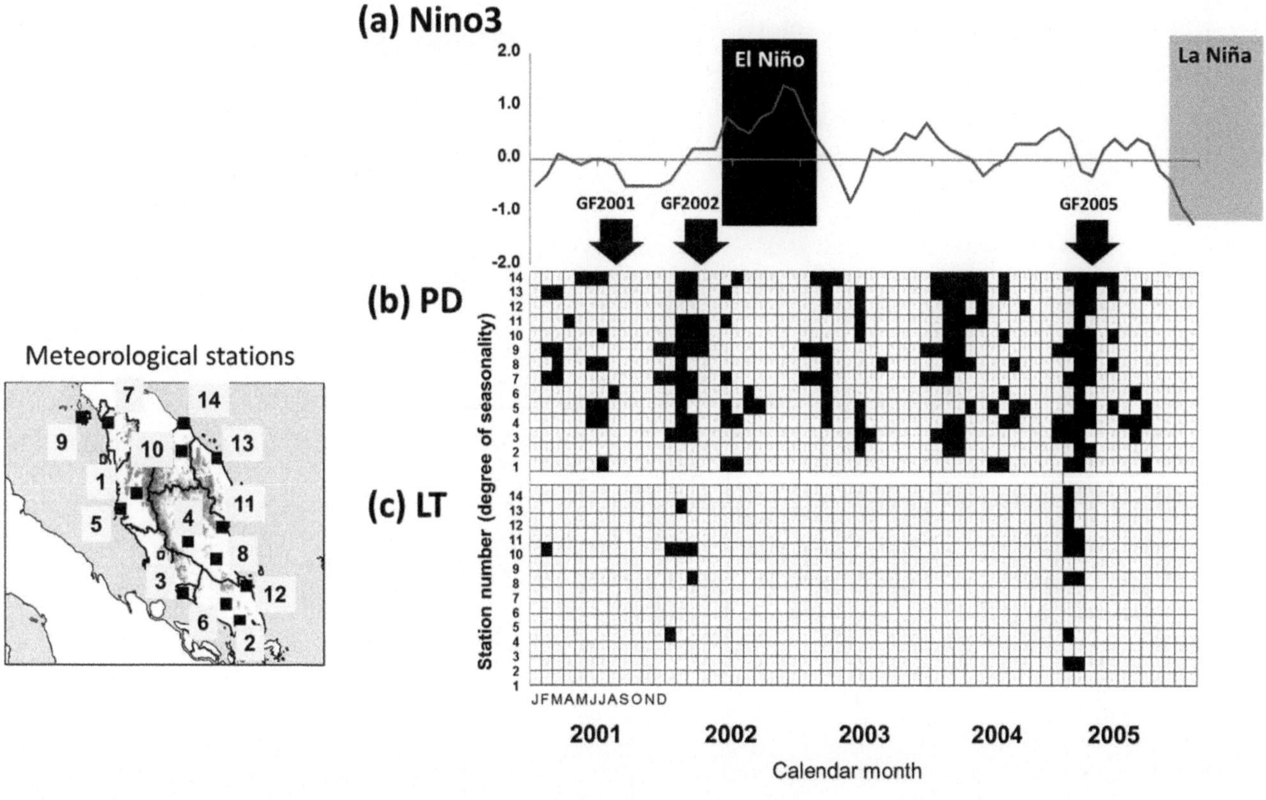

Figure 2. Occurrences of ENSO, PD and LT in Peninsular Malaysia from 2001 to 2005. (a) Niño 3 values as ENSO index, (b) months with PD (30-day moving total rainfall <40 mm) and (c) months with LT (daily minimum temperature <20°C). An El Niño episode occurred from summer 2002 to winter 2002–03, and La Niña from winter 2005 to spring 2006. Arrows indicate regional-scale GF episodes (GF2001, GF2002 and GF2005). Station names are identified in Table 2.

Table 2. Summary of climate at 14 meteorological stations in Peninsular Malaysia.

Meteorological station	Long. (E)	Lat. (N)	Alt (m)	Ann. rainfall (mm) (1981–2008)	Rainfall seasonality (CV of mo. rainfall)	Frequency of LT(day^{-1})		Frequency of PD(day^{-1})		Rank correlation between mo lowest 30dMTR and min temperatures (r_s)
						2001–2005	1981–2008	2001–2005	1981–2008	
1. Ipoh	101° 06′	4° 34′	40.1	2583	0.49	0	0.0001	0.0296	0.0181	0.068
2. Senai	103° 40′	1° 38′	37.8	2501	0.52	0.0022	0.0047	0.0197	0.0207	0.248***
3. Malacca	102° 15′	2° 16′	8.5	2024	0.54	0	0.0006	0.0947	0.0669	0.02
4. Temerloh	100° 23′	3° 28′	39.1	1949	0.58	0.0011	0.0054	0.0942	0.0682	0.290***
5. Sitiawan	100° 42′	4° 13′	7	1810	0.59	0	0.0008	0.0805	0.078	0.039
6. Kluang	103° 19′	2° 01′	88.1	2196	0.61	0	0.0001	0.0279	0.0336	0.155**
7. Alor Star	100° 24′	6° 12′	3.9	1994	0.65	0	0.0038	0.1218	0.1356	0.514***
8. Muadzam Shaw	103° 05′	3° 03′	33.3	2432	0.71	0.0027	0.0057	0.0723	0.0403	0.176**
9. Langkawi	99° 44′	6° 20′	6.4	2404	0.75	0	0.0004	0.1999	0.1784	0.133*
10. Kuala Krai	102° 12′	5° 32′	68.3	2486	0.85	0.011	0.0205	0.0756	0.0767	0.128*
11. Kuantan	103° 13′	3° 47′	15.3	2969	0.35	0.0033	0.0037	0.0581	0.0565	−0.001
12. Mersing	103° 50′	2° 27′	43.6	2660	0.39	0.0011	0.0003	0.0323	0.0459	−0.038
13. Kulala Terengganu	103° 06′	5° 23′	5.2	2593	1.16	0.0016	0.0018	0.1057	0.1262	0.022
14. Kota Bharu	102° 17′	6° 10′	4.6	2532	1.2	0.0005	0.0015	0.1528	0.1533	0.038

*$p<0.05$,
**$p<0.01$,
***$p<0.001$.

was recorded only in the east, but PD was recorded at all stations. In contrast, no clear geographical overlaps between LT and GF were found in any of the three regional-scale episodes. We examined statistically whether GF was associated with last GF status and intensities of preceding occurrences of PD and LT for forests within 50 km of meteorological stations. The results of GLMM analysis suggested that the intensity of PD during the putative trigger period significantly positively influenced GF status in GF2001 and GF2002, but not in GF2005 (Table 3, Figure 4). In contrast, significant effects of last GF status and the intensity of preceding LT on GF scores were not found in any three regional-scale GF episodes (Table 3).

Discussion

Geographical Pattern of General Flowering

GF varied both spatially and temporally. From 2001 to 2005, three regional-scale GF episodes were observed throughout Peninsular Malaysia, although GF was not fully synchronized among forests within the region during these episodes. The regional-scale GF episodes in Peninsular Malaysia would be linked to inter regional synchronization of GF between west and east Malaysia in Malesia. In GF2001, GF2002 and GF2005, GF forests were observed in not only Peninsular Malaysia, but also from Bornean dipterocarp forests [8,26].

Our results revealed that geographical patterns of GF forests differed between the three regional-scale episodes. The geographical patterns may be partly explained by geographical floristic variation. There are several main forest types with different elevational ranges and edaphic formations where dipterocarp species dominate in the peninsula [27]. The significant positive effect of elevation on GF scores in GF2005 probably resulted from the greater GF scores in hill forest, where hill species such as *Shorea curtisii* dominate, than in lowland forest. In addition, the floristic composition of dominant dipterocarp species in the same forest type may also cause geographical variation, because some dipterocarp species flower more frequently than others [2,27,28]; for example, *Dryobalanops aromatica* flowers annually in Malaysia [21].

Factors Affecting General Flowering Distribution Pattern

In 2001–2005, GF tended to occur in regions with preceding PD (Figure 3), and the intensity of PD significantly positively influenced the following GF status in GF2001 and GF2002 (Table 3). Therefore, preceding occurrence of PD might be one factor that shapes the spatial pattern of GF. However, the almost annual occurrence of PD throughout Peninsular Malaysia was too frequent to explain the occurrence of GF, suggesting that PD alone does not explain the spatial and temporal patterns of GF. The coarse relationships between GF and the proposed climatic cues

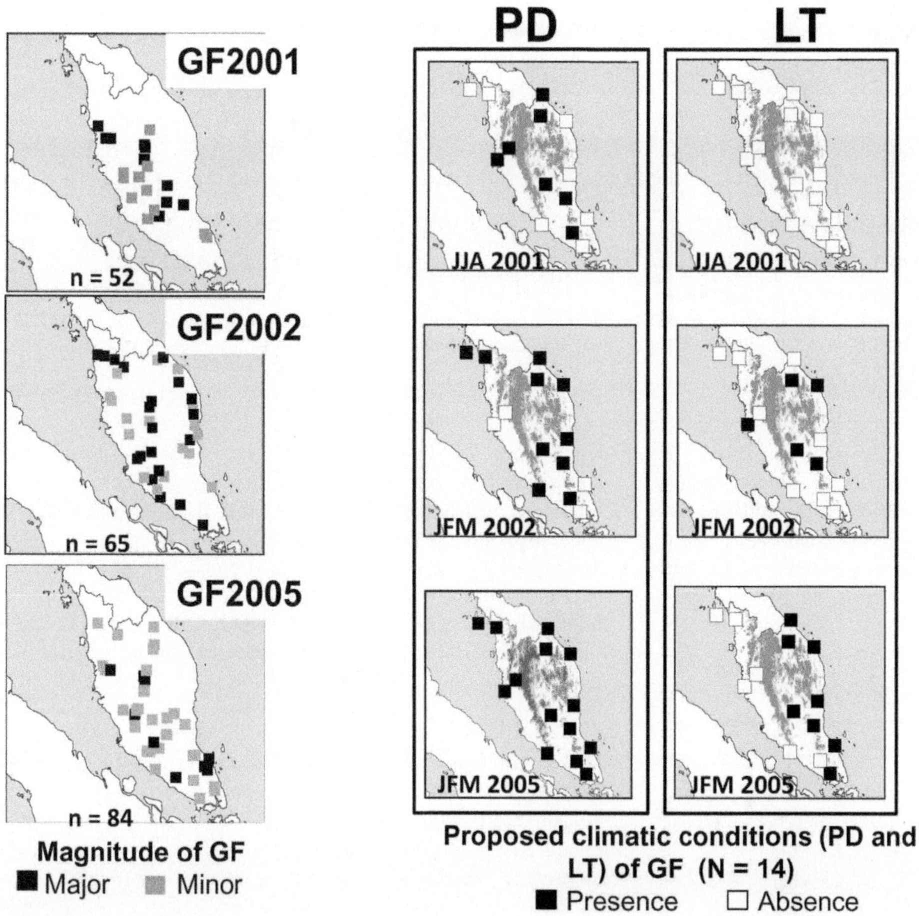

Figure 3. Distribution patterns of major and minor GF (left), and meteorological stations with PD (centre) and LT (right) during the three regional-scale GF episodes. Plots show the presence or absence of PD and LT during the putative floral induction season: June–August (GF2001) or January–March (GF2002 and GF2005).

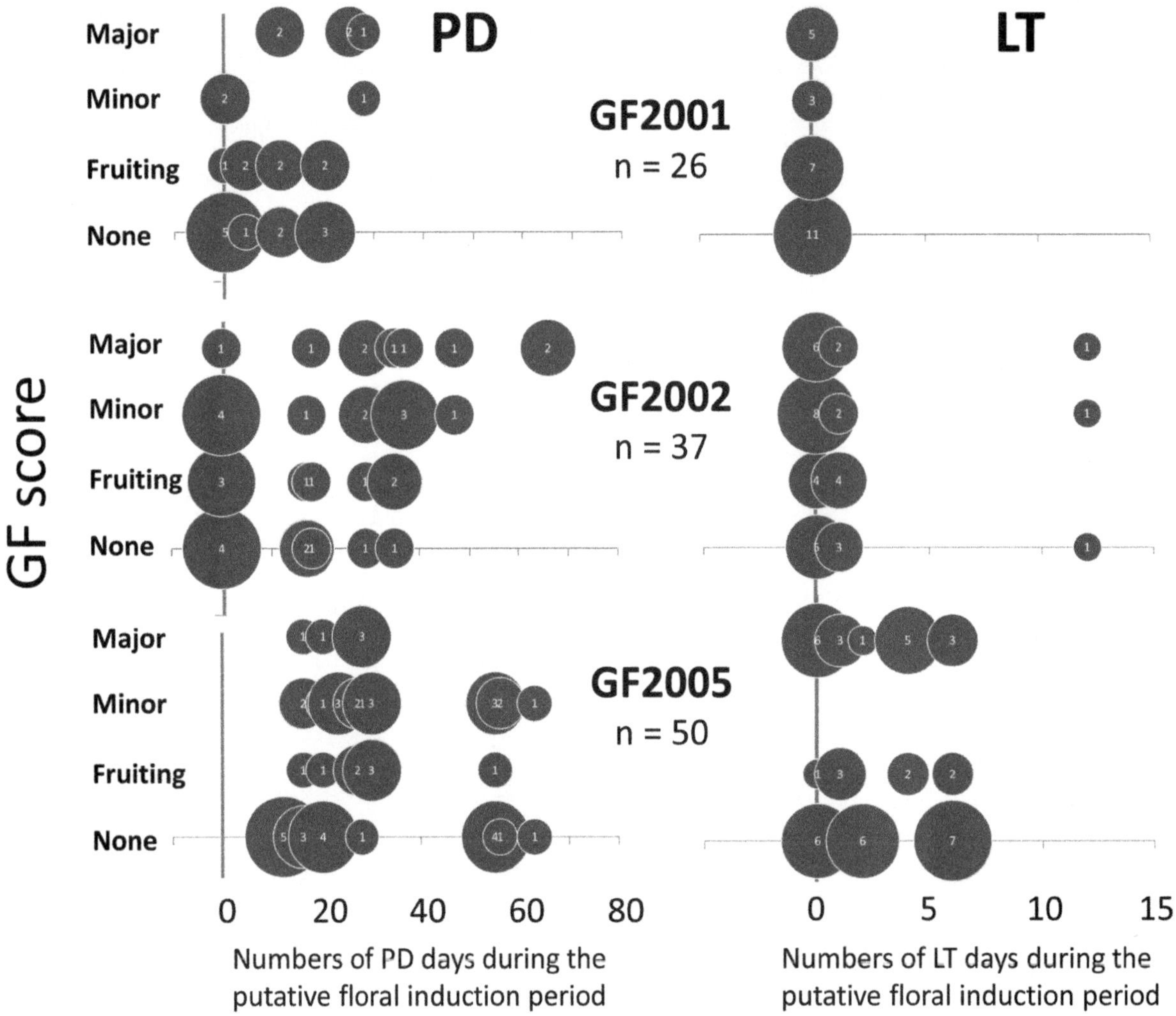

Figure 4. Relationships between GF scores and numbers of PD (left) or LT (right) during the putative trigger periods in GF2001, GF2002 and GF2005. The numbers in the bubbles and bubble's volumes mean numbers of forests used for the analyses. These analyses used data from meteorological stations within 50 km of sites.

might be partly due to the geographical variation in proximate cues for GF. In future studies, we need to examine the definition of PD in terms of climate. For example, previous studies used different definition of PD: Sakai et al. (2006) defined PD as 30-day moving total rainfall <40 mm in a Bornean rainforest (average annual rainfall in 1985–2003: 2722 mm) [8], but Brearly (2007) defined it as in <60 mm in a rainforest of Central Kalimantan (annual rainfall: around 3800 mm) [12].

Table 3. Effects of last GF status and the two plausible meteorological cues (PD and LT) on GF status.

	GF2001	GF2002	GF2005
(Intercept)	−3.45e+00	−1.31e+00	−1.42e+e00
Last GF scores	–	−3.56e-01	8.41e-01
No. PD days	1.59e-01*	8.40e-02*	1.85e-02
No. LT days	–	−1.22e-01	−1.58e-01

*p<0.05.
Results from GLMM using with binomial error structure were shown. The forest id was used as a random effect in the models.

LT was widely recorded only before GF2002 and GF2005. Therefore, LT is likely associated with regional-scale GF. However, LT was not observed before GF2001 and any other local and sporadic episodes. We also found that LT did not explain the distribution of GF in GF2002 and 2005 (Table 3). Therefore, LT as estimated here does not determine the occurrence and spatial distribution of GF. It has been believed that it is difficult to distinguish between PD and LT because of its correlation [10]. However, the decreasing frequency of LT during the 2000s might have obscured any spatial relationship between PD and LT in the three regional-scale episodes.

Sakai et al. (2006) implied that magnitude of GF may be influenced the level of accumulated resources in addition to the intensity of the trigger [8]. However, no significant effects of last GF status suggest that the resource condition of trees would not be more decisive than climatic cues even if it has an effect. In Japanese beech trees (*Fagus crenata*), for example, it is likely that climatic variation is more critical to controlling the amount of flowering than resource condition, while the quantity of seed production in a year is negatively correlated with that in the preceding year [29].

The coarse relationships between GF and the previously proposed climatic cues may be also due to the poor spatial representativeness of the meteorological stations. Rainfall and temperature are spatially and temporally heterogeneous even in the peninsula [30,31], and rainfall spatial variability generally increases with distance in the peninsula [32,33]. Furthermore, differences in observations between flowering and fruiting may influence the association between GF and climatic cues, because factors affecting the density of flowering differ from those affecting the density of fruiting [29]. Percentages of flowering and fruit production are generally highly correlated, even in dipterocarp species [8,13], but more data on flowering are needed to settle the dispute on climatic cues of GF.

Trends in Meteorological Cues: Past and Future

There were also considerable geographical variations in the occurrences and frequencies of the proposed climatic cues. The frequencies of PD were significantly correlated with the seasonality of rainfall among the 14 meteorological stations. Peninsular Malaysia lies to the south of the Kangar–Pattani line in the Indo-Sundaic region of South-East Asia, and its vegetation is thought to be aseasonal rain forest [34]. In general, frequent flowering and fruiting in a region with strong rainfall seasonality is well known, because most species, including dipterocarps, flower and fruit annually in seasonal forests in South-East Asia [21,27]. Comparison of frequencies of PD between seasonal and aseasonal rainforests would be important to understand latitudinal transition from annual flowering and GF as well as its mechanism.

The correlation between PD and LT was discussed in many studies [1,4,8,10,13], but its geographical variation has not been well considered. In this study, significant positive correlations between monthly lowest minimum temperatures and monthly lowest 30-day running totals of rainfall were found at only seven of the 14 meteorological stations, in central and north-western parts of Peninsular Malaysia. Therefore, contribution of LT to cues for GF might geographically vary even if LT is responsible for GF.

The frequencies of PD and LT decreased from the 1980s to the 2000s. If PD and LT is associated with regional-scale GF, projected increases in temperature (1.1–3.6°C) in Peninsular Malaysia may reduce the frequency of LT events for both seasonal types of GF [35]; in contrast, rainfall variability in Peninsular Malaysia is likely dependent on model choice, and there is no clear trend, owing to the high variability in the precipitation-modulating factor. Therefore, if LT is associated with specific PD-related conditions for GF, increases in temperature would result in decreases in regional-scale GF episodes regionally. Further study is needed to understand how climate change affects frequency and magnitude of PD and LT as well as relationship between LT and PD.

ENSO-related droughts in Borneo may have serious simultaneous impacts on the regeneration of dipterocarp forests by GF episodes under global climate change [8,22]. In earlier studies of association of GF with ENSO, no simple association was found [1,4,5,20,22]. However, our results reveal that regional-scale GF in Peninsular Malaysia was triggered during the absence of El Niño and La Niña. This result is consistent with previous studies suggesting that GF was induced at the transition between El Niño and La Niña [8,10,36]. Our results also explain why the ENSO-neutral condition important for regional-scale GF episodes: the frequency of PD was higher under El Niño than under neutral and La Niña conditions, but the frequency of LT was lower. Therefore, GF would be favoured under ENSO-neutral conditions in a region where LT is associated with climatic cues for GF. Complex Coupled Global Circulation Models suggest that ENSO frequency and magnitude under global warming are still unclear [37]. In addition, the ENSO–rainfall relationship may vary geographically, even within a region [38,39]. Further study is required to examine the combined effect of global climate change and ENSO on GF.

Concluding Remarks for Future Study

Although PD and LT do not fully explain the spatial and temporal patterns of GF in Peninsular Malaysia, PD was likely to trigger GF during the study period. However, it is still unclear why PD can explain only a part of GF occurrences in Peninsular Malaysia. The coarse relationships between GF and the proposed climatic cues may be due to the geographical variation in proximate cues for GF. Thus, information on floristic and phylogeographic regions should also be included to explain the spatial and temporal pattern of GF. Furthermore, intra-specific variation in the response of trees to the environment may also cause different geographical pattern in responses to climate [40–42]; for example, geographical variation in masting behaviour occurs in Japanese beech, and its populations in northern Japan can be classified into five [43] or seven regional groups [29]. Furthermore, considering the two different monsoon activities [10], there may be different mechanisms for triggering GF between the spring and autumn types. Both experimental and theoretical approaches focusing on internal and external proximate cues may be required to explore the mechanisms of inter- and intra-regional reproductive synchronization in Southeast Asian dipterocarp forests.

Supporting Information

File S1 Combined Supplementary file containing the following items: Table S1, Table S2, Table S3, Table S4, Table S5 and Figure S1. Table S1 Summary of minimum temperature and 30-day moving total rainfall at the 14 meteorological stations. Table S2 Frequencies of LT in 1980's, 1990's, 2000's and the entire period (1981–2008). Table S3 Frequencies of PD in 1980's, 1990's, 2000's and the entire period (1981–2008). Table S4 Frequencies of LT in neutral, El Niño and La Niña periods. Table S5 Frequencies of PD in neutral, El Niño and La Niña periods. Figure S1. Relationships between frequencies of the proposed climatic cues (LT and PD), annual rainfall, and rainfall seasonality across the 14 meteorological stations.

Acknowledgments

We thank the Malaysian Meteorological Department for providing climatological data of Malaysia. We thank T. Okuda, S. Nishimura, K. Yoshida, J. Matsumoto, N. Uchida, A. A. Aminudin, Shahri, and E. S. Quah for their help in the field work. We also thank S. Sakai and P. S. Ashton for their valuable comments and suggestions on a draft of the manuscript.

Author Contributions

Conceived and designed the experiments: SN MY. Performed the experiments: SN MY TH NSMD CDF MH. Analyzed the data: SN ROS TH. Wrote the paper: SN MY ROS TH NSMD MH.

References

1. Ashton PS, Givnish TJ, Appanah S (1988) Staggered flowering in the Dipterocarpaceae: New insights into floral induction and the evolution of mast fruiting in the aseasonal tropics. Am Nat 132: 44–66.
2. Appanah S (1993) Mass flowering of dipterocarp forests in the aseasonal tropics. J Biosci 18: 457–474.
3. Sakai S, Momose K, Yumoto T, Nagamitsu T, Nagamasu H, et al. (1999) Plant reproductive phenology over four years including an episode of general flowering in a lowland dipterocarp forest, Sarawak, Malaysia. Am J Bot 86: 1414–1436.
4. Yasuda M, Matsumoto J, Osada N, Ichikawa S, Kachi N, et al. (1999) The mechanism of general flowering in Dipterocarpaceae in the Malay Peninsula. J Trop Ecol 15: 437–449.
5. Sakai S (2002) General flowering in lowland mixed dipterocarp forests of South-east Asia. Biol J Linn Soc 75: 233–247.
6. Numata S (2004) Mass flowering of dipterocarps in Peninsular Malaysia. Flowering Newsletter 37: 32–38.
7. Cannon CH, Curran LM, Marshall AJ, Leighton M (2007) Long-term reproductive behaviour of woody plants across seven Bornean forest types in the Gunung Palung National Park (Indonesia): suprannual synchrony, temporal productivity and fruiting diversity. Ecol Lett 10: 956–969.
8. Sakai S, Harrison RD, Momose K, Kuraji K, Nagamasu H, et al. (2006) Irregular droughts trigger mass flowering in aseasonal tropical forests in Asia. Am J Bot 93: 1134–1139.
9. Whitmore TC, Burnham CP (1984) Tropical rain forests of the far east. New York: Oxford University Press.
10. Numata S, Yasuda M, Okuda T, Kachi N, Nur Supardi MN (2003) Temporal and spatial patterns of mass flowerings on the Malay Peninsula. Am J Bot 90: 1025–1031.
11. Foxworthy FW (1932) Dipterocarpaceae of the Malay Peninsula. Kepong. 289 p.
12. Medway FLSL (1972) Phenology of a tropical rainforest in Malaya. Biol J Linn Soc 4: 117–146.
13. Brearley FQ, Proctor J, Suriantata, Nagy L, Dalrymple G, et al. (2007) Reproductive phenology over a 10-year period in a lowland evergreen rain forest of central Borneo. J Ecol 95: 828–839.
14. Sork VL (1993) Evolutionary ecology of mast-seeding in temperate and tropical oaks (*Quercus* spp.). Vegetatio 107/108: 133–147.
15. Isagi Y, Sugimura K, Sumida A, Ito H (1997) How Does Masting Happen and Synchronize? J Theor Biol 187: 231–239.
16. Satake A, Iwasa Y (2002) Spatially limited pollen exchange and long-range synchronization of trees. Ecology 83: 993–1005.
17. Ichie T, Nakagawa M (2013) Dynamics of mineral nutrient storage for mast reproduction in the tropical emergent tree *Dryobalanops aromatica*. Ecol Res: 1–8.
18. Sasaki S, Tan CH, Zulfatah AR (1979) Some observations of unusual flowering and fruiting of dipterocarps. Malay Forest 42: 38–65.
19. Yap SK, Chan HT (1990) Phenological behaviour of some Shorea species in peninsular Malaysia. In: Bawa KS, Hadley M, editors. Reproductive ecology of tropical forest plants. Paris, France: the United Nations Educational Scientific and Culutual Organization.
20. Wich SA, van Schaik CP (2000) The impact of El Niño on mast-fruiting in Sumatra and elsewhere in Malesia. J Trop Ecol 16: 563–577.
21. Burgess PF (1972) Studies on the regeneration of the hill forests of the Malay peninsula. Malay Forest 35: 103–123.
22. Curran LM, Caniago I, Paoli GD, Astianti D, Kusneti M, et al. (1999) Impact of El Niño and logging on canopy tree recruitment in Borneo. Science 286: 2184–2188.
23. Curran LM, Trigg SN, McDoland AK, Astiani D, Hardiono YM, et al. (2004) Lowland forest loss in protected areas of Indonesian Borneo. Science 303: 1000–1003.
24. Kettle CJ, Ghazoul J, Ashton PS, Cannon CH, Chong L, et al. (2010) Mass Fruiting in Borneo: A Missed Opportunity. Science 330: 584.
25. Numata S, Suzuki R, Nishimura S, Naito Y, Konuma A, et al. (2012) Fruiting behavior of dipterocarps in two consecutive episodes of general flowering in a Malaysian lowland rain forest. J For Res 17: 378–387.
26. Kishimoto-Yamada K, Itioka T (2008) Survival of Flower-visiting chrysomelids during non general-flowering periods in Bornean dipterocarp Forests. Biotropica 40: 600–606.
27. Symington CF, Ashton PS, Appanah S (2004) Forester's manual of dipterocarps; Barlow HS, editor. Kuala Lumpur: Forest Research Institute Malaysia and Malaysian Nature Society. 519 p.
28. Ashton PS (1988) Dipterocarp biology as a window to the understanding of tropical forest structure. Ann Rev Ecol Syst 19: 347–370.
29. Masaki T, Oka T, Osumi K, Suzuki W (2008) Geographical variation in climatic cues for mast seeding of *Fagus crenata*. Popul Ecol 50: 357–366.
30. Desa MMN, Niemczynowicz J (1996) Spatial variability of rainfall in Kuala Lumpur, Malaysia: long and short term characteristics. Hydro Sci J 41: 345–362.
31. Tangang FT, Juneng L (2004) Mechanisms of Malaysian Rainfall Anomalies. J Climate 17: 3616–3622.
32. Dale WL (1959) The rainfall of Malaya. J Trop Geograp 13: 23–37.
33. Dale WL (1960) The rainfall of Malaya, Part II. J Trop Geograp 14: 11–28.
34. Corlett RT (2009) The ecology of tropical east Asia New York: Oxford University Press. 262 p.
35. Malaysian Meteorological Department (2009) Climate Change Scenarios For Malaysia 2001–2099. 68 p.
36. Hamann A (2004) Flowering and fruiting phenology of Philippine submontane rain forest: climatic factors as proximate and ultimate causes. J Ecol 92: 24–31.
37. Collins M, An S-I, Cai W, Ganachaud A, Guilyardi E, et al. (2010) The impact of global warming on the tropical Pacific Ocean and El Niño. Nat Geosci 3: 391–397.
38. Cheang B-k (1993) Interannual variability of monsoons in Malaysia and its relationship with ENSO. J Earth Syst Sci 102: 219–239.
39. Gomyo M, Kuraji K (2009) Spatial and temporal variations in rainfall and the ENSO-rainfall relationship over Sarawak, Malaysian Borneo. SOLA 5: 41–44.
40. Menzel A, Sparks TH, Estrella N, Koch E, Aasa A, et al. (2006) European phenological response to climate change matches the warming pattern. Glob Change Biol 12: 1969–1976.
41. Doi H, Takahashi M (2008) Latitudinal patterns in the phenological responses of leaf colouring and leaf fall to climate change in Japan. Glob Ecol Biogeograp 17: 556–561.
42. Doi H, Takahashi M, Katano I (2010) Genetic diversity increases regional variation in phenological dates in response to climate change. Glob Change Biol 16: 373–379.
43. Suzuki W (2005) Mast seeding and its spatial scale in *Fagus crenata* in northern Japan. For Ecol Manage 205: 105–116.

Permissions

All chapters in this book were first published in PLOS ONE, by The Public Library of Science; hereby published with permission under the Creative Commons Attribution License or equivalent. Every chapter published in this book has been scrutinized by our experts. Their significance has been extensively debated. The topics covered herein carry significant findings which will fuel the growth of the discipline. They may even be implemented as practical applications or may be referred to as a beginning point for another development.

The contributors of this book come from diverse backgrounds, making this book a truly international effort. This book will bring forth new frontiers with its revolutionizing research information and detailed analysis of the nascent developments around the world.

We would like to thank all the contributing authors for lending their expertise to make the book truly unique. They have played a crucial role in the development of this book. Without their invaluable contributions this book wouldn't have been possible. They have made vital efforts to compile up to date information on the varied aspects of this subject to make this book a valuable addition to the collection of many professionals and students.

This book was conceptualized with the vision of imparting up-to-date information and advanced data in this field. To ensure the same, a matchless editorial board was set up. Every individual on the board went through rigorous rounds of assessment to prove their worth. After which they invested a large part of their time researching and compiling the most relevant data for our readers.

The editorial board has been involved in producing this book since its inception. They have spent rigorous hours researching and exploring the diverse topics which have resulted in the successful publishing of this book. They have passed on their knowledge of decades through this book. To expedite this challenging task, the publisher supported the team at every step. A small team of assistant editors was also appointed to further simplify the editing procedure and attain best results for the readers.

Apart from the editorial board, the designing team has also invested a significant amount of their time in understanding the subject and creating the most relevant covers. They scrutinized every image to scout for the most suitable representation of the subject and create an appropriate cover for the book.

The publishing team has been an ardent support to the editorial, designing and production team. Their endless efforts to recruit the best for this project, has resulted in the accomplishment of this book. They are a veteran in the field of academics and their pool of knowledge is as vast as their experience in printing. Their expertise and guidance has proved useful at every step. Their uncompromising quality standards have made this book an exceptional effort. Their encouragement from time to time has been an inspiration for everyone.

The publisher and the editorial board hope that this book will prove to be a valuable piece of knowledge for researchers, students, practitioners and scholars across the globe.

LIST OF CONTRIBUTORS

Pamela M. Willis and Michael J. Ryan
Section of Integrative Biology, University of Texas at Austin, Austin, Texas, United States of America

Gil G. Rosenthal
Department of Biology, Texas A & M University, College Station, Texas, United States of America
Centro de Investigacio´n Cientı´fica de las Huastecas "Aguazarca", Calnali, Hidalgo, Mexico

Anja Vogel
Institute of Ecology, Friedrich Schiller University Jena, Jena, Germany

Michael Scherer-Lorenzen
Michael Scherer-Lorenzen, Faculty of Biology - Geobotany, University of Freiburg, Freiburg, Germany

Alexandra Weigelt
Alexandra Weigelt, Institute of Biology, University of Leipzig, Leipzig, Germany

Michael J. O'Brien, Christopher D. Philipson and Andy Hector
Institute of Evolutionary Biology and Environmental Studies, University of Zurich, Zurich, Switzerland

John Tay
School of International Tropical Forestry, University Malaysia Sabah, Kota Kinabalu, Sabah, Malaysia

Sadia E. Ahmed and Robert M. Ewers
Department of Life Sciences, Imperial College London, Silwood Park Campus, Ascot, Berkshire, United Kingdom

Assaf Anyamba
National Aeronautics and Space Administration, Goddard Space Flight Center, Biospheric Sciences Laboratory, Greenbelt, Maryland, United States of America
Universities Space Research Association, Columbia, Maryland, United States of America

Jennifer L. Small and Edwin W. Pak
National Aeronautics and Space Administration, Goddard Space Flight Center, Biospheric Sciences Laboratory, Greenbelt, Maryland, United States of America
Science Systems and Applications Incorporated, Lanham, Maryland, United States of America

Seth C. Britch and Kenneth J. Linthicum
United States Department of Agriculture, Agricultural Research Service, Center for Medical, Agricultural, & Veterinary Entomology, Gainesville, Florida, United States of America

Compton J. Tucker and Curt A. Reynolds
National Aeronautics and Space Administration, Goddard Space Flight Center, Biospheric Sciences Laboratory, Greenbelt, Maryland, United States of America

James Crutchfield
United States Department of Agriculture, Foreign Agricultural Service, International Production & Assessment Division, Washington, District of Columbia, United States of America

Christina L. Tague
Bren School of Environmental Science and Management, University of California Santa Barbara, Santa Barbara, California, United States of America

Nathan G. McDowell
Earth and Environmental Sciences Division, Los Alamos National Laboratory, Los Alamos, New Mexico, United States of America

Craig D. Allen
United States Geological Survey, Fort Collins Science Center, Jemez Mountains Field Station, Los Alamos, New Mexico, United States of America

Vanessa Nessner Kavamura and Milena Duarte Lançoni
Laboratory of Environmental Microbiology, Embrapa Environment, Jaguariúna, São Paulo, Brazil
Department of Soil Science, "Luiz de Queiroz" College of Agriculture, University of São Paulo, Piracicaba, São Paulo, Brazil

Rodrigo Gouvêa Taketani, Rodrigo Mendes and Itamar Soares de Melo
Laboratory of Environmental Microbiology, Embrapa Environment, Jaguariúna, São Paulo, Brazil

Fernando Dini Andreote
Department of Soil Science, "Luiz de Queiroz" College of Agriculture, University of São Paulo, Piracicaba, São Paulo, Brazil

Anja Vogel and Nico Eisenhauer
Institute of Ecology, Friedrich-Schiller-University Jena, Jena, Germany

Thomas Fester
Department Umweltmikrobiologie, Helmholtz-Zentrum für Umweltforschung - UFZ, Leipzig, Germany

Michael Scherer-Lorenzen
Faculty of Biology, Geobotany, University of Freiburg, Freiburg, Germany

Bernhard Schmid
Institute of Evolutionary Biology and Environmental Studies, University of Zurich, Zurich, Switzerland

Wolfgang W. Weisser
Terrestrial Ecology Research Group, Department of Ecology and Ecosystem Management, Center for Food and Life Sciences Weihenstephan, Technische Universität München, Freising, Germany

Alexandra Weigelt
Institute of Ecology, Friedrich-Schiller-University Jena, Jena, Germany

Institute of Biology, University of Leipzig, Leipzig, Germany

Jeanne L. Nel, David C. Le Maitre, Belinda Reyers Greg G. Forsyth, Andre K. Theron, Patrick J. O'Farrell, Lara van Niekerk and Laurie Barwell
Natural Resources and the Environment, Council for Scientific and Industrial Research (CSIR), Stellenbosch, Western Cape, South Africa

Deon C. Nel
World Wide Fund for Nature (WWF), Cape Town, Western Cape, South Africa

Sally Archibald
School of Animal, Plant and Environmental Sciences, University of the Witwatersrand, Johannesburg, Gauteng, South Africa

Natural Resources and the Environment, Council for Scientific and Industrial Research (CSIR), Pretoria, Gauteng, South Africa

Brian W. van Wilgen
Natural Resources and the Environment, Council for Scientific and Industrial Research (CSIR), Stellenbosch, Western Cape, South Africa
Centre for Invasion Biology, Department of Botany and Zoology, Stellenbosch University, Stellenbosch, Western Cape, South Africa

Jean-Marc Mwenge Kahinda, Francois A. Engelbrecht and Evison Kapangaziwiri
Natural Resources and the Environment, Council for Scientific and Industrial Research (CSIR), Pretoria, Gauteng, South Africa

Alicia Carrillo-Bastos, Gerald A. Islebe and Nuria Torrescano-Valle
El Colegio de la Frontera Sur, Unidad Chetumal, Herbario, Chetumal, Quintana Roo, México

Marius Warg Næss
Center for International Climate and Environmental Research - Oslo (CICERO), Fram Centre, Tromsø, Norway

Bård-Jørgen Bårdsen
Norwegian Institute for Nature Research (NINA), Arctic Ecology Department, Fram Centre, Tromsø, Norway

Karin Aharonson-Raz, Eyal Klement, Amir Steinman and Anat Lichter-Peled
Koret School of Veterinary Medicine, The Robert H. Smith Faculty of Agriculture, Food and Environment, The Hebrew University of Jerusalem, Rehovot, 76100, Israel

Shlomit Tal and Daniel Cohen
School of Public Health, Sackler Faculty of Medicine, Tel Aviv University, Tel Aviv, 39040, Israel

Boris Gelman
Kimron Veterinary Institute, Bet-Dagan, 50250, Israel

Selena Ahmed
Sustainable Food and Bioenergy Systems Program, Department of Health and Human Development, Montana State University, Bozeman, Montana, United States of America
Department of Biology, Tufts University, Medford, Massachusetts, United States of America
College of Life and Environmental Sciences, Minzu University of China, Beijing, China

John Richard Stepp
College of Life and Environmental Sciences, Minzu University of China, Beijing, China
Department of Anthropology, University of Gainesville, Gainesville, Florida, United States of America

Colin Orians and Sarabeth Buckley
Department of Biology, Tufts University, Medford, Massachusetts, United States of America

Timothy Griffin and Sean Cash
Friedman School of Nutrition Science and Policy, Tufts University, Boston, Massachusetts, United States of America

Corene Matyas
Department of Geography, University of Gainesville, Gainesville, Florida, United States of America

Albert Robbat
Department of Chemistry, Tufts University, Medford, Massachusetts, United States of America

Dayuan Xue and Chunlin Long
College of Life and Environmental Sciences, Minzu University of China, Beijing, China

Uchenna Unachukwu
Department of Biochemistry, The Graduate Center of the City University of New York, New York, New York, United States of America

David Small
School of Engineering, Tufts University, Medford, Massachusetts, United States of America

Edward Kennelly
College of Life and Environmental Sciences, Minzu University of China, Beijing, China
Department of Biochemistry, The Graduate Center of the City University of New York, New York, New York, United States of America

Zhuolin Li, Yuting Zhang, Dafu Yu, Na Zhang, Jinwei Zhang, Jiahong Tang, Junfeng Wang and Chunsheng Mu
Key Laboratory of Vegetation Ecology, Ministry of Education, Institute of Grassland Science, Northeast Normal University, Changchun, China

Jixiang Lin
Key Laboratory of Salinealkali Vegetation Ecology Restoration in Oil Field Ministry of Education, Alkali Soil Nature Environmental Science Center, Northeast Forestry University, Harbin, China

Marcos D. Robles, Robert M. Marshall and Jeanmarie A. Haney
The Nature Conservancy Center for Science and Public Policy, Tucson, Arizona, United States of America

Frances O'Donnell
Northern Arizona University, Flagstaff, Arizona, United States of America

Edward B. Smith
The Nature Conservancy in California, Sacramento, California, United States of America

David F. Gori
The Nature Conservancy in New Mexico, Santa Fe, New Mexico, United States of America

Dolors Armenteras
Landscape Ecology and Ecosystem Modelling Laboratory, Department of Biology, Sciences Faculty, Colombia National University, Bogotá, Colombia

Javier Retana
CREAF i Unitat d'Ecologı´a, Universitat Auto`noma de Barcelona, Barcelona, Spain

Thaís Helena Ferreira, Agustina Gentile, Lara Isys Dias and Marcelo Menossi
Laborató rio de Genoma Funcional, Departamento de Genética, Evolução e Bioagentes, Instituto de Biologia, Universidade Estadual de Campinas. Campinas, São Paulo, Brazil

Romel Duarte Vilela and Laurı´cio Endres
Centro de Ciências Agrárias, Universidade Federal de Alagoas. Rio Largo, Alagoas, Brazil

Gustavo Gilson Lacerda Costa
Laborató rio Central de Tecnologias de alto desempenho, Universidades Estadual de Campinas. Campinas, São Paulo, Brazil

André Kost
University of Bonn, Dept. of Geography, Bonn, Germany

Peter Läderach and Lorena Gómez
International Center for Tropical Agriculture, Managua, Nicaragua

Myles Fisher
International Center for Tropical Agriculture, Cali, Colombia

Simon Cook
International Water Management Institute, Sunil Mawatha, Pelawatte, Battaramulla, Sri Lanka

Geetha Ramaswami
Centre for Ecological Sciences, Indian Institute of Science, Bangalore, India

Raman Sukumar
Centre for Ecological Sciences, Indian Institute of Science, Bangalore, India
Divecha Centre for Climate Change, Indian Institute of Science, Bangalore, India

Haitao Shi, Yanping Wang and Zhulong Chan
Key Laboratory of Plant Germplasm Enhancement and Specialty Agriculture, Wuhan Botanical Garden, Chinese Academy of Sciences, Wuhan, China

Zhangmin Cheng and Tiantian Ye
Key Laboratory of Plant Germplasm Enhancement and Specialty Agriculture, Wuhan Botanical Garden, Chinese Academy of Sciences, Wuhan, China
University of Chinese Academy of Sciences, Beijing, China

Alvaro G. Gutiérrez
Department of Ecological Modeling, Helmholtz Centre for Environmental Research (UFZ), Leipzig, Germany
Forest Ecology Group, Institute of Terrestrial Ecosystems, Department of Environmental Sciences, Swiss Federal Institute of Technology (ETH Zürich), Zürich, Switzerland
nstituto de Conservación Biodiversidad y Territorio, Facultad de Ciencias Forestales y Recursos Naturales, Universidad Austral de Chile, Valdivia, Chile

Juan J. Armesto
Instituto de Ecología y Biodiversidad (IEB), Santiago, Chile
Departamento de Ecología, Facultad de Ciencias Biológicas, Universidad Catolica de Chile, Santiago, Chile

M. Francisca Díaz
Departamento de Ciencias Biológicas, Facultad de Ciencias Biológicas, Universidad Andrés Bello, Santiago, Chile

Andreas Huth
Department of Ecological Modeling, Helmholtz Centre for Environmental Research (UFZ), Leipzig, Germany

Mahendar Thudi, Hari D. Upadhyaya, Abhishek Rathore, Pooran Mal Gaur, Lakshmanan Krishnamurthy, Spurthi N. Nayak and Rajeev Kumar Varshney
International Crops Research Institute for the Semi-Arid Tropics (ICRISAT), Hyderabad, Andhra Pradesh, India

Manish Roorkiwal
International Crops Research Institute for the Semi-Arid Tropics (ICRISAT), Hyderabad, Andhra Pradesh, India
Guru Gobind Singh Indraprastha University, Delhi, India

Sushil Kumar Chaturvedi and Partha Sarathi Basu
Indian Institute of Pulses Research (IIPR), Kanpur, Uttar Pradesh, India

N. V. P. R. Gangarao
International Crops Research Institute for the Semi-Arid Tropics (ICRISAT), Nairobi, Kenya

Asnake Fikre
Ethiopian Institute of Agricultural Research (EIAR), Debre Zeit, Ethiopia

Paul Kimurto
Egerton University (EU), Egerton, Kenya

Prakash C. Sharma
Guru Gobind Singh Indraprastha University, Delhi, India

M. S. Sheshashayee
University of Agricultural Sciences- Bangalore Bangalore, Karnataka, India

Satoshi Tobita
Japan International Research Center for Agricultural Sciences (JIRCAS), Tsukuba, Japan

Junichi Kashiwagi
Hokkaido University, Sapporo, Japan

Osamu Ito
United Nations University, Yokohama, Japan

Andrzej Killian
DArT Pty. Ltd., Yarralumla, Australia

Nicole Ashley Davies and Adrian Bradley
School of Biomedical Sciences, The University of Queensland, St Lucia, Queensland, Australia

Galina Gramotnev, Clive McAlpine, Leonie Seabrook, Greg Baxter and Jonathan R. Rhodes
Landscape Ecology and Conservation Group, School of Geography, Planning and Environmental Management, The University of Queensland, St Lucia, Queensland, Australia

Daniel Lunney
Office of Environment and Heritage New South Wales, Hurstville, New South Wales, Australia

School of Veterinary and Life Sciences, Murdoch University, Perth, Western Australia, Australia

Shinya Numata and Tetsuro Hosaka
Department of Tourism Science, Tokyo Metropolitan University, Hachiouji, Tokyo, Japan

Masatoshi Yasuda
Kyushu Research Center, Forestry and Forest Products Research Institute, Kumamoto, Kumamoto, Japan

Ryo O. Suzuki
Sugadaira Montane Research Center, University of Tsukuba, Ueda, Nagano, Japan

Nur Supardi Md. Noor and Christine D. Fletcher
Forestry Division, Forest Research Institute Malaysia, Kepong, Selangor, Malaysia

Mazlan Hashim
Institute of Geospatial Science and Technology, Universiti Teknologi Malaysia, UTM, Skudai, Johor Bahru, Malaysia

Index